D1070781

STRESS CONCENTRATION FACTORS

STRESS CONCENTRATION FACTORS

Charts and relations useful in making strength calculations for machine parts and structural elements

R. E. PETERSON

Consultant

Westinghouse Research Laboratories

A WILEY-INTERSCIENCE PUBLICATION

JOHN WILEY & SONS, New York • London • Sydney • Toronto

Library of Congress Cataloging in Publication Data

Peterson, Rudolph Earl, 1901–
Stress concentration factors.

"A Wiley-Interscience publication."
Includes bibliographical references.
1. Strength of materials—Tables, calculations, etc. 2. Strains and stresses—Tables, calculations, etc. 3. Mechanical engineering—Tables, calculations, etc. I. Title.

TA407.2.P47 620.1'12 73-9834
ISBN 0-471-68329-9

Printed in the United States of America

10 9 8 7 6 5 4 3 2 1

PREFACE

In the interval of two decades since publication of *Stress Concentration Design Factors*,[1] a large amount of material on stress concentration has become available, partly because of the growing capability of modern computer technology. This large increase in available data has necessitated the development of a system of classification; the outline in the Contents (vii) will help the reader to become familiar with the system and to expedite usage in design applications.

The variety of applications extends to nearly every field of design of structures and machines: nuclear equipment (Chapter 4, sections A.f, A.l, A.z, A.z4); deep sea vehicles (Chapter 4, Section A.f); aircraft details (Chapter 4, Sections A.k, A.q, A.v, A.x); space applications (Chapter 2, Sections A.h and A.i); solid propellant rocket grains (Chapter 4, Section A.y); underground tunnels (Chapter 4, Section A.z7); turbines (Chapter 5, Section, F and M); and so on. Emphasis has been placed on design use in the format of the charts and the selection of parameters and their ranges.

Some of the material included in the preceding book[1] has been omitted* in the present volume, but even with the omissions the book has become rather thick.

The importance of stress concentration in design, particularly with respect to fatigue, is emphasized by examples in the preface of the preceding book;[1] to this can be added similar references[2,3,3a,316a].

In the development of fracture mechanics the stress intensity factor is used. For limiting conditions this factor is related directly to the stress concentration factor; three charts are given for width correction factors (Figs. 20, 39b, and 131).

It will be noted that the full-page charts are dimensionless; with the sketch of the member on each chart, this means that the book is unique as a technical reference and can be used internationally by those with very limited English language capability.

In the earlier book,[1] ring binding was used to allow the pages of the open book to lie flat, a convenience when making calculations. However, page failures occurred at the corners of the rectangular holes owing to stress concentration (a subject, incidentally, covered in Chapter 4, Section A.v). The present book has a more durable binding that will permit the pages to lie reasonably flat at any selected book opening.

A "quick-finder" feature has been provided for locating certain much-used charts.

*The text and charts for K_t', a factor combining stress concentration and Mises strength theory are not included (see Chapter 1, Section B, Mises Theory, for comments). The relations for limited life (see Chapter 1, Section G for comments) are also not included. Relations for combined stress based on an energy consideration (Chapter 1, Section F)[1] and the Appendices[1], giving mainly derivations, are omitted.

These key charts have a gray screen on the right margin. By placing the book on the bound edge, grasping the pages with the right hand, and slowly releasing the pages with the thumb one can quickly find the desired chart. One can also fix additional edges in this way for personal use.

One of the objectives of application of more accurate factors is to achieve better balanced designs* of structures and machines—conserving material, obtaining cost reduction, and achieving lighter and more efficient apparatus.

Acknowledgment is given to numerous technical consultations with M. M. Leven. He and R. L. Johnson have read the manuscript and have made many valuable recommendations. Discussions were held with W. K. Wilson with regard to stress intensity factors. Some of the newer material in this book has been issued in Westinghouse Research Reports. The encouragement and support of L. B. Kramer and A. C. Hagg in the development of these reports and the subsequent incorporation of the data into this book is much appreciated.

Even though the manuscript has been checked and rechecked, experience indicates that with so much material it is almost impossible to completely eliminate errors. I shall appreciate being informed of any errors that are found.

The dedication of this volume to my wife Marie will be understood by all who know her.

RUDOLPH EARL PETERSON

Churchill Borough
Pittsburgh, Pennsylvania
April 1973

---

*Balanced design is delightfully phrased in the poem[4] "The Deacon's Masterpiece, or the Wonderful One Hoss Shay" by Oliver Wendell Holmes (1858)

"Fur", said the Deacon, " 't 's mighty plain
That the weakes' place mus' stan' the strain,
'N the way t' fix it, uz I maintain,
Is only jest
T' make that place uz strong uz the rest"

After "one hundred years to the day" the shay failed "all at once, and nothing first."

CONTENTS

LIST OF SYMBOLS

α, θ, ϕ = angles.
a = hole radius, minimum width of ellipse.
A = area.
b = maximum dimension of ellipse,
b = pitch, spacing between notches or holes.
b = width of keyway.
c = spring index (mean coil diameter divided by wire diameter).
c = distance spanning a row of notches.
c = distance from neutral axis to extreme fiber.
c = distance from center of hole to the near edge of bar or plate.
C_w = Wahl factor, helical spring.
d = diameter (width).
d = smaller diameter (or width) of a stepped shaft or bar.
D = larger diameter (or width) of a stepped shaft or bar.
e = distance from center of hole to farther edge of bar.
E = modulus of elasticity.
E' = modulus of elasticity of inclusion material.
g = gravitational acceleration.
γ = weight per unit volume.
h = thickness.
h = moment arm for gear tooth.
h = ligament width (hole pattern).
I = moment of inertia.
J = polar moment of inertia.
K_t = stress concentration factor for normal stress, $\sigma_{\max}/\sigma_{\text{nom}}$. ($K_t$ is a *theoretical* factor.) Various expanded subscripts are used, such as K_{te} for elliptical notch, K_{th} for hyperbolic notch, K_{tA}, K_{t2}, $K_{t\alpha}$, etc., for special cases.
K_{tg} = stress concentration factor based on gross stress.
K_{tn} = stress concentration factor based on nominal stress.
K_t' = combined factor taking account of stress concentration (normal stress) and Mises criterion of failure. (K_t' is a *theoretical* factor.) Similar subscript variations as indicated under K_t.
K_{ts} = stress concentration factor for shear stress, $\tau_{\max}/\tau_{\text{nom}}$ (K_{ts} is a *theoretical* factor.) Various expanded subscripts, K_{tse}, etc.
K_{tf} (or K_{tf}') = estimated fatigue notch factor for normal stress. (K_{tf}, or K_{tf}', is a *calculated* factor.)
K_{tsf} = estimated fatigue notch factor for shear stress. (K_{tsf} is a *calculated* factor.)
K_f = fatigue notch factor for axial or bending load. (K_f is a factor determined by *fatigue tests.*)
K_{fs} = fatigue notch factor for torsion. (K_{fs} is a factor determined by *fatigue tests.*)
L = axial length of shoulder.
L = limit design factor.
L_b = limit design factor for bending.
L_s = limit design factor for shear (torsion).
l = moment arm.
M = moment.

m = height of T-head.
n = factor of safety.
ν = Poisson's ratio.
p = pressure.
P = load.
P_d = diametral pitch (gearing).
q = notch sensitivity factor $= \dfrac{K_f - 1}{K_t - 1}$
R_1 = radius of hole in disk, inside radius of ring.
R_2 = outside radius of disk or ring.
R_0 = radius to non-central hole in disk.
r = minimum notch radius, radius of hole in plate or bar.
s = ligament width (two holes).
σ = normal stress.
$\sigma_1, \sigma_2, \sigma_3$ = principal normal stresses.
σ_a = alternating normal stress (completely reversed).
σ_o = steady (or static) normal stress.
σ_y = yield strength (tension).
σ_u, σ_{ut} = tensile strength (ultimate).*
σ_{uc} = compressive strength (ultimate).
σ_f = fatigue strength (endurance limit), for axial load or bending, of unnotched specimen.
σ_{nf} = fatigue strength, for axial loading or bending, of notched specimen.
$\sigma_{\max}$ = maximum normal stress.
σ_{nom} = nominal normal stress computed from P/A or Mc/I or an elementary formula which does not take account of the stress concentration "peak." In the case of a member with a hole, the net section is used.
τ = shear stress.
τ_a = alternating shear stress (completely reversed).
τ_o = steady (or static) shear stress.
τ_y = yield strength (shear).
τ_f = fatigue strength in torsion (shear endurance limit) of unnotched specimen.
τ_{nf} = fatigue strength in torsion of notched specimen.
$\tau_{\max}$ = maximum shear stress.
τ_{nom} = nominal shear stress computed from Tc/J.
t = base width of gear tooth.
t = notch depth, or shoulder height, or keyway depth, or plate thickness.
T = torque.
v = peripheral velocity.
V = volume.
w = width.
W = load per inch of face (gearing).

*Use σ_{ut} when tensile and compressive strengths are involved.

STRESS CONCENTRATION FACTORS

CHAPTER 1

DEFINITIONS AND DESIGN RELATIONS

(A) STRESS CONCENTRATION

The elementary formulas used in design are based on members having a constant section or a section with gradual change of contour (Fig. 1). Such conditions, however, are hardly ever attained throughout the highly stressed region of actual machine parts or structural members. The presence of shoulders, grooves, holes, keyways, threads, and so on, results in a modification of the simple stress distributions of Fig. 1, so that localized high stresses occur as shown in Fig. 2. This localization of high stresses is known as *stress concentration*, measured by the *stress concentration factor*, defined as

$$\left.\begin{aligned} K_t &= \frac{\sigma_{\max}}{\sigma_{\text{nom}}} \quad \text{for normal stress (tension or bending)} \\ K_{ts} &= \frac{\tau_{\max}}{\tau_{\text{nom}}} \quad \text{for shear stress (torsion)} \end{aligned}\right\} \qquad [1]$$

where σ_{nom} and τ_{nom} are defined in accordance with the elementary formulas of Fig. 1.

The subscript t indicates that stress concentration factor is a theoretical factor (i.e., based on usual assumptions in theory of elasticity—Hooke's law, homogeneity, etc.).* The basic subscript t distinguishes theoretical factors from experimentally determined factors, such as "fatigue notch factor" K_f, to be described later.†

Stress concentration factors are obtained mathematically or experimentally[6] by such means as photoelasticity, precision strain gage, membrane analogy for torsion, or electrical analogy for torsion. When the experimental work is conducted with sufficient precision, excellent agreement is obtained with well-established mathematical stress concentration factors.

*This book deals only with elastic stress concentration factors; in the plastic range one must consider separate stress and strain concentration factors which depend on the shape of the stress-strain curve and the stress or strain level.[4a]

†Since by definition stress concentration factor is always a theoretical factor, it is strictly speaking, not necessary to use the term "theoretical stress concentration factor"; sometimes, however, this is done to make sure there is no misunderstanding. Such terms as "fatigue stress concentration factor" or "effective stress concentration factor" which apply to test results are likely to be misleading. For this reason terms that do not involve the words "stress concentration" are thought to be more appropriate for test results.[5]

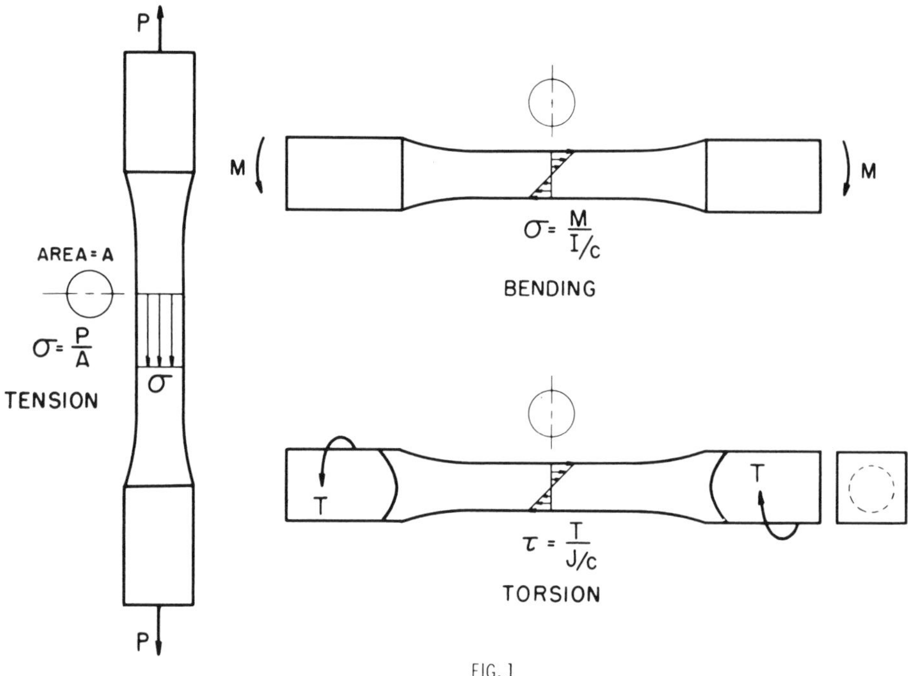

FIG. 1

ELEMENTARY STRESS CASES
(CONSTANT CROSS-SECTION OR GRADUAL CROSS-SECTIONAL CHANGE)

Unfortunately, the application phase of this work, as treated in the remainder of Chapter 1, is not on as precise a basis as the theoretical phase. Many more data are necessary, particularly because, for the required precision in materials tests, statistical procedures often are necessary. Directional effects in materials must also be carefully taken into account. It is hardly necessary to point out that the designer cannot wait for exact answers to all these questions. As always, existing information must be reviewed and judgment used in developing reasonable approximate procedures for design, tending toward the safe side in doubtful cases. In time, the application is certain to advance and revisions will need to be made accordingly. On the other hand, it can be said that our limited experience in using these methods has been satisfactory.

(B) STRENGTH THEORY

If our design problems involved only uniaxial stress systems, we would need to give only limited consideration to strength theory. However, even very simple load conditions may result in biaxial stress systems. An example is a thin spherical vessel subjected to internal pressure, resulting in biaxial tension acting on an element of the vessel. Another example is a round bar subjected to torsion, resulting in biaxial tension and compression acting on a surface element oriented at 45°.

From the standpoint of stress concentration cases, it should be noted that such simple loading as an axial pull produces biaxial surface stresses in a grooved bar, (Fig. 3). Axial load P results in axial tension σ_1 and circumferential tension σ_2 acting on a surface element of the groove.

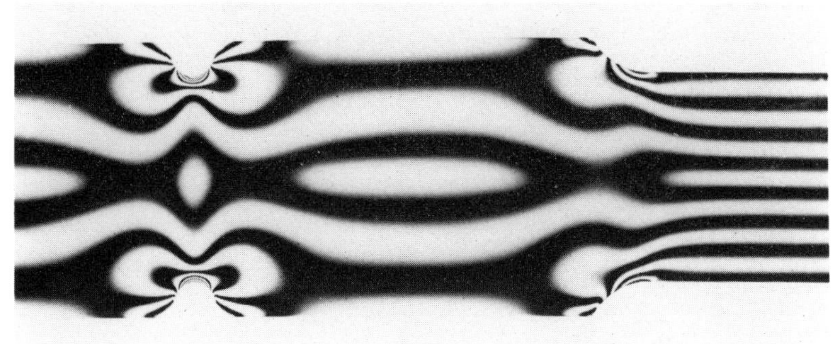

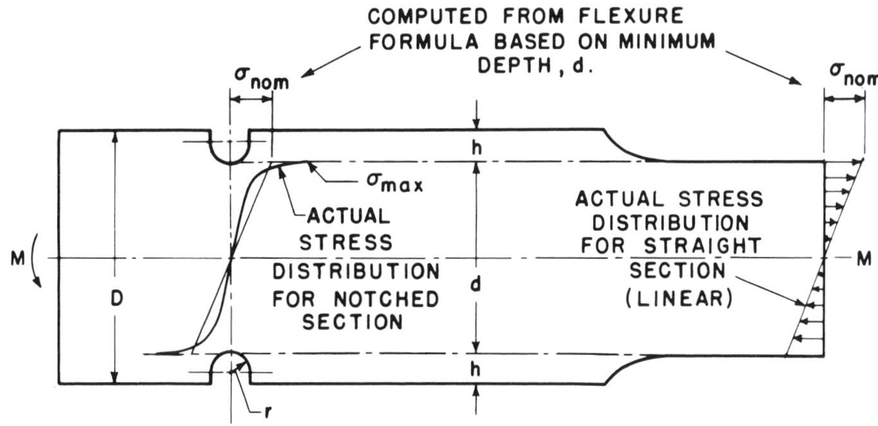

FIG. 2.

STRESS CONCENTRATION INTRODUCED BY NOTCH.

(OR IN GENERAL BY CROSS SECTIONAL CHANGE WHICH IS NOT GRADUAL)

PHOTOELASTIC FRINGE PHOTOGRAPH – LEVEN

A considerable number of theories have been proposed relating uniaxial and biaxial or triaxial stress systems;[7] only the theories ordinarily utilized for design purposes[8] are considered here. These are: *for brittle materials:* * normal stress criterion, Mohr theory; *for ductile materials:* maximum shear theory, Mises criterion.

(a) Normal Stress Criterion

This criterion can be stated as follows: failure occurs in a multiaxial system when either a principal tensile stress reaches the uniaxial tensile strength σ_{ut}, or a principal compressive stress reaches the uniaxial compressive strength σ_{uc}. For a brittle material σ_{uc} is usually considerably greater than σ_{ut}. In Fig. 4, which represents biaxial conditions (σ_1 and σ_2 principal stresses, $\sigma_3 = 0$), the normal stress criterion is represented by the square *CFHJ*. For $\sigma_{ut} = \sigma_{uc}$, the normal stress criterion is known as the maximum normal stress theory.

*The distinction between brittle and ductile materials is arbitrary, sometimes an elongation of 5% is considered to be an arbitrary dividing line.[9]

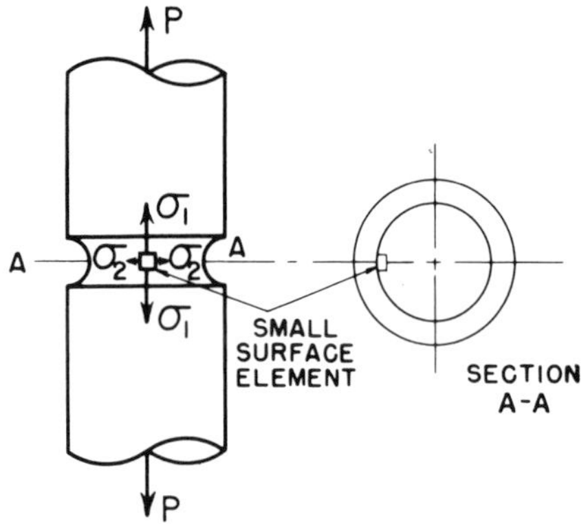

FIG. 3
BI-AXIAL STRESS IN NOTCHED TENSILE MEMBER.

The strength of a bar under uniaxial tension σ_{ut} is OB in Fig. 4. Note that the presence of an additional stress σ_2 at right angles does not affect the strength.

For torsion, $\sigma_2 = -\sigma_1$ (line AOE), and, since these principal stresses are equal in magnitude to the applied stress τ, the normal stress condition of failure for torsion (MA, Fig. 4) is

$$\tau_u = \sigma_{ut} \qquad [2]$$

In other words, according to the normal stress criterion the torsion and tension strength values should be equal.

(b) Mohr Theory

The condition of failure of brittle materials according to the Mohr theory is illustrated in Fig. 5. Circles of diameters σ_{ut} and σ_{uc} are drawn as shown. Any other stress state, for which the Mohr circle (which envelops circles of diameter σ_1 and σ_2 as shown) is tangent to the line of tangency* of the σ_{ut} and σ_{uc} circles, represents a condition of failure.[10] The resultant plot for biaxial conditions is shown[10] in Fig. 4. As will be seen later (Fig. 6) this is similar to the representation for the maximum shear theory, except for nonsymmetry. The following condition holds[10] for the second, or northwest, quadrant of Fig. 4:

$$\frac{\sigma_1}{\sigma_{ut}} - \frac{\sigma_2}{\sigma_{uc}} = 1 \qquad [3]$$

Certain tests of brittle materials seem to substantiate the normal stress criterion,[11] whereas other tests and reasoning lead to a preference for the Mohr theory, [10,12] The normal stress criterion gives the same result in the first (northeast) and third (southwest) quadrants. For the torsion case, use of the Mohr theory is on the "safe side," since the limiting strength value used is $M'A'$ instead of MA (Fig. 4). The following can be shown for $M'A'$ of Fig. 4:

$$\tau_u = \frac{\sigma_{ut}}{1 + (\sigma_{ut}/\sigma_{uc})} \qquad [4]$$

*The straight line is a special case of the more general Mohr theory, which is based on a curved envelope.

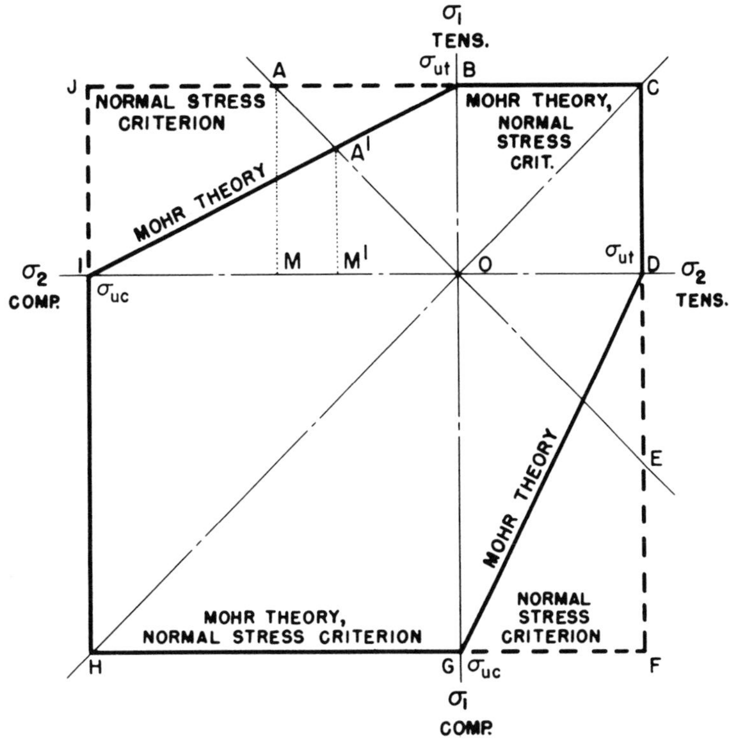

FIG. 4
BI-AXIAL CONDITIONS FOR STRENGTH THEORIES FOR BRITTLE MATERIALS.

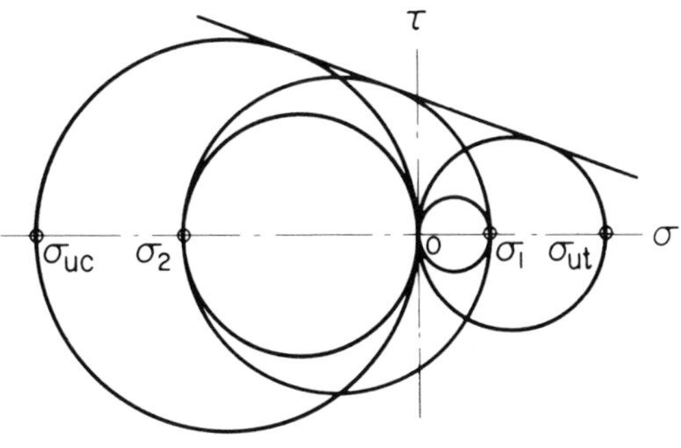

FIG. 5
MOHR THEORY OF FAILURE OF BRITTLE MATERIALS.

(c) Maximum Shear Theory

The maximum shear theory was developed as a criterion for yield failure, but it has also been applied to fatigue failure, which in ductile materials is initiated by maximum resolved shear stress.[13] According to the maximum shear theory, failure occurs when the maximum shear stress in a multiaxial system reaches the value of the shear stress in a uniaxial bar at failure. In Fig. 6, the maximum shear theory is represented by the six-sided figure. Suppose we are considering fatigue failure and represent the uniaxial fatigue limit in alternating tension-compression by σ_f. For principal stresses σ_1, σ_2, and σ_3 the maximum shear stresses are

$$\frac{\sigma_1 - \sigma_2}{2} \qquad \frac{\sigma_1 - \sigma_3}{2} \qquad \frac{\sigma_2 - \sigma_3}{2}$$

For the biaxial case $\sigma_3 = 0$, and for σ_1 greater than σ_2 but both tension, failure occurs when

$$\frac{\sigma_1 - 0}{2} = \frac{\sigma_f}{2}$$

$$\sigma_1 = \sigma_f \qquad [5]$$

This is the condition represented in the first (northeast) quadrant of Fig. 6. However, in the second (northwest) and fourth (southeast) quadrants, where the biaxial stresses are of

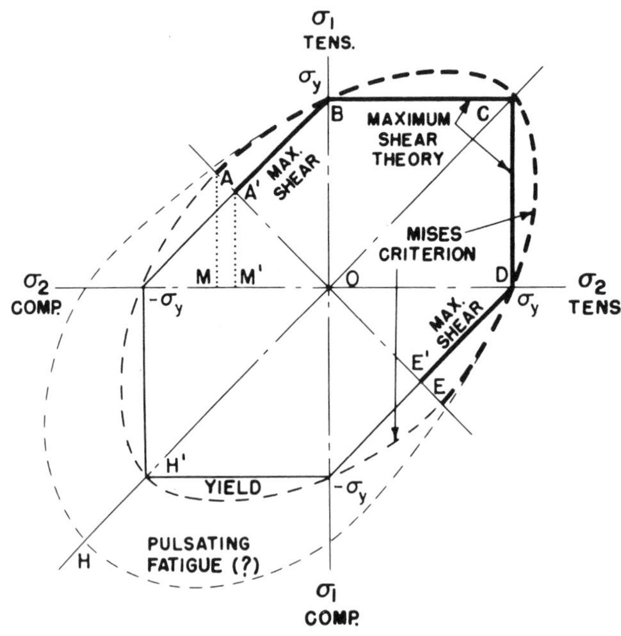

FIG. 6
BI-AXIAL CONDITIONS FOR STRENGTH THEORIES FOR DUCTILE MATERIALS.

opposite sign, a different result is obtained. For $\sigma_2 = -\sigma_1$, represented by line AE of Fig. 6, fatigue failure occurs in accordance with the maximum shear theory when

$$\frac{\sigma_1 - (-\sigma_1)}{2} = \frac{\sigma_f}{2} \qquad \sigma_1 = \frac{\sigma_f}{2} \qquad M'A' = \frac{OB}{2} \text{ in Fig. 6}$$

In the torsion test $\sigma_2 = -\sigma_1 = \tau$

$$\tau_f = \frac{\sigma_f}{2} \tag{6}$$

This is just half the value corresponding to the normal stress criterion.

(d) Mises Criterion

The following mathematical expression was proposed by R. von Mises[14] as representing a criterion of failure by yielding:

$$\sigma_y = \sqrt{\frac{(\sigma_1 - \sigma_2)^2 + (\sigma_2 - \sigma_3)^2 + (\sigma_1 - \sigma_3)^2}{2}} \tag{7}$$

where σ_y = yield strength in uniaxially loaded bar. If $\sigma_3 = 0$,

$$\sigma_y = \sqrt{\sigma_1^2 - \sigma_1\sigma_2 + \sigma_2^2} \tag{8}$$

This relationship is shown by the dashed ellipse of Fig. 6 where $OB = \sigma_y$ in this case. Unlike the six-sided figure, it does not have the discontinuities which seem unrealistic in a physical sense. Hencky[15] later showed that if shear energy is made a criterion of failure, one also obtains [7]. Eichinger[16] and Nadai[17] obtained equivalent results by using octahedral shear stress. However, close agreement with the results predicted by [7] is obtained if one considers the statistical behavior of a randomly oriented aggregate of crystals, as shown by Sachs[18] and by Cox and Sopwith.[19] Beeching[20] has suggested that a term other than "shear energy theory" was needed to avoid confusion. The brief designation "Mises criterion" has been coming into increased use[21,22] and will be used in this book.

Considering the torsion case, with $\sigma_2 = -\sigma_1 = \tau$, the Mises criterion results in

$$\tau_y = \frac{\sigma_y}{\sqrt{3}} = 0.577\sigma_y \qquad MA = (0.577)\, OB \text{ in Fig. 6} \tag{9}$$

where τ_y = yield strength of plain torsion bar.

Note from Figs. 4 and 6 that all the foregoing theories are in agreement at C, representing equal tensions, but they differ along AE, representing tension and compression of equal magnitudes (torsion).

Yield tests of ductile materials have shown that the Mises criterion interprets well the results of a variety of biaxial conditions. It has been pointed out by Prager and Hodge[22] that, although the agreement must be regarded as fortuitous, the Mises criterion would still be of practical interest because of its mathematical simplicity even if the agreement with test results had been less satisfactory.

There is evidence[23–29] that for ductile materials the Mises criterion also gives a reasonably good interpretation of fatigue results in the upper right half $ABCDE$ of the ellipse of

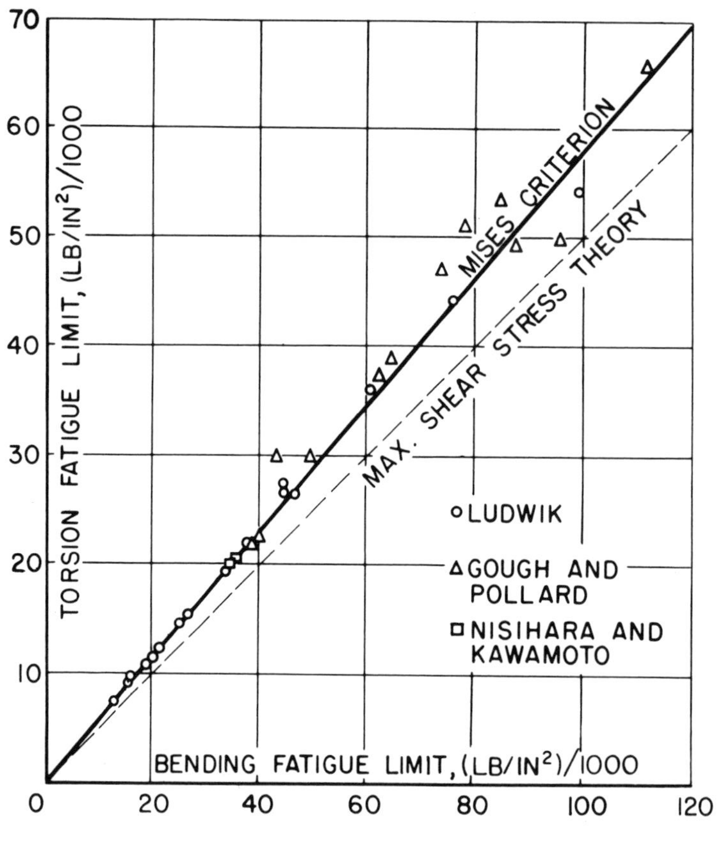

FIG. 7

COMPARISON OF TORSION AND BENDING FATIGUE LIMITS FOR DUCTILE MATERIALS

Fig. 6 for completely alternating or pulsating tension cycling. As shown in Fig. 7, results from alternating tests are in better agreement with the Mises criterion (upper line) than with the maximum shear theory (lower line). If yielding is considered the criterion of failure, the ellipse of Fig. 6 is symmetrical about AE. With regard to the region below AE (compression side), there is evidence that for pulsating compression (for example, 0 to maximum compression) this area is considerably enlarged.[30–32] For the cases treated here we deal primarily with the upper right area.*

The earlier edition of this book[1] presents a method for taking account of the biaxiality in a grooved member in simple tension or bending by means of a factor K_t', which combines stress concentration and Mises strength theory.[23] Since K_t' is lower than K_t, generally less than 10% but not exceeding 15%, it seems that the usual design practice is to use K_t, which is on the safe side and removes any reservation concerning the validity of the combined factor. Therefore, the K_t' charts are not included here. As far as the writer knows, no inter-

*It will be noted that all representations in Figs. 4 and 6 are symmetrical about line HC. In materials as produced, forgings, bars, etc., strong directional effects can exist (i.e., transverse strength can be considerably less than longitudinal strength). Findley[33] gives methods for taking anisotropy into account in applying strength theories.

vening research has invalidated the K_t' concept. For further research on stress concentration effect in fatigue, the K_t' values of the earlier edition[1] should be consulted.

(C) NOTCH SENSITIVITY

It is well known that the effect of a notch on the fatigue strength of a part varies considerably with material and notch geometry and is usually* less than the effect that would be predicted by use of the stress concentration factor, which is a theoretical factor. This general phenomenon is denoted *notch sensitivity.* Notch sensitivity may be considered as a measure of the degree to which the theoretical effect is obtained:

$$q = \frac{K_f - 1}{K_t - 1} \qquad [10]$$

or

$$q = \frac{K_{fs} - 1}{K_{ts} - 1} \qquad [11]$$

where q = fatigue notch sensitivity,† a measure of the agreement between K_f and K_t
K_t = stress concentration factor for normal stress, theoretical factor
K_{ts} = stress concentration factor (shear stress)
K_f = fatigue notch factor (normal stress) =

$$\frac{\sigma_f}{\sigma_{nf}} = \frac{\text{fatigue limit of unnotched specimen (axial or bending)}}{\text{fatigue limit of notched specimen (axial or bending)}}$$

K_{fs} = fatigue notch factor (shear stress) =

$$\frac{\tau_f}{\tau_{nf}} = \frac{\text{fatigue limit of unnotched specimen (torsion)}}{\text{fatigue limit of notched specimen (torsion)}}$$

The preceding definition of q has the merit of providing a scale of notch sensitivity, which, as can be seen from [10], varies from $q = 0$ or *no notch effect* (when $\sigma_{nf} = \sigma_f$, $K_f = 1$) to $q = 1$, or *full theoretical effect* (when $K_f = K_t$).

Relations 10 and 11 can be put in the following form for design use:

$$K_{tf} = q(K_t - 1) + 1 \qquad [12]$$

$$K_{tsf} = q(K_{ts} - 1) + 1 \qquad [13]$$

where K_{tf} = *estimated* fatigue notch factor (normal stress), a calculated factor using an average q value obtained from Fig. 8 or a similar curve.
K_{tsf} = *estimated* fatigue notch factor (shear stress)

Where no q data exist, for example, with newly developed materials, it is suggested that the full theoretical factor, K_t or K_{ts} be used. It should be noted in this connection that if notch sensitivity is not taken into consideration at all in design ($q = 1$), the error will be on the safe side ($K_{tf} = K_t$ in [12]).

*Under certain conditions full theoretical effect is, for all practical purposes obtained (to be discussed later).
†See reference 5 for definitions.

In plotting K_f for geometrically similar specimens, it was found that for a typical resulting curve K_f decreased as the specimen size decreased;[34–36] for this reason, it is not possible to obtain reliable comparative q values for different materials by making tests of a standardized specimen of fixed dimensions.[37] Since the local stress distribution (stress gradient,* volume at peak stress) is more dependent on the notch radius, r, than on other geometrical variables,[38–41] it was apparent that it would be more logical to plot[42] q versus r, rather than q versus d (for the geometrically similar specimens, the curve shapes are, of course, the same). Plotted q versus r curves[43,44] based on available data[45–59] were found to be within reasonable scatter bands.

A q versus r chart for design purposes is given in Fig. 8, which averages the previously mentioned plots.[43,44] Note that the chart is not verified for notches having a depth greater than four times the notch radius because data are not available. Also note that the curves are to be considered as approximate (see shaded band).

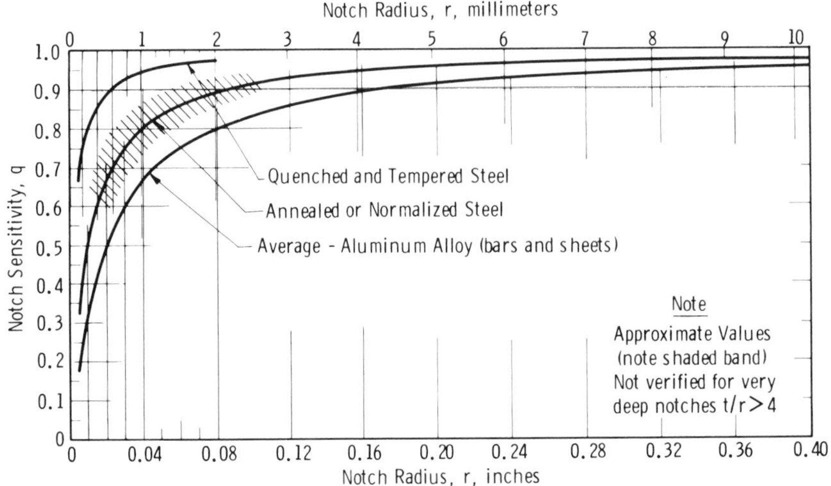

Fig. 8—Average fatigue notch sensitivity

Notch sensitivity values for radii approaching zero still must be studied. It is, however, well known that tiny holes and scratches do not result in a strength reduction corresponding to geometric stress concentration factors; in fact, in steels of low tensile strength the effect is often quite small. However, in higher strength steels the effect of tiny holes or scratches is more pronounced. Much more data are needed, preferably obtained from statistically planned investigations; until better information is available, it is believed that Fig. 8 will provide reasonable values for design use.

Several expressions have been proposed for the q versus r curve. Such a formula could be useful in setting up a complete computer design program. Since it would be unrealistic to expect failure at a volume corresponding to the point of peak stress,[38,40] formulations for K_f are based on failure over a distance below the surface, ϵ or $2\rho'$ for a sharp notch[39] and δ for a rounded notch.[60] From the K_f formulations, q versus r relations are obtained;[60] these

*The stress is approximately linear in the peak stress region.[40,40a]

and other variations[61–65] are found in the literature. All of the formulas yield acceptable results for design purposes; one must, however, always remember the approximate nature of the relations. In Fig. 8 the following simple formula[44,64] is used:*

$$q = \frac{1}{1 + a/r} \qquad [14]$$

where a = material constant
r = notch radius

In Fig. 8, $a = 0.0025$ for quenched and tempered steel, $a = 0.01$ for annealed or normalized steel, $a = 0.02$ for aluminum alloy sheets and bars (avg.). In Ref. 44 more detailed values are given, including the following approximate design values for steels as a function of tensile strength:

$\sigma_u/1000$	a
50	0.015
75	0.010
100	0.007
125	0.005
150	0.0035
200	0.0020
200	0.0020
250	0.0013

where σ_u = tensile strength, lb/in.^{2}. In using the foregoing a values, one must keep in mind that the curves represent averages (see shaded band in Fig. 8).

A method has been proposed by Neuber[65a] wherein an equivalent larger radius is used to provide a lower K factor; the increment to the radius is dependent on the stress state, the kind of material, and its tensile strength. Application of this method gives results which are in reasonably good agreement with Fig. 8a of Ref. 1.

(D) DESIGN RELATIONS FOR STEADY STRESS

(a) Ductile Materials

Under ordinary conditions a ductile member when loaded with a "steady stress" (i.e., a steadily increasing uniaxial stress) does not suffer loss of strength due to presence of a notch. If the member is loaded statically (steady load) but may also be subjected to shock loading, or if the part is to be subjected to high[66] or low temperature, or if the part contains sharp discontinuities, a ductile material may behave in the manner of a brittle material, a consideration to be analyzed by fracture mechanics methods.[67–69]

*The corresponding Kuhn-Hardrath formula,[62] based on Neuber relations, is

$$q = \frac{1}{1 + \sqrt{\rho'/r}}$$

Either formula may be used for design purposes;[60] a or ρ' is determined by test data.

These are special cases, and if there is doubt, K_t should be applied; ordinarily, for static (steady) stress a stress concentration factor is not applied.*

The following factor of safety n, [15], [16], [18], and [19], are based on the Mises criterion of failure, as discussed in Section B.

For axial loading (normal, or direct, stress):

$$n = \frac{\sigma_y}{\sigma_{od}} \tag{15}$$

where σ_y = yield strength
σ_{od} = steady normal stress

For bending:

$$n = \frac{L_b \sigma_y}{\sigma_{ob}} \tag{16}$$

where L_b = limit design factor for bending
σ_{ob} = steady bending stress

In general, the limit design factor L is the ratio of the load (or moment) needed to cause complete yielding throughout the section of a bar over the load (or moment) needed to cause initial yielding at the "extreme fiber,"[70] assuming no stress concentration. For tension $L = 1$; for bending of a rectangular bar $L_b = 3/2$; for bending of a round bar, $L_b = 16/(3\pi) = 1.70$; for torsion of a round bar $L_s = 4/3$; for a tube† (Fig. 9) it can readily be shown that for bending and torsion, respectively,

$$L_b = \frac{16}{3\pi}\left[\frac{1 - (di/d_o)^3}{1 - (di/d_o)^4}\right] \qquad L_s = \frac{4}{3}\left[\frac{1 - (di/d_o)^3}{1 - (di/d_o)^4}\right] \tag{17}$$

where d_i = inside diameter of tube
d_o = outside diameter of tube

Criteria other than complete yielding can be used. For a rectangular bar in bending L_b values have been calculated:[71] yielding to $\frac{1}{4}$ depth, $L_b = 1.22$; yielding to $\frac{1}{2}$ depth, $L_b = 1.375$; 0.1% inelastic strain in steel with yield point of 30,000 psi, $L_b = 1.375$. For a circular bar in bending: yielding to $\frac{1}{4}$ depth, $L_b = 1.25$; yielding to $\frac{1}{2}$ depth, $L_b = 1.5$. For a tube $d_i/d_o = \frac{3}{4}$: yielding $\frac{1}{4}$ depth, $L_b = 1.23$; yielding $\frac{1}{2}$ depth, $L_b = 1.34$.

All the foregoing L values are based on the assumption that the stress-strain diagram becomes horizontal after the yield point is reached. This is a reasonable assumption for low- or medium-carbon steel. For other stress-strain diagrams which can be represented by a sloping line or curve beyond the elastic range, a value of L closer to 1.0 should be used.[70] For design $L\sigma_y$ should not exceed the tensile strength σ_{ut}.

---

*This consideration is on the basis of strength only. Stress concentration does not usually reduce the strength of a notched member in a static test, but usually it does reduce total deformation to rupture. This means lower "ductility," or, expressed in a different way, less area under the stress-strain diagram (less energy expended in producing complete failure). It is often of major importance to have as much energy-absorption capacity as possible (cf. metal versus plastic for automobile body). However, this is a consideration depending on consequence of failure, and so on, and is not within the scope of this book, which deals only with strength factors. Plastic behavior is involved in a limited way in the use of the factor L, as will be explained in the following paragraphs.

†For thin tubes, failure by buckling must be considered.

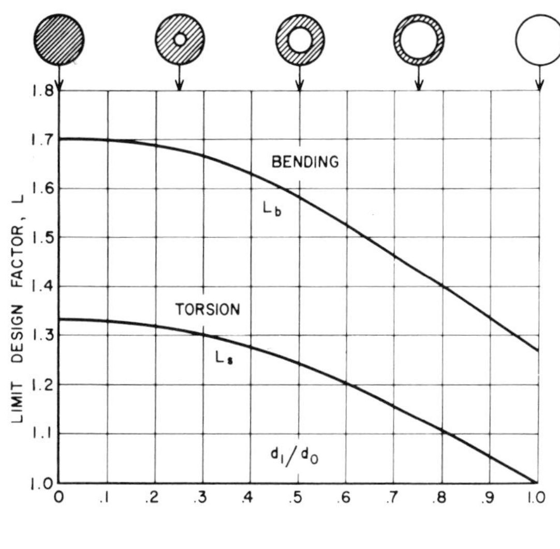

FIG. 9
LIMIT DESIGN FACTORS FOR TUBULAR MEMBERS

For torsion of a round bar (shear stress):

$$n = \frac{L_s \tau_y}{\tau_o} = \frac{L_s \sigma_y}{\sqrt{3}\tau_o} \qquad [18]$$

where τ_y = yield strength in torsion
τ_o = steady shear stress

For combined normal (axial and bending) and shear stress:

$$n = \frac{\sigma_y}{\sqrt{(\sigma_{od} + \sigma_{ob}/L_b)^2 + 3\,(\tau_o/L_s)^2}} \qquad [19]$$

(b) Brittle Materials

It is customary to apply the full K_t factor in the design of members of brittle materials. The use of the full K_t factor for cast iron may be considered, in a sense, as penalizing this material unduly, since experiments show that the full effect is usually not obtained.[72] The use of the full K_t factor may be partly justified as compensating, in a way, for the poor shock resistance of brittle materials. Since it is not possible to design rationally for shock or mishandling in transportation and installation, the larger sections obtained by the preceding rule may be a means of preventing some failures that might otherwise occur. However, notable designs of cast iron members have been made (large paper-mill rolls, etc.) involving rather high stresses where full application of stress concentration factors would rule out this material. Such designs should be carefully made and may be viewed as exceptions to the rule. For ordinary design, it seems wise to proceed cautiously in the treatment of notches in brittle materials, especially in critical load-carrying members.

The following factors of safety, [20] and [21], are based on the *normal stress criterion* of failure (Fig. 4).

For axial tension or bending (normal stress):

$$n = \frac{\sigma_{ut}}{K_t \sigma_o} \qquad [20]$$

where σ_{ut} = tensile ultimate strength
K_t = stress concentration factor for normal stress

For torsion of a round bar (shear stress):

$$n = \frac{\sigma_{ut}}{K_{ts} \tau_o} \qquad [21]$$

where K_{ts} = stress concentration factor for shear stress.

The following factors of safety, [22] and [23], are based on the *Mohr theory* of failure (Fig. 4).

For axial tension or bending: same as [20].

For torsion of a round bar (shear stress) from [4]:

$$n = \frac{\sigma_{ut}}{K_{ts} \tau_o} \left[\frac{1}{1 + (\sigma_{ut}/\sigma_{uc})} \right] \qquad [22]$$

where σ_{ut} = tensile ultimate strength
σ_{uc} = compressive ultimate strength

Since the factors based on the Mohr theory are on the "safe side" compared to those based on the normal stress criterion, [22] and [23] are suggested for design use.[9]

For combined normal and shear stress:

$$n = \frac{2\sigma_{ut}}{K_t \sigma_o + \sqrt{(K_t \sigma_o)^2 + 4 \{K_{ts} \tau_o [1 + (\sigma_{ut}/\sigma_{uc})]\}^2}} \qquad [23]$$

(E) DESIGN RELATIONS FOR ALTERNATING STRESS

(a) Ductile Materials

For alternating (completely reversed cyclic) stress, stress concentration effect must be considered. As explained in Section C, the fatigue notch factor K_f is usually less than the stress concentration factor K_t. The factor K_{tf} represents a calculated estimate of the actual fatigue notch factor K_f. Naturally, if K_f is available from tests one uses this, but a designer is very seldom in such a fortunate position. The expression for K_{tf} and K_{tsf}, [12] and [13], respectively, are repeated here:

$$K_{tf} = q(K_t - 1) + 1$$

$$K_{tsf} = q(K_{ts} - 1) + 1$$

The following expressions for factors of safety, [24] through [26], are based on the Mises criterion of failure as discussed in Section B.

For axial or bending loading (normal stress):

$$n = \frac{\sigma_f}{K_{tf} \sigma_a} = \frac{\sigma_f}{[q(K_t - 1) + 1]\sigma_a} \qquad [24]$$

where σ_f = fatigue limit (endurance limit) in axial or bending test (normal stress)
σ_a = alternating stress amplitude (normal stress)

For torsion of a round bar (shear stress):

$$n = \frac{\tau_f}{K_{tsf}\tau_a} = \frac{\sigma_f}{\sqrt{3}\ K_{tsf}\tau_a} = \frac{\sigma_f}{\sqrt{3}\ [q(K_{ts} - 1) + 1]\tau_a} \qquad [25]$$

where τ_f = fatigue limit in torsion
τ_a = alternating stress amplitude (shear stress)

For combined normal stress and shear stress:

$$n = \frac{\sigma_f}{\sqrt{(K_{tf}\sigma_a)^2 + 3(K_{tsf}\tau_a)^2}} \qquad [26]$$

By rearranging [26], the following equation for an ellipse is obtained:

$$\frac{\sigma_a^2}{(\sigma_f/nK_{tf})^2} + \frac{\tau_a^2}{(\sigma_f/n\sqrt{3}\ K_{tsf})^2} = 1 \qquad [27]$$

where σ_f/nK_{tf} and $\sigma_f/n\sqrt{3}\ K_{tsf}$ are the major and minor semiaxes. Fatigue tests of unnotched specimens by Gough and Pollard[28] and by Nisihara and Kawamoto[29] are in excellent agreement with the elliptical relation. Fatigue tests of notched specimens[73] are not in as good agreement with the elliptical relation as are the unnotched, but for design purposes the elliptical relation seems reasonable for ductile materials.

(b) Brittle Materials

Since our knowledge in this area is very limited, it is suggested that unmodified K_t factors be used. The Mohr theory (see Section B of this chapter and Appendix B, Ref. 1) is suggested for design purposes for brittle materials subjected to alternating stress.

For axial or bending loading (normal stress):

$$n = \frac{\sigma_f}{K_t\sigma_a} \qquad [28]$$

For torsion of a round bar (shear stress), using [116], Appendix B, Ref. 1:

$$n = \frac{\tau_f}{K_{ts}\tau_a} = \frac{\sigma_f}{K_{ts}\tau_a}\left[\frac{1}{1 + (\sigma_{ut}/\sigma_{uc})}\right] \qquad [29]$$

For combined normal stress and shear stress:

$$n = \frac{2\sigma_f}{K_t\sigma_a + \sqrt{(K_t\sigma_a)^2 + 4\ \{K_{ts}\tau_a\ [1 + (\sigma_{ut}/\sigma_{uc})]\}^2}} \qquad [30]$$

(F) DESIGN RELATIONS FOR COMBINED ALTERNATING AND STEADY STRESSES

The majority of important strength problems comprises neither simple static nor alternating cases, but involves fluctuating stress, which is a combination of both.[8] A cyclic fluctuating

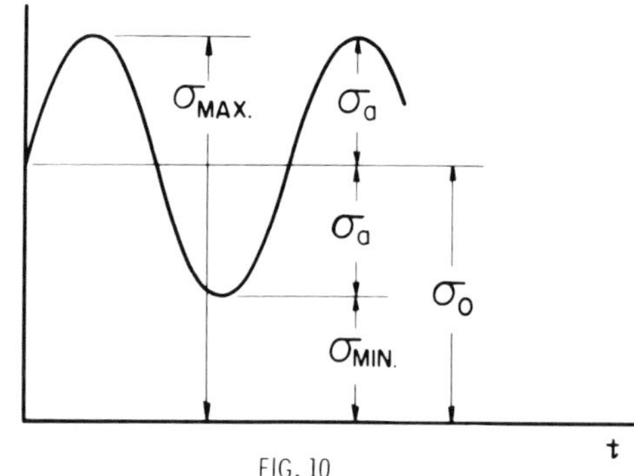

FIG. 10

COMBINED ALTERNATING AND STEADY STRESSES

stress (Fig. 10) having a maximum value σ_{max} and minimum value σ_{min} can be considered as having an *alternating component* of amplitude

$$\sigma_a = \frac{\sigma_{max} - \sigma_{min}}{2} \tag{31}$$

and a *steady component*

$$\sigma_o = \frac{\sigma_{max} + \sigma_{min}}{2} \tag{32}$$

(a) Ductile Materials

It is the usual practice, in designing parts to be made of ductile materials for normal temperature use, to apply the stress concentration factor to the alternating component but not to the steady component.[8] This appears to be a reasonable procedure and is in conformity with test data[74] such as that shown in Fig. 11*a* (see limitations stated in Section D).

By plotting minimum and maximum limiting stresses (Fig. 11*a*), the relative positions of the static properties, such as yield strength and tensile strength, are clearly shown. However, one can also use a simpler representation such as that of Fig. 11*b*, with alternating component as ordinate.

If, in Fig. 11*a*, the curved lines are replaced by straight lines connecting the end points σ_f and σ_u, σ_f/K_{tf} and σ_u, we have a simple approximation which is on the safe side for steel members.* From Fig. 11*b* we can obtain the following simple rule for factor of safety:

$$n = \frac{1}{\sigma_o/\sigma_u + K_{tf}\sigma_a/\sigma f} \tag{33}$$

This is the same as the Soderberg rule, [34], except that σ_u is used instead of σ_y; that is the Soderberg rule is based on the yield strength[8] (see lines in Fig. 11 connecting σ_f and σ_y, σ_f/K_{tf} and σ_y; see also Appendix D, Ref. 1):

$$n = \frac{1}{\sigma_o/\sigma_y + K_{tf}\sigma_a/\sigma_f} \tag{34}$$

*For steel members, a cubic relation[75,76] fits available data fairly well, $\sigma_{Ka} = \{\sigma_f/7K_f\}\ \{8 - [(\sigma_o/\sigma_u) + 1]^3\}$; this is the equation for the lower full curve of Fig. 11*b*. For certain aluminum alloys, the σ_a, σ_o curve has a shape[77] which is concave slightly below the σ_f/K_f, σ_u line at the upper end and is above the line at the lower end.

By referring to Fig. 11*b* it can be shown that $n = (OB/OA)$. Note that in Fig. 11*a*, the pulsating (0 to max) condition corresponds to $\tan^{-1} 2$, or approximately $63\frac{1}{2}°$; in Fig. 11*b*, 45°.

Expression [34] may be further modified to be in conformity with [15] and [16], which means applying limit design for yielding, with the factors and considerations as stated following [16]:

$$n = \frac{1}{(\sigma_{od}/\sigma_y) + (\sigma_{ob}/L_b\sigma_y) + (K_{tf}\,\sigma_a/\sigma_f)} \qquad [35]$$

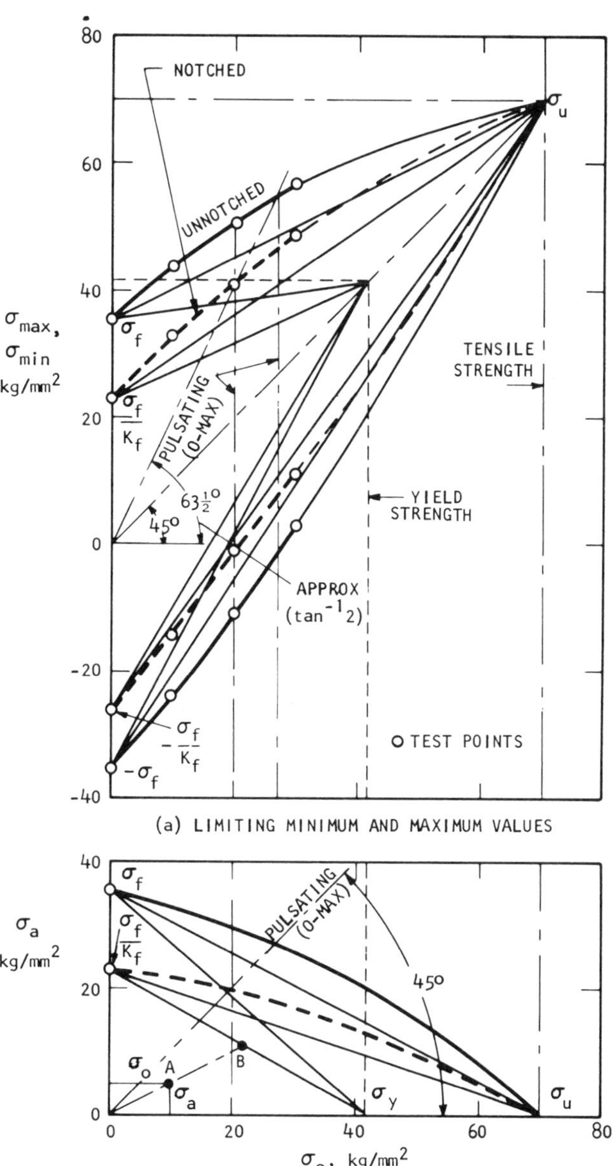

FIG. 11—LIMITING VALUES OF COMBINED ALTERNATING AND STEADY STRESSES FOR PLAIN AND NOTCHED SPECIMENS. (Data of Schenck, 0.7% C Steel)

As mentioned previously $L_b\sigma_y$ must not exceed σ_u, that is, the factor of safety n from [35] must not exceed n from [33].

For torsion, the same assumptions and use of the Mises criterion result in the following:

$$n = \frac{1}{\sqrt{3}\,[(\tau_o/L_s\sigma_y) + (K_{tsf}\,\tau_a/\sigma_f)]} \qquad [36]$$

For notched specimens, [36] represents a reasonable design relation, being on the safe edge of test data.[78] It is interesting to note that, for unnotched torsion specimens, steady torsion (up to a maximum stress equal to the yield strength in torsion) does not lower the limiting alternating torsional range.[78] It is apparent that further research is needed in the torsion region; however, since notch effects are involved in design (almost without exception), the use of [36] is indicated. Even in the absence of stress concentration, [36] would be on the "safe side," but by a large margin for relatively large values of steady torque.

For a combination of *steady and alternating normal stresses plus steady and alternating shear stresses* (alternating components in phase) the following relation, derived by Soderberg,[9] is based on expressing the shear stress on an arbitrary plane in terms of steady and alternating components, assuming failure is governed by the maximum shear theory and a "straight-line" relation similar to [34], and finding the plane which gives a minimum factor of safety, n (see Appendix D, Ref. 1):

$$n = \frac{1}{\sqrt{[(\sigma_o/\sigma_y) + (K_t\sigma_a/\sigma_f)]^2 + 4[(\tau_o/\sigma_y) + (K_{ts}\tau_a/\sigma_f)]^2}} \qquad [37]$$

The following modifications are made to correspond to the end conditions represented by [15], [16], [18], [24], and [25]. Then [37] becomes

$$n = \frac{1}{\sqrt{[(\sigma_{od}/\sigma_y) + (\sigma_{ob}/L_b\sigma_y) + (K_{tf}\,\sigma_a/\sigma_f)]^2 + 3[(\tau_o/L_s\sigma_y) + (K_{tsf}\,\tau_a/\sigma_f)]^2}} \qquad [38]$$

For steady stress only, [38] reduces to [19].
For alternating stress only, [38] reduces to [26].
For normal stress only, [38] reduces to [35].
For torsion only, [38] reduces to [36].

In tests by Ono[79] and by Lea and Budgen[80] the alternating bending fatigue strength was found not to be affected by the addition of a steady torque (less than yield torque). Other tests reported in a discussion by Davies[81] indicate a lowering of the bending fatigue strength by the addition of steady torque. Hohenemser and Prager[82] found that a steady tension lowered the alternating torsional fatigue strength; Gough and Clenshaw[73] found that steady bending lowered the torsional fatigue strength of plain specimens but that the effect was smaller for specimens involving stress concentration. Further experimental work is needed in this area of special combined stress combinations, especially in the region involving the additional effect of stress concentration. In the meantime, while it appears that use of [38] may be overly "safe" in certain cases of alternating bending plus steady torque, it is believed that [38] provides a reasonable general design rule.

(b) Brittle Materials

A "straight-line" simplification similar to that of Fig. 11 and [33] can be made for brittle material, except that stress concentration effect is considered to apply also to the steady component:

$$n = \frac{1}{K_t\,[(\sigma_o/\sigma_{ut}) + (\sigma_a/\sigma_f)]} \qquad [39]$$

As previously mentioned, unmodified K_t factors are used for the brittle material cases.

For *combined shear and normal stresses* data are very limited. For combined alternating bending and steady torsion, Ono[79] reported a decrease of the bending fatigue strength of cast iron as steady torsion was added. By use of the Soderberg method[9] and basing failure on the normal stress criterion (Appendix D, Ref. 1), we obtain

$$n = \frac{2}{K_t\,[(\sigma_o/\sigma_{ut}) + (\sigma_a/\sigma_f)] + \sqrt{K_t^2\,[(\sigma_o/\sigma_{ut}) + (\sigma_a/\sigma_f)]^2 + 4K_{ts}^2\,[(\tau_o/\sigma_{ut}) + (\tau_a/\sigma_f)]^2}} \qquad [40]$$

To correspond to the Mohr theory (end conditions [20], [22], [28], and [29]), [40] is modified as follows:

$$n = \frac{2}{K_t\,[(\sigma_o/\sigma_{ut}) + (\sigma_a/\sigma_f)] + \sqrt{K_t^2\,[(\sigma_o/\sigma_{ut}) + (\sigma_a/\sigma_f)]^2 + 4K_{ts}^2\,\{[(\tau_o/\sigma_{ut}) + (\tau_a/\sigma_f)]\,[1 + (\sigma_{ut}/\sigma_{uc})]\}^2}} \qquad [41]$$

For steady stress only, [41] reduces to [23].
For alternating stress only, [41] reduces to [30].
For normal stress only, [41] reduces to [39].
For torsion only, [41] reduces to

$$n = \frac{1}{K_{ts}\,[(\tau_o/\sigma_{ut}) + (\tau_a/\sigma_f)]\,[1 + (\sigma_{ut}/\sigma_{uc})]} \qquad [42]$$

This in turn can be reduced to the component cases [22] and [29].

(G) LIMITED NUMBER OF CYCLES OF ALTERNATING STRESS

The previous edition of this book[1] presented formulas for a limited number of cycles (upper branch of the S-N diagram). These relations were based on an average of available test data and therefore apply to polished test specimens 0.2 to 0.3 in. diameter. If the member being designed is not too far from this size range, the formulas may be useful as a rough guide, but otherwise the formulas are questionable since the number of cycles required for a crack to propagate to rupture of a member depends on the size of the member.

Fatigue failure consists of three stages: crack initiation, crack propagation, and rupture. Crack initiation is thought to be not strongly dependent on size, although from statistical consideration of number of "weak spots," one would expect some effect. In recent years, important progress has been made in the area of knowledge about crack propagation under cyclic stress.[83,84] Although this information has not yet been converted to convenient design use for the types of stress distribution found in practice, it is believed that reasonable estimates can be made for a number of problems.

CHAPTER 2

NOTCHES AND GROOVES

The U-shaped notch or circumferential groove (of which the semicircular shape is a special case) is a geometrical shape of considerable interest in engineering. It occurs in machine elements: in turbine rotors between blade rows and at seals; in a variety of shafts (Fig. 12) as a shoulder relief groove or as a retainer for a spring washer; and in numerous other design configurations.

The round-bottomed V-shaped notch or circumferential groove (and to a lesser extent, the U-shaped notch) is a conventional contour shape for stress-concentration test pieces in the areas of fatigue, creep-rupture, and brittle fracture.

A threaded part may be considered an example of a multi-grooved member.

Two basic K_t factors may be defined: K_{tg}, based on the larger (gross) section of width D; and K_{tn}, based on the smaller (net) section of width d (see sketch on Fig. 16). For tension (Fig. 16), $K_{tg} = \sigma_{\max}/\sigma$, where $\sigma = P/hD$; $K_{tn} = \sigma_{\max}/\sigma_{\text{nom}}$, where $\sigma_{\text{nom}} = P/hd$. Since design calculations are usually based on the critical section (width d) where $\sigma_{\max}$ is located, K_{tn} is the generally used factor; unless otherwise specified K_t refers to K_{tn}.

The difference between K_{tg} and K_{tn} is illustrated in Fig. 16. Assuming a constant width D and a constant force P, as notches are cut deeper (increasing $2r/D$), K_{tn} decreases, reflecting a decreasing stress concentration (peak stress divided by average stress across d) until as $2r/D \rightarrow 1$, $K_t \rightarrow 1$, in effect a uniform stress tension specimen. K_{tg} increases as $2r/D$ increases, reflecting the increase in $\sigma_{\max}$ owing to the loss of section.

In the earlier book,[1] Neuber's method[85] was used for the stress factors for the flat bar with single opposite notches and for the round bar with a single circumferential groove. The Neuber method makes use of the exact values of the deep hyperbolic notch and the shallow elliptical notch (in infinitely wide members)[86] and modifies these for finite width members by using the following ingenious simple relation, which has the correct end conditions:

$$K_t = 1 + \sqrt{\frac{(K_{te} - 1)^2 (K_{th} - 1)^2}{(K_{te} - 1)^2 + (K_{th} - 1)^2}} \qquad [43]$$

where K_{te} = stress concentration factor for a shallow elliptical notch (with same t/r as U-notch) in a semi-infinitely wide member

K_{th} = stress concentration factor for a deep hyperbolic notch (with same r/d as U-notch) in an infinitely wide member

t = notch depth

r = notch radius (minimum contour radius)

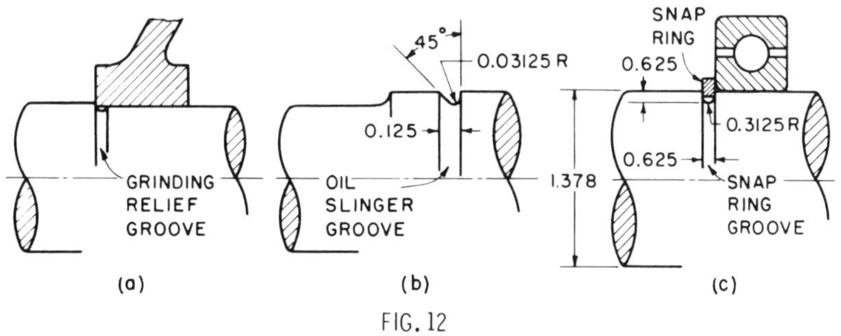

FIG. 12

EXAMPLES OF GROOVED SHAFTS
(Dimensions in inches)

d = minimum diameter or width of member
D = maximum diameter or width of member

Since use of the exponent 2 is arbitrary, relation [43] is not exact. Recent investigations have provided more accurate values for the parameter ranges covered by the investigations, as will be presented in the following sections. If the actual member being designed has a notch or groove which is either very deep or shallow, the Neuber approximation will be close, but for values of d/D in the region of 1/2, the Neuber K_t can be as much as 12% too low, which is on the unsafe side from a design standpoint. More accurate values have been obtained over the most used ranges of parameters; these form the basis of new charts presented here. However, when a value for an extreme condition such as a very small or large r/d is sought, the Neuber method is the only means of obtaining a useful factor (some charts covering extreme ranges are also included in this book).

Another use of the charts of Neuber factors is in designing a test piece for maximum K_t, as detailed in Section D of this chapter.

The K_t factors for the flat members covered in this chapter are for two-dimensional states of stress (plane stress) and apply strictly to very thin sheets, or more strictly to where $t/r \to 0$, where t = plate thickness and r = notch radius. As t/r increases, a state of plane strain is approached, in which case the stress at the notch surface at the middle of the plate thickness increases and the stress at the plate surface decreases. Some guidance may be obtained by referring to the introductory remarks at the beginning of Chapter 4.

The K_t factor for a notch can be lowered by use of a reinforcing bead.[86a]

(A) TENSION (AXIAL LOADING)

(a) Opposite Deep Hyperbolic Notches in an Infinite Plate; Shallow Elliptical, Semicircular, U-Shaped or Keyhole-Shaped Notch in a Semi-Infinite Plate; Equivalent Elliptical Notch

In Fig. 14, K_t values are given for the deep ($d/D \to 0$) hyperbolic notch in an infinite plate.[85,86] In Fig. 15, K_t values are given for an elliptical or U-shaped notch in a semi-infinite plate.[87–89] For the higher t/r values, K_t for the U-notch is up to 1% higher than for the elliptical notch. For practical purposes the solid curve of Fig. 15 covers both cases.

The semicircular notch ($t/r = 1$) in a semi-infinite plate has been studied by a number of investigators; the following summary of their resulting K_t factors is given by Ling:[90]

1936	Maunsell	3.05
1940	Isibasi	3.06
1941	Weinel	3.063
1948	Ling	3.065
1965	Yeung	3.06
1965	Mitchell	3.08

In many reports the writer has used $K_t = 3.065$, which is satisfactory for design use.

Similar to the "equivalent elliptical hole" in an infinite plate (see Chapter 4, Section A.t), an "equivalent elliptical notch" in a semi-infinite plate may be defined as an elliptical notch that has the same t/r and envelops the notch geometry under consideration. All such notches, U-shaped, keyhole (circular hole connected to edge by saw-cut), and so on have very nearly the same K_t as the equivalent elliptical notch. The "equivalent elliptical notch" applies for tension; it is not applicable for shear.

Factors have been approximated by splitting a plate with a central hole axially through the middle of the hole and using the K_t for the hole to represent the resulting notches.[91] The elliptical hole factors[231]

$$K_t = 1 + 2\sqrt{\frac{t}{r}} \tag{44}$$

are shown on Fig. 15. The factors for the U-shaped slot[232] are practically the same. A comparison of the curves for notches and holes in Fig. 15 shows that the preceding approximation can be in error by as much as 10% for the larger values of t/r.

(b) Opposite Single Semicircular Notches in a Finite-Width Plate

For the tension case of opposite semicircular notches in a finite-width plate, K_t factors[92–95] are given in Fig. 16.

Slot[95a] found that with $r/D = \frac{1}{4}$ in a strip of length $1.5D$, good agreement was obtained with the stress distribution for σ applied at infinity.

(c) Opposite Single U-Shaped Notches in a Finite-Width Plate

Strain gage tests,[96] photoelastic tests,[97] and mathematical analysis[98] provide consistent data for the flat bar (two-dimensional case) of Fig. 17. An important check is provided by including in Fig. 17 the curve representing mathematical results[92,93] for the semicircular notch (Fig. 16), a special case of the U-notch; the agreement is excellent for values of $D/d \lesseqqgtr 2$. Recent photoelastic results[98a] for $D/d = 1.05$ are in good agreement.

Barrata[98b] has compared empirical formulas for K_t with experimentally determined values and concludes that the following two formulas are satisfactory for predictive use.

Barrata and Neal:[89]

$$K_{tn} = \left(0.780 + 2.243\sqrt{\frac{t}{r}}\right)\left[0.993 + 0.180\left(\frac{2t}{D}\right) - 1.060\left(\frac{2t}{D}\right)^2 + 1.710\left(\frac{2t}{D}\right)^3\right]\left(1 - \frac{2t}{D}\right) \tag{44a}$$

Heywood:[98c]

$$K_{tn} = 1 + \left[\frac{t/r}{1.55\,(D/d) - 1.3}\right]^n \qquad [44b]$$

$$n = \frac{D/d - 1 + 0.5\sqrt{t/r}}{D/d - 1 + \sqrt{t/r}}$$

Referring to Fig. 17, formula [44a] gives values in good agreement with the solid curves for $r/d < 0.25$; formula [44b] is in better agreement for $r/d > 0.25$. For the dashed curves (not the dot-dash curve for semicircular notches), formula [44a] gives lower values as r/d increases. The tests on which the formulas are based do not include parameter values corresponding to the dashed curves, which are uncertain owing to their determination by interpolation of r/d curves having D/d as abscissae. In the absence of better basic data, the dashed curves, representing higher values, should be used for design.

In Fig. 17 the values of r/d are from 0 to 0.3 and the values of D/d are from 1 to 2, covering the most widely used parameter ranges. There is considerable evidence[96–98] that for greater values of r/d and D/d, the K_t versus D/d curve for a given r/d does not flatten out but reaches a peak value and then decreases slowly toward a somewhat lower K_t value as $D/d \to \infty$. The effect is small and is not shown on Fig. 17.

In Fig. 17 the range of parameters corresponds to the recent investigations.[96–98] For smaller and larger r/d values, the Neuber values (Figs. 18 and 19), although approximate, are the only wide-range values available and are useful for certain problems. The largest errors are at the mid-region of d/D; for shallow or deep notches, the error becomes progressively smaller. Some specific photoelastic tests[98d] with $d/D \backsim 0.85$ and r/D varying from $\backsim 0.001$ to 0.02 gave higher K_t values than Fig. 18.

(d) "Finite-Width Correction Factors" for Opposite Narrow Single Elliptical Notches in a Finite-Width Plate

For the very narrow elliptical notch, approaching a crack (Fig. 20), "finite-width correction" formulas have been proposed by Dixon,[99] Westergaard,[100] Irwin,[101] Bowie,[102] Brown and Srawley,[103] and Koiter.[104]

The formula[89,103] based on Bowie's results is satisfactory for values of $2l/w < 0.5$:

$$\frac{K_{tg}}{K_{t\infty}} = 0.993 + 0.180\left(\frac{2l}{w}\right) - 1.060\left(\frac{2l}{w}\right)^2 + 1.710\left(\frac{2l}{w}\right)^3 \qquad [45]$$

where l = crack length

$$\frac{K_{tn}}{K_{t\infty}} = \frac{K_{tg}}{K_{t\infty}}\left(1 - \frac{2l}{w}\right)$$

The Koiter[104] formula [46]* covers the entire $2l/w$ range (Fig. 20); for the lower $2l/w$ range agreement with [45] is good; for the mid-$2l/w$ range, [46] gives somewhat higher values:

$$\frac{K_{tg}}{K_{t\infty}} = \left[1 - 0.50\left(\frac{2l}{w}\right) - 0.0134\left(\frac{2l}{w}\right)^2 + 0.081\left(\frac{2l}{w}\right)^3\right]\left[1 - \frac{2l}{w}\right]^{-1/2} \qquad [46]$$

*Reference 104, formula [9] divided by $K_{t\infty} = 1.122$ and $\sqrt{1 - 2l/w}$.

Formulas [45] and [46] represent the ratio of stress-intensity factors; in the small-radius, narrow-notch limit, the ratio is valid for stress concentration.[105,106]

As shown in Fig. 20, the results for the semicircular notch[92,93] are consistent with the results for the "crack"; for $2l/w < 0.5$ the curves are nearly in agreement.

Note that the endpoints for the $K_{tn}/K_{t\infty}$ curves are 1.0 at $2l/w = 0$ and $1/K_{t\infty}$ at $2l/w = 1$. The $1/K_{t\infty}$ values at $2l/w = 1$ for elliptical notches are obtained from $K_{t\infty}$ of Fig. 15; these $1/K_{t\infty}$ values at $2l/w = 1$ are useful in sketching in approximate values for elliptical notches.

(e) Opposite Single V-Shaped Notches in a Finite-Width Plate

$K_{t\alpha}$ values have been obtained[98] for the flat tension bar with opposite V notches as a function of the V angle, α (Fig. 21). The Leven-Frocht method[107] of relating $K_{t\alpha}$ to the K_t of a corresponding U notch, as used in Fig. 21, shows that for $D/d = 1.66$ the angle has little effect up to 90°; for $D/d = 3$ it has little effect up to 60°. In comparing these results with Fig. 38, where the highest $D/d = 1.82$, the agreement is good, even though the two cases are different (symmetrical notches, in tension; notch on one side, in bending).

(f) Single Notch on One Side of a Plate or Bar

Neuber[107a] has obtained approximate K_t values for a semi-infinite plate with a deep hyperbolic notch, wherein tension loading is applied along a midline through the minimum section (Fig. 22). Figure 22*a* presents K_t curves based on photoelastic tests;[108] corresponding Neuber K_t factors obtained by use of Fig. 22 and relation [43] are on the average 18% lower than the K_t factors of Fig. 22*a*.

The curve for the semicircular notch is obtained by noting that for this case $D/d = 1 + r/d$ and that $K_t = 3.065$ at $r/d \rightarrow 0$.

(g) Multiple Notches in a Plate

It has long been recognized that a single notch represents a higher degree of stress concentration than a series of closely spaced notches of the same kind as the single notch. Considered from the standpoint of flow analogy (Fig. 13), a smoother flow is obtained in Figs. 13*b* and 13*c*.

For the infinite row of semicircular edge notches, factors have been obtained mathematically[109] as a function of notch spacing and relative width of bar, with results summarized in Figs. 23 and 24. For infinite notch spacing, the K_t factors are in agreement with the single notch factors of Isida and Ling (Fig. 16).

For a specific case[95a] with $r/D = \frac{1}{4}$ and $b/a = 3$, good agreement was obtained with the corresponding Atsumi[109] value.

An analysis[110] of a semi-infinite plate with an edge of wave form of depth t and minimum radius r gave $K_t = 2.13$ for $t/r = 1$ and $b/a = 2$, which is in agreement with Fig. 23.

K_t factors are available for the case of an infinite row of circular holes in a plate stressed in tension in the direction of the row.[111,111a] If we consider the plate as split along the axis of

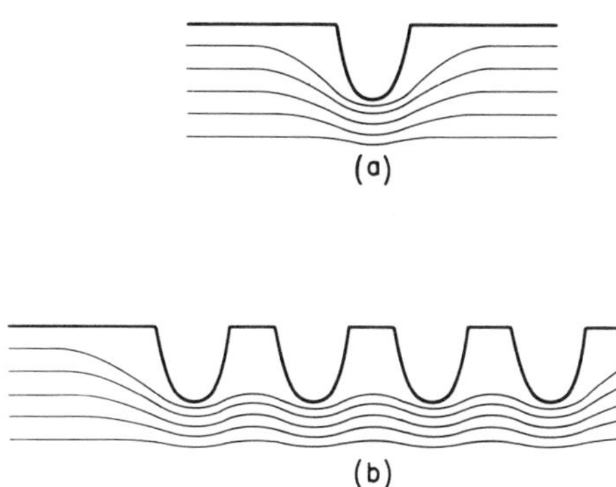

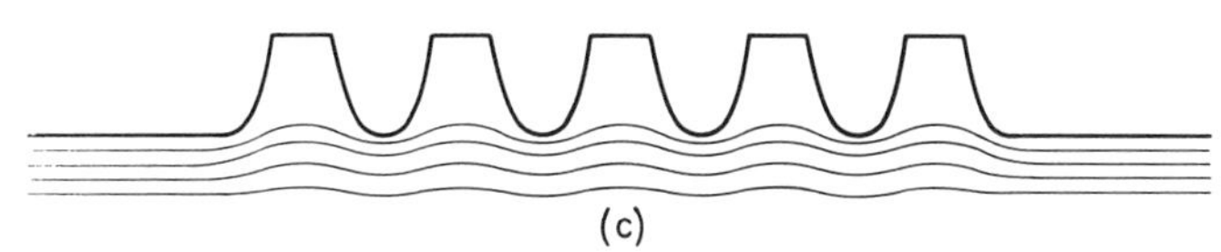

FIG. 13

MULTIPLE NOTCHES

the holes, the K_t values should be nearly the same (for the single hole, $K_t = 3$; for the single notch, $K_t = 3.065$). The K_t curve for the holes as a function of b/a fits well (slightly below) the top curve of Fig. 23.

Use of the "load relieving factor" of Neuber[112] gives K_t factors that are, for this case of semicircular notches, too high.

For a finite number of multiple notches (Fig. 13*b*), the stress concentration of the intermediate notches is considerably reduced. The maximum stress concentration occurs at the end notches[113] (Figs. 26 and 27), and this is also reduced (as compared to a single notch) but to a lesser degree than for the intermediate notches. Sometimes a member can be designed as in Fig. 13*c*, resulting in a reduction of stress concentration as compared to Fig. 13*b*.

Factors for two pairs of notches[114] in a square pattern ($b/D = 1$) are included on Fig. 24.

Photoelastic tests have been made for various numbers (up to six) of semicircular notches.[113,115] The results (Figs. 25, 26, and 27) are consistent with the recent mathematical factors for the infinite row.[109] In Fig. 26 K_t for a groove which corresponds to a lower K_t limit for any number of semicircular notches with overall edge dimension c is shown for comparison.

For the Aero thread shape (semicircular notches with $b/a = 1.33$), two-dimensional photoelastic tests[115] of six notches gave K_t values of 1.94 for the intermediate notch and 2.36 for the end notch. For the Whitworth thread shape (V notch with rounded bottom), the corresponding photoelastic tests[115] gave K_t values of 3.35 and 4.43, respectively.

Fatigue tests[116] of threaded specimens and specimens having a single groove of the same dimensions showed considerably higher strength for the threaded specimens.

Bolt and nut combinations are discussed in Chapter 5.

(h) Hemispherical Depression (Pit) in the Surface of a Semi-Infinite Body

For equal biaxial stress, $K_t = 2.23$ was found[117] for $\nu = \frac{1}{4}$. This is about 7% higher than for the corresponding case of a spherical cavity ($K_t = 2.09$ for $\nu = \frac{1}{4}$). Moreover, the semicircular edge notch in tension ($K_t = 3.065$) is about 2% higher than the circular hole in tension ($K_t = 3$).

(i) Hyperboloid Depression (Pit) in the Surface of a Finite-Thickness Plate

The hyperboloid depression simulates the type of pit caused by meteoroid impact of an aluminum plate.[118] For equal biaxial stress, values of K_t in the range of 3.4 to 3.8 were obtained for individual specific geometries as detailed in the report.[119] The authors point out that the K_t values are higher than for complete penetration in the form of a circular hole ($K_t = 2$).

(j) Opposite Shallow Spherical Depressions (Dimples) in a Plate

This particular geometry has been suggested for a test piece in which a crack forming at the thinnest location can progress only into a region of lower stress.[120] "Dimpling" is often used to remove a small surface defect; if the depth is small relative to the thickness (h_o/h approaching 1.0 in Fig. 28), the stress increase is small. The $K_{tg} = \sigma_{\max}/\sigma$ values are shown for uniaxial stressing[120] in Fig. 28; these values also apply for equal biaxial stressing.[120]

The calculated values of Fig. 28 are for a shallow spherical depression having a diameter greater than four or five times the thickness of the plate. In terms of the variables given in Fig. 28 the spherical radius is

$$r = \frac{1}{4}\left[\frac{a^2}{(h - h_o)} + (h - h_o)\right] \tag{47}$$

For $a \gtreqless 5h$, $r/h_o > 25$. For such a relatively large radius, the stress increase for thin section^{S} ($h_o/h \rightarrow 0$) is due to the thinness of section rather than stress concentration per se (i.e., stress gradient is not steep).

For comparison, K_{tg} for a groove having the same sectional contour as the dimple is shown by the dashed line on Fig. 28, the K_{tg} values being calculated from the K_{tn} values of Fig. 19. Note that in Fig. 19 the K_{tn} values for $r/d = 25$ represent a stress concentration of about 1%; the K_{tg} factors therefore essentially represent the loss of section.

Removal of a surface defect in a thick section by means of creating a relatively shallow spherical depression results in negligible stress concentration, on the order of 1%.

(k) Deep Hyperbolic Groove in an Infinite Member (Circular Net Section)

Exact K_t values for Neuber's solution[85,86] are given in Fig. 29. Note that Poisson's ratio has only a relatively small effect.

(l) U-Shaped Circumferential Groove in a Round Bar

The K_t value for the round bar with a U groove (Fig. 30) is obtained by multiplying the K_t of Fig. 17 by the ratio of the corresponding Neuber three-dimensional K_t over two-dimensional K_t values. This is an approximation (as indicated on Fig. 30), because of lack of research on this particular case; however, after comparison with the bending and torsion cases, the results seem reasonable.

Cheng[120a] has by a photoelastic test obtained $K_t = 1.85$ for $r/d = 0.209$ and $D/d = 1.505$. The corresponding K_t from Fig. 30 is 1.92, which agrees fairly well with Cheng's value, which he believes to be somewhat low.

Approximate K_t factors are given in Fig. 31 for smaller r/d values and in Fig. 32 for larger r/d values (e.g., test specimens).

(B) BENDING

(a) Opposite Deep Hyperbolic Notches in an Infinite Plate

Exact K_t values of Neuber's solution[85,86] are presented in Fig. 33.

(b) Opposite Single Semicircular Notches in a Finite-Width Plate

Figure 34 provides K_t values determined mathematically.[90,92]

Slot[95a] found that with $r/D = \frac{1}{4}$ a strip of length $1.5D$, good agreement was obtained with the stress distribution for M applied at infinity.

(c) Opposite Single U-Shaped Notches in a Finite-Width Plate

The curves of Fig. 35 are obtained by increasing the photoelastic values[121] which as in tension are known to be low, to agree with the semicircular notch mathematical values of Fig. 34, which are assumed to be accurate. Recent photoelastic[98a] and numerical results[121a] are in good agreement.

Approximate K_t values for extended r/d values are given in Figs. 36 and 37.

(d) V-Shaped Notches in a Plate

The effect of notch angle on the K_t value is presented in Fig. 38 for a bar in bending with a notch on one side.[107] The K_t value is for a U notch; $K_{t\alpha}$ is for a notch with inclined sides having an included angle α but with all other dimensions the same as for the corresponding K_t case. The curves of Fig. 38 are based on data from specimens covering a D/d range up to 1.82; any effect of D/d up to this value is sufficiently small that single α curves are adequate. For larger D/d values, the α curves may be lower (see Fig. 21).

(e) Notch on One Side of a Plate or Bar; Finite-Width Correction Factor for Crack on One Side of Bar

Figure 39 provides K_t for bending of a semi-infinite plate with a deep hyperbolic notch.[121b] In Fig. 39*a*, K_t curves based on photoelastic tests[107] are given; corresponding Neuber K_t factors obtained by use of Fig. 39 and relation [43] are on the average 6% higher than the K_t factors of Fig. 39*a*.

The curve for the semicircular notch is obtained by noting that for this case $D/d = 1 + r/d$ and that $K_t = 3.065$ at $r/d \to 0$.

In Fig. 39*b*, finite-width correction factors are given for a bar with a notch on one side; the full curve represents a crack[121c] and the dashed curved a semicircular notch.[107] The correction factor for the crack is the ratio of stress-intensity factors; in the small-radius, narrow-notch limit, the ratio is valid for stress concentration.[105,106] Note that the endpoints of the curves are 1.0 at $l/w = 0$ and $1/K_{t\infty}$ at $l/w = 1$. The $1/K_{t\infty}$ values at $l/w = 1$ for elliptical notches are obtained from $K_{t\infty}$ of Fig. 15; these $1/K_t$ values at $l/w = 1$ are useful in sketching in approximate values for elliptical notches.

If $K_{tg}/K_{t\infty}$ (not shown in Fig. 39*b*) is plotted, the curves start at 1.0 at $l/w = 0$ and then dip below 1.0, reach a minimum in the $l/w = 0.10$ to 0.15 range, and then turn upward to go to infinity at $l/w = 1.0$. This means that for bending the effect of the nominal stress gradient is to cause σ_{max} to decrease slightly as the notch is cut into the surface, but beyond a depth of $l/w = 0.25$ to 0.3 the maximum stress is greater than for the infinitely deep bar. The same effect, only of slight magnitude, was obtained by Isida[92] for the bending case of a bar with opposite semicircular notches (Fig. 34; K_{tg} not shown). In tension, since there is no nominal stress gradient, this effect is not obtained.

Fig. 40 gives K_t for various impact specimens.

(f) Single or Multiple Notches with Semicircular or Semielliptical Notch Bottoms

From work on propellant grains, it is known[122] "that invariably the stress concentration factor for an optimized semielliptic notch is significantly lower than that for the more easily formed semicircular notch."

Photoelastic tests[122–124] were made on beams in bending, with variations of beam and notch depth, notch spacing, and semielliptical notch bottom shape. The ratio of beam depth to notch depth D/t (notch on one side only) varied from 2 to 10; Fig. 41 provides results for $D/t = 5$.

For the single notch with $t/(b/2) = 2.666$, the ratio of K_t for the semicircular bottom to the K_t for the optimum semielliptical bottom, $b/a = 2.4$, is 1.25 (see Fig. 41). In other words, a considerable stress reduction (20% in this case) results from using a semielliptical notch bottom instead of a semicircular notch bottom. As can be found from Fig. 41, even larger stress reductions can be obtained for multiple notches.

Although these results are for a specific case of a beam in bending, it is reasonable to expect that, in general, a considerable stress reduction can be obtained by use of the semi-elliptic notch bottom.

The optimum b/a of the semi-ellipse varies from 1.8 to greater than 3, with the single notch and the wider spaced multiple notches averaging at about 2 and the closer spaced notches increasing toward 3 and greater.

Other uses of the elliptical contour are: the slot-end of Fig. 136, where the optimum b/a is about 3; and the shoulder fillet (see Fig. 77).

(g) Various Edge Notches in an Infinite Sheet in Transverse Bending

K_t factors for opposite deep hyperbolic notches[125,126] are given in Fig. 42. The factors were obtained for simple bending ($M_1 = 1$, $M_2 = 0$). The applicability of simple and cylindrical bending ($M_1 = 1$, $M_2 = \nu$) is discussed by Lee.[126]

Simple bending cases of a semi-infinte plate with a V-shaped notch or a rectangular notch with rounded corners[127] are covered in Fig. 43. At $r/t = 1$, both curves have the same K_t (semicircular notch). Note that the curve for the rectangular notch has a minimum K_t value at about $r/t = \frac{1}{2}$.

In Fig. 44, K_t factors are given for the elliptical notch.[128] For comparison, the corresponding curve for the tension case is shown; note that the K_t factors are considerably higher.

In Fig. 45, the K_t factor for an infinite row of semicircular notches is given as a function of the notch spacing.[129] As the notch spacing increases, the K_t value for the single notch is approached asymptotically.

(h) Notches in Finite-Width Sheet in Transverse Bending

Approximate values have been obtained by the Neuber method[85,86] which makes use of the exact values for the deep hyperbolic notch[125,126] and the shallow elliptical nothch[128] in infinitely wide members and modifies these for finite-width members by using a second-power relation that has the correct end condtions. The results are shown for the thin sheet in Fig. 46.

No direct results are available for intermediate thicknesses. If we consider the tension case as representing maximum values for a thick plate in bending, we can use Fig. 17 for $t/h \rightarrow o$. For the thin plate ($t/h \rightarrow \infty$) use is made of Fig. 46, as described in the preceding paragraph. For intermediate thickness ratios, some guidance can be obtained from Fig. 160 (see also values on Fig. 162 in the region of $b/a = 1$.

(i) Deep Hyperbolic Groove in an Infinite Member (Circular Net Section)

K_t values for Neuber's exact solution[85,86] are given in Fig. 47.

(j) U-Shaped Circumferential Groove in a Round Bar

The K_t values of Fig. 48 are obtained by the method used in the tension case (see Section A.1). Approximate K_t factors for small r/d values are given in Fig. 49 and for large r/d values (e.g., test specimens) in Fig. 50.

(C) TORSION (SHEAR)

(a) Deep Hyperbolic Notches in an Infinite Plate in Shear

The factors given in Fig. 51 are from Neuber.[130] Shearing forces are applied parallel to the notch axis* as shown in Fig. 51.

*For equilibrium, the shear force couple $2bV$ must be counterbalanced by an equal couple symmetrically applied remotely from the notch.[130] To avoid possible confusion with the combined shear and bending case, the countercouple is not shown in Fig. 51.

The location of σ_{max} is at

$$x = \frac{r}{\sqrt{1 + (2r/d)}} \qquad [48]$$

The location of τ_{max} along the line corresponding to the minimum section is at

$$y = \frac{d}{2}\sqrt{\frac{(d/2r) - 2}{d/2r}} \qquad [48a]$$

The value of $K_t/2$, τ at 45° to σ_{max}, is greater than the K_{ts} value for the minimum section shown in Fig. 51.

For combined shear and bending, Neuber[130] shows that for large d/r values it is a good approximation to add the two K_t factors, even though the maxima do not occur at the same location along the notch surface.

The case of a twisted sheet with hyperbolic notches has been analyzed by Lee.[126]

(b) Deep Hyperbolic Groove in an Infinite Member (Circular Net Section)

The K_t values for Neuber's exact solution[85,86] are presented in Fig. 52.

(c) U-Shaped Circumferential Groove in a Round Bar

Figure 53 is based on electric analog results,[131] using a technique that has also provided results in agreement with the exact values for the hyperbolic notch in the parameter range of present interest. Mathematical results for semicircular grooves[132,133] are in reasonably good agreement with Fig. 53. The K_{ts} values of Fig. 53 are somewhat higher (up to 6%) than previously used Neuber values (Fig. 46 of Ref. 1), and they are also higher (average 4½%) than photoelastic values.[134] The present values[131] are not in agreement with certain other published values.[135,136]

Figure 54 shows a leveling of the K_{ts} curve at a D/d value of about 2 (or less for high r/d values). Approximate K_{ts} factors beyond the r/d range of Fig. 53 are given for small r/d values in Fig. 55 and for large r/d values (e.g., test specimens) in Fig. 56.

(d) V-Shaped Circumferential Groove in a Round Bar

Figure 57 shows $K_{ts\alpha}$ for the V groove,[131] with variable angle α, using the chart style as in Figs. 21 and 38. For $\alpha \lesseqgtr 90°$ the curves are approximately independent of r/d. For 135°, separate curves are needed for $r/d = 0.005$, 0.015, and 0.05. The effect of the V angle may be compared with Figs. 21 and 38.

(D) TEST SPECIMEN DESIGN FOR MAXIMUM K_t FOR A GIVEN r/D

In designing a test piece, suppose that we have a given outside diameter,* D, and for a particular notch bottom radius,† r, we wish to know the notch depth (or the d/D ratio) that gives a maximum K_t.

*Often dictated by available bar size.

†Minimum notch bottom radius often dictated by ability of shop to produce accurate, smooth, small radius.

From the curves of Figs. 18, 31, 36, 49, and 55, (see Ref. 1) maximum K_t values are plotted in Fig. 58 with r/D and d/D as variables. Although these values are approximate, in that the Neuber approximation is involved (as detailed in the introductory remarks at the beginning of this chapter), the maximum region is quite flat and therefore K_t is not highly sensitive to variations in d/D in the maximum region.

From Fig. 58 it can be seen that a rough guide for obtaining maximum K_t in a specimen in the most-used r/D range is to make the smaller diameter, or width, about three-fourths of the larger diameter, or width (assuming that one is working with a given r and D).

Another specimen design problem occurs when r and d are given (the smaller diameter d may, in some cases, be determined by the testing machine capacity). In this case K_t increases with increase of D/d, reaching a "knee" at a D/d value which depends on the r/d value (see Fig. 54). For the smaller r/d values a value of $d/D = \frac{1}{2}$ would be indicated, and for the larger r/d values $d/D = \frac{3}{4}$ would be appropriate.

(E) DESIGN EXAMPLE FOR A SHAFT WITH A CIRCUMFERENTIAL GROOVE

Suppose we wish to estimate the bending fatigue strength of the shaft shown in Fig. 12*c* for two materials: an axle steel (normalized 0.40% C), and a heat-treated 3½% nickel steel (SAE 2345). These materials will have fatigue strengths (endurance limits) of approximately 30,000 and 70,000 lb/in.^{2}, respectively, when tested in the conventional manner, with no stress concentration effect, in a rotating beam machine.

First we determine K_t. From Fig. 12*c*, $D = 1.378$ in., $d = 1.253$ in., $r = 0.03125$ in. We calculate $D/d = 1.10$ and $r/d = 0.025$; from Fig. 48 we find $K_t = 2.90$.

From Fig. 8 we obtain, for $r = 0.03125$ in., a q value of 0.76 for the axle steel and 0.93 for the heat-treated alloy steel.

Substituting in [12]:

$$K_{tf} = 1 + 0.76\,(2.90 - 1) = 2.44$$

$$\sigma_{tf} = \frac{30{,}000}{2.44} = 12{,}300 \text{ lb/in.}^2 \qquad \text{for the axle steel}$$

$$K_{tf} = 1 + 0.93(2.90 - 1) = 2.77$$

$$\sigma_{tf} = \frac{70{,}000}{2.77} = 25{,}200 \text{ lb/in.}^2 \qquad \text{for the heat-treated alloy steel}$$

This tells us that we can expect strength values of approximately 12,000 and 25,000 lb/in.^{2} under fatigue conditions for the shaft of Fig. 12*c* when the shaft is made of normalized axle steel and quenched-and-tempered alloy steel (as specified), respectively. These are not working stresses, since a factor of safety must be applied which depends on type of service, consequences of failure, and so on. Different factors of safety are used throughout industry depending on service and experience. The strength of a member, however, is not, in the same sense, a matter of opinion or judgment and should be estimated in accordance with the best methods available. Naturally, a test of the member is desirable whenever possible, but in any event an initial calculation is made and this should be done carefully and should include all known factors.

FIG. 14

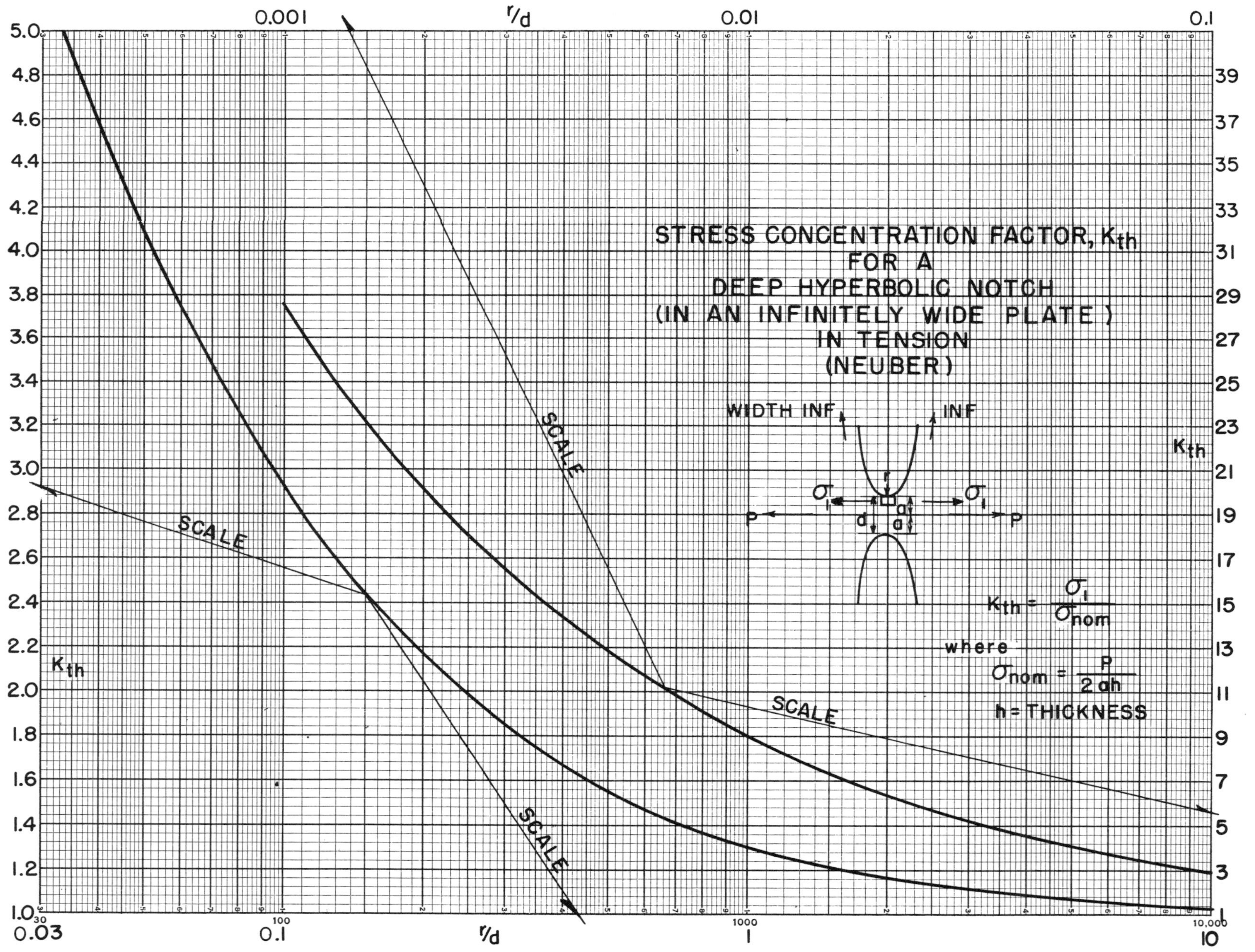
STRESS CONCENTRATION FACTOR, K_{th}
FOR A
DEEP HYPERBOLIC NOTCH
(IN AN INFINITELY WIDE PLATE)
IN TENSION
(NEUBER)
WIDTH INF
INF
$K_{th} = \frac{\sigma_1}{\sigma_{nom}}$
where
$\sigma_{nom} = \frac{P}{2ah}$
h = THICKNESS
SCALE
SCALE
SCALE
SCALE
r/d
r/d
K_{th}
K_{th}

FIG. 15

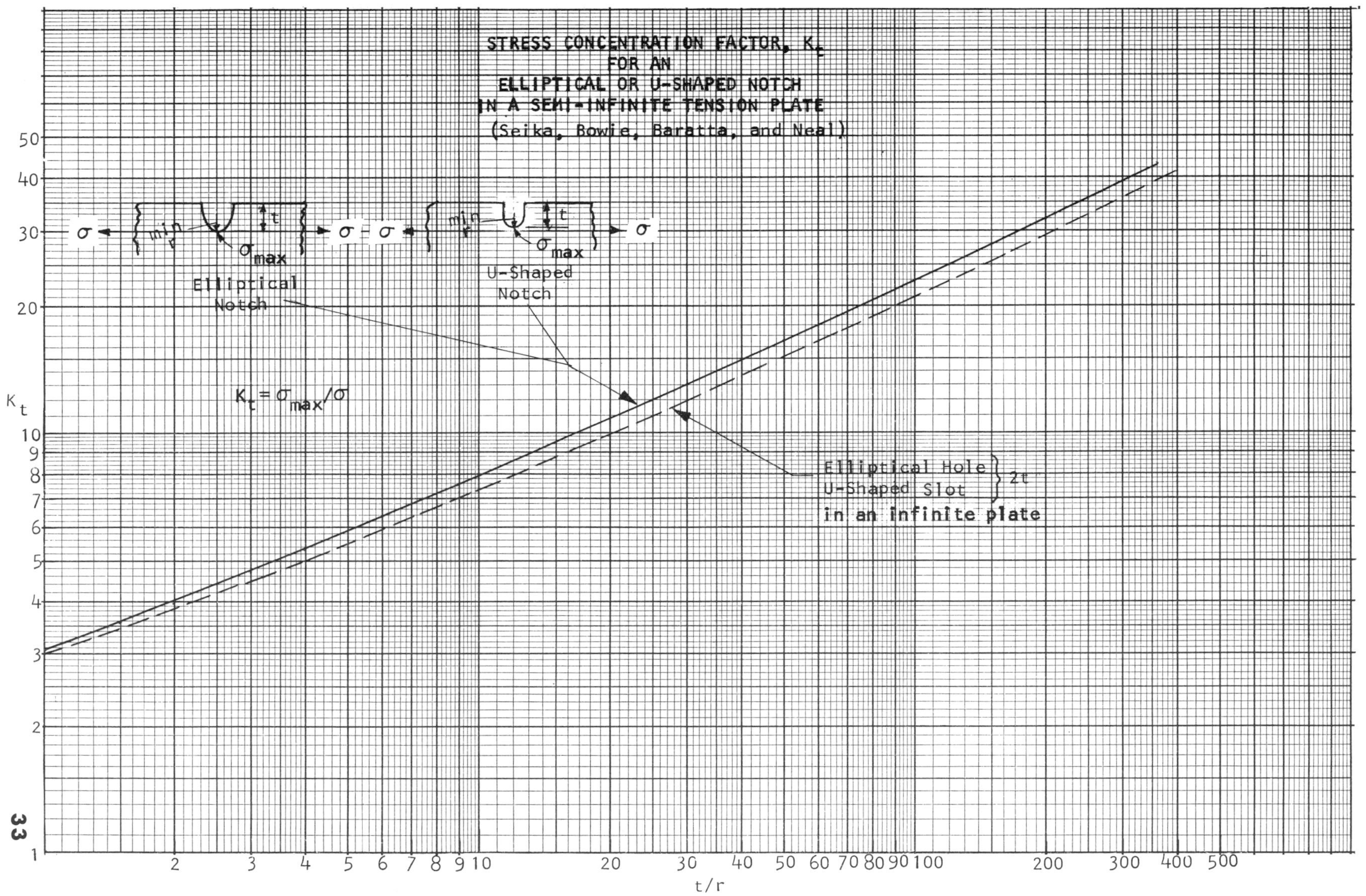

STRESS CONCENTRATION FACTOR, K_t
FOR AN
ELLIPTICAL OR U-SHAPED NOTCH
IN A SEMI-INFINITE TENSION PLATE
(Seika, Bowie, Baratta, and Neal)
σ
t
r
σ_{max}
Elliptical Notch
U-Shaped Notch
$K_t = \sigma_{max}/\sigma$
Elliptical Hole
U-Shaped Slot
2t
in an infinite plate
K_t
t/r
1
2
3
4
5
6
7
8
9
10
20
30
40
50
60
70
80
90
100
200
300
400
500

FIG. 16

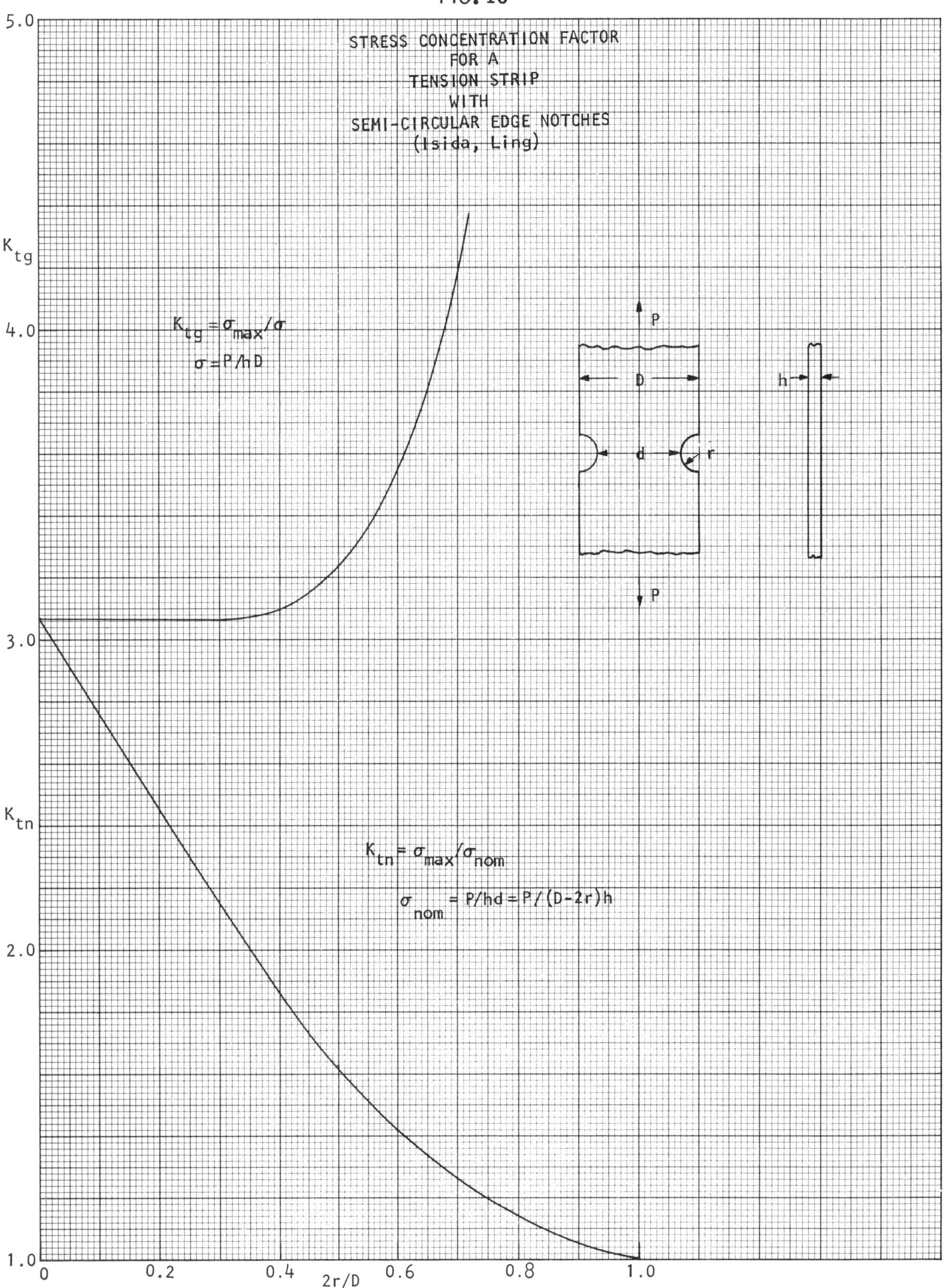
STRESS CONCENTRATION FACTOR
FOR A
TENSION STRIP
WITH
SEMI-CIRCULAR EDGE NOTCHES
(Isida, Ling)
K_{tg}
$K_{tg} = \sigma_{max}/\sigma$
$\sigma = P/hD$
K_{tn}
$K_{tn} = \sigma_{max}/\sigma_{nom}$
$\sigma_{nom} = P/hd = P/(D-2r)h$
P
D
d
r
h
5.0
4.0
3.0
2.0
1.0
0
0.2
0.4
0.6
0.8
1.0
2r/D

FIG. 17

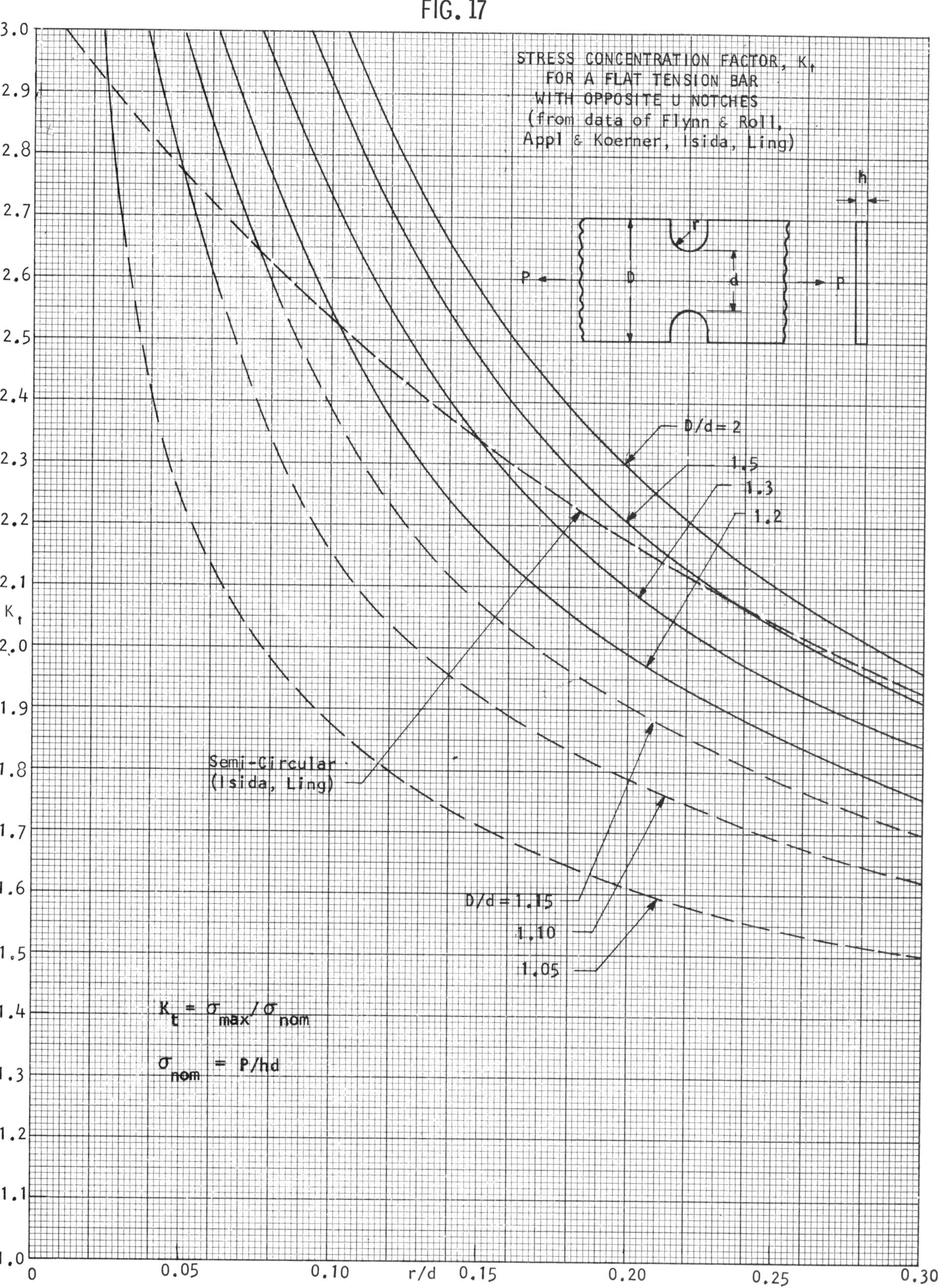
STRESS CONCENTRATION FACTOR, K_t
FOR A FLAT TENSION BAR
WITH OPPOSITE U NOTCHES
(from data of Flynn & Roll,
Appl & Koerner, Isida, Ling)
h
r
P
D
d
P
D/d = 2
1.5
1.3
1.2
Semi-Circular
(Isida, Ling)
D/d = 1.15
1.10
1.05
$K_t = \sigma_{max}/\sigma_{nom}$
$\sigma_{nom} = P/hd$
K_t
r/d
3.0
2.9
2.8
2.7
2.6
2.5
2.4
2.3
2.2
2.1
2.0
1.9
1.8
1.7
1.6
1.5
1.4
1.3
1.2
1.1
1.0
0
0.05
0.10
0.15
0.20
0.25
0.30

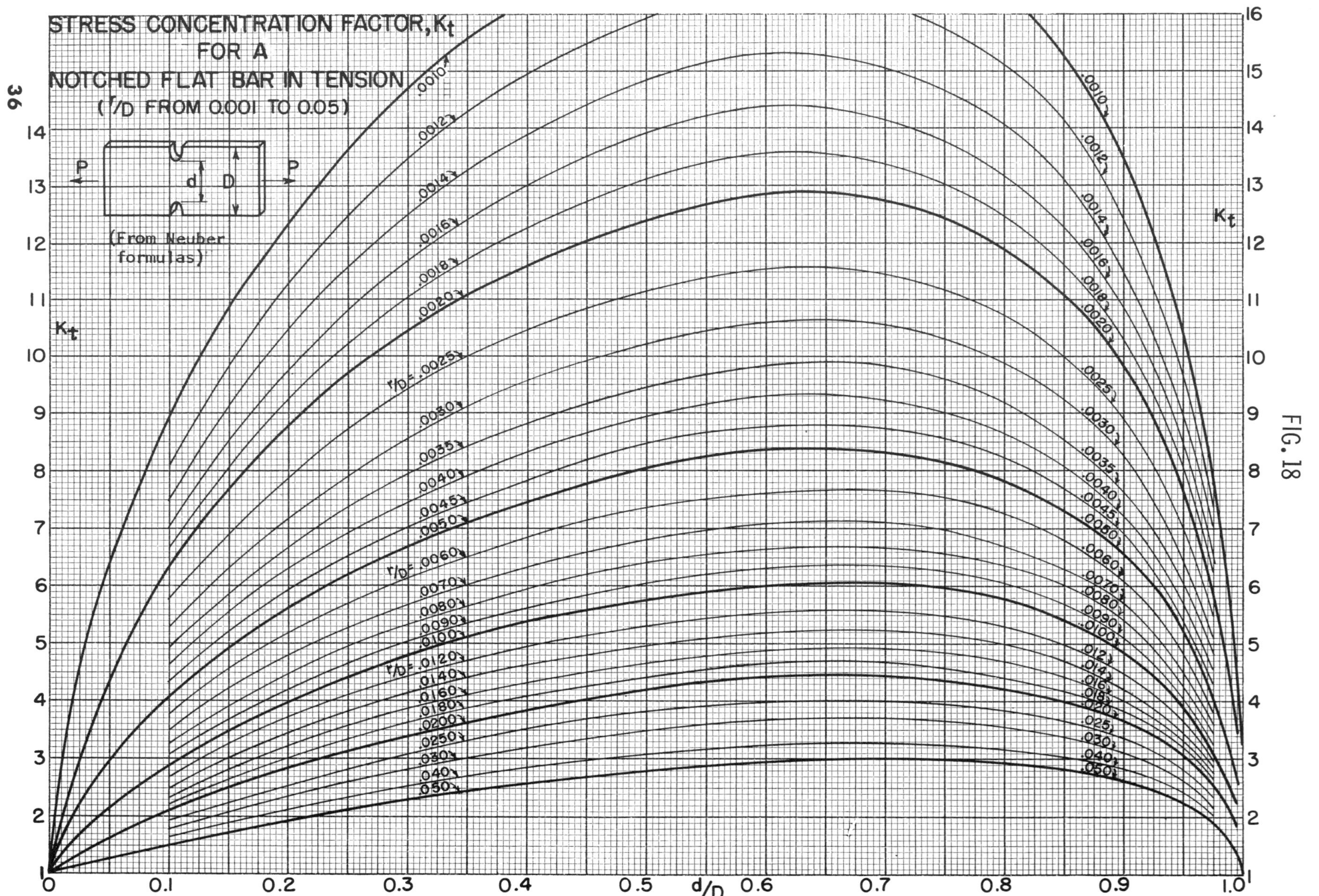

STRESS CONCENTRATION FACTOR, K_t
FOR A
NOTCHED FLAT BAR IN TENSION
(r/D FROM 0.001 TO 0.05)
P
d
D
P
(From Neuber formulas)
K_t
d/D

FIG. 18

FIG. 19

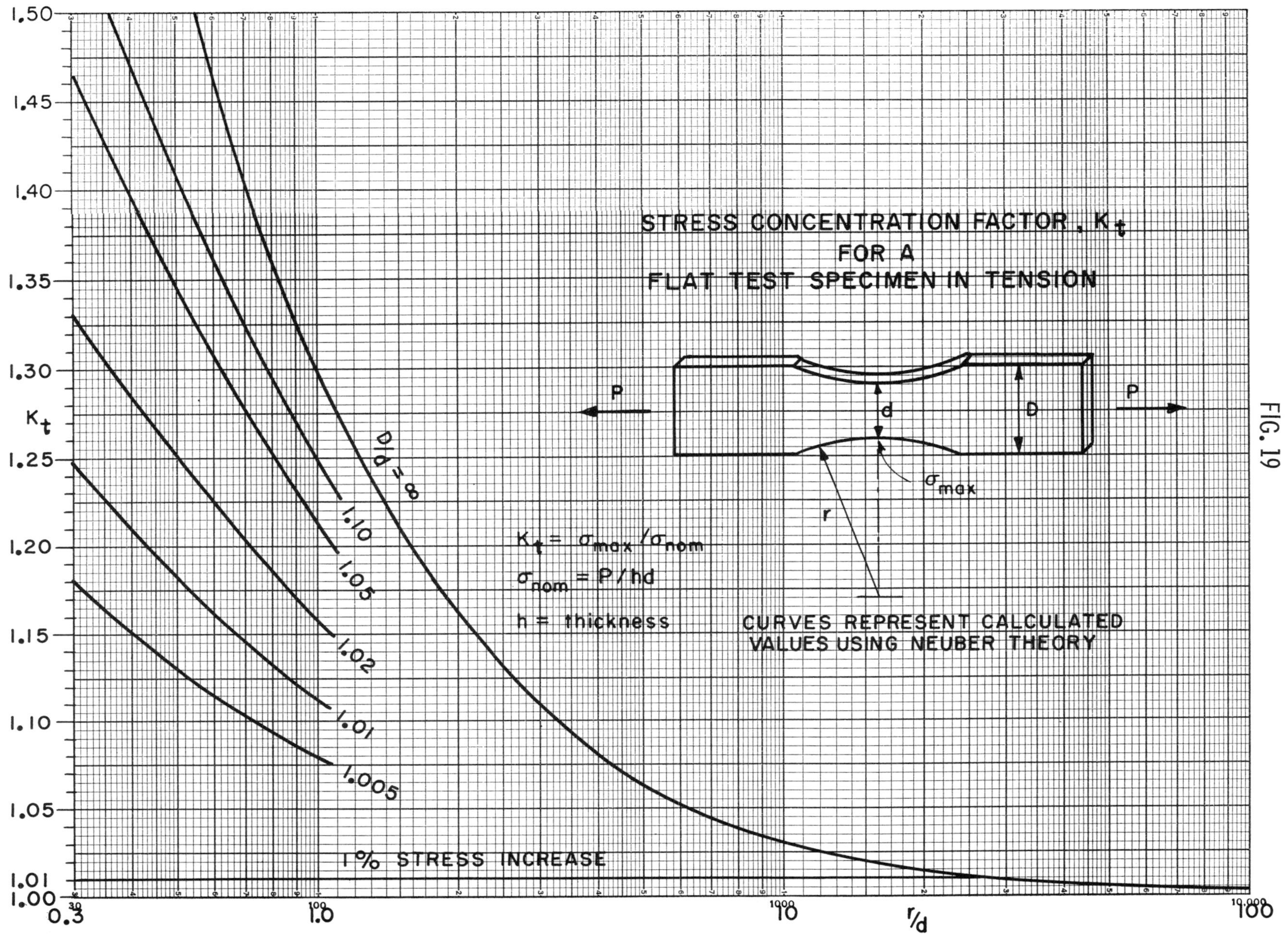

STRESS CONCENTRATION FACTOR, K_t
FOR A
FLAT TEST SPECIMEN IN TENSION
P
P
d
D
σ_{max}
r
$K_t = \sigma_{max} / \sigma_{nom}$
$\sigma_{nom} = P/hd$
h = thickness
CURVES REPRESENT CALCULATED
VALUES USING NEUBER THEORY
$D/d = \infty$
1.10
1.05
1.02
1.01
1.005
1% STRESS INCREASE
K_t
1.50
1.45
1.40
1.35
1.30
1.25
1.20
1.15
1.10
1.05
1.01
1.00
0.3
1.0
10
100
r/d

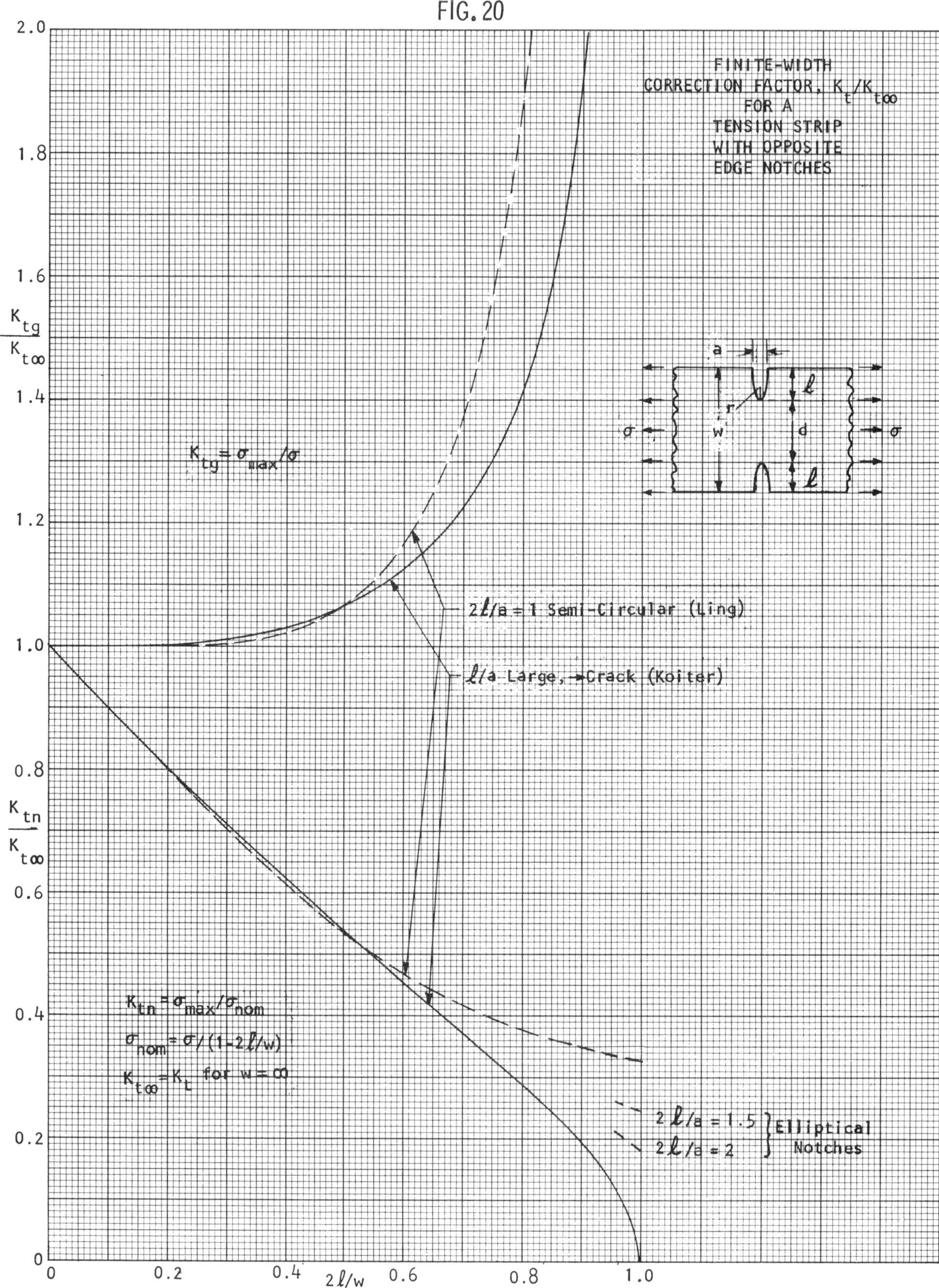

FIG. 20
FINITE-WIDTH CORRECTION FACTOR, $K_t/K_{t\infty}$ FOR A TENSION STRIP WITH OPPOSITE EDGE NOTCHES
$K_{tg}/K_{t\infty}$
$K_{tg} = \sigma_{max}/\sigma$
$2\ell/a = 1$ Semi-Circular (Ling)
ℓ/a Large, → Crack (Koiter)
$K_{tn}/K_{t\infty}$
$K_{tn} = \sigma_{max}/\sigma_{nom}$
$\sigma_{nom} = \sigma/(1-2\ell/w)$
$K_{t\infty} = K_t$ for $w = \infty$
$2\ell/a = 1.5$
$2\ell/a = 2$
Elliptical Notches
$2\ell/w$
a
r
w
d
ℓ
σ

FIG. 21

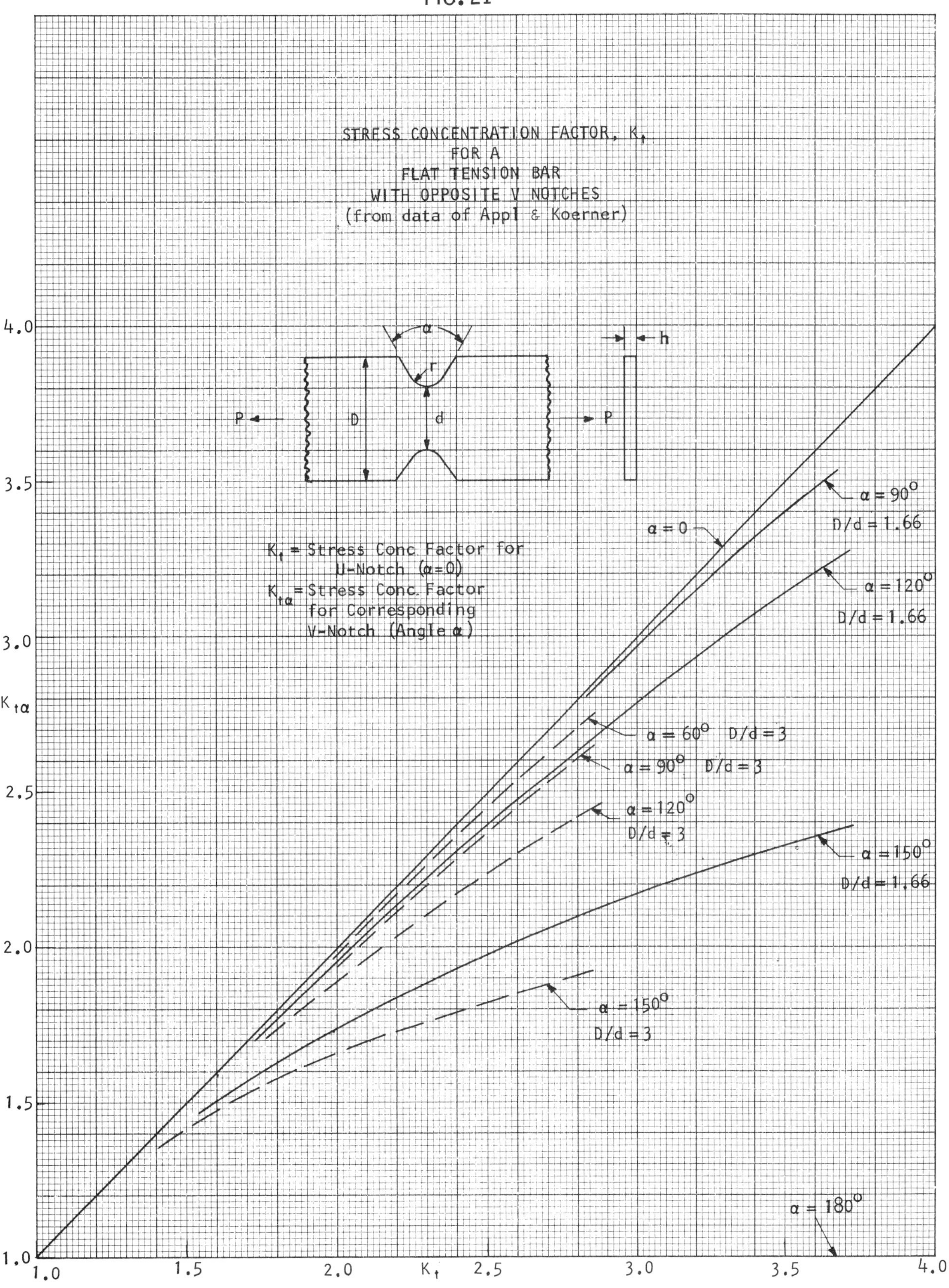

STRESS CONCENTRATION FACTOR, K_t
FOR A
FLAT TENSION BAR
WITH OPPOSITE V NOTCHES
(from data of Appl & Koerner)
α
r
h
P
D
d
P
K_t = Stress Conc Factor for U-Notch (α=0)
$K_{tα}$ = Stress Conc. Factor for Corresponding V-Notch (Angle α)
α = 0
α = 90° D/d = 1.66
α = 120° D/d = 1.66
α = 60° D/d = 3
α = 90° D/d = 3
α = 120° D/d = 3
α = 150° D/d = 1.66
α = 150° D/d = 3
α = 180°
$K_{tα}$
K_t
4.0
3.5
3.0
2.5
2.0
1.5
1.0
1.0
1.5
2.0
2.5
3.0
3.5
4.0

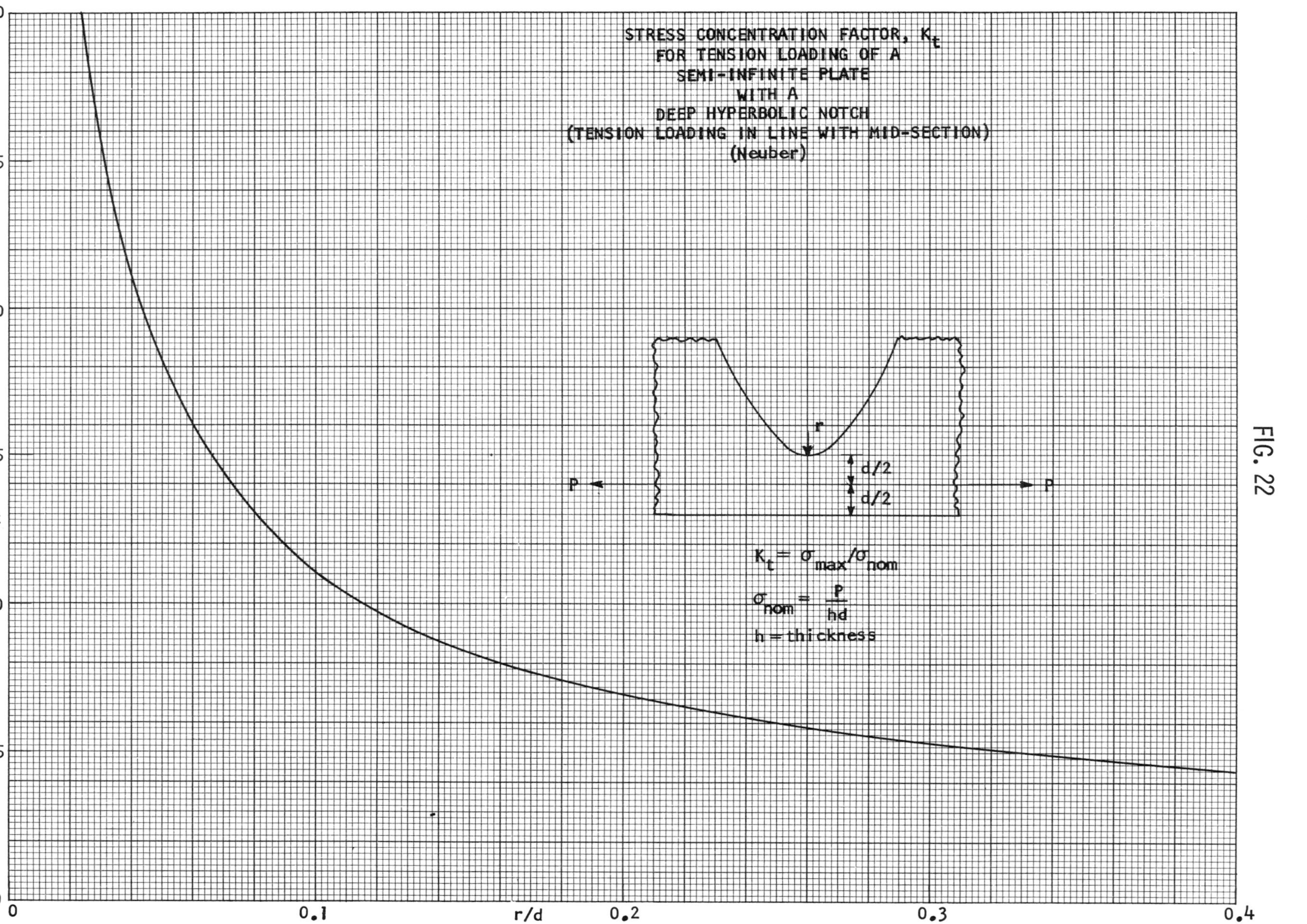

STRESS CONCENTRATION FACTOR, K_t
FOR TENSION LOADING OF A
SEMI-INFINITE PLATE
WITH A
DEEP HYPERBOLIC NOTCH
(TENSION LOADING IN LINE WITH MID-SECTION)
(Neuber)
r
d/2
d/2
P
P
$K_t = \sigma_{max}/\sigma_{nom}$
$\sigma_{nom} = \frac{P}{hd}$
h = thickness
K_t
4.0
3.5
3.0
2.5
2.0
1.5
1.0
0
0.1
r/d
0.2
0.3
0.4

FIG. 22

FIG. 22a

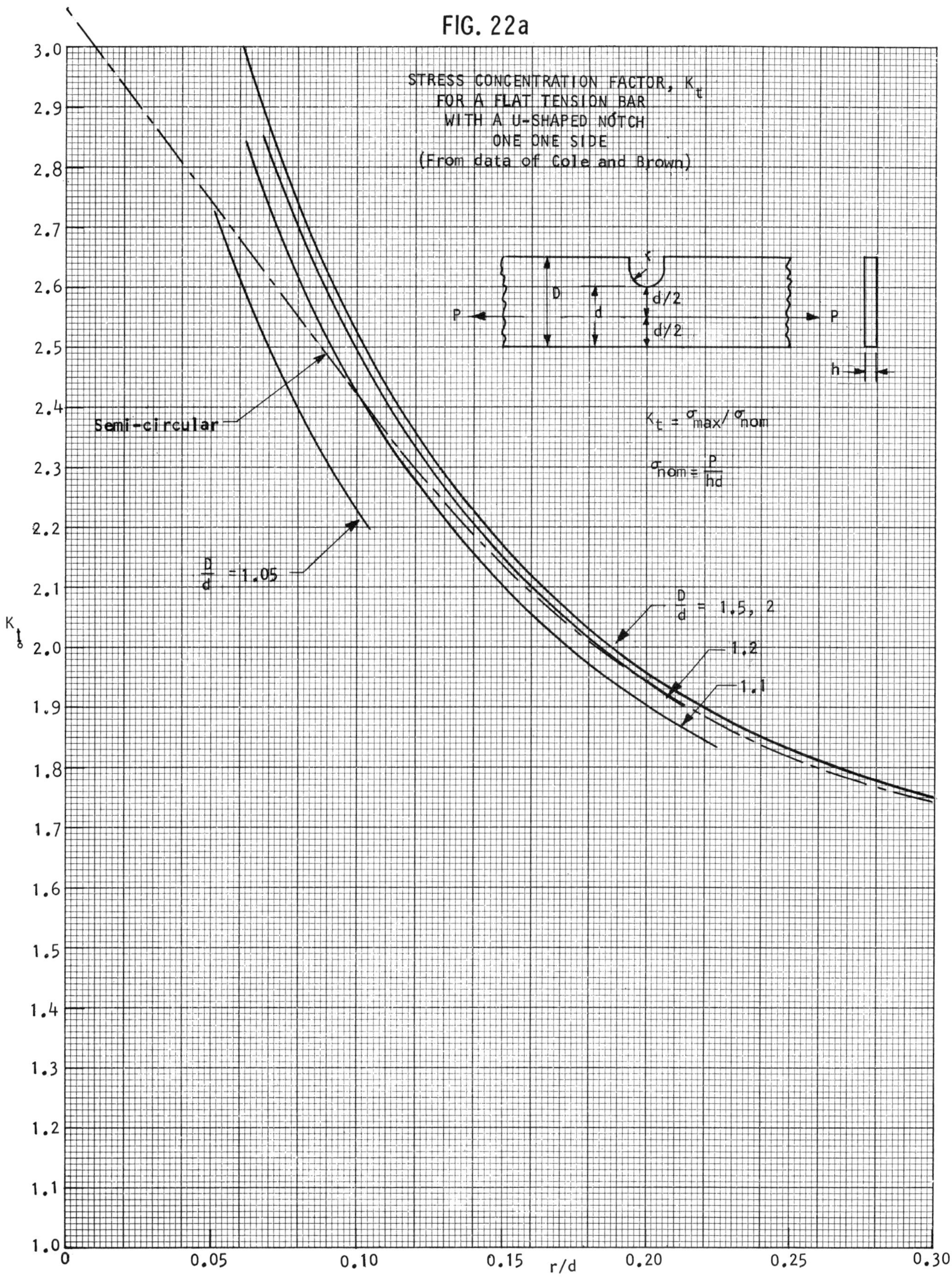
STRESS CONCENTRATION FACTOR, K_t
FOR A FLAT TENSION BAR
WITH A U-SHAPED NOTCH
ONE ONE SIDE
(From data of Cole and Brown)
Semi-circular
$\frac{D}{d}$ = 1.05
$\frac{D}{d}$ = 1.5, 2
1.2
1.1
r
D
d
d/2
d/2
P
P
h
$K_t = \sigma_{max}/\sigma_{nom}$
$\sigma_{nom} = \frac{P}{hd}$
K_t
r/d
3.0
2.9
2.8
2.7
2.6
2.5
2.4
2.3
2.2
2.1
2.0
1.9
1.8
1.7
1.6
1.5
1.4
1.3
1.2
1.1
1.0
0
0.05
0.10
0.15
0.20
0.25
0.30

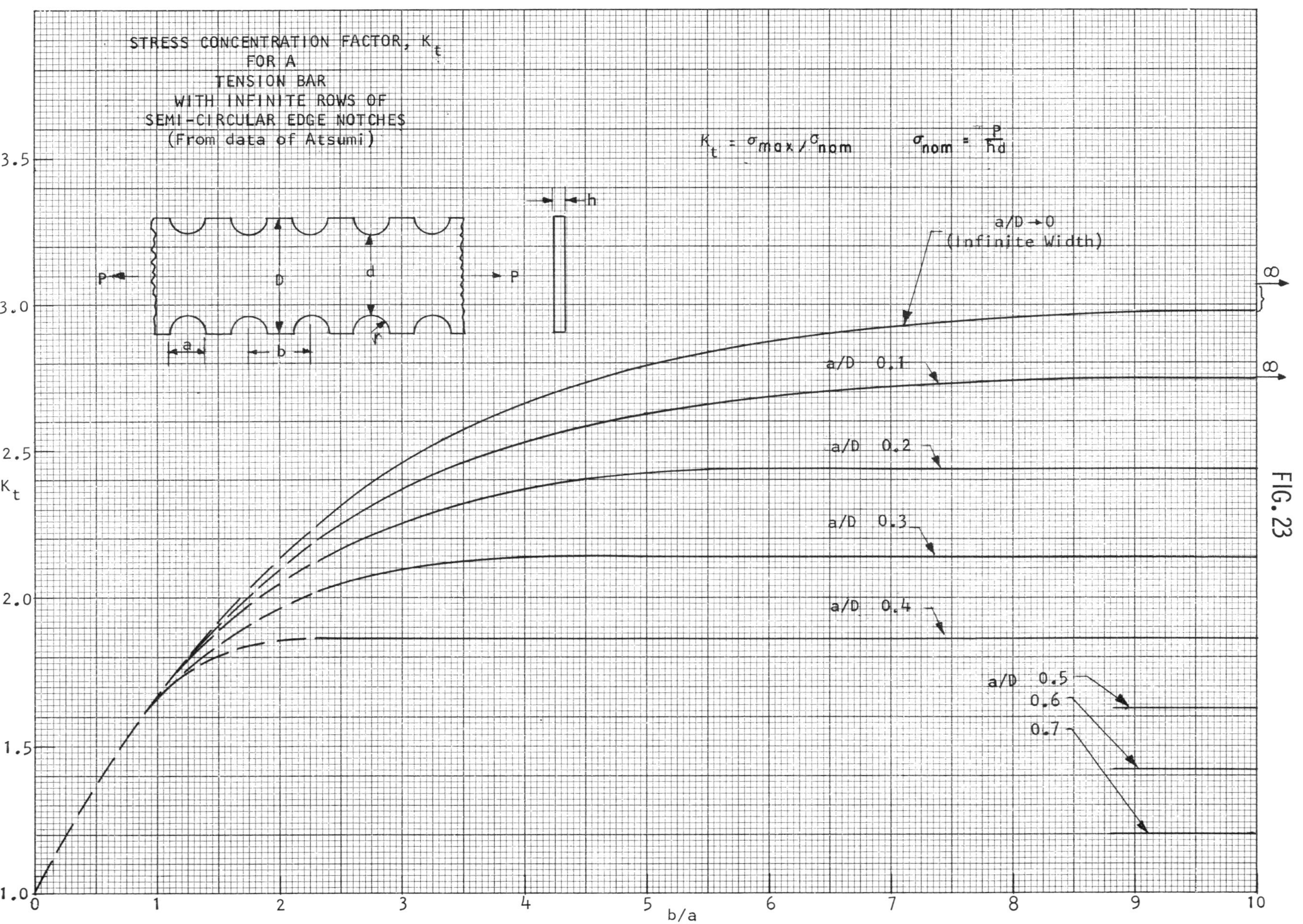

STRESS CONCENTRATION FACTOR, K_t
FOR A
TENSION BAR
WITH INFINITE ROWS OF
SEMI-CIRCULAR EDGE NOTCHES
(From data of Atsumi)
$K_t = \sigma_{max} / \sigma_{nom}$
$\sigma_{nom} = \frac{P}{hd}$
P
P
D
d
a
b
r
h
a/D → 0
(Infinite Width)
∞
∞
a/D 0.1
a/D 0.2
a/D 0.3
a/D 0.4
a/D 0.5
0.6
0.7
K_t
1.0
1.5
2.0
2.5
3.0
3.5
0
1
2
3
4
5
6
7
8
9
10
b/a

FIG. 23

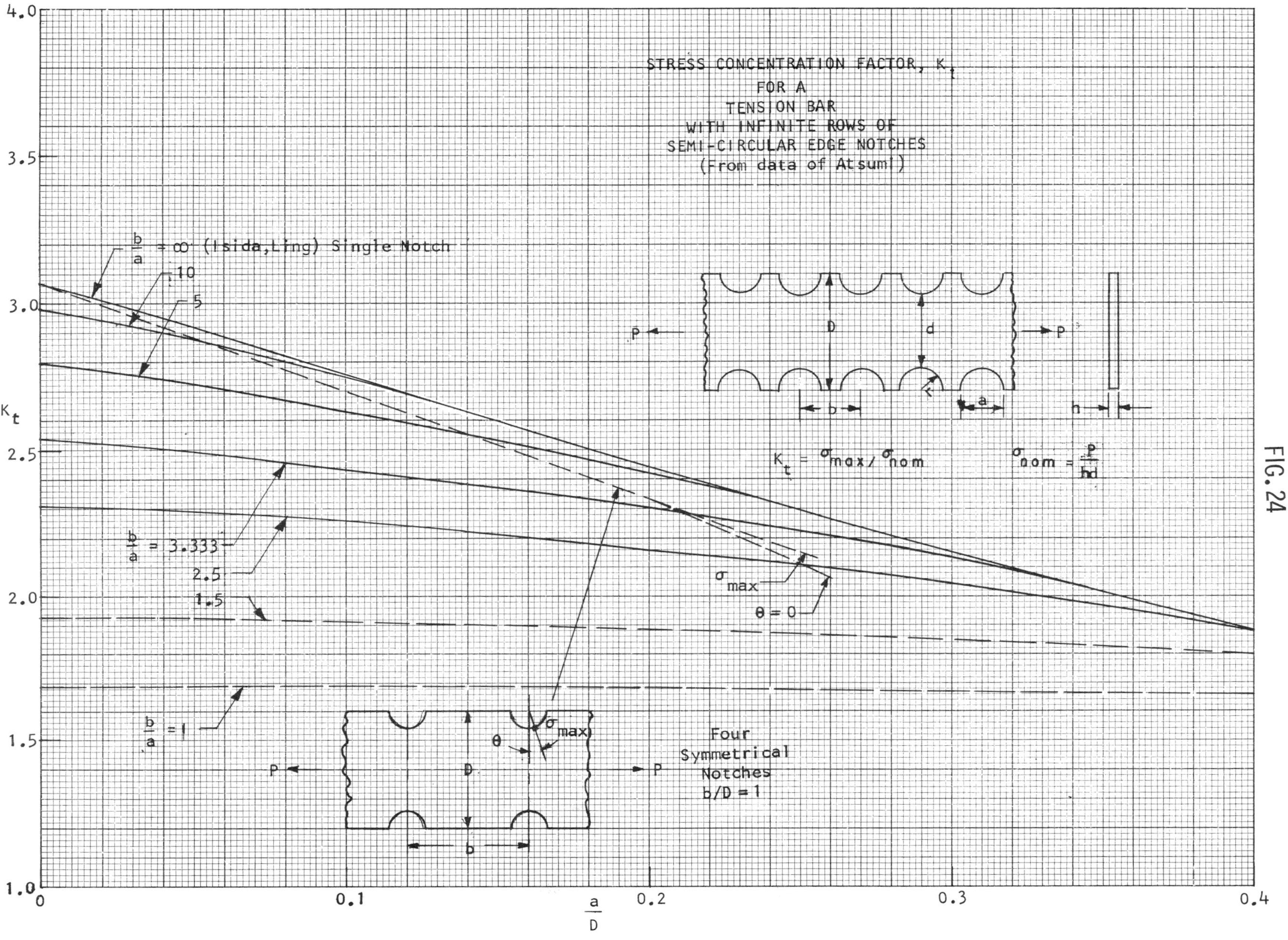

STRESS CONCENTRATION FACTOR, K_t
FOR A
TENSION BAR
WITH INFINITE ROWS OF
SEMI-CIRCULAR EDGE NOTCHES
(From data of Atsumi)
$\frac{b}{a} = \infty$ (Isida, Ling) Single Notch
10
5
$\frac{b}{a} = 3.333$
2.5
1.5
$\frac{b}{a} = 1$
σ_{max}
$\theta = 0$
$K_t = \sigma_{max}/\sigma_{nom}$
$\sigma_{nom} = \frac{P}{hd}$
Four Symmetrical Notches b/D = 1
K_t
$\frac{a}{D}$

FIG. 24

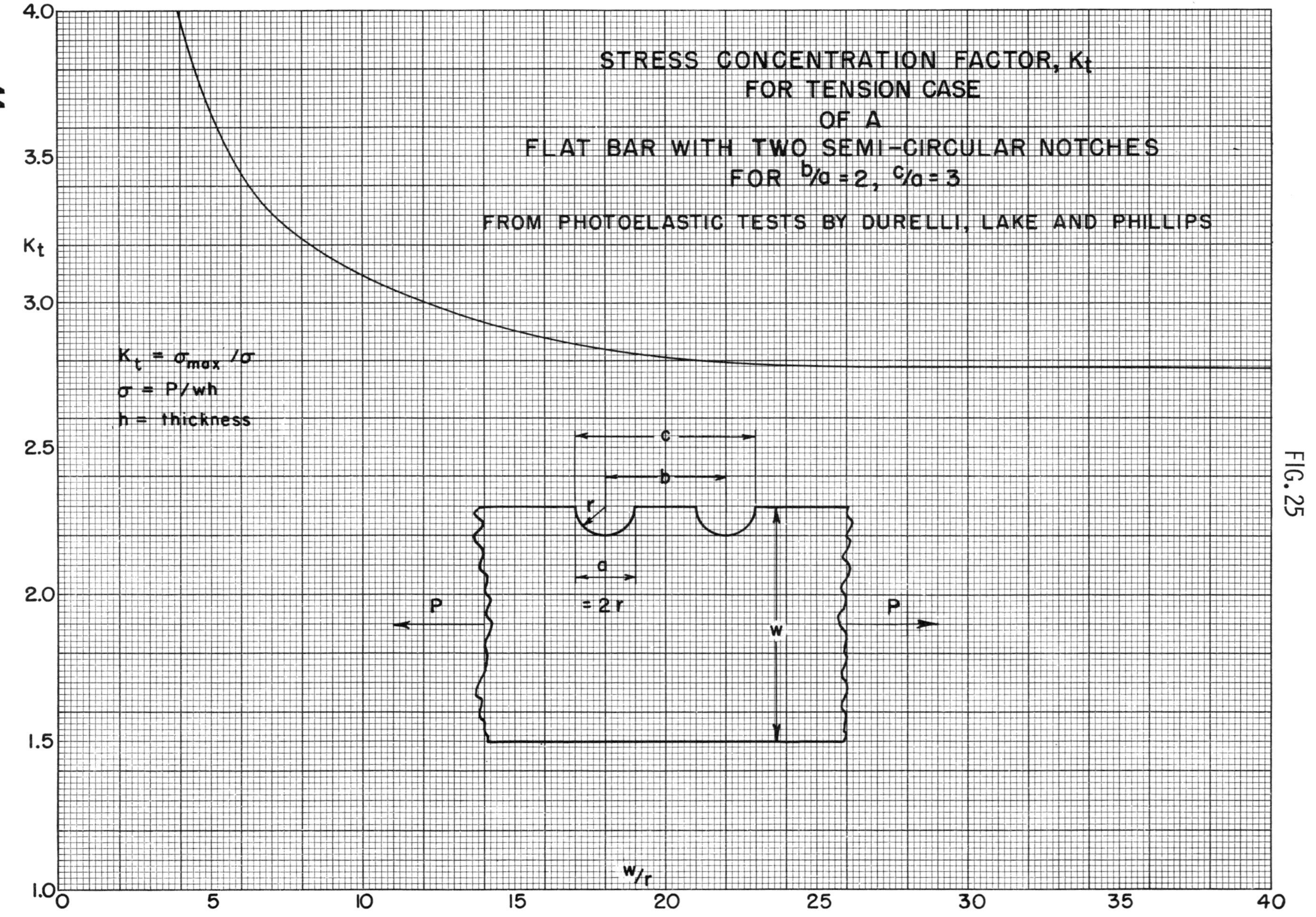
STRESS CONCENTRATION FACTOR, K_t
FOR TENSION CASE
OF A
FLAT BAR WITH TWO SEMI-CIRCULAR NOTCHES
FOR $b/a = 2$, $c/a = 3$
FROM PHOTOELASTIC TESTS BY DURELLI, LAKE AND PHILLIPS
$K_t = \sigma_{max} / \sigma$
$\sigma = P/wh$
h = thickness
c
b
r
a
= 2r
w
P
P
K_t
4.0
3.5
3.0
2.5
2.0
1.5
1.0
w/r
0
5
10
15
20
25
30
35
40

FIG. 25

FIG. 26

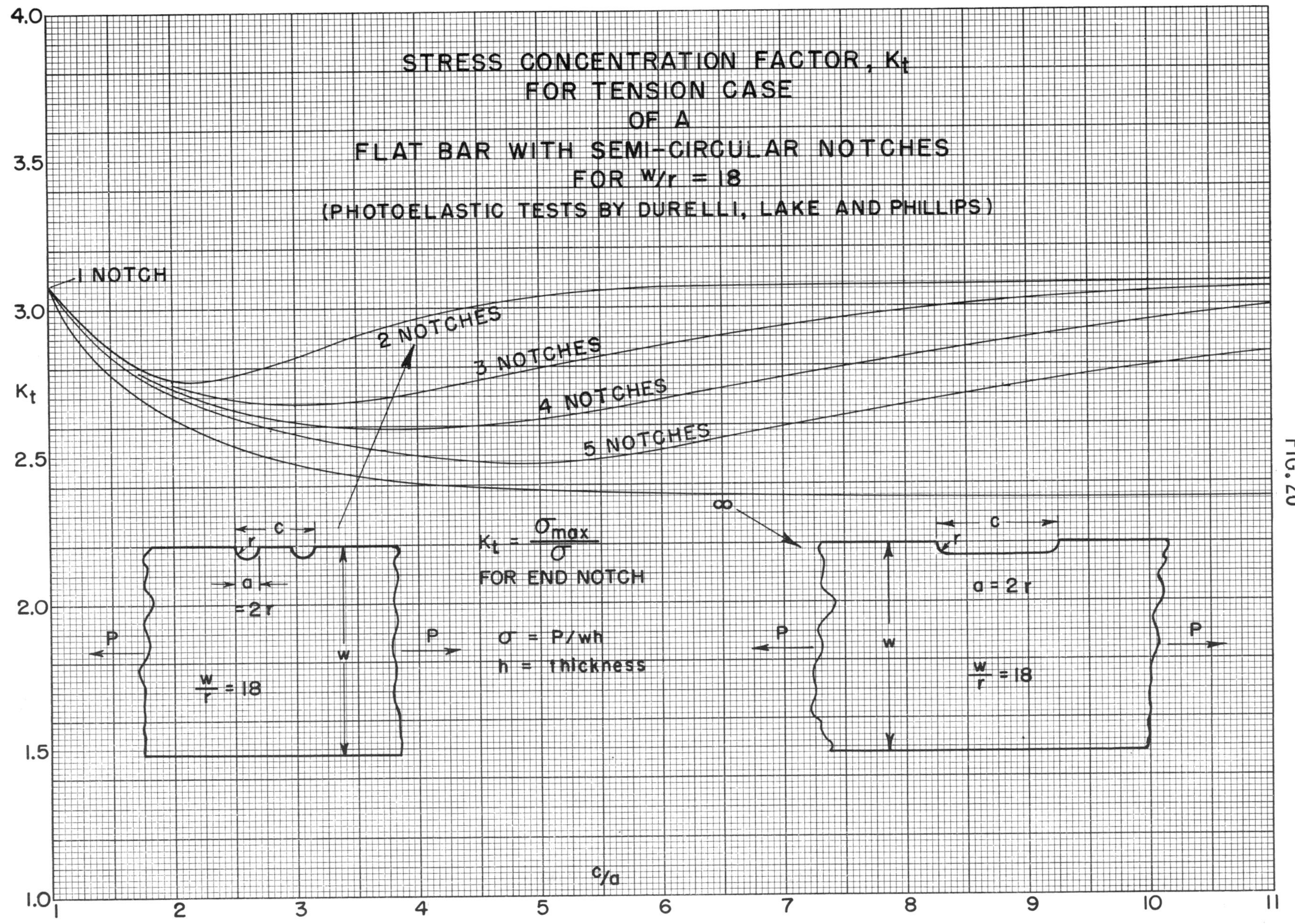

STRESS CONCENTRATION FACTOR, K_t
FOR TENSION CASE
OF A
FLAT BAR WITH SEMI-CIRCULAR NOTCHES
FOR $w/r = 18$
(PHOTOELASTIC TESTS BY DURELLI, LAKE AND PHILLIPS)
1 NOTCH
2 NOTCHES
3 NOTCHES
4 NOTCHES
5 NOTCHES
∞
$K_t = \frac{\sigma_{max}}{\sigma}$
FOR END NOTCH
$\sigma = P/wh$
h = thickness
P
w
c
r
a
$= 2r$
$a = 2r$
$\frac{w}{r} = 18$
K_t
4.0
3.5
3.0
2.5
2.0
1.5
1.0
c/a
1
2
3
4
5
6
7
8
9
10
11

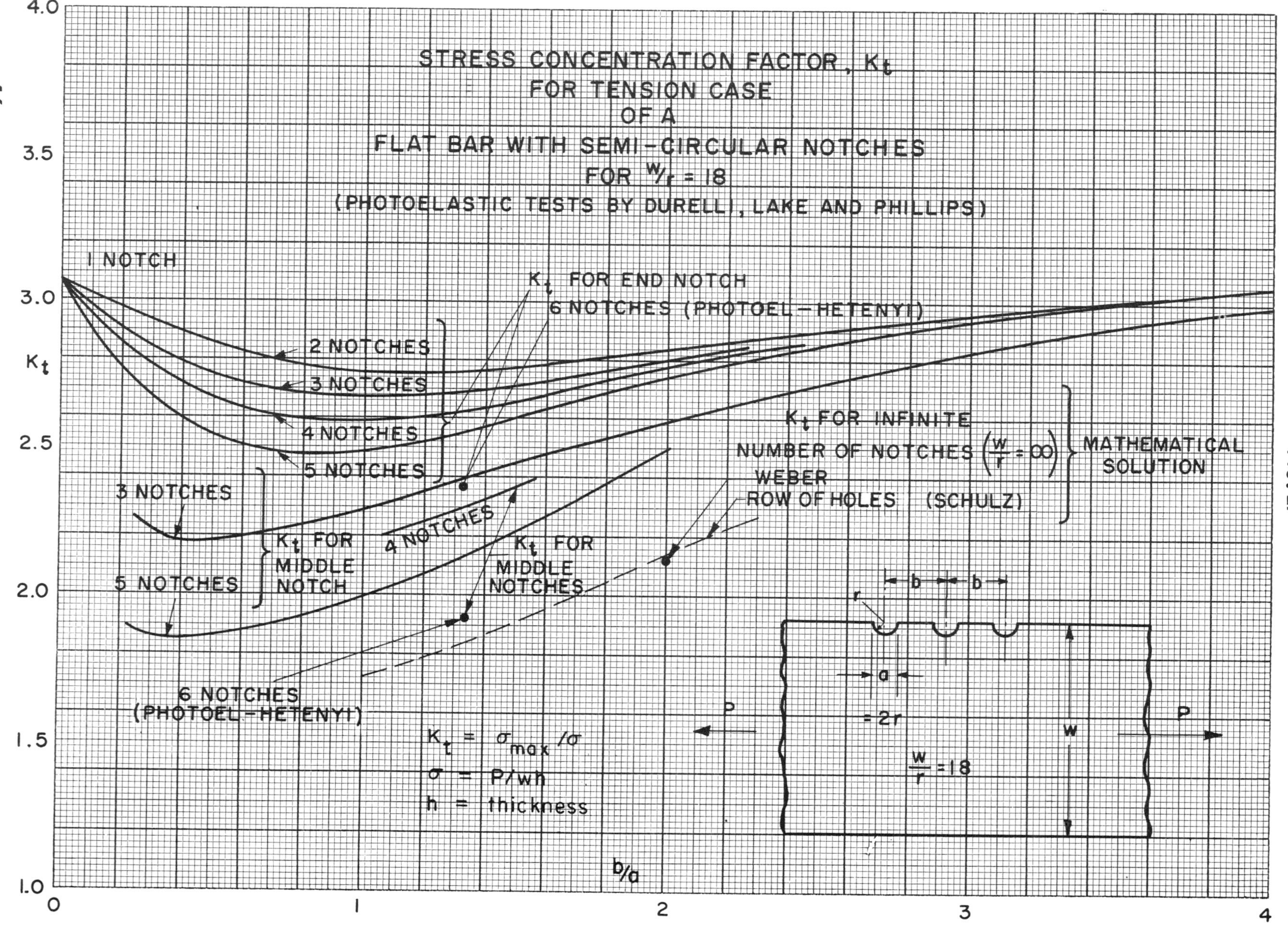

STRESS CONCENTRATION FACTOR, K_t
FOR TENSION CASE
OF A
FLAT BAR WITH SEMI-CIRCULAR NOTCHES
FOR $w/r = 18$
(PHOTOELASTIC TESTS BY DURELLI, LAKE AND PHILLIPS)
1 NOTCH
K_t FOR END NOTCH
6 NOTCHES (PHOTOEL-HETENYI)
2 NOTCHES
3 NOTCHES
4 NOTCHES
5 NOTCHES
K_t FOR INFINITE
NUMBER OF NOTCHES $\left(\frac{w}{r} = \infty\right)$
WEBER
ROW OF HOLES (SCHULZ)
MATHEMATICAL SOLUTION
3 NOTCHES
5 NOTCHES
K_t FOR MIDDLE NOTCH
4 NOTCHES
K_t FOR MIDDLE NOTCHES
6 NOTCHES
(PHOTOEL-HETENYI)
$K_t = \sigma_{max} / \sigma$
$\sigma = P/wh$
h = thickness
b/a
K_t
4.0
3.5
3.0
2.5
2.0
1.5
1.0
0
1
2
3
4
r
b
b
a
= 2r
P
P
w
$\frac{w}{r} = 18$

FIG. 27

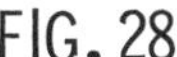

FIG. 28

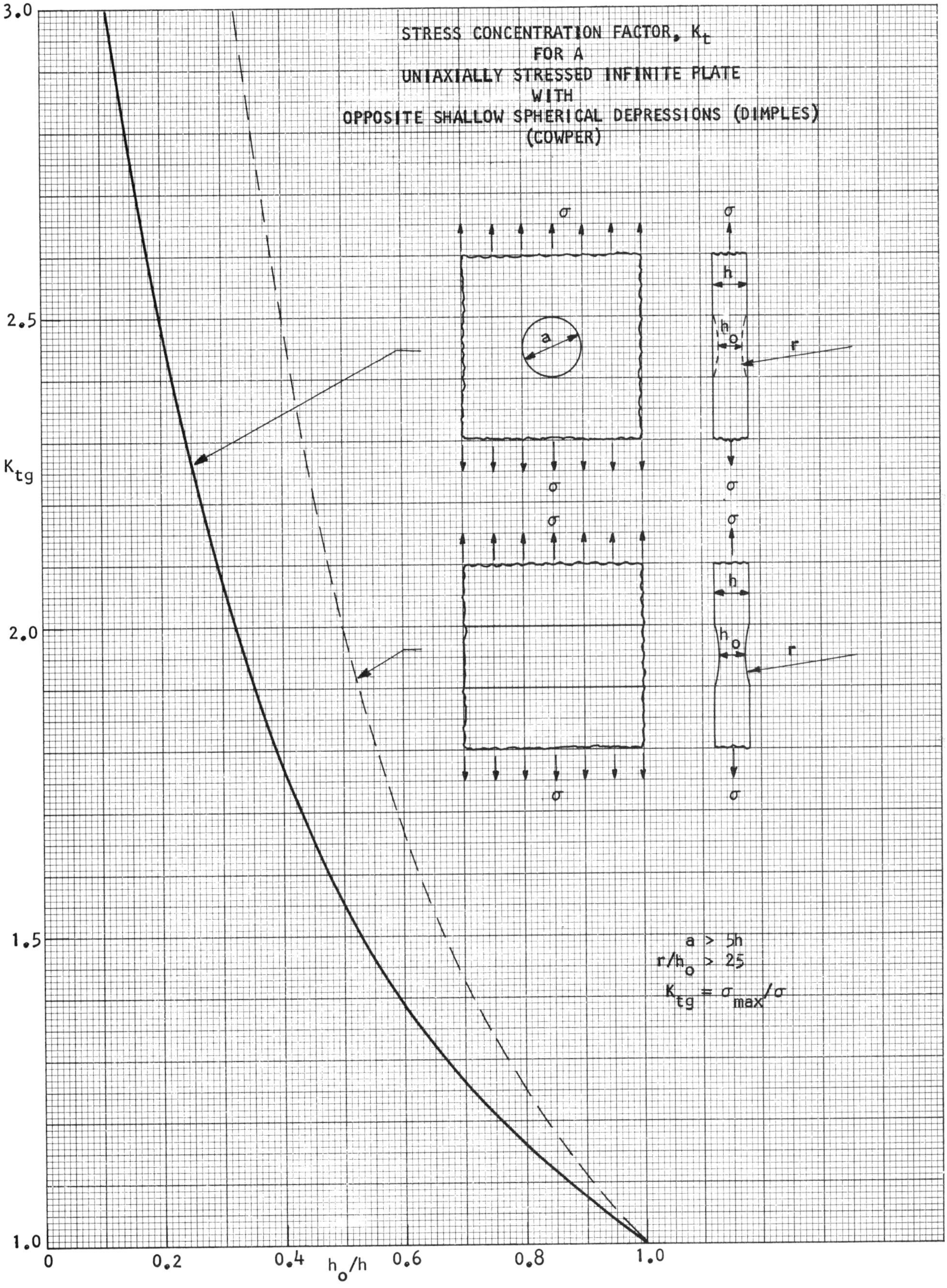
STRESS CONCENTRATION FACTOR, K_t
FOR A
UNIAXIALLY STRESSED INFINITE PLATE
WITH
OPPOSITE SHALLOW SPHERICAL DEPRESSIONS (DIMPLES)
(COWPER)
K_{tg}
h_o/h
3.0
2.5
2.0
1.5
1.0
0
0.2
0.4
0.6
0.8
1.0
σ
a
h
h_o
r
$a > 5h$
$r/h_o > 25$
$K_{tg} = \sigma_{max}/\sigma$

STRESS CONCENTRATION FACTOR, K_{th}
FOR A
DEEP HYPERBOLIC GROOVE
(IN AN INFINITELY WIDE MEMBER)
3 DIMENSIONAL CASE,
TENSION
(NEUBER SOLUTION)
INF
P
σ_1
σ_2
r
a
d
$K_{th} = \frac{\sigma_1}{\sigma_{nom}}$
where $\sigma_{nom} = \frac{P}{\pi a^2}$
EFFECT OF
POISSON'S RATIO
$\nu = 0.2$
$\nu = 0.3$
$\nu = 0.5$
SCALE
r/d
K_{th}

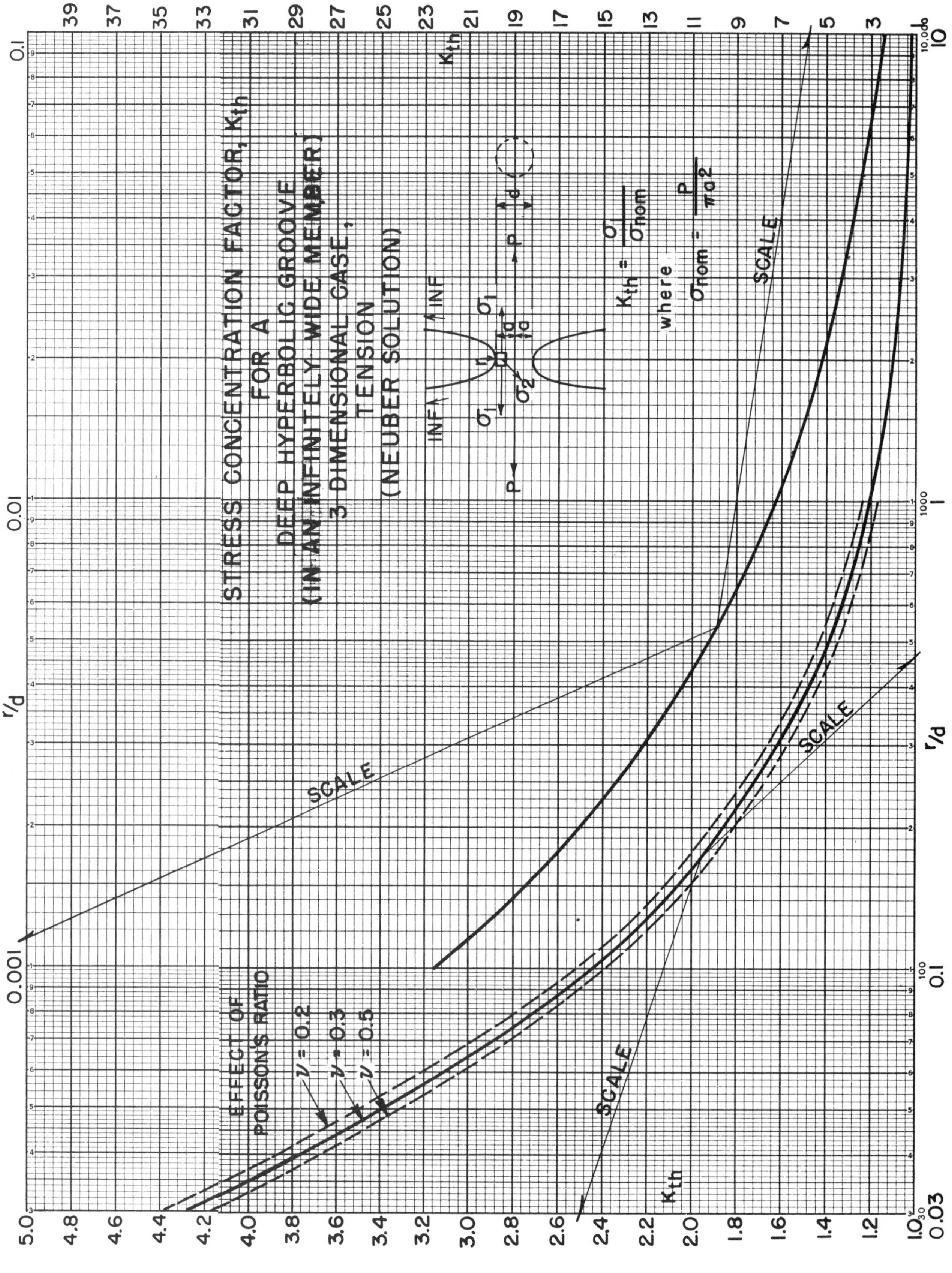

FIG. 29

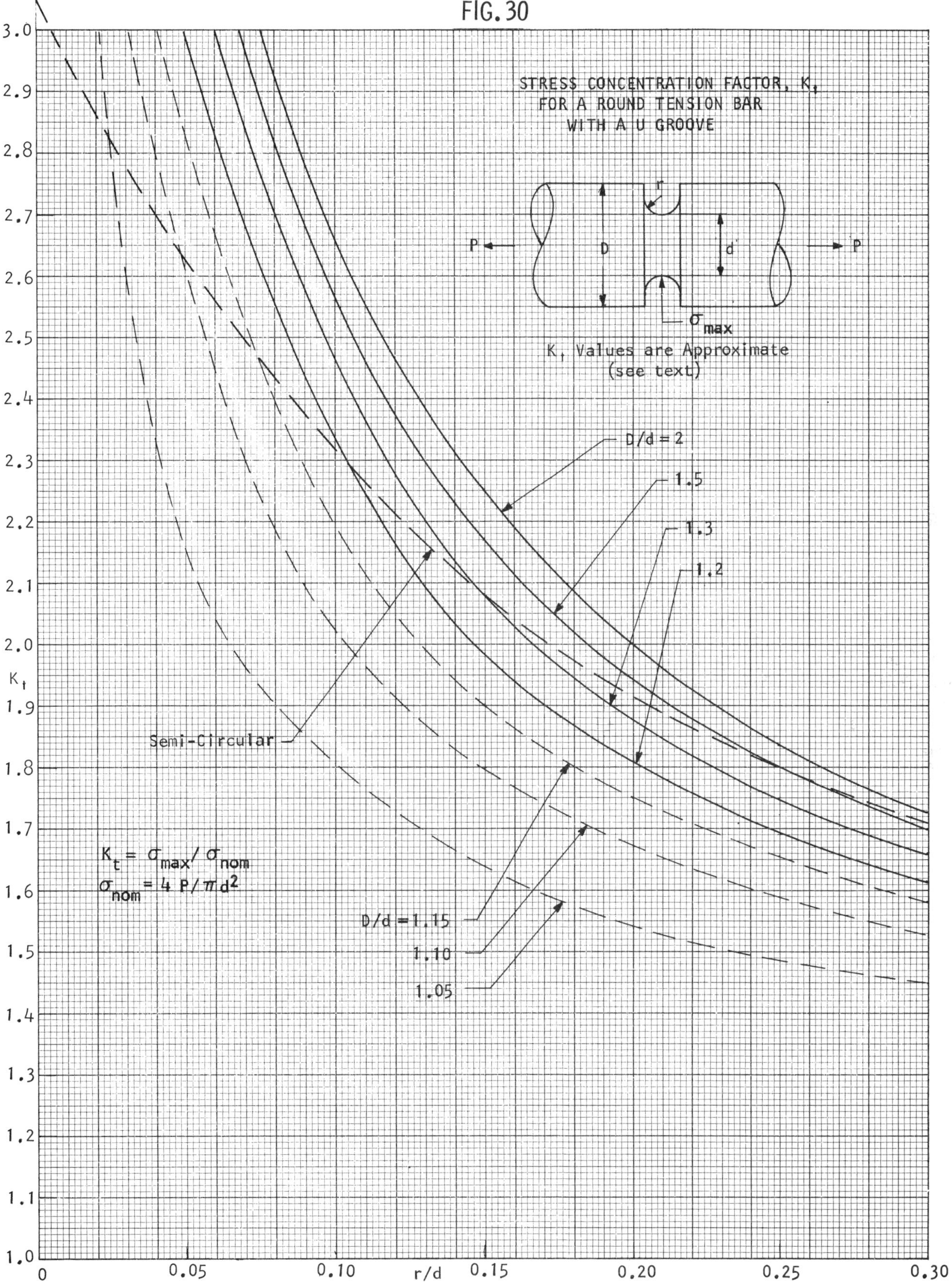
FIG. 30
STRESS CONCENTRATION FACTOR, K_t
FOR A ROUND TENSION BAR
WITH A U GROOVE
P
D
r
d
P
σ_{max}
K_t Values are Approximate
(see text)
D/d = 2
1.5
1.3
1.2
Semi-Circular
D/d = 1.15
1.10
1.05
$K_t = \sigma_{max}/\sigma_{nom}$
$\sigma_{nom} = 4P/\pi d^2$
K_t
3.0
2.9
2.8
2.7
2.6
2.5
2.4
2.3
2.2
2.1
2.0
1.9
1.8
1.7
1.6
1.5
1.4
1.3
1.2
1.1
1.0
0
0.05
0.10
r/d
0.15
0.20
0.25
0.30

FIG. 31

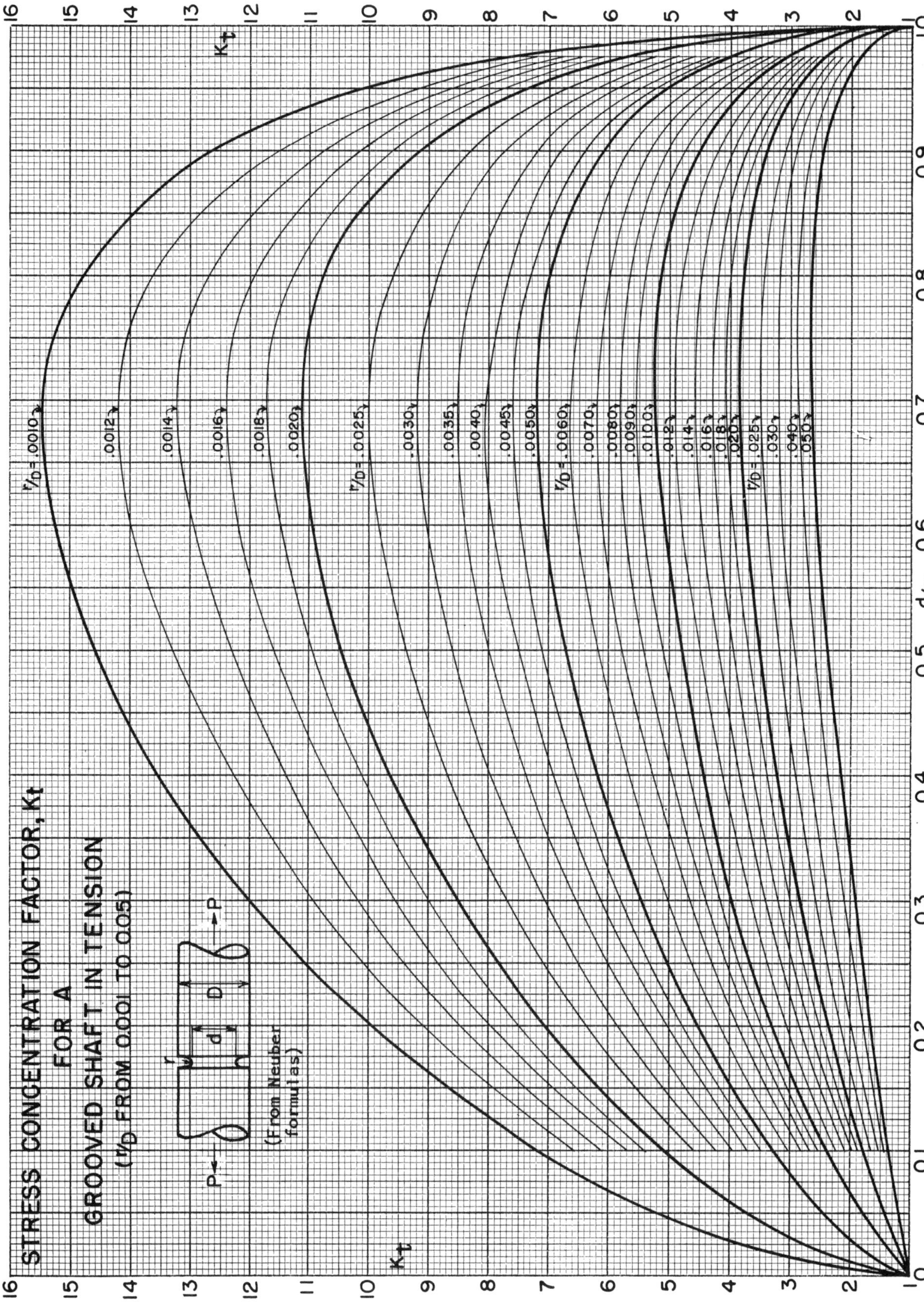

STRESS CONCENTRATION FACTOR, K_t FOR A GROOVED SHAFT IN TENSION

(r/D FROM 0.001 TO 0.05)

(From Neuber formulas)

K_t: 1, 2, 3, 4, 5, 6, 7, 8, 9, 10, 11, 12, 13, 14, 15, 16

d/D: 0, 0.1, 0.2, 0.3, 0.4, 0.5, 0.6, 0.7, 0.8, 0.9, 1.0

P, r, d, D, P

r/D = .0010, .0012, .0014, .0016, .0018, .0020

r/D = .0025, .0030, .0035, .0040, .0045, .0050

r/D = .0060, .0070, .0080, .0090, .0100, .012, .014, .016, .018, .020

r/D = .025, .030, .040, .050

FIG. 32

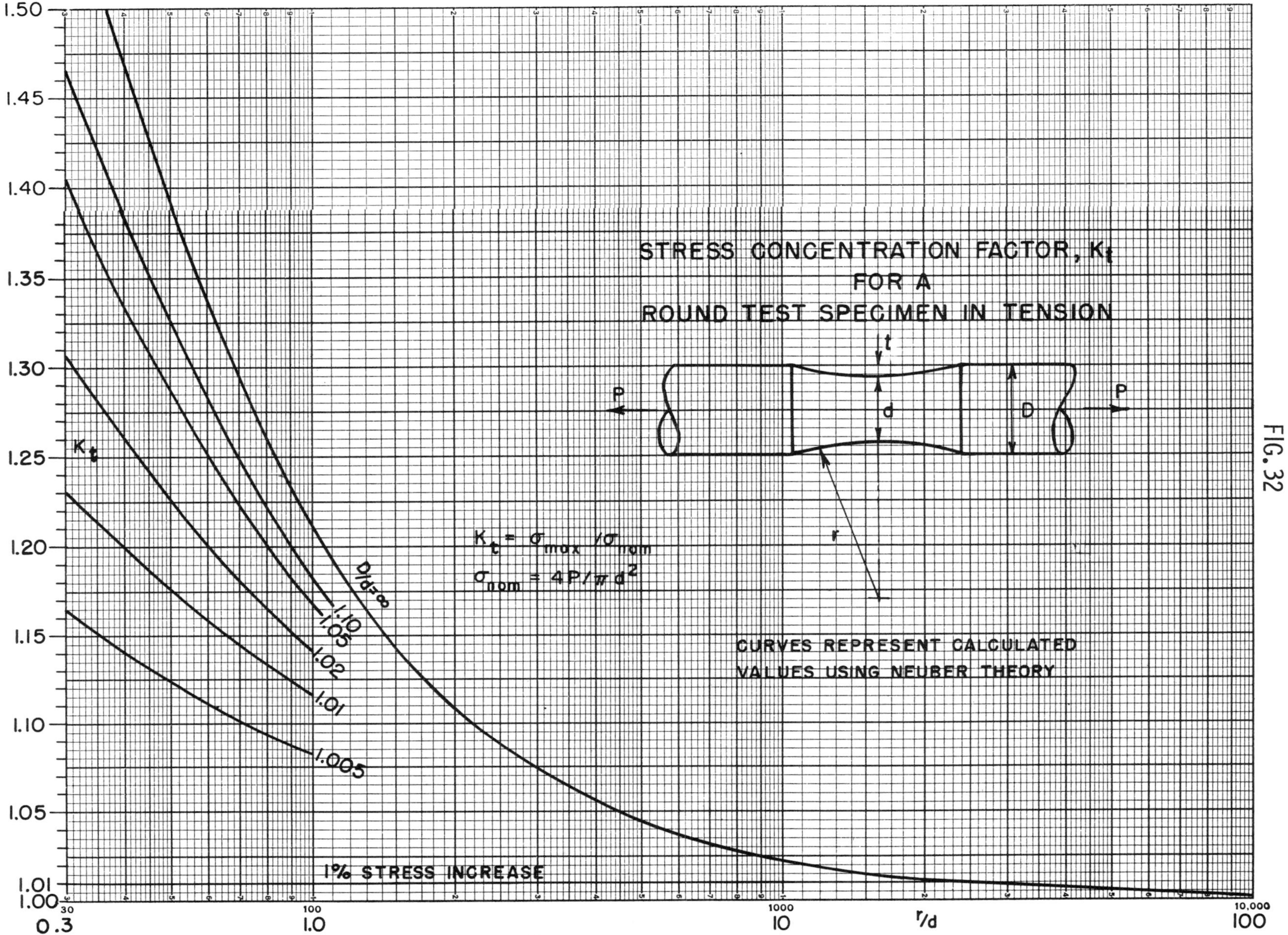

STRESS CONCENTRATION FACTOR, K_t
FOR A
ROUND TEST SPECIMEN IN TENSION
P
P
t
d
D
r
$K_t = \sigma_{max} / \sigma_{nom}$
$\sigma_{nom} = 4P/\pi d^2$
CURVES REPRESENT CALCULATED
VALUES USING NEUBER THEORY
D/d=∞
1.10
1.05
1.02
1.01
1.005
1% STRESS INCREASE
K_t
1.50
1.45
1.40
1.35
1.30
1.25
1.20
1.15
1.10
1.05
1.01
1.00
0.3
1.0
10
r/d
100

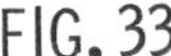

FIG. 33

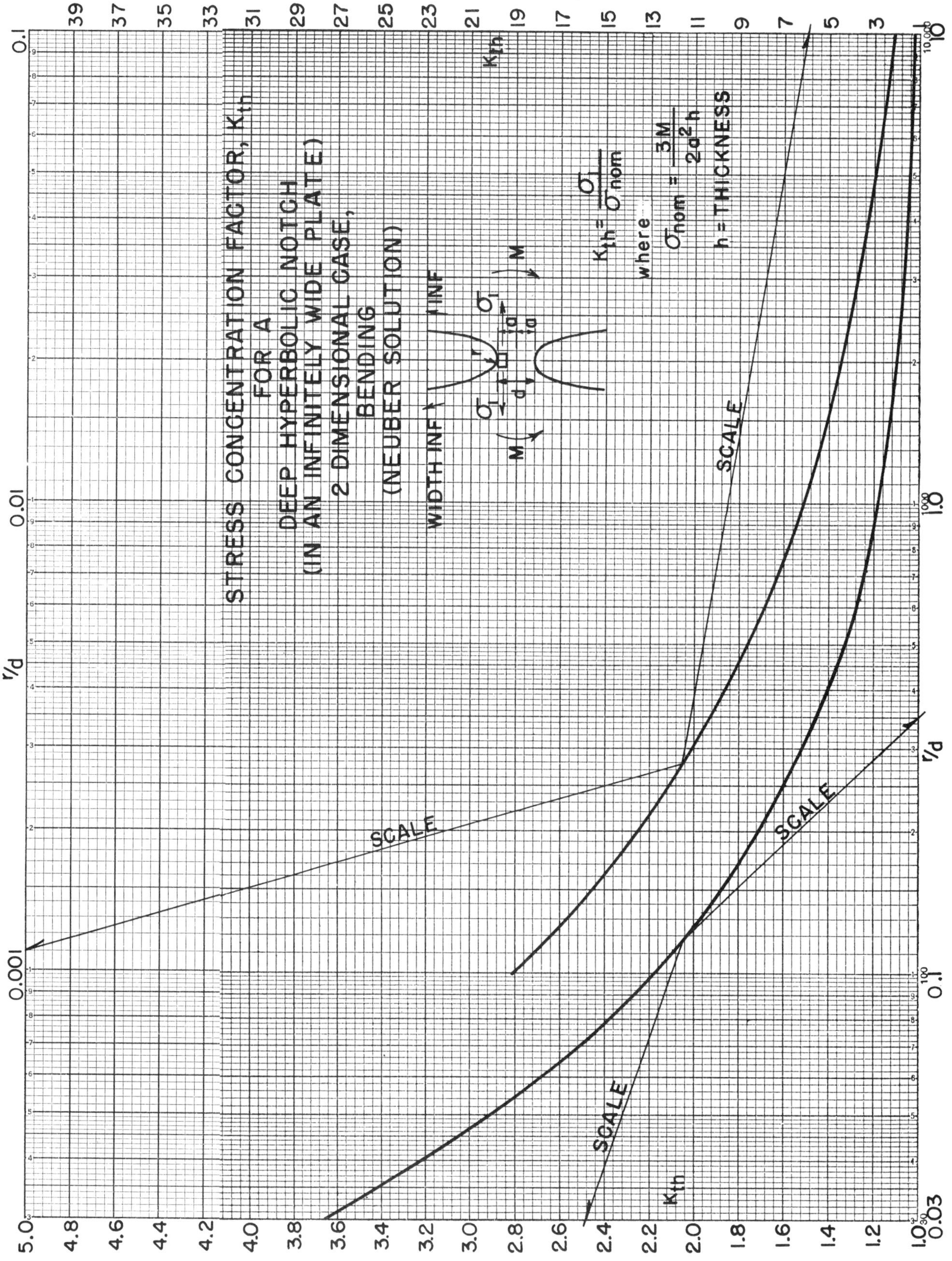

STRESS CONCENTRATION FACTOR, K_{th}
FOR A
DEEP HYPERBOLIC NOTCH
(IN AN INFINITELY WIDE PLATE)
2 DIMENSIONAL CASE,
BENDING
(NEUBER SOLUTION)
WIDTH INF
INF
$K_{th} = \frac{\sigma_1}{\sigma_{nom}}$
where
$\sigma_{nom} = \frac{3M}{2a^2 h}$
h = THICKNESS
SCALE
SCALE
SCALE
SCALE
K_{th}
K_{th}
r/d
r/d
0.001
0.01
0.1
0.03
0.1
1.0
10

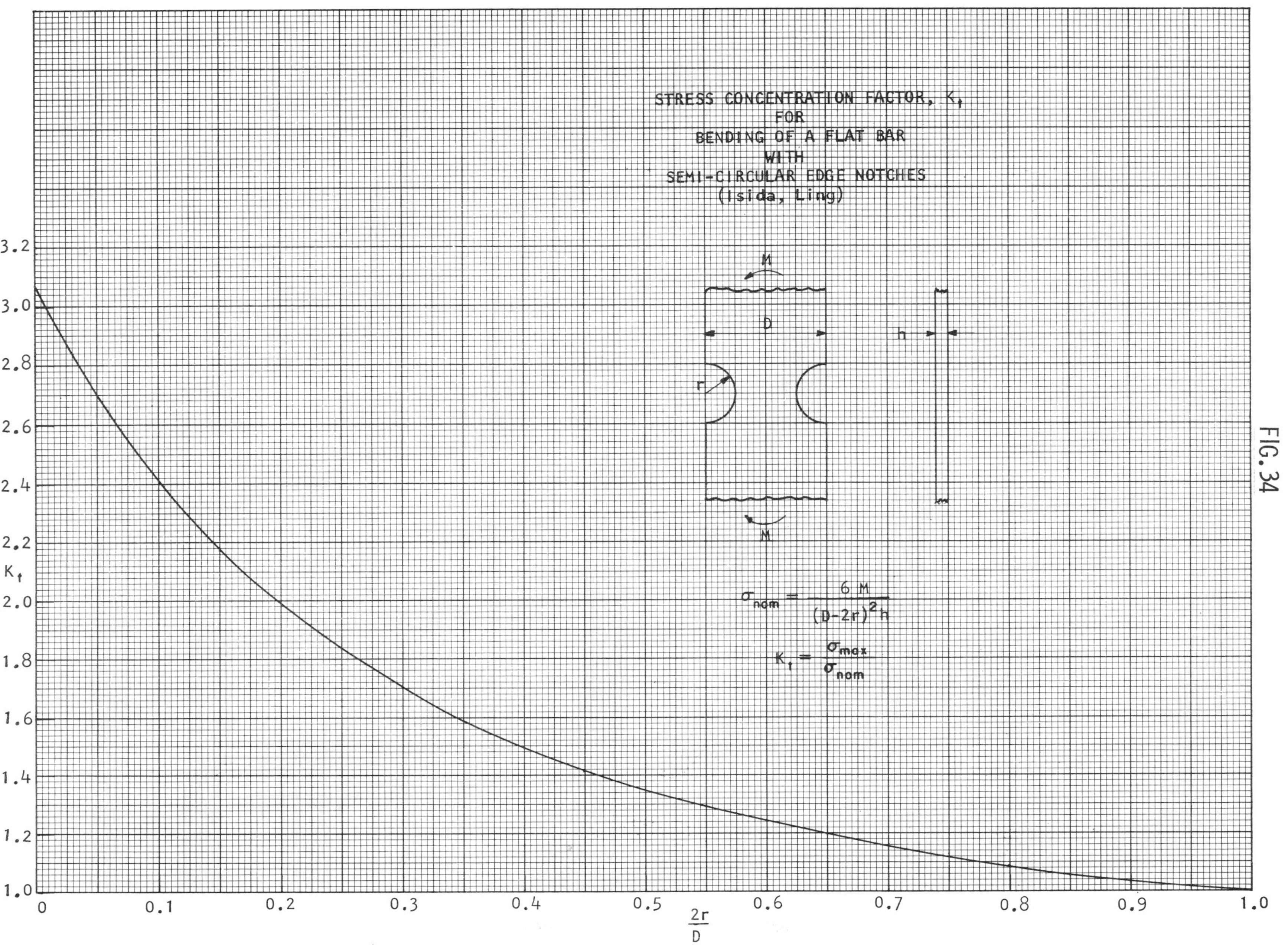

STRESS CONCENTRATION FACTOR, K_t
FOR
BENDING OF A FLAT BAR
WITH
SEMI-CIRCULAR EDGE NOTCHES
(Isida, Ling)
M
D
h
r
M
$\sigma_{nom} = \frac{6\,M}{(D-2r)^2 h}$
$K_t = \frac{\sigma_{max}}{\sigma_{nom}}$
K_t
3.2
3.0
2.8
2.6
2.4
2.2
2.0
1.8
1.6
1.4
1.2
1.0
0
0.1
0.2
0.3
0.4
0.5
0.6
0.7
0.8
0.9
1.0
$\frac{2r}{D}$

FIG. 34

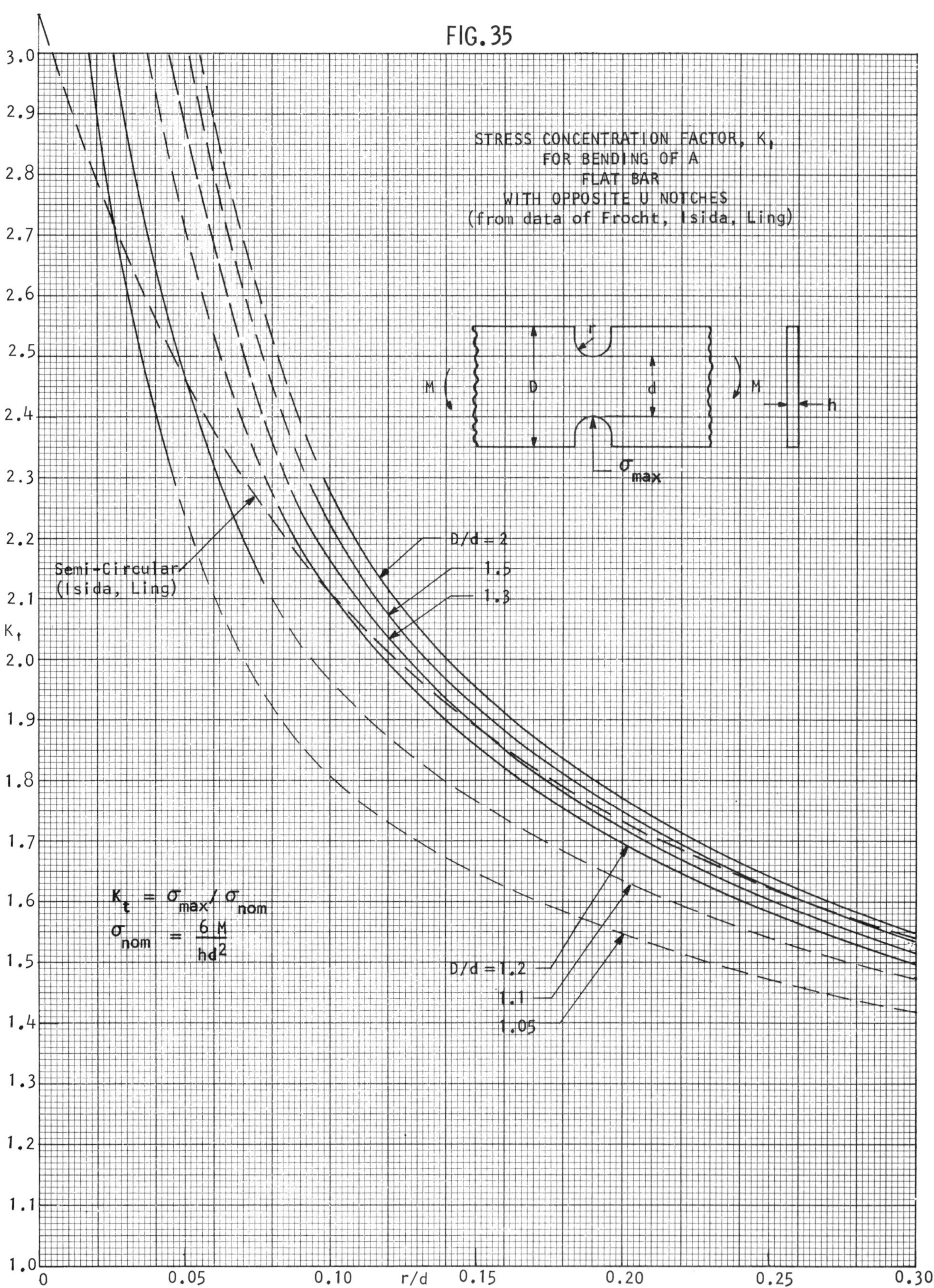
FIG. 35
STRESS CONCENTRATION FACTOR, K_t
FOR BENDING OF A
FLAT BAR
WITH OPPOSITE U NOTCHES
(from data of Frocht, Isida, Ling)
M
D
r
d
M
h
σ_{max}
D/d = 2
1.5
1.3
Semi-Circular
(Isida, Ling)
D/d = 1.2
1.1
1.05
$K_t = \sigma_{max}/\sigma_{nom}$
$\sigma_{nom} = \frac{6M}{hd^2}$
K_t
r/d
3.0
2.9
2.8
2.7
2.6
2.5
2.4
2.3
2.2
2.1
2.0
1.9
1.8
1.7
1.6
1.5
1.4
1.3
1.2
1.1
1.0
0
0.05
0.10
0.15
0.20
0.25
0.30

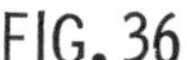

FIG. 36

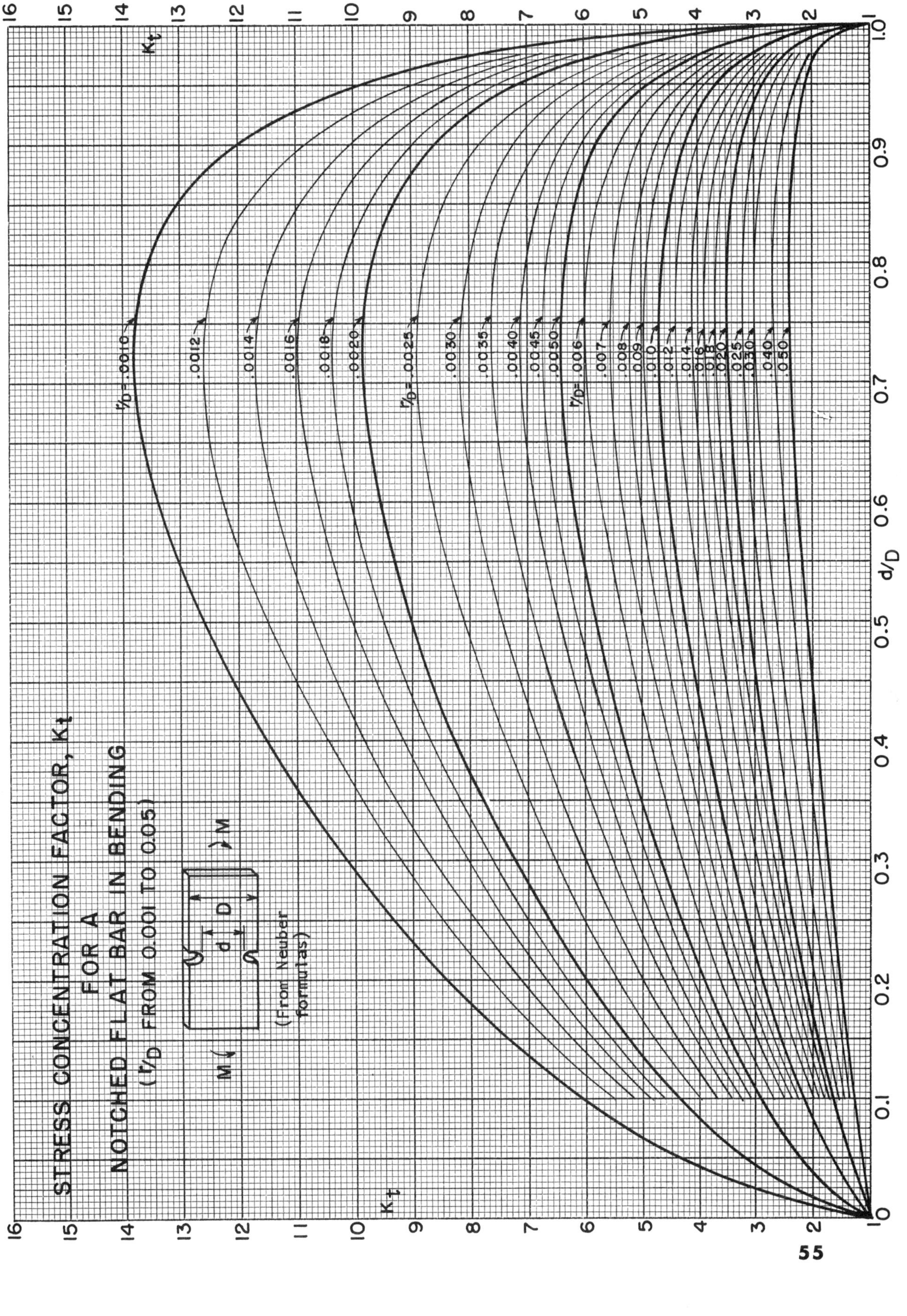

STRESS CONCENTRATION FACTOR, Kt
FOR A
NOTCHED FLAT BAR IN BENDING
(r/D FROM 0.001 TO 0.05)
M
d
D
M
(From Neuber formulas)
Kt
d/D
r/D = .0010
.0012
.0014
.0016
.0018
.0020
r/D = .0025
.0030
.0035
.0040
.0045
.0050
r/D = .006
.007
.008
.009
.010
.012
.014
.016
.018
.020
.025
.030
.040
.050
0
0.1
0.2
0.3
0.4
0.5
0.6
0.7
0.8
0.9
1.0
1
2
3
4
5
6
7
8
9
10
11
12
13
14
15
16

FIG. 37

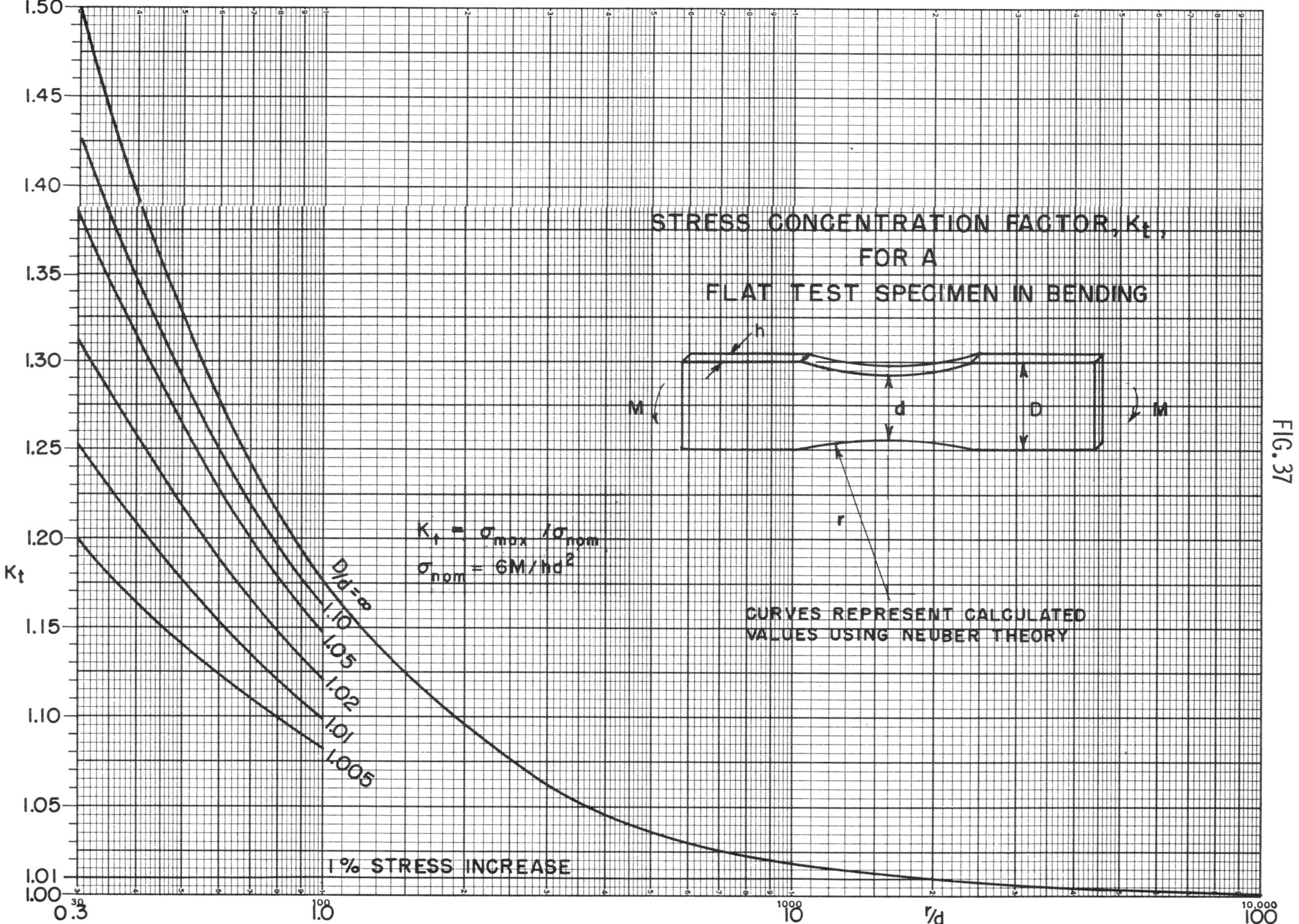
STRESS CONCENTRATION FACTOR, K_t,
FOR A
FLAT TEST SPECIMEN IN BENDING
h
M
d
D
M
r
$K_t = \sigma_{max} / \sigma_{nom}$
$\sigma_{nom} = 6M/hd^2$
CURVES REPRESENT CALCULATED
VALUES USING NEUBER THEORY
D/d = ∞
1.10
1.05
1.02
1.01
1.005
1 % STRESS INCREASE
K_t
1.50
1.45
1.40
1.35
1.30
1.25
1.20
1.15
1.10
1.05
1.01
1.00
0.3
1.0
10
100
r/d

FIG. 38

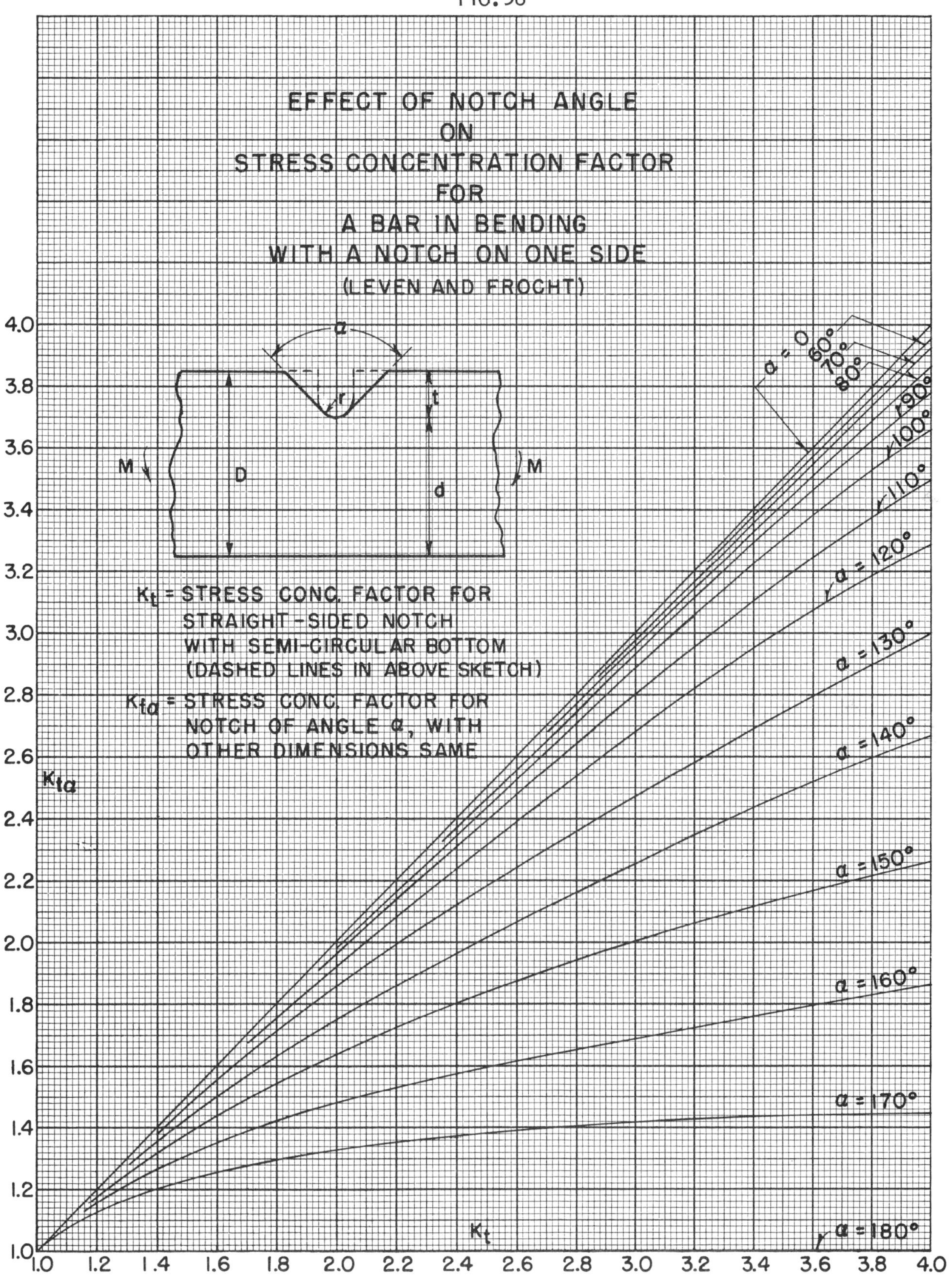

EFFECT OF NOTCH ANGLE
ON
STRESS CONCENTRATION FACTOR
FOR
A BAR IN BENDING
WITH A NOTCH ON ONE SIDE
(LEVEN AND FROCHT)
α
r
t
M
D
d
M
K_t = STRESS CONC. FACTOR FOR
STRAIGHT-SIDED NOTCH
WITH SEMI-CIRCULAR BOTTOM
(DASHED LINES IN ABOVE SKETCH)
$K_{t\alpha}$ = STRESS CONC. FACTOR FOR
NOTCH OF ANGLE α, WITH
OTHER DIMENSIONS SAME
$K_{t\alpha}$
K_t
α = 0
60°
70°
80°
90°
100°
110°
α = 120°
α = 130°
α = 140°
α = 150°
α = 160°
α = 170°
α = 180°
4.0
3.8
3.6
3.4
3.2
3.0
2.8
2.6
2.4
2.2
2.0
1.8
1.6
1.4
1.2
1.0
1.0
1.2
1.4
1.6
1.8
2.0
2.2
2.4
2.6
2.8
3.0
3.2
3.4
3.6
3.8
4.0

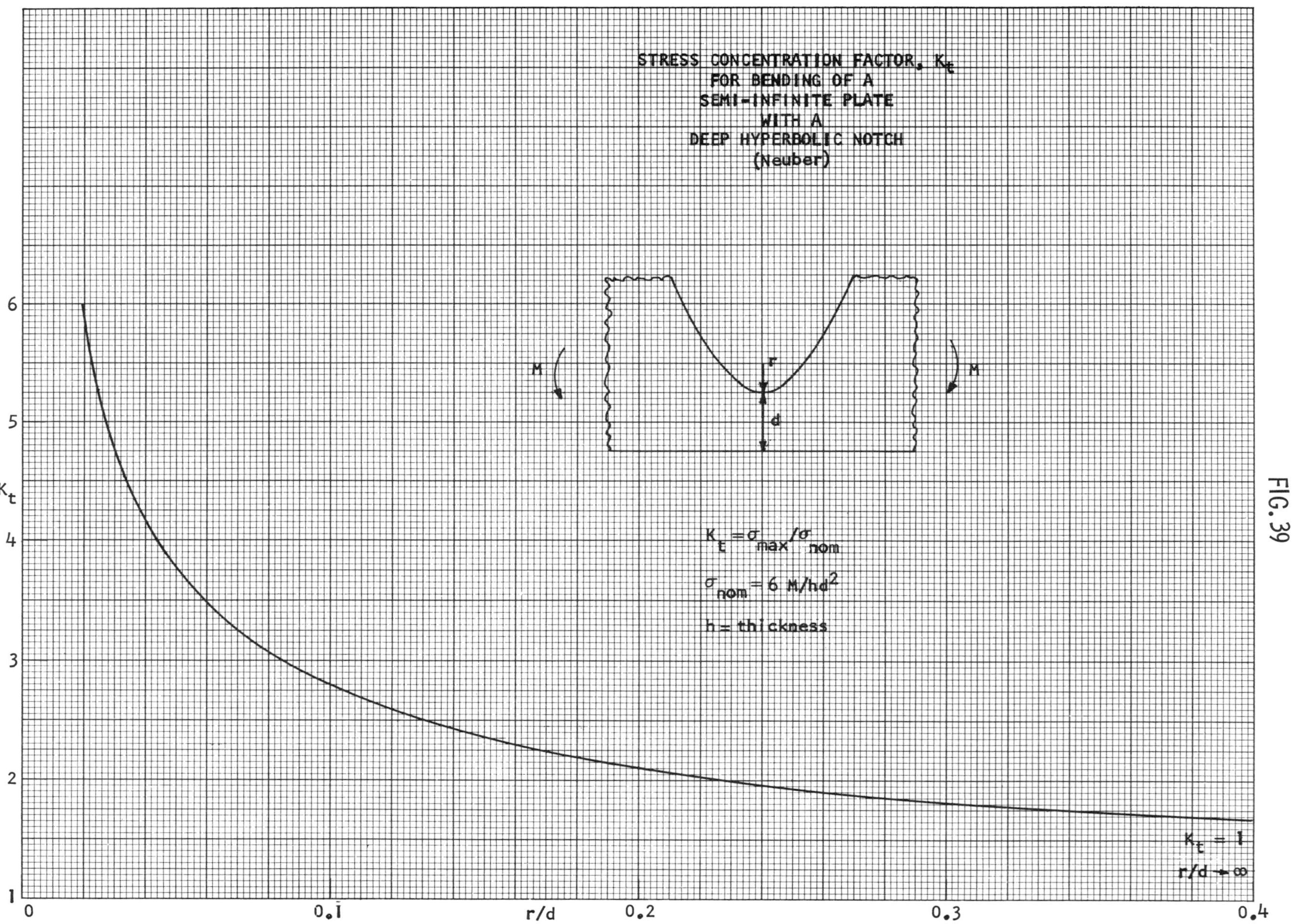

STRESS CONCENTRATION FACTOR, K_t
FOR BENDING OF A
SEMI-INFINITE PLATE
WITH A
DEEP HYPERBOLIC NOTCH
(Neuber)
M
M
r
d
$K_t = \sigma_{max}/\sigma_{nom}$
$\sigma_{nom} = 6\ M/hd^2$
h = thickness
$K_t = 1$
$r/d \to \infty$
K_t
1
2
3
4
5
6
0
0.1
0.2
0.3
0.4
r/d

FIG. 39

FIG. 39a

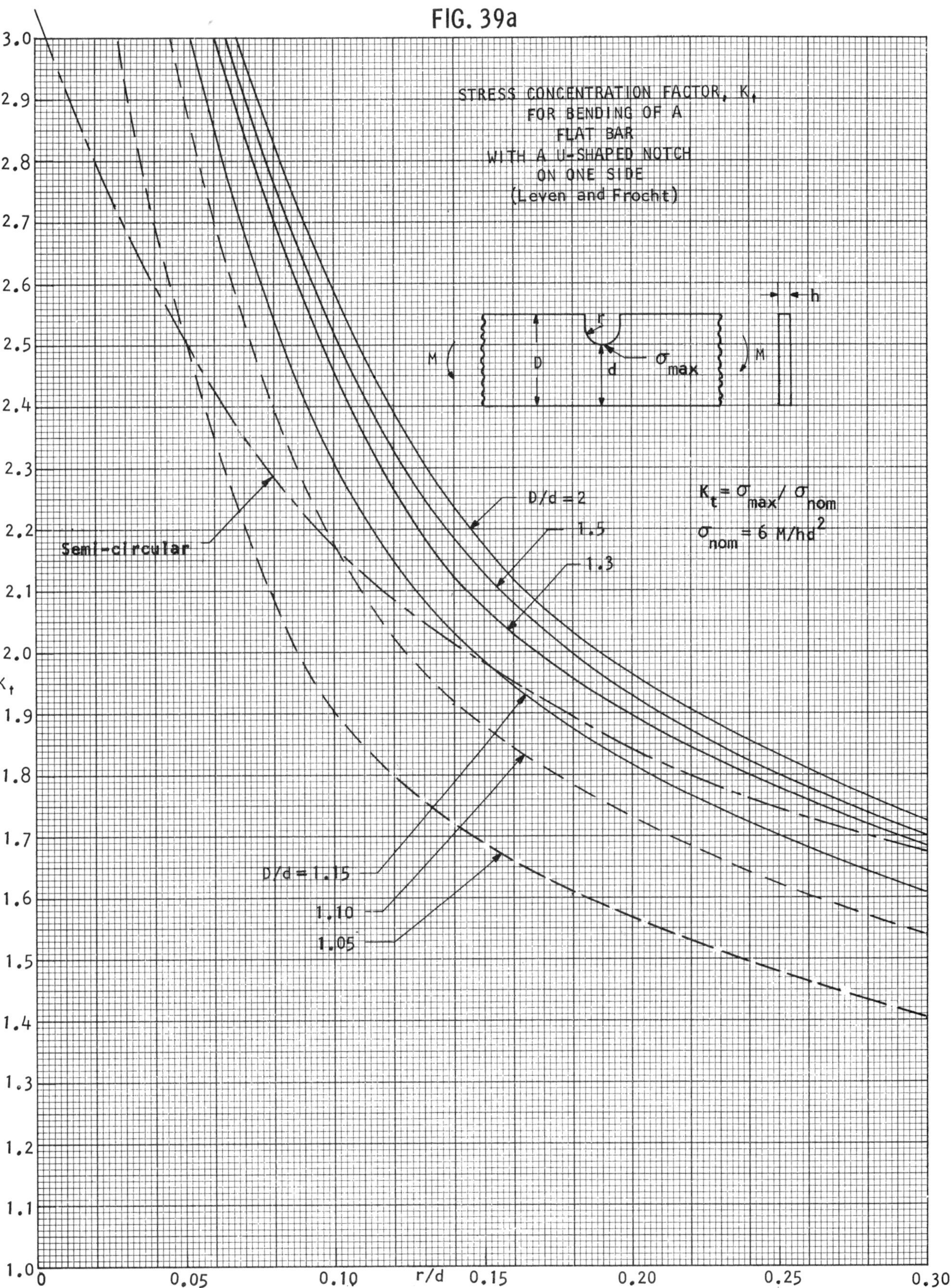
STRESS CONCENTRATION FACTOR, K_t
FOR BENDING OF A
FLAT BAR
WITH A U-SHAPED NOTCH
ON ONE SIDE
(Leven and Frocht)
h
r
M
D
d
σ_{max}
M
$K_t = \sigma_{max} / \sigma_{nom}$
$\sigma_{nom} = 6\ M/hd^2$
D/d = 2
1.5
1.3
Semi-circular
D/d = 1.15
1.10
1.05
K_t
r/d
3.0
2.9
2.8
2.7
2.6
2.5
2.4
2.3
2.2
2.1
2.0
1.9
1.8
1.7
1.6
1.5
1.4
1.3
1.2
1.1
1.0
0
0.05
0.10
0.15
0.20
0.25
0.30

FIG. 39b

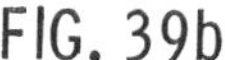

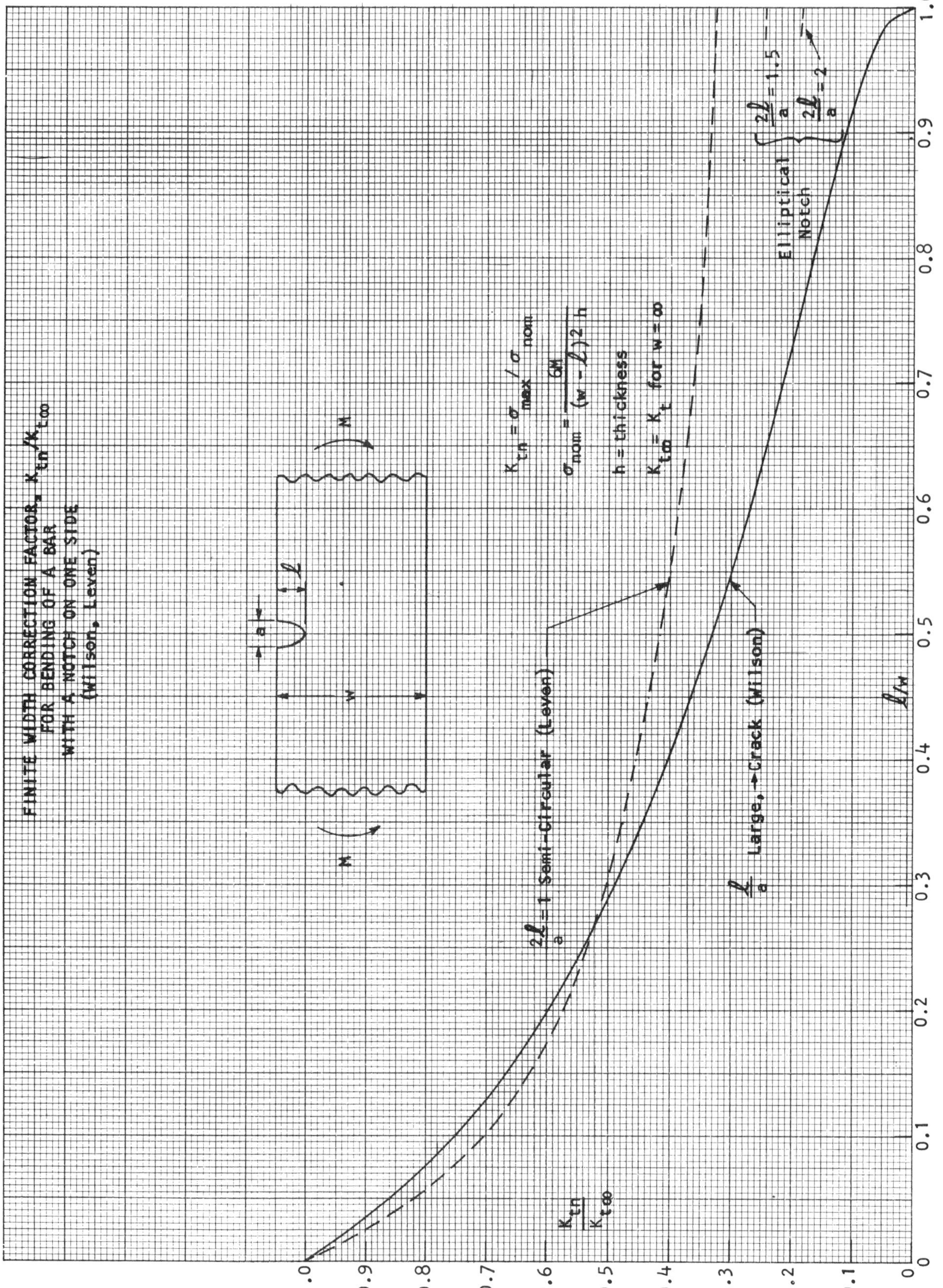
FINITE WIDTH CORRECTION FACTOR, $K_{tn}/K_{t\infty}$
FOR BENDING OF A BAR
WITH A NOTCH ON ONE SIDE
(Wilson, Leven)
M
M
a
ℓ
w
$K_{tn} = \sigma_{max} / \sigma_{nom}$
$\sigma_{nom} = \frac{6M}{(w - \ell)^2 h}$
h = thickness
$K_{t\infty} = K_t$ for $w = \infty$
$\frac{2\ell}{a} = 1$ Semi-Circular (Leven)
$\frac{\ell}{a}$ Large, → Crack (Wilson)
Elliptical Notch
$\frac{2\ell}{a} = 1.5$
$\frac{2\ell}{a} = 2$
$\frac{K_{tn}}{K_{t\infty}}$
ℓ/w
1.0
0.9
0.8
0.7
0.6
0.5
0.4
0.3
0.2
0.1
0

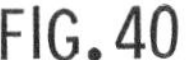

FIG. 40

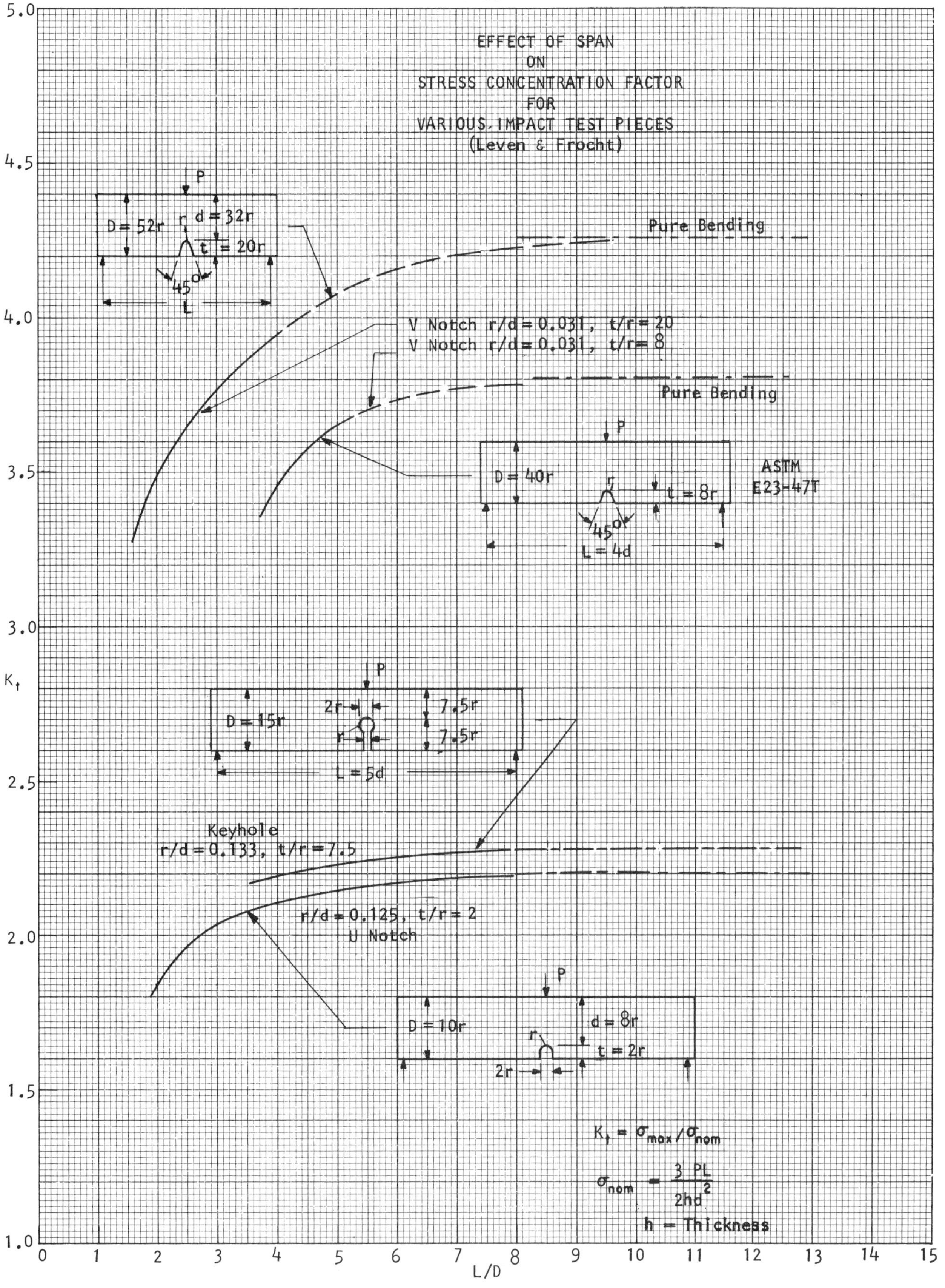
EFFECT OF SPAN
ON
STRESS CONCENTRATION FACTOR
FOR
VARIOUS IMPACT TEST PIECES
(Leven & Frocht)
5.0
4.5
4.0
3.5
3.0
2.5
2.0
1.5
1.0
K_t
0 1 2 3 4 5 6 7 8 9 10 11 12 13 14 15
L/D
P
D = 52r
d = 32r
r
t = 20r
45°
L
Pure Bending
V Notch r/d = 0.031, t/r = 20
V Notch r/d = 0.031, t/r = 8
Pure Bending
P
D = 40r
r
t = 8r
45°
L = 4d
ASTM
E23-47T
P
D = 15r
2r
7.5r
7.5r
r
L = 5d
Keyhole
r/d = 0.133, t/r = 7.5
r/d = 0.125, t/r = 2
U Notch
P
D = 10r
d = 8r
r
t = 2r
2r
$K_t = \sigma_{max}/\sigma_{nom}$
$\sigma_{nom} = \frac{3\ PL}{2hd^2}$
h = Thickness

FIG. 41

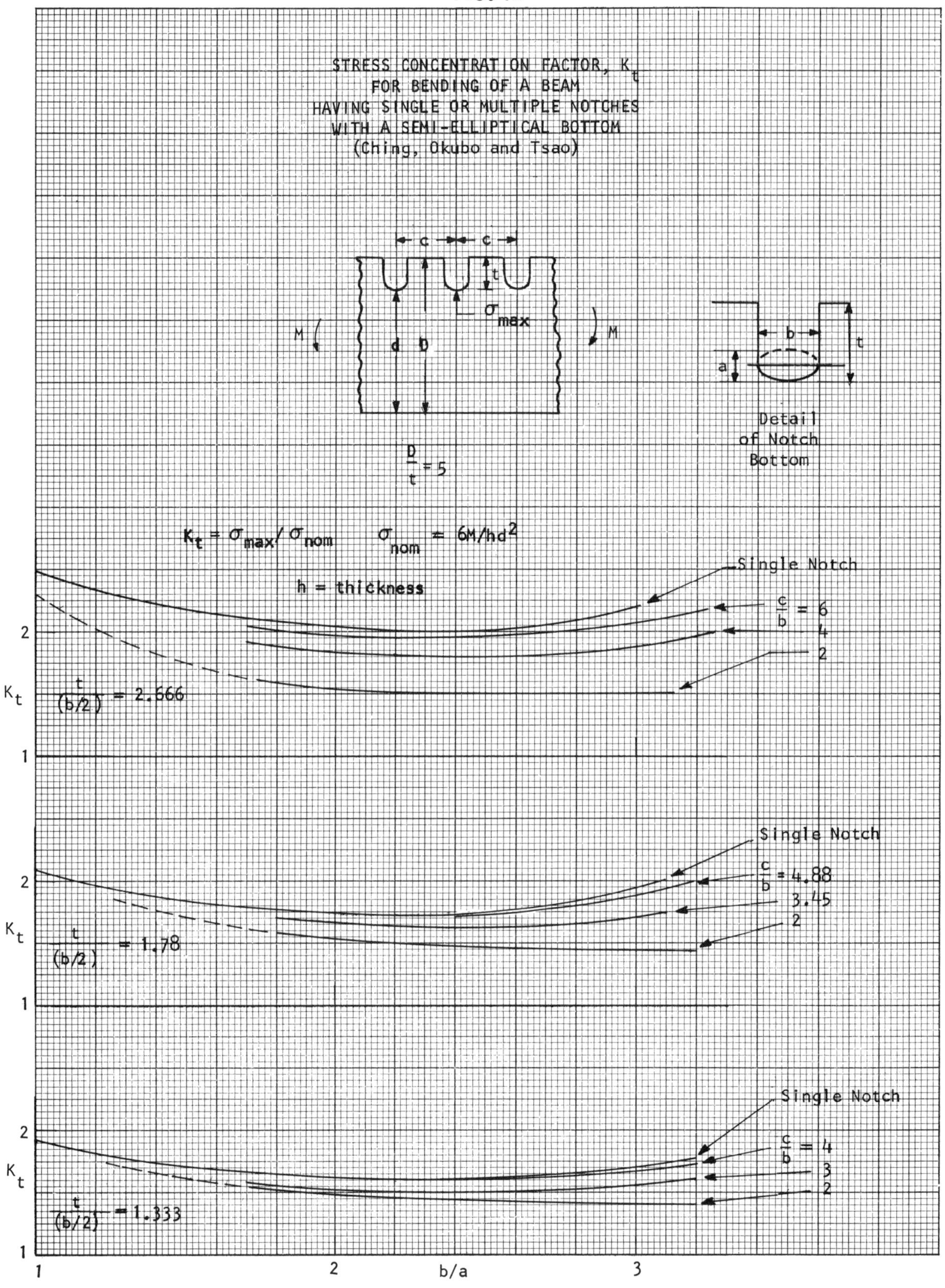

STRESS CONCENTRATION FACTOR, K_t
FOR BENDING OF A BEAM
HAVING SINGLE OR MULTIPLE NOTCHES
WITH A SEMI-ELLIPTICAL BOTTOM
(Ching, Okubo and Tsao)
c
c
t
σ_{max}
M
M
d
D
b
a
t
Detail of Notch Bottom
$\frac{D}{t} = 5$
$K_t = \sigma_{max}/\sigma_{nom}$ $\sigma_{nom} = 6M/hd^2$
h = thickness
Single Notch
$\frac{c}{b} = 6$
4
2
K_t
$\frac{t}{(b/2)} = 2.666$
Single Notch
$\frac{c}{b} = 4.88$
3.45
2
K_t
$\frac{t}{(b/2)} = 1.78$
Single Notch
$\frac{c}{b} = 4$
3
2
K_t
$\frac{t}{(b/2)} = 1.333$
1
2
3
b/a

FIG. 42

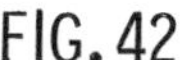

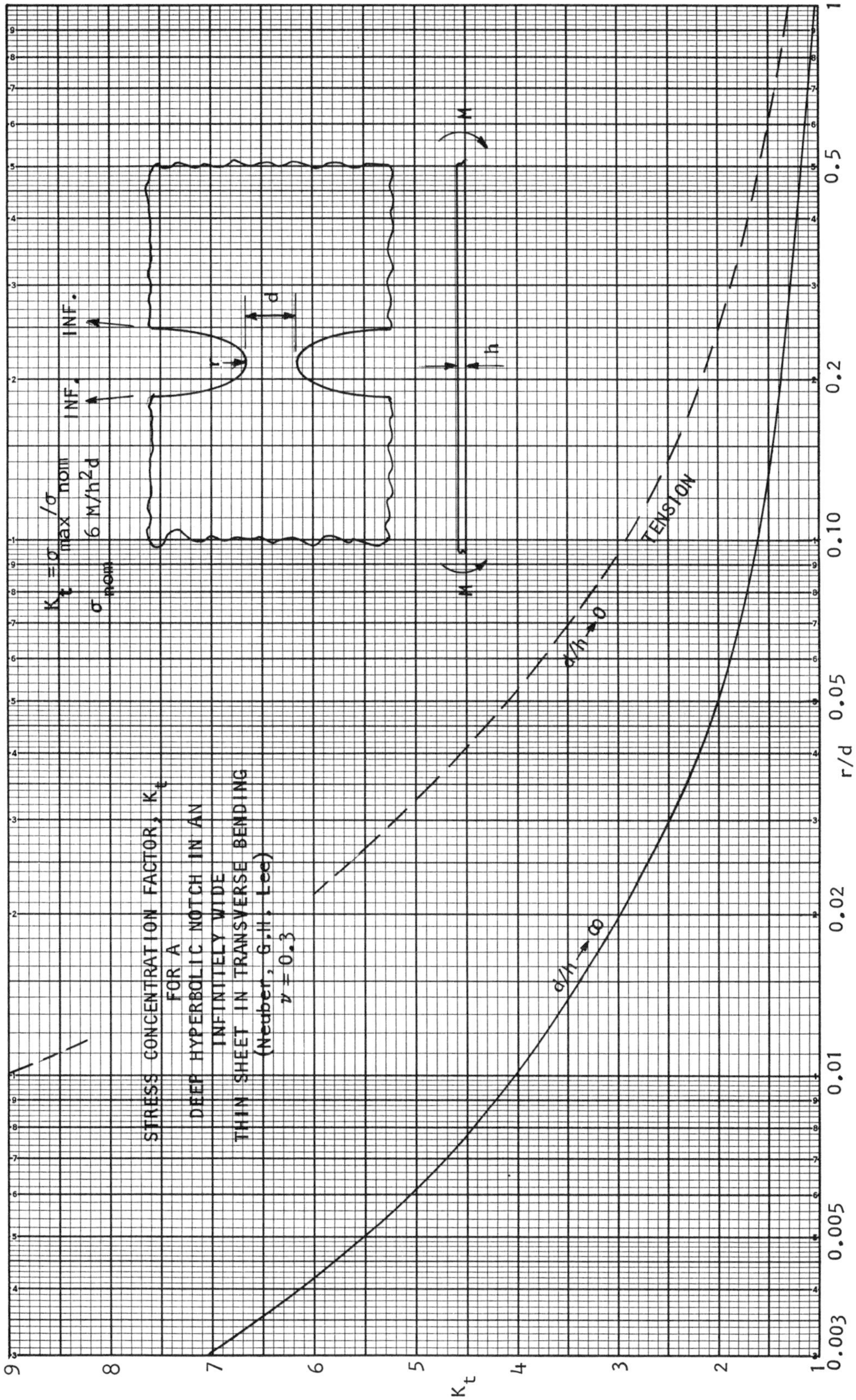

STRESS CONCENTRATION FACTOR, K_t
FOR A
DEEP HYPERBOLIC NOTCH IN AN
INFINITELY WIDE
THIN SHEET IN TRANSVERSE BENDING
(Neuber, G.H. Lee)
$\nu = 0.3$
$K_t = \sigma_{max}/\sigma_{nom}$
$\sigma_{nom} = 6\ M/h^2d$
INF.
INF.
d
r
h
M
M
TENSION
d/h → 0
d/h → ∞
K_t
1
2
3
4
5
6
7
8
9
r/d
0.003
0.005
0.01
0.02
0.05
0.10
0.2
0.5
1

FIG. 43

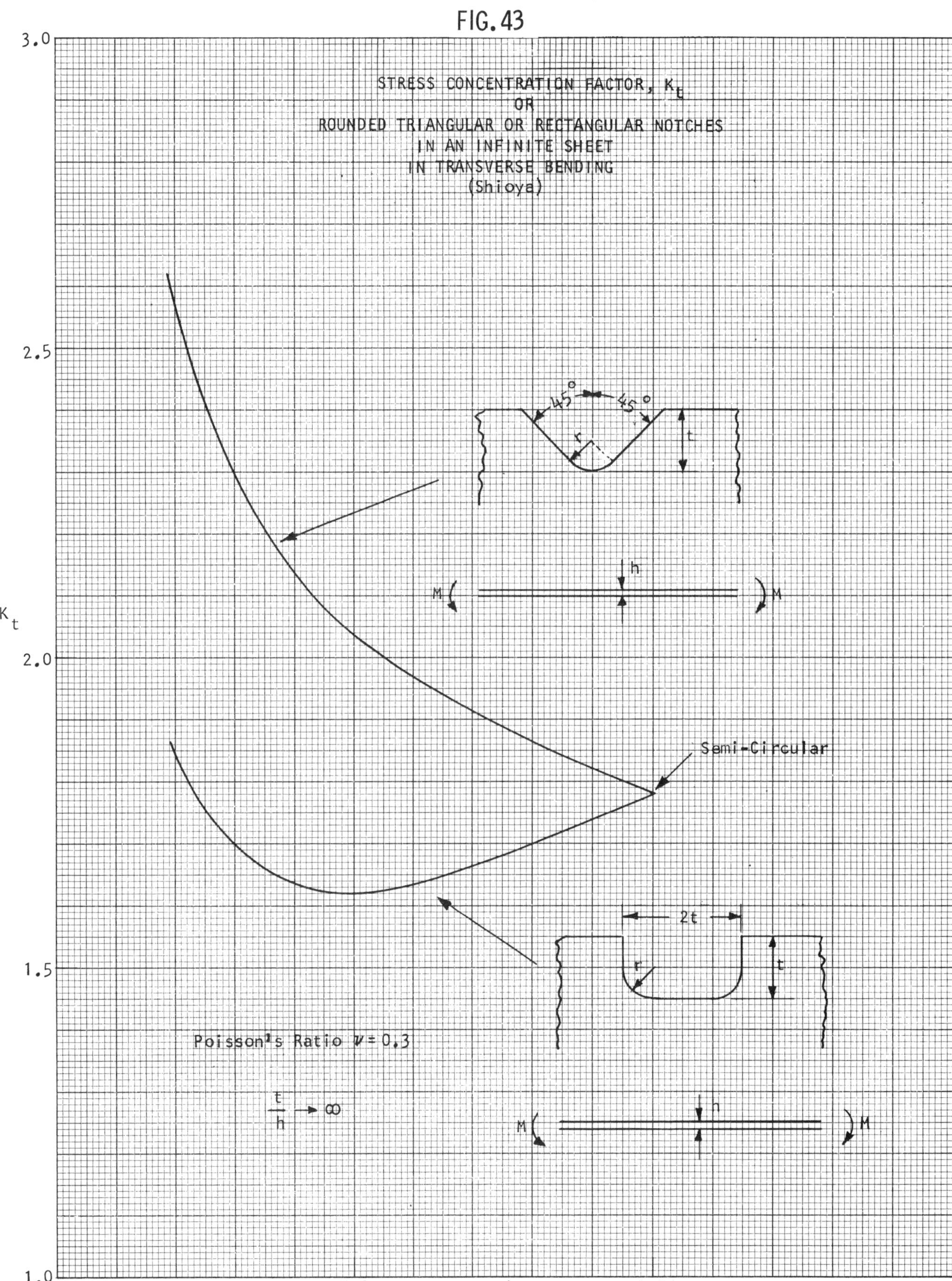

STRESS CONCENTRATION FACTOR, K_t
OR
ROUNDED TRIANGULAR OR RECTANGULAR NOTCHES
IN AN INFINITE SHEET
IN TRANSVERSE BENDING
(Shioya)
3.0
2.5
2.0
1.5
1.0
K_t
0
0.5
r/t
1.0
45°
45°
r
t
h
M
M
Semi-Circular
2t
r
t
h
M
M
Poisson's Ratio $\nu = 0.3$
$\frac{t}{h} \rightarrow \infty$

FIG. 44

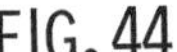

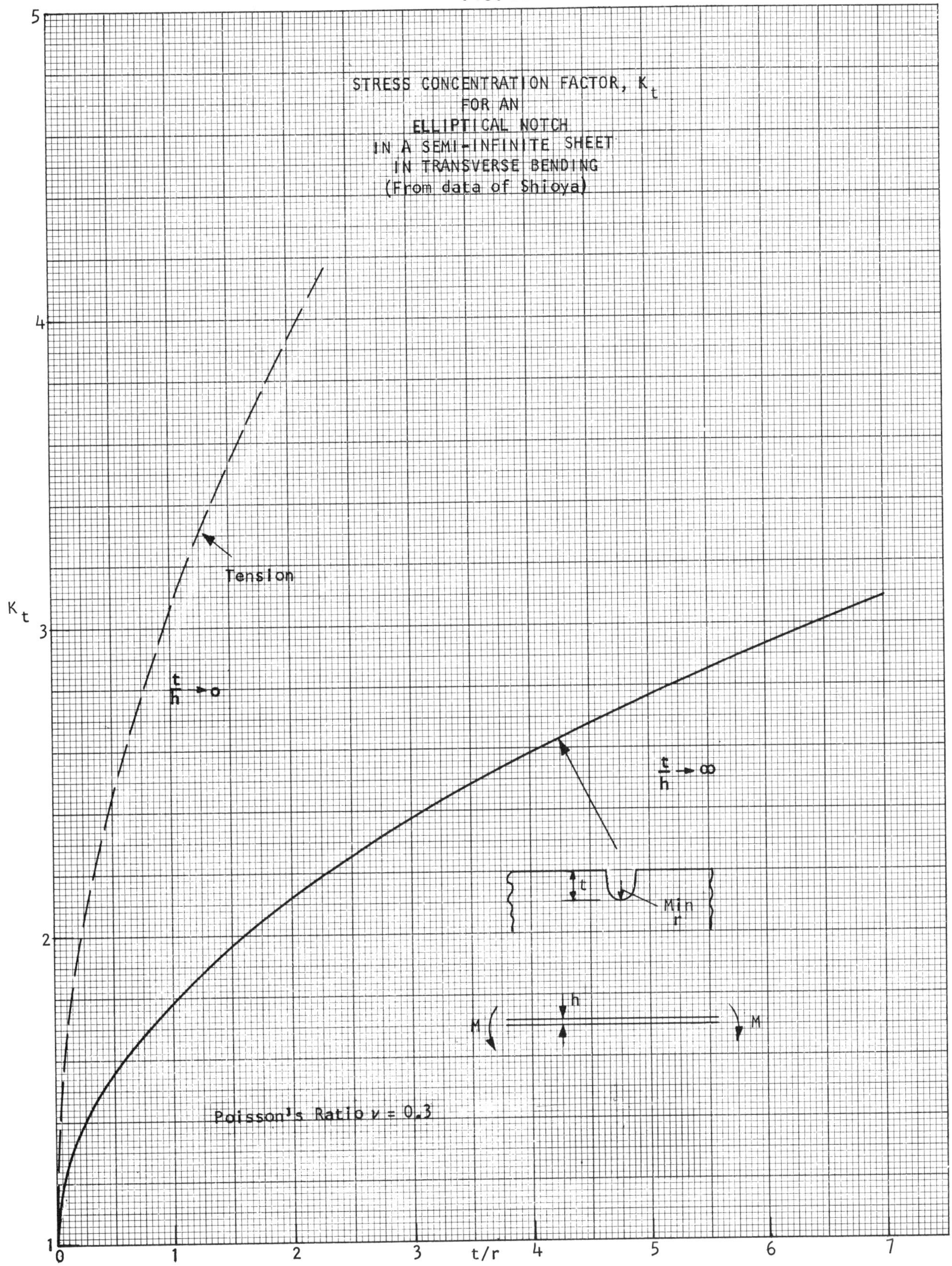
STRESS CONCENTRATION FACTOR, K_t
FOR AN
ELLIPTICAL NOTCH
IN A SEMI-INFINITE SHEET
IN TRANSVERSE BENDING
(From data of Shioya)
Tension
$\frac{t}{h} \rightarrow 0$
$\frac{t}{h} \rightarrow \infty$
t
Min r
h
M
M
Poisson's Ratio ν = 0.3
K_t
t/r
1
2
3
4
5
0
1
2
3
4
5
6
7

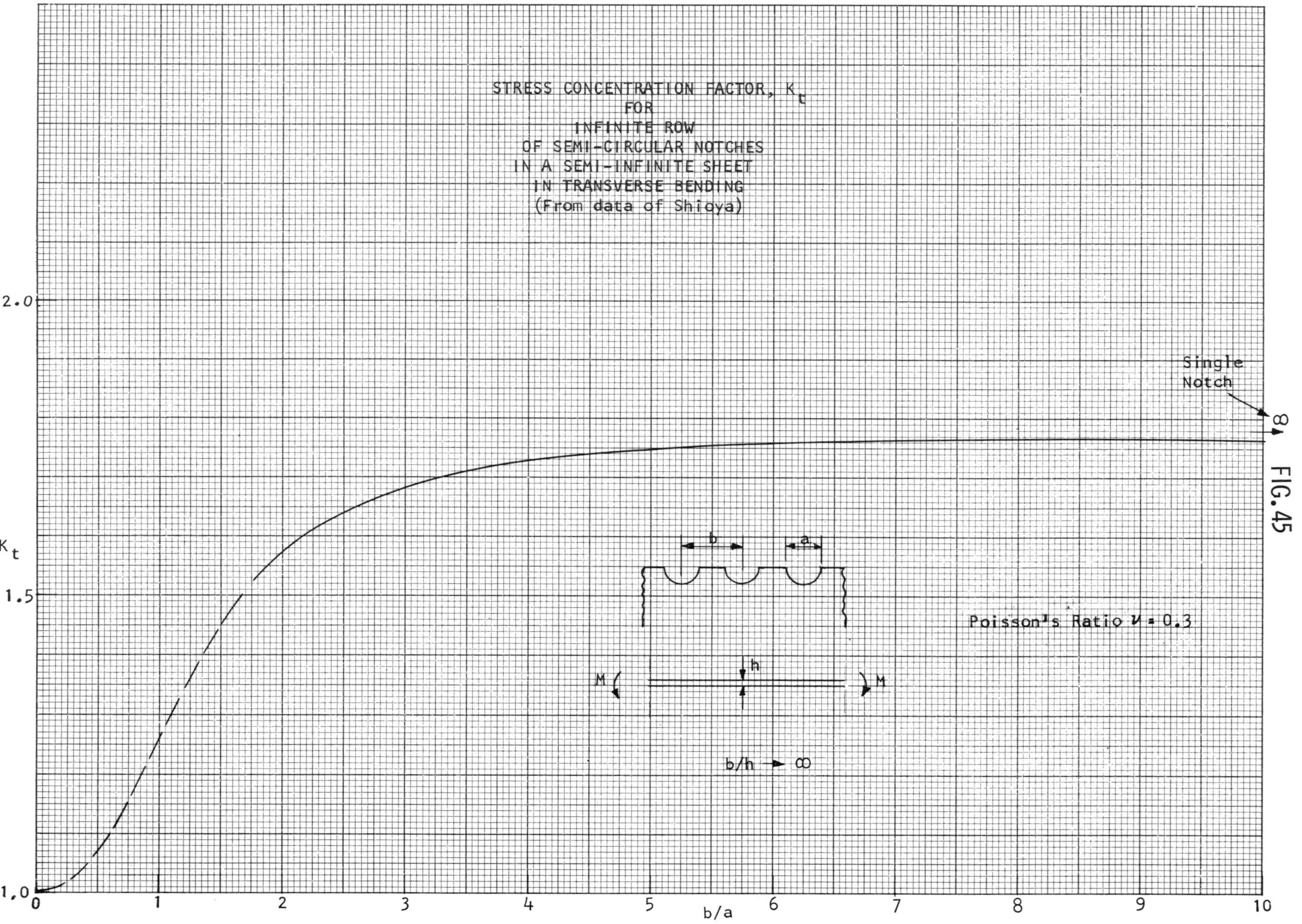
STRESS CONCENTRATION FACTOR, K_t
FOR
INFINITE ROW
OF SEMI-CIRCULAR NOTCHES
IN A SEMI-INFINITE SHEET
IN TRANSVERSE BENDING
(From data of Shioya)
K_t
2.0
1.5
1.0
0
1
2
3
4
5
6
7
8
9
10
b/a
Single
Notch
∞
b
a
h
M
M
b/h → ∞
Poisson's Ratio ν = 0.3

FIG. 45

FIG. 46

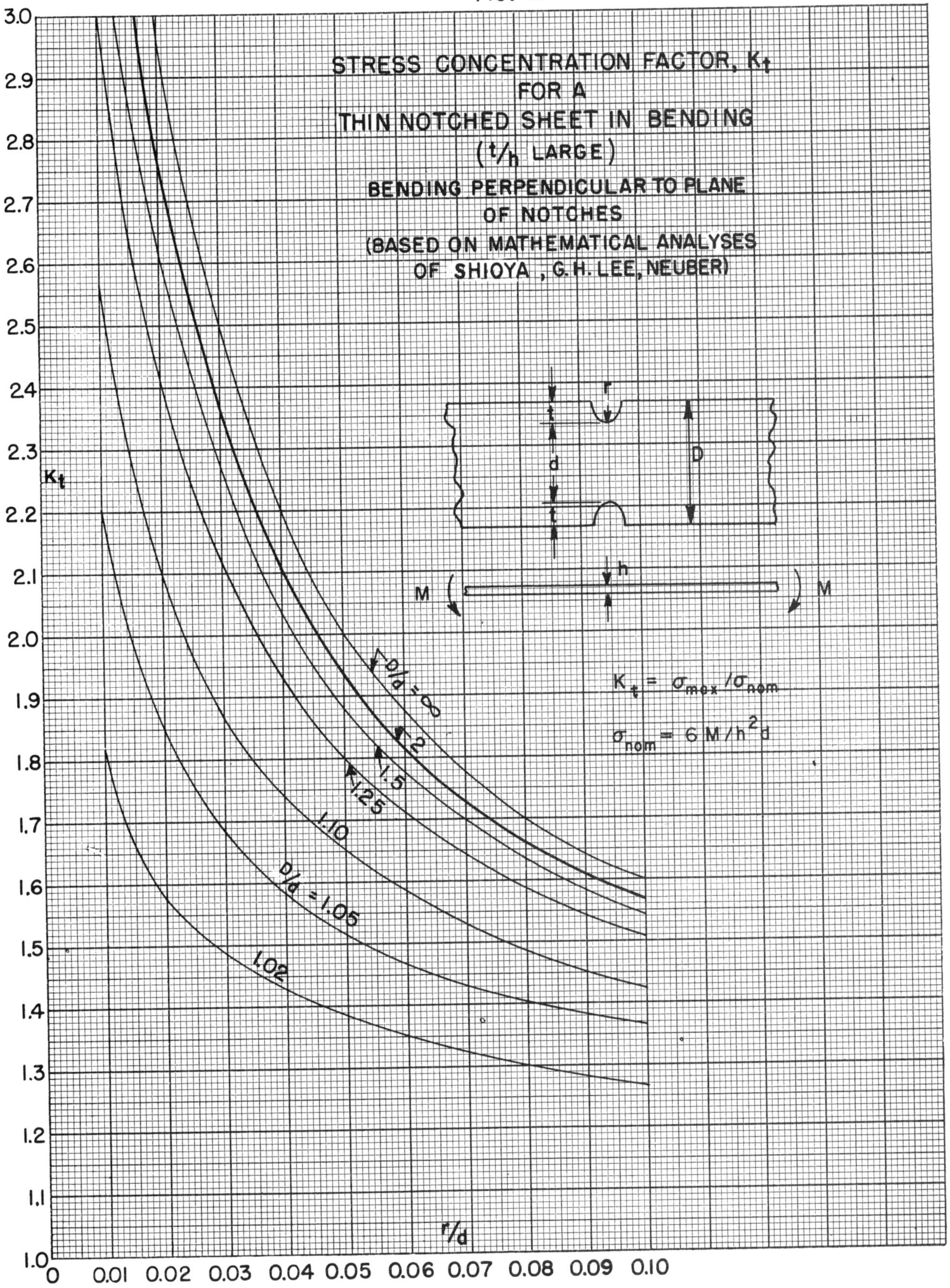

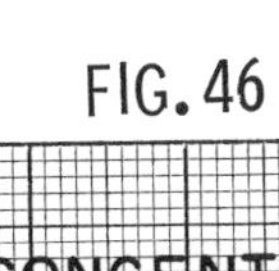
STRESS CONCENTRATION FACTOR, K_t
FOR A
THIN NOTCHED SHEET IN BENDING
(t/h LARGE)
BENDING PERPENDICULAR TO PLANE
OF NOTCHES
(BASED ON MATHEMATICAL ANALYSES
OF SHIOYA, G. H. LEE, NEUBER)
$K_t = \sigma_{max} / \sigma_{nom}$
$\sigma_{nom} = 6\, M/h^2 d$
$D/d = \infty$
2
1.5
1.25
1.10
$D/d = 1.05$
1.02
r
t
d
D
h
M
M
K_t
r/d
3.0
2.9
2.8
2.7
2.6
2.5
2.4
2.3
2.2
2.1
2.0
1.9
1.8
1.7
1.6
1.5
1.4
1.3
1.2
1.1
1.0
0
0.01
0.02
0.03
0.04
0.05
0.06
0.07
0.08
0.09
0.10

FIG. 47

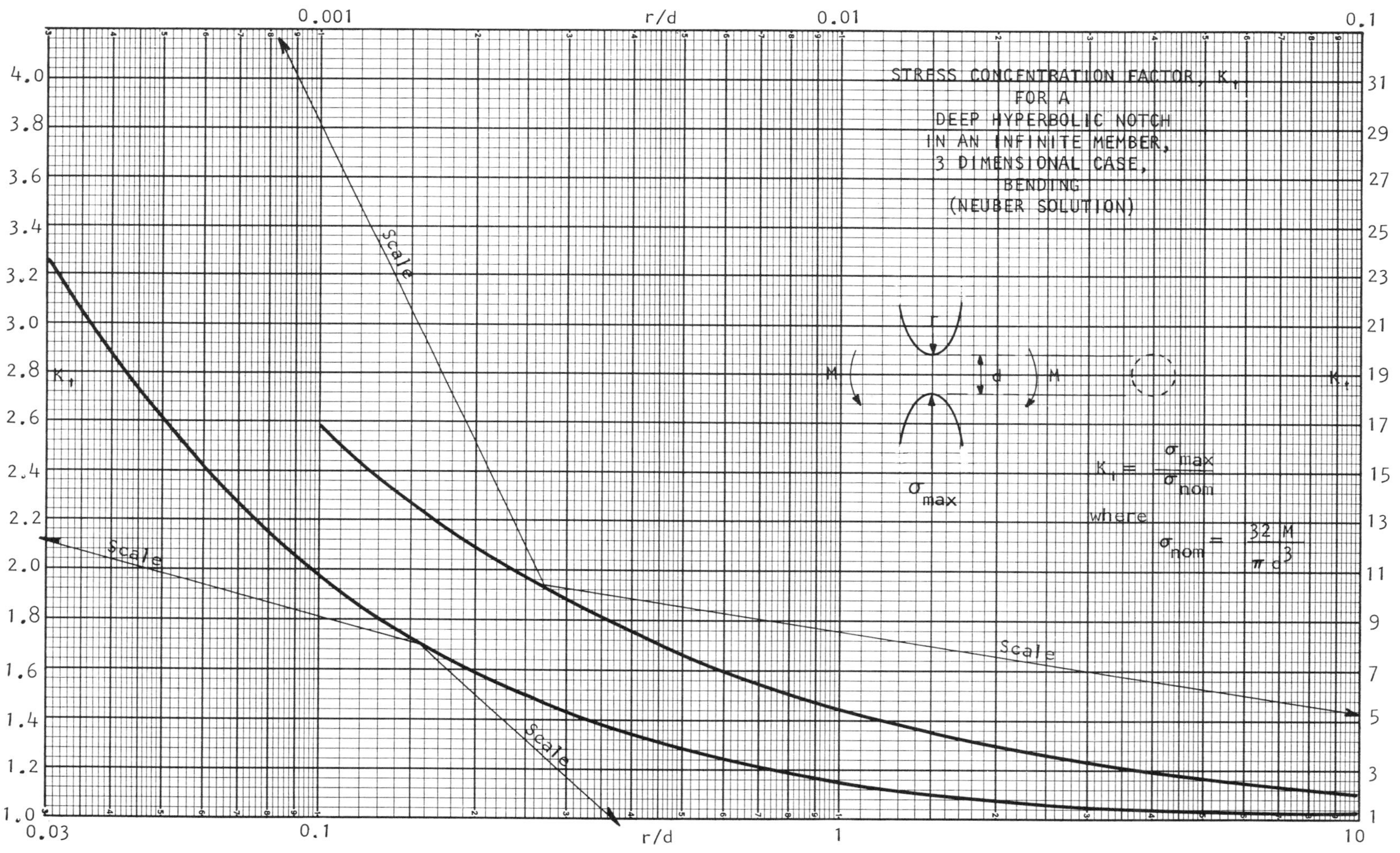

STRESS CONCENTRATION FACTOR, K_t
FOR A
DEEP HYPERBOLIC NOTCH
IN AN INFINITE MEMBER,
3 DIMENSIONAL CASE,
BENDING
(NEUBER SOLUTION)
$K_t = \frac{\sigma_{max}}{\sigma_{nom}}$
where
$\sigma_{nom} = \frac{32\,M}{\pi d^3}$
σ_{max}
M
M
d
r
r/d
r/d
K_t
K_t
Scale
Scale
Scale
Scale

FIG. 48

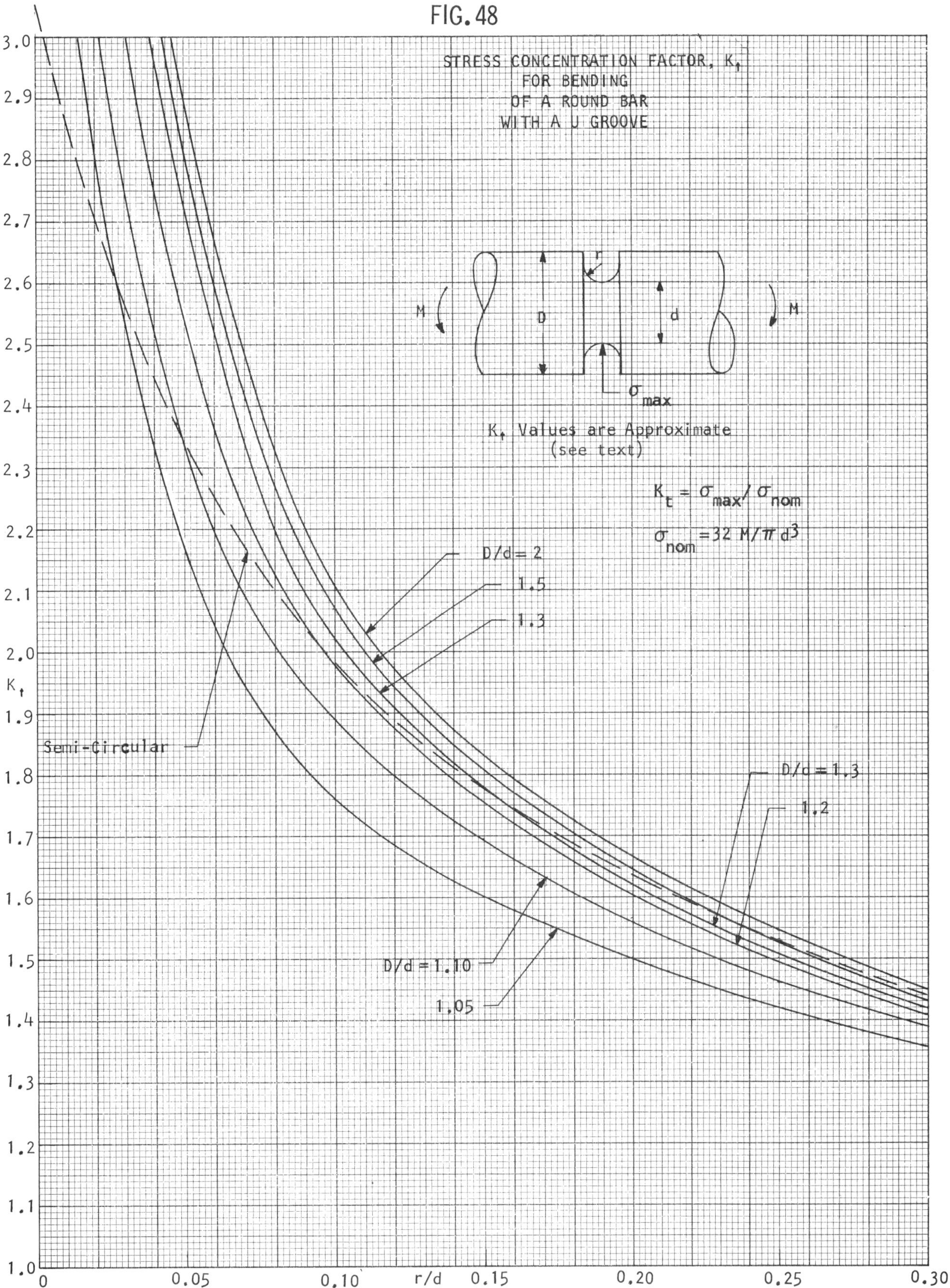
STRESS CONCENTRATION FACTOR, K_t
FOR BENDING
OF A ROUND BAR
WITH A U GROOVE
M
D
r
d
M
σ_{max}
K_t Values are Approximate
(see text)
$K_t = \sigma_{max}/\sigma_{nom}$
$\sigma_{nom} = 32\ M/\pi d^3$
D/d = 2
1.5
1.3
Semi-Circular
D/d = 1.3
1.2
D/d = 1.10
1.05
K_t
r/d
3.0
2.9
2.8
2.7
2.6
2.5
2.4
2.3
2.2
2.1
2.0
1.9
1.8
1.7
1.6
1.5
1.4
1.3
1.2
1.1
1.0
0
0.05
0.10
0.15
0.20
0.25
0.30

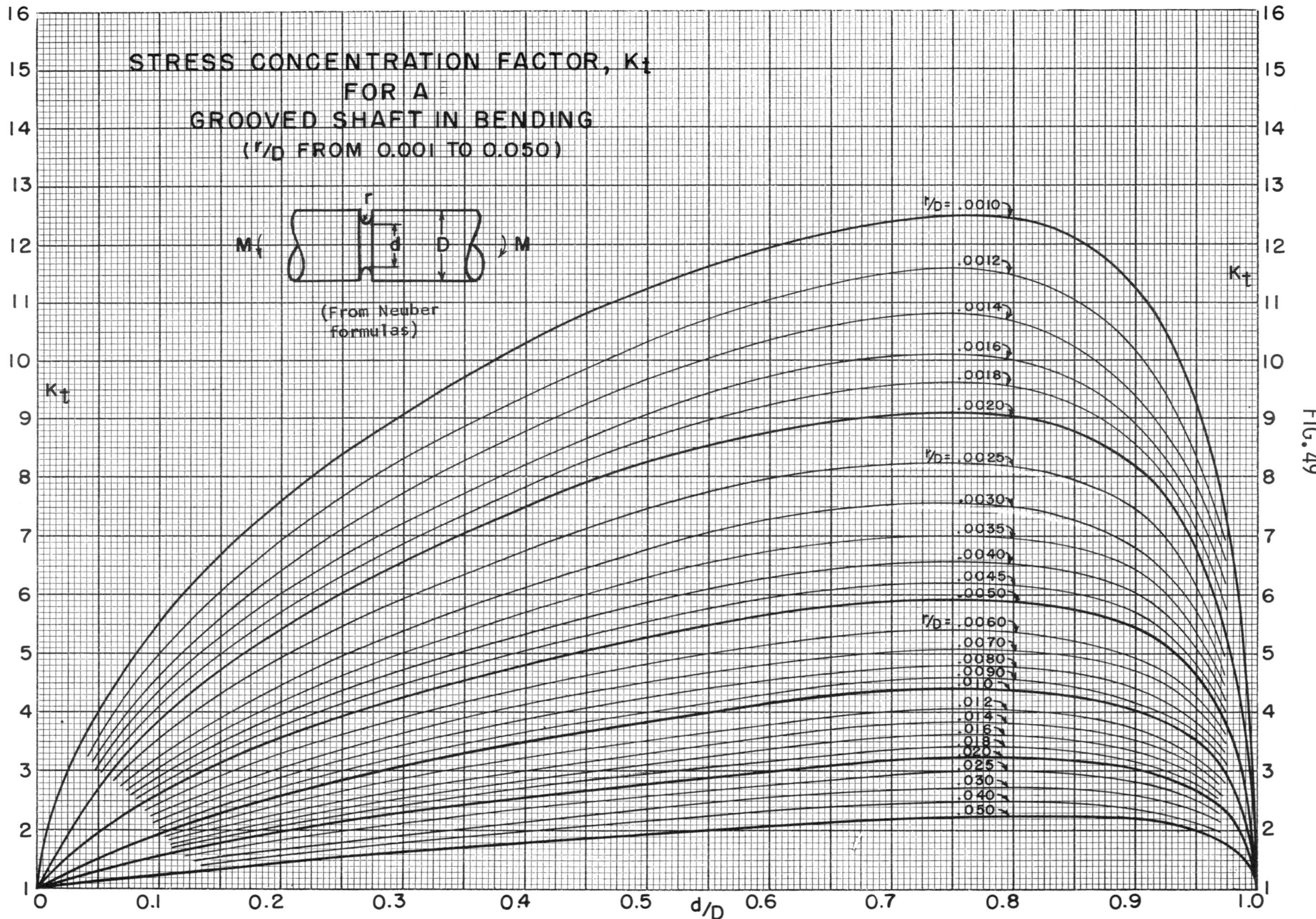

STRESS CONCENTRATION FACTOR, K_t
FOR A
GROOVED SHAFT IN BENDING
(r/D FROM 0.001 TO 0.050)
M
r
d
D
(From Neuber formulas)
K_t
d/D
r/D = .0010
.0012
.0014
.0016
.0018
.0020
r/D = .0025
.0030
.0035
.0040
.0045
.0050
r/D = .0060
.0070
.0080
.0090
.010
.012
.014
.016
.018
.020
.025
.030
.040
.050

FIG. 49

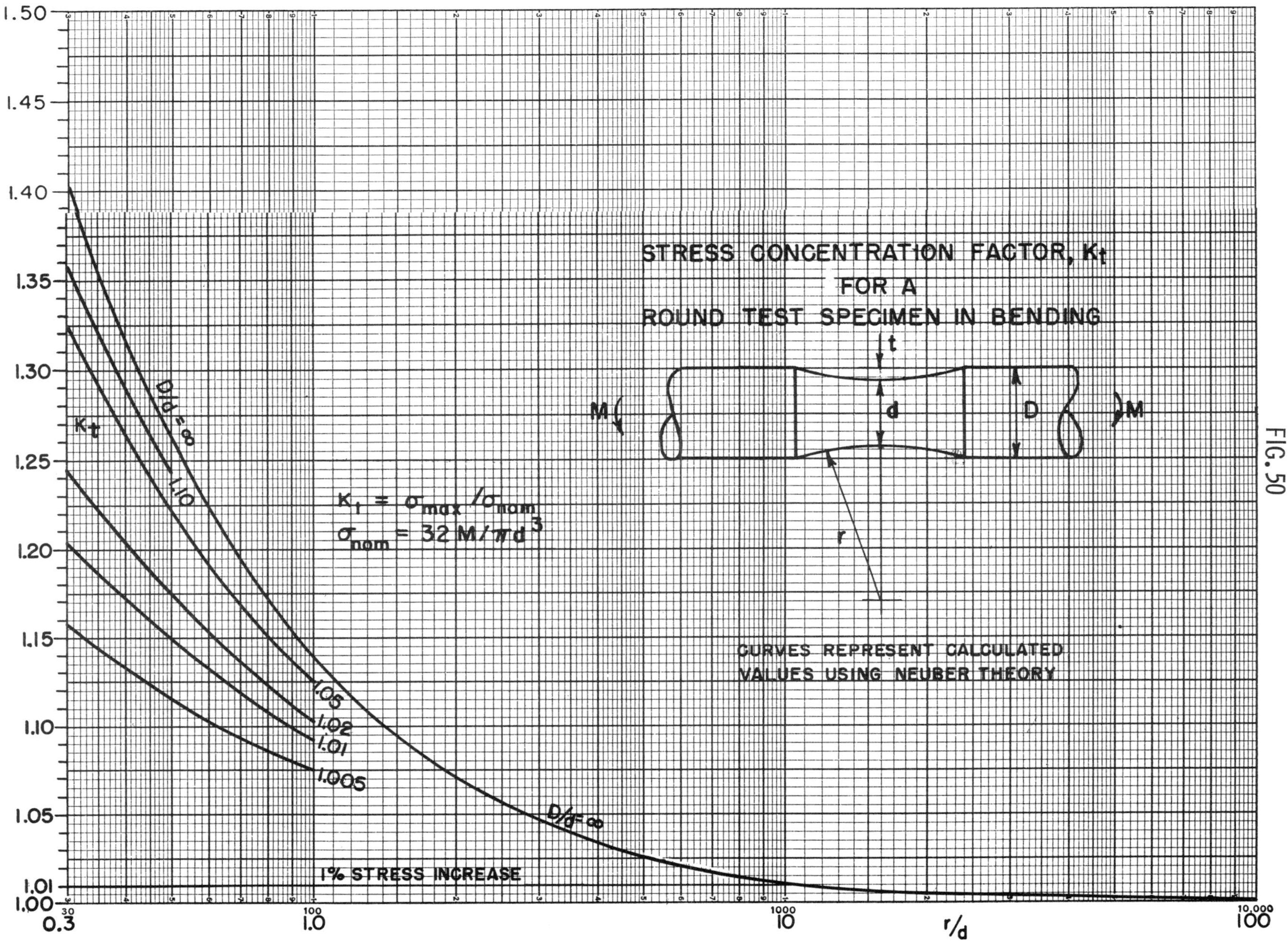

STRESS CONCENTRATION FACTOR, Kt
FOR A
ROUND TEST SPECIMEN IN BENDING
CURVES REPRESENT CALCULATED
VALUES USING NEUBER THEORY
$K_t = \sigma_{max} / \sigma_{nom}$
$\sigma_{nom} = 32\,M/\pi d^3$
1% STRESS INCREASE
M
D
d
t
r
D/d = ∞
1.10
1.05
1.02
1.01
1.005
Kt
r/d
1.50
1.45
1.40
1.35
1.30
1.25
1.20
1.15
1.10
1.05
1.01
1.00
0.3
1.0
10
100

FIG. 50

Fig. 51

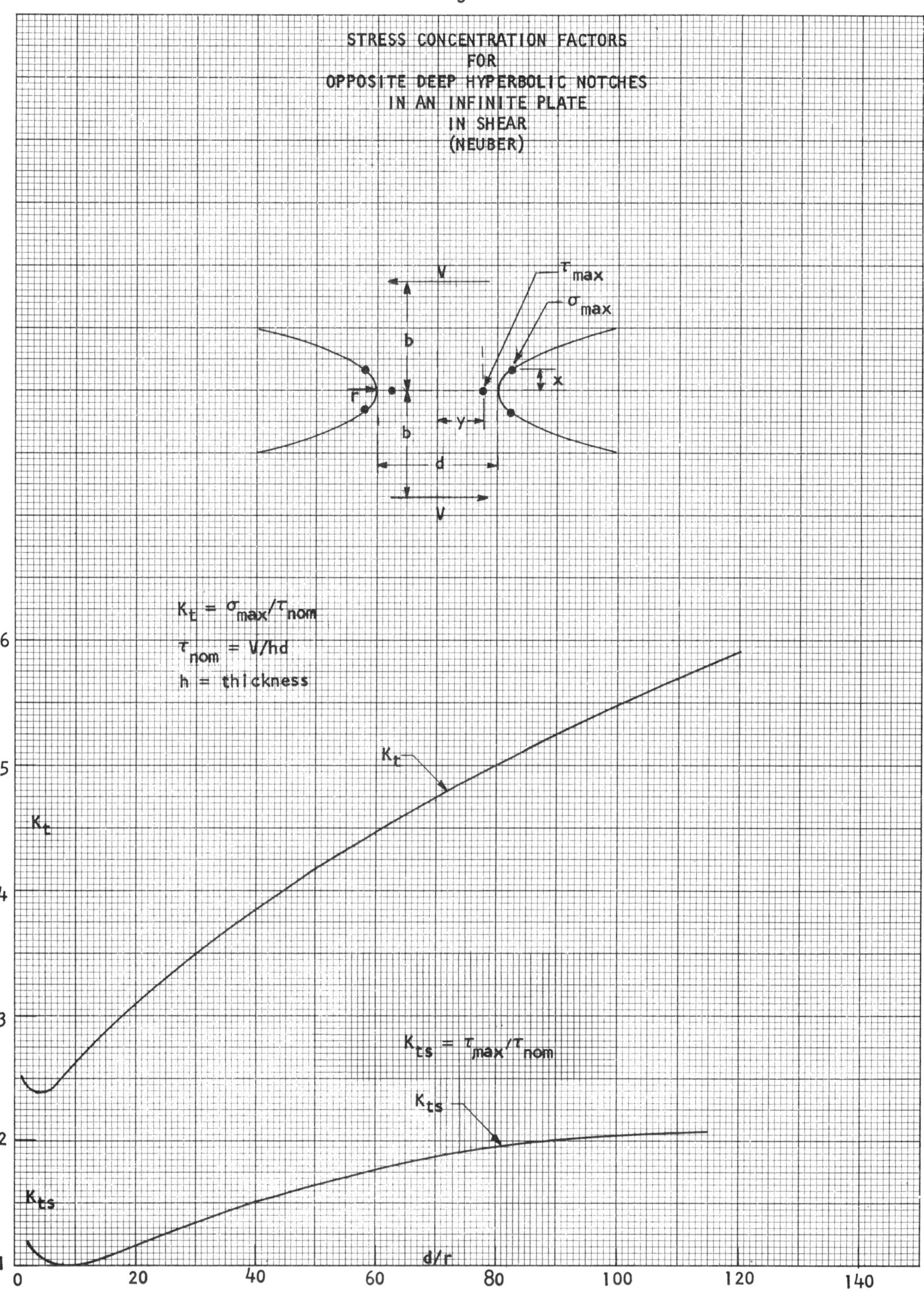

STRESS CONCENTRATION FACTORS
FOR
OPPOSITE DEEP HYPERBOLIC NOTCHES
IN AN INFINITE PLATE
IN SHEAR
(NEUBER)
V
b
b
r
y
d
x
V
τ_{max}
σ_{max}
$K_t = \sigma_{max}/\tau_{nom}$
$\tau_{nom} = V/hd$
h = thickness
K_t
$K_{ts} = \tau_{max}/\tau_{nom}$
K_{ts}
K_t
K_{ts}
d/r
1
2
3
4
5
6
0
20
40
60
80
100
120
140

FIG.52

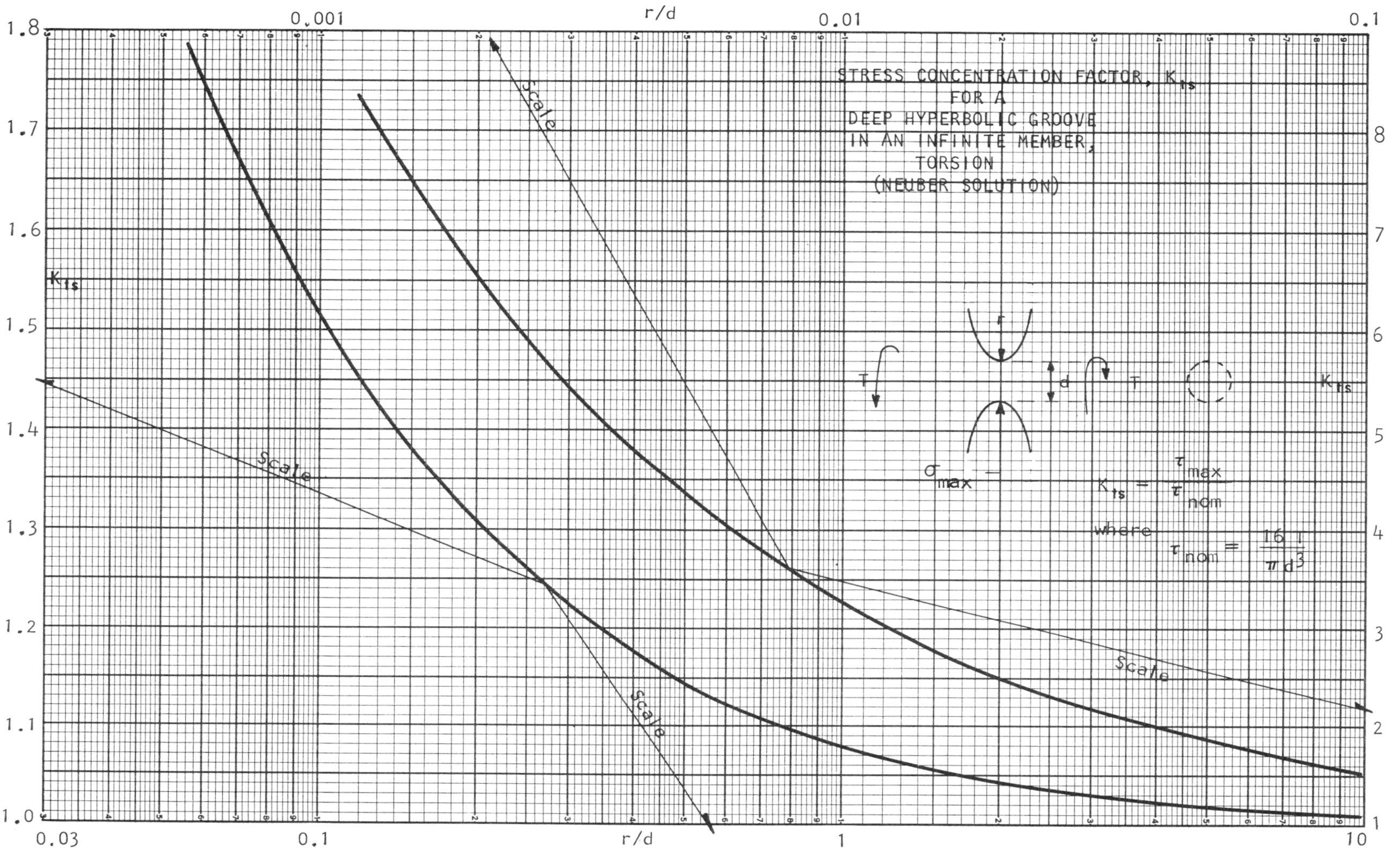

STRESS CONCENTRATION FACTOR, K_{ts}
FOR A
DEEP HYPERBOLIC GROOVE
IN AN INFINITE MEMBER,
TORSION
(NEUBER SOLUTION)
$K_{ts} = \frac{\tau_{max}}{\tau_{nom}}$
where $\tau_{nom} = \frac{16T}{\pi d^3}$
σ_{max}
r
d
T
K_{ts}
r/d
Scale
Scale
Scale
Scale

FIG. 53

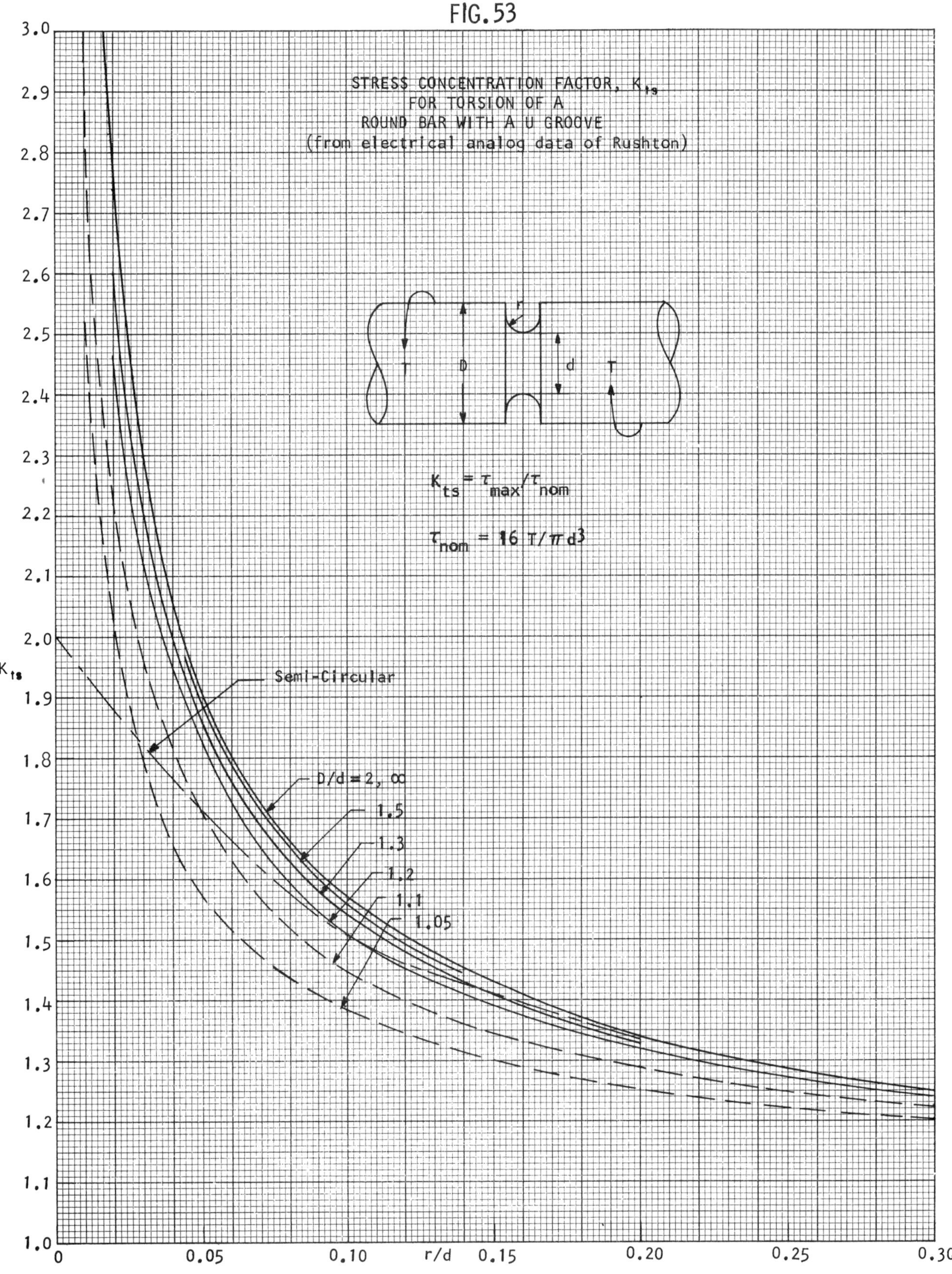
STRESS CONCENTRATION FACTOR, K_{ts}
FOR TORSION OF A
ROUND BAR WITH A U GROOVE
(from electrical analog data of Rushton)
T
D
r
d
T
$K_{ts} = \tau_{max}/\tau_{nom}$
$\tau_{nom} = 16\ T/\pi d^3$
Semi-Circular
D/d = 2, ∞
1.5
1.3
1.2
1.1
1.05
K_{ts}
r/d
3.0
2.9
2.8
2.7
2.6
2.5
2.4
2.3
2.2
2.1
2.0
1.9
1.8
1.7
1.6
1.5
1.4
1.3
1.2
1.1
1.0
0
0.05
0.10
0.15
0.20
0.25
0.30

FIG. 54

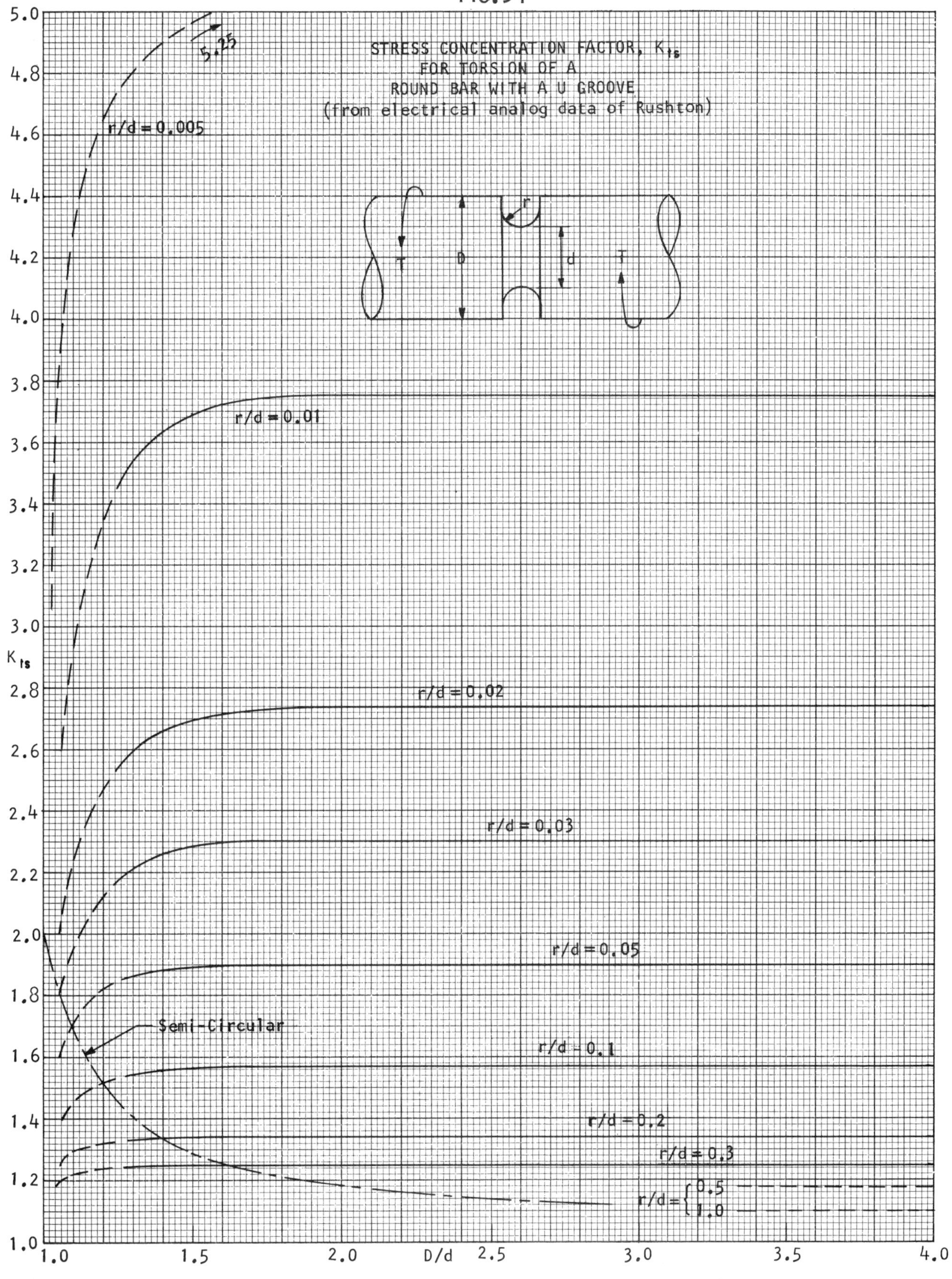

STRESS CONCENTRATION FACTOR, K_{ts}
FOR TORSION OF A
ROUND BAR WITH A U GROOVE
(from electrical analog data of Rushton)
5.25
r/d = 0.005
r/d = 0.01
r/d = 0.02
r/d = 0.03
r/d = 0.05
r/d = 0.1
r/d = 0.2
r/d = 0.3
r/d = {0.5, 1.0}
Semi-Circular
T
D
r
d
T
K_{ts}
5.0
4.8
4.6
4.4
4.2
4.0
3.8
3.6
3.4
3.2
3.0
2.8
2.6
2.4
2.2
2.0
1.8
1.6
1.4
1.2
1.0
1.0
1.5
2.0
D/d
2.5
3.0
3.5
4.0

FIG. 55

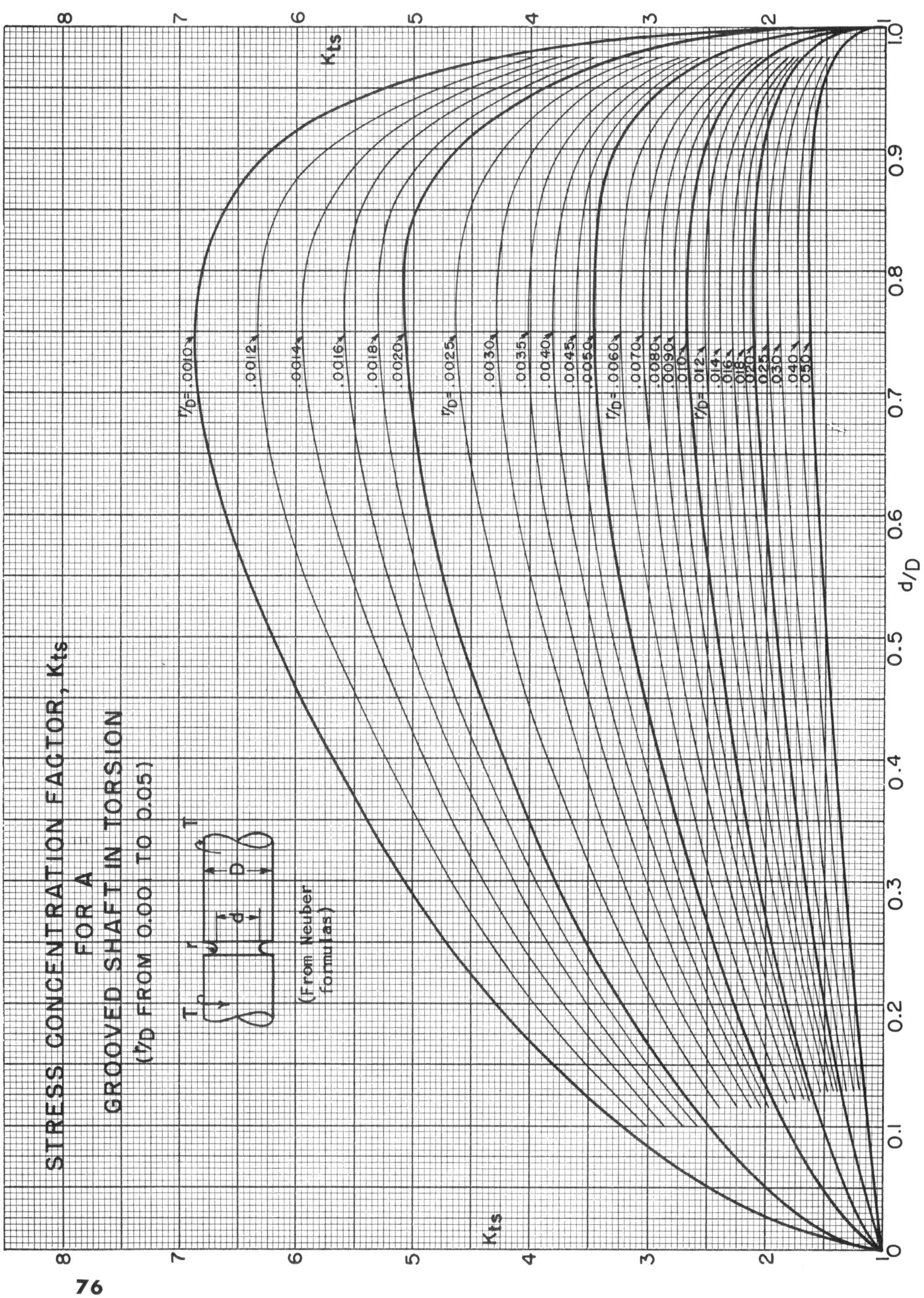

STRESS CONCENTRATION FACTOR, K_{ts}
FOR A
GROOVED SHAFT IN TORSION
(r/D FROM 0.001 TO 0.05)

T r d D T

(From Neuber formulas)

r/D = .0010, .0012, .0014, .0016, .0018, .0020, r/D = .0025, .0030, .0035, .0040, .0045, .0050, r/D = .0060, .0070, .0080, .0090, .010, r/D = .012, .014, .016, .018, .020, .025, .030, .040, .050

K_{ts}

1, 2, 3, 4, 5, 6, 7, 8

d/D

0, 0.1, 0.2, 0.3, 0.4, 0.5, 0.6, 0.7, 0.8, 0.9, 1.0

FIG. 56

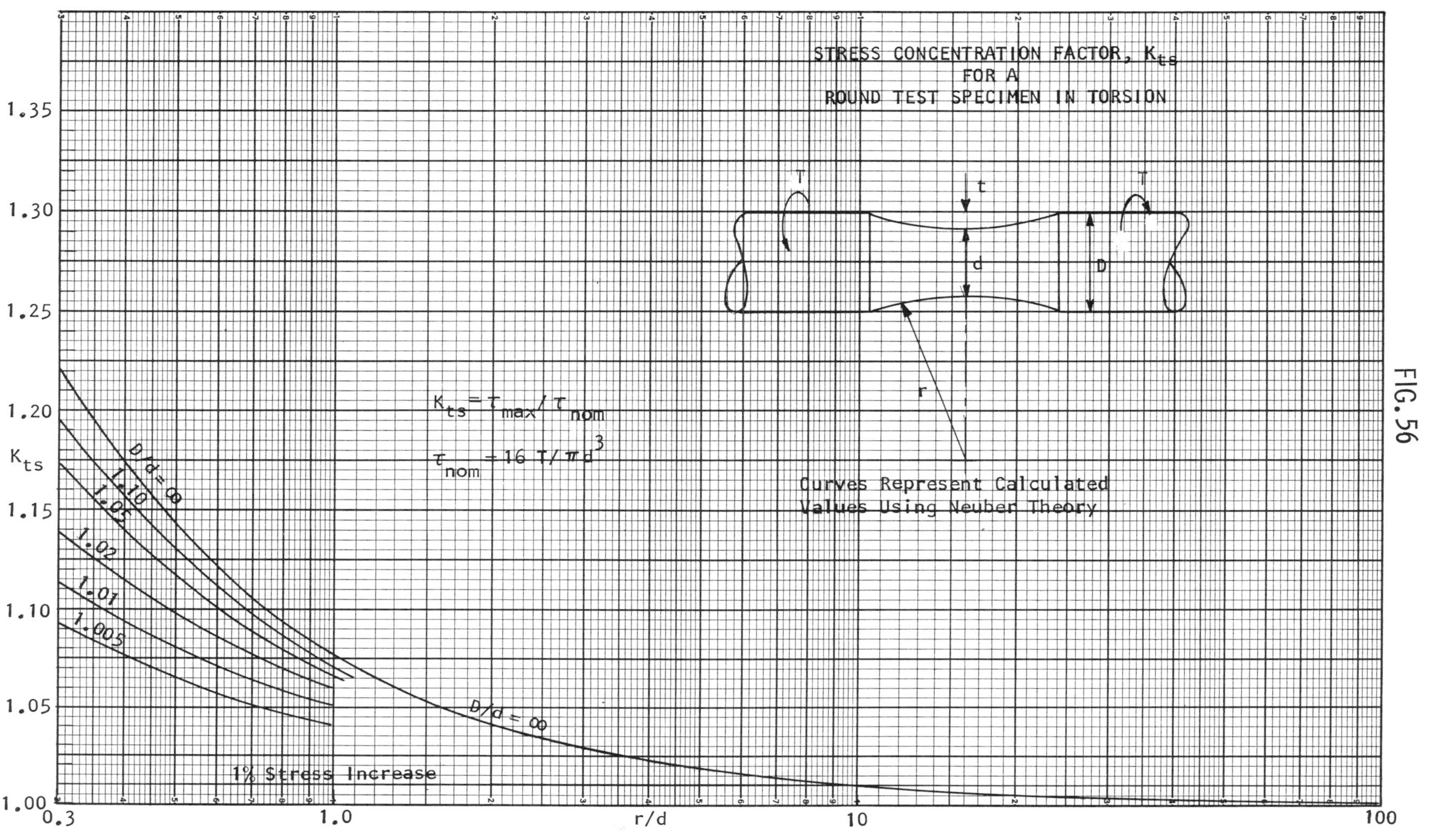

STRESS CONCENTRATION FACTOR, K_{ts}
FOR A
ROUND TEST SPECIMEN IN TORSION
T
T
t
d
D
r
$K_{ts} = \tau_{max} / \tau_{nom}$
$\tau_{nom} = 16\ T / \pi d^3$
Curves Represent Calculated
Values Using Neuber Theory
D/d = ∞
1.10
1.05
1.02
1.01
1.005
D/d = ∞
1% Stress Increase
K_{ts}
1.35
1.30
1.25
1.20
1.15
1.10
1.05
1.00
0.3
1.0
r/d
10
100

FIG. 57

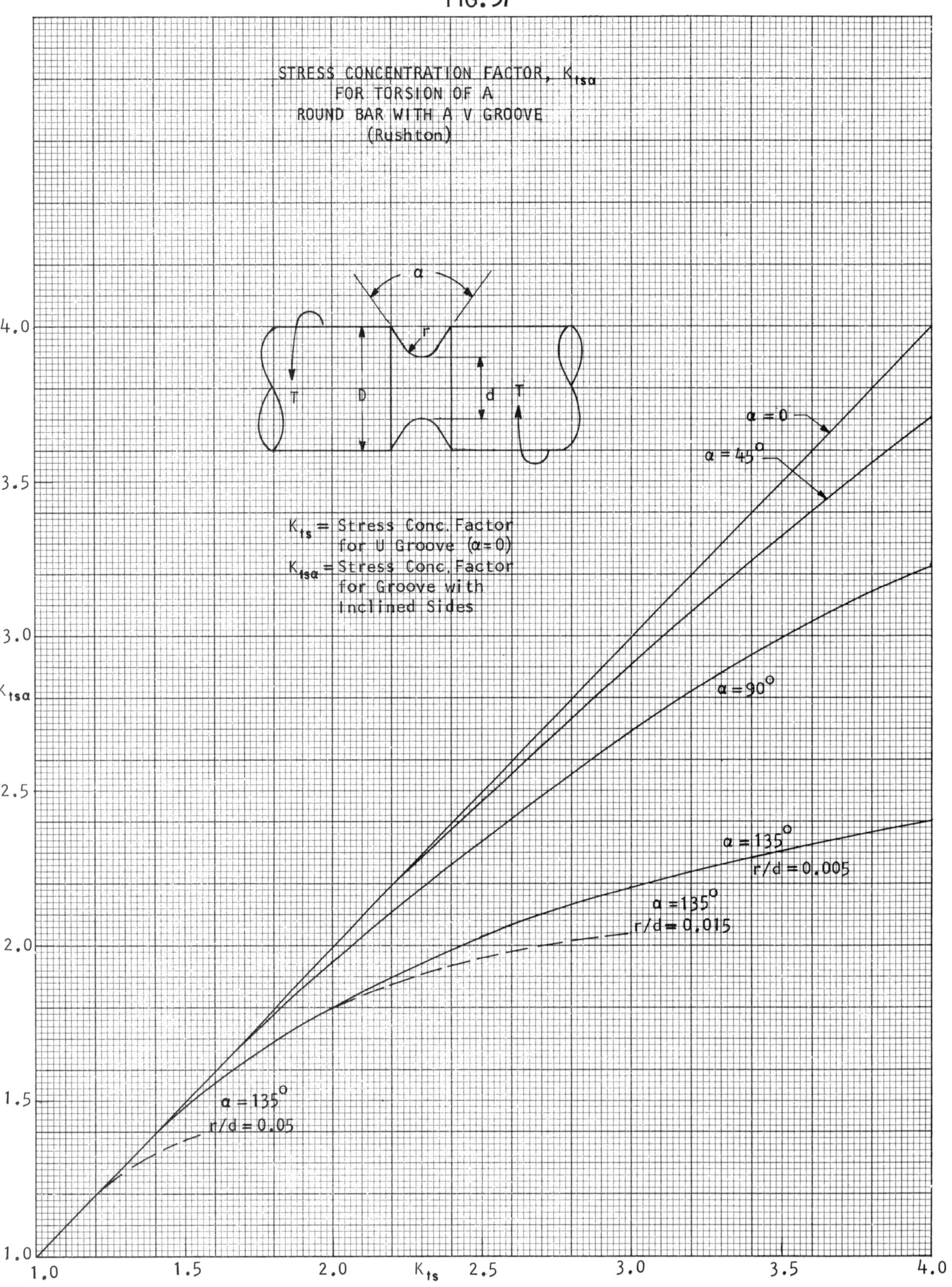

STRESS CONCENTRATION FACTOR, $K_{ts\alpha}$
FOR TORSION OF A
ROUND BAR WITH A V GROOVE
(Rushton)
α
r
T
D
d
T
K_{ts} = Stress Conc. Factor for U Groove (α = 0)
$K_{ts\alpha}$ = Stress Conc. Factor for Groove with Inclined Sides
α = 0
α = 45°
α = 90°
α = 135° r/d = 0.005
α = 135° r/d = 0.015
α = 135° r/d = 0.05
$K_{ts\alpha}$
K_{ts}
1.0
1.5
2.0
2.5
3.0
3.5
4.0

FIG. 58

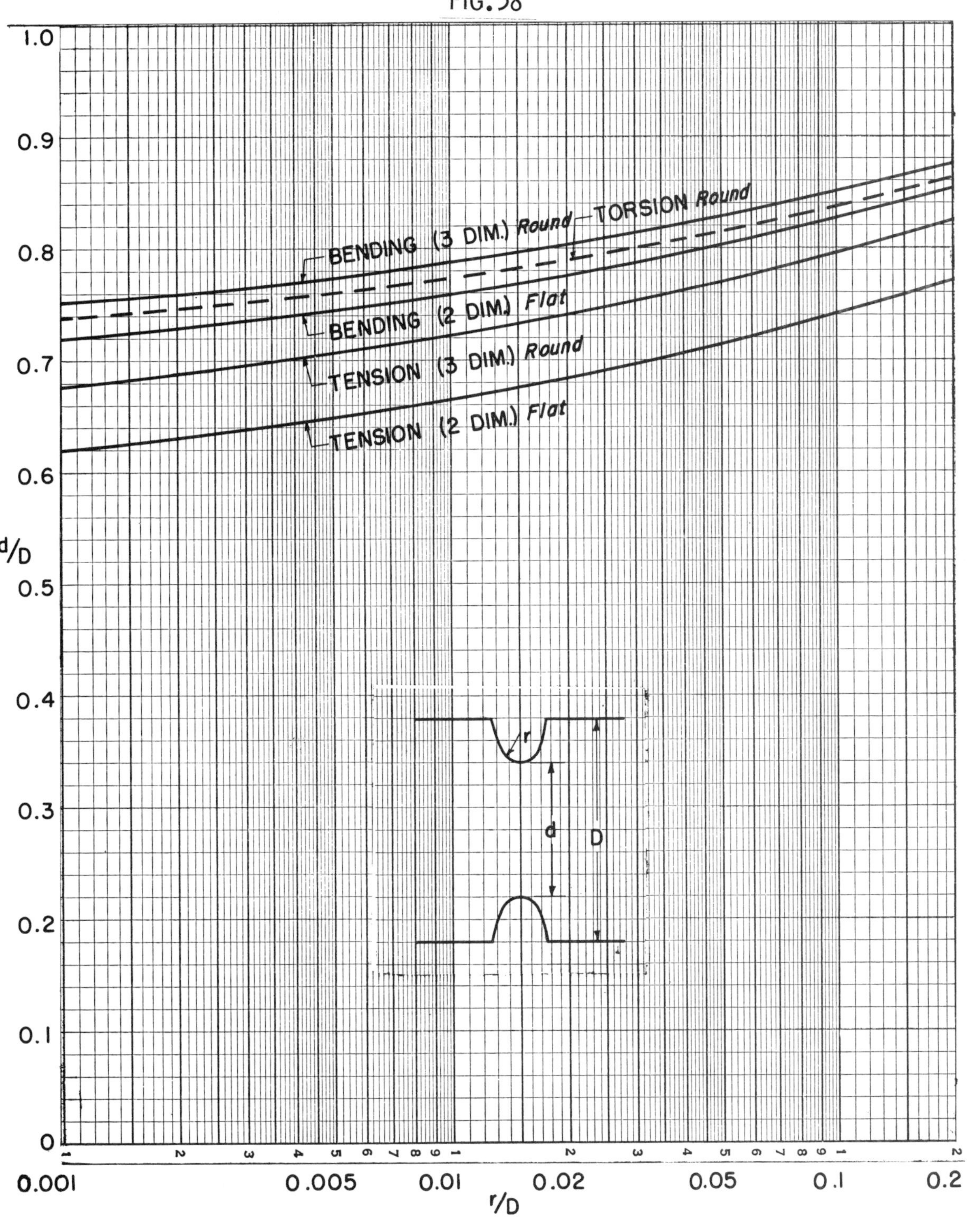

APPROXIMATE
GEOMETRIC RELATIONS FOR MAXIMUM STRESS CONCENTRATION FOR GROOVED SPECIMENS
(Based on Neuber Relations)

CHAPTER 3

SHOULDER FILLETS

The shoulder fillet (Fig. 59) is the type of stress concentration that is more frequently encountered in machine design practice than any other. Shafts, axles, spindles, rotors, and so forth, usually involve, in each case, a number of diameters connected by shoulders with rounded fillets replacing the sharp corners that were often used in former years.

K_t is based on the smaller width or diameter, d; in tension (Fig. 65), $K_t = \sigma_{max}/\sigma_{nom}$, where $\sigma_{nom} = P/hd$.

The fillet factors for tension and bending are based on photoelastic values; for torsion the fillet factors are from a mathematical analysis. In the earlier book[1] a method was given for obtaining approximate K_t values for smaller r/d values. The present charts extend well into the small r/d range, owing to use of recently published results.

The K_t factors for the flat members considered in this chapter are for two-dimensional states of stress (plane stress) and apply only to very thin sheets or, more strictly, to where $t/r \to 0$, where t = plate thickness and r = fillet radius. As t/r increases a state of plane strain is approached; here the stress at the fillet surface at the middle of the plate thickness increases and the stress at the plate surface decreases. The introductory remarks at the beginning of Chapter 4 may be helpful in explaining this.

(A) TENSION (AXIAL LOADING)

(a) Opposite Shoulder Fillets in a Flat Bar or Plate

Figure 65 presents recently determined K_t factors for a stepped flat tension bar. The curves are modifications of the K_t factors previously determined through photoelastic tests,[121] whose values have been found to be too low, owing probably to the small size of the models and to possible edge effects. The curves in the r/d range of 0.03 to 0.3 have been obtained as follows:

$$K_t, \text{Fig. } 65 = K_t, \text{Fig. } 57, \text{Ref. } 1 \left[\frac{K_t, \text{Fig. } 17}{K_t \text{ notch, Ref. } 121} \right] \qquad [48b]$$

Recently the r/d range has been extended to lower values by photoelastic tests[98a]; these data fit well with the above results from [48b] for $D/d > 1.1$.

Other recent photoelastic tests[137] give K_t values which agree reasonably well with the $D/d = 1.5$ and 2 curves of Fig. 65.

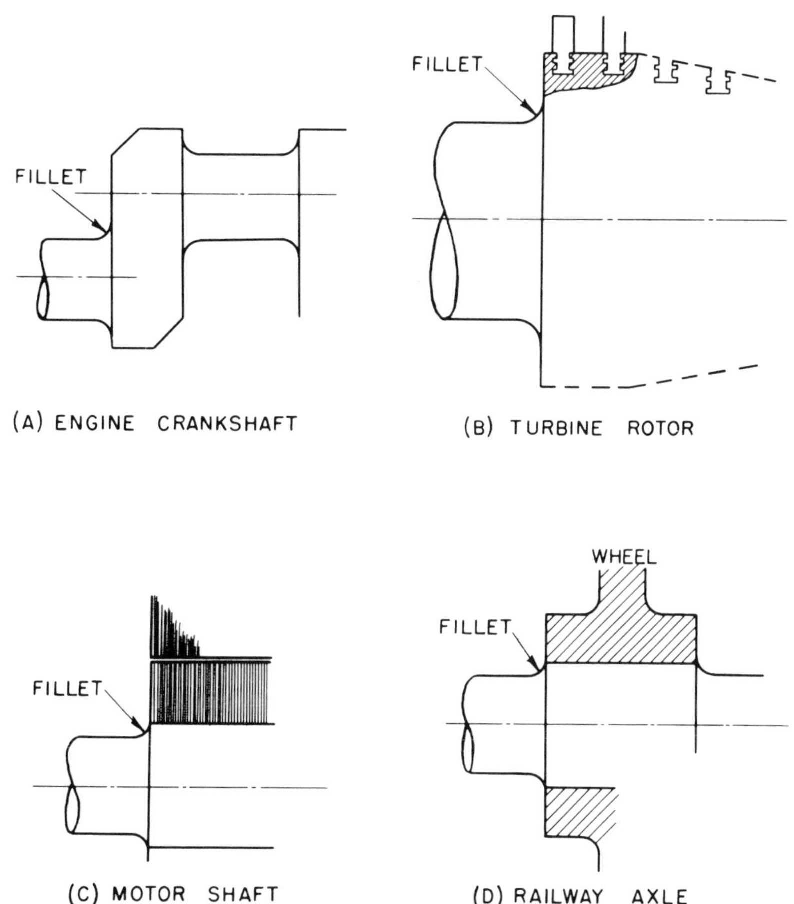

FIG. 59

EXAMPLES OF FILLETED MEMBERS

(b) Effect of Shoulder Geometry in a Flat Member

The factors of Fig. 65 are for the case where the large diameter D extends back from the shoulder a relatively great distance. Frequently one encounters a case in design where this shoulder width L (Fig. 60) is relatively narrow.

In one of the early investigations in the photoelasticity field, Baud[138] noted that in the case of a narrow shoulder the outer part is unstressed, and he proposed the formula

$$D_x = d + 0.3L \qquad [49]$$

where D_x is the depth (Fig. 60) of a wide shoulder member which has the same K_t factor.

The same result can be obtained graphically by drawing intersecting lines at an angle Θ of 17° (Fig. 60). Sometimes a larger angle Θ is used, up to 30°. The Baud rule, which was proposed as a rough approximation, has been quite useful.

Although the K_t factors for bending of plates with narrow shoulders (Figs. 74 to 76) were published[139] in 1951, it is only recently that the tension case has been systematically investigated[140] (Figs. 66 to 69). Referring to Figs. 68 and 69, note that at $L/d = 0$ a cusp

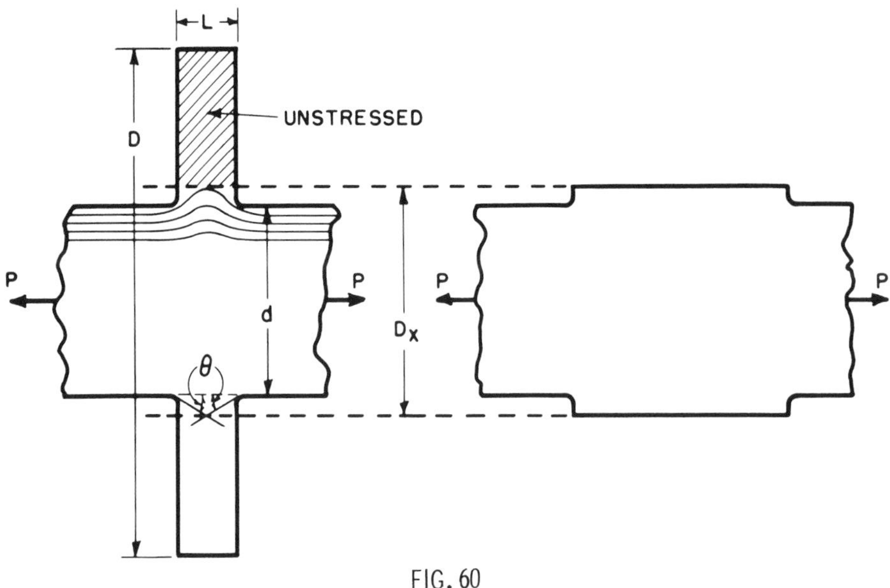

FIG. 60

EFFECT OF NARROW SHOULDER.

remains; $K_t = 1$ at $L/d = -2r/d$, (see dashed lines in Figs. 68 and 69 for extrapolation to $K_t = 1$). The formula gives the exact L/d value for $K_t = 1$ for $D/d = 1.8$ (Fig. 68) when $r/d \leqq 0.4$, and for $D/d = 5$ (Fig. 69) when $r/d \leqq 2$. The authors[140] state that their results are consistent with previous data[141,142] obtained for somewhat different geometries. Empirical formulas are developed to cover the authors results.[140]

Round bar values are not available; it is suggested that the approximation method given in Ref. 1, p. 61, be used.

(c) Effect of a Trapezoidal Protuberance on the Edge of a Flat Bar

An example of such a geometry is an approximation to a weld bead. The geometrical configuration is shown in the sketch on Fig. 70. A finite difference method was used for a tension bar with equivalent surface shear loading.[143] The resulting K_t factors for $\Theta = 30°$ and 60° are given in Figs. 70 and 71, respectively. The dashed curve corresponds to a proturberance height where the radius is exactly tangent to the angular side (i.e., below the dashed curve there are no straight sides, only segments of a circle).

A comparison[143] of K_t factors with corresponding (large w/t) factors obtained from Ref. 1, Figs. 36 and 62, for filleted members with angle correction, showed the latter to be about 7% higher on the average, with variations from 2 to 15%. A similar comparison by the writer, using the increased K_t fillet values of Fig. 65, showed these values (corrected for angle) to be about 17% higher (varying between 14 and 22%) than the Derecho-Munse values.

Strain gage measurements[143] resulted in K_t factors 32, 23, and 31% higher, with one value (for the lowest K_t) 2.3% lower than the computed values.

The authors[143] comment: "the above comparisons suggest that the values.* . . . may be slightly lower than they should be. In may be noted here that had a further refinement

*In figures corresponding to Figs. 70 and 71.

of the grid spacing been possible in the previously discussed finite-difference solution, slightly higher values of the stress concentration factor could have been obtained." It is possible that the factors may be more than slightly lower.

A typical weld bead would correspond to a geometry of small t/w, with D/d near 1.0. Referring, for example, to Fig. 70, we see that for $t/w = 0.1$ and $r/w = 0.1$, K_t is surprisingly low, 1.55. Even if we increase this by 17%, to be on the safe side in design, we still have a relatively low factor, $K_t = 1.8$.

(d) Fillet of Noncircular Contour in a Flat Stepped Bar

Circular fillets are usually used for simplicity in drafting and machining. The circular fillet does not correspond to minimum stress concentration.

The variable radius fillet is often found in old machinery (using many cast-iron parts) where the designer or builder apparently produced the result intuitively. Sometimes the variable radius fillet is approximated by two radii, resulting in the compound fillet illustrated in Fig. 61.

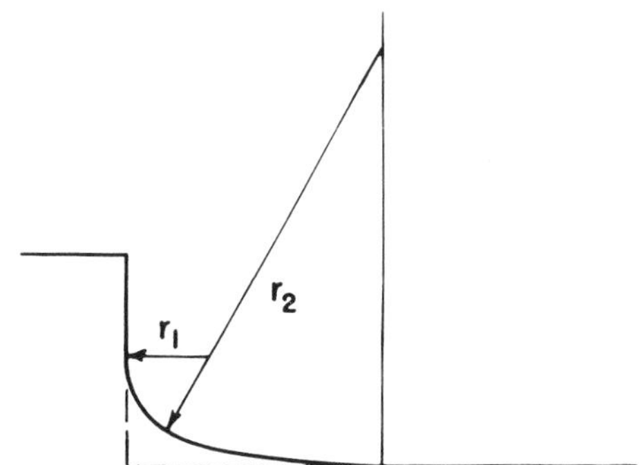

FIG. 61 COMPOUND FILLET

In 1934 Baud[144] proposed a fillet form with the same contour as that given mathematically for an ideal, frictionless liquid flowing by gravity from an opening in the bottom of a tank (Fig. 62):

$$x = 2\frac{d}{\pi}\sin^2\frac{\theta}{2} \qquad [50]$$

$$y = \frac{d}{\pi}\left[\log\tan\left(\frac{\theta}{2} + \frac{\pi}{4}\right) - \sin\theta\right] \qquad [51]$$

Baud noted that in this case the liquid at the boundary has constant velocity and reasoned that the same boundary may also be the contour of constant stress for a tension member. By means of a photoelastic test in tension Baud observed that no appreciable stress concentration occurred with a fillet of streamline form.

For bending and torsion Thum and Bautz[145] applied a correction in accordance with the cube of the diameter, resulting in a shorter fillet than for tension. (See Table I.) Thum and Bautz[145] also demonstrated by means of fatigue tests in bending and in torsion that,

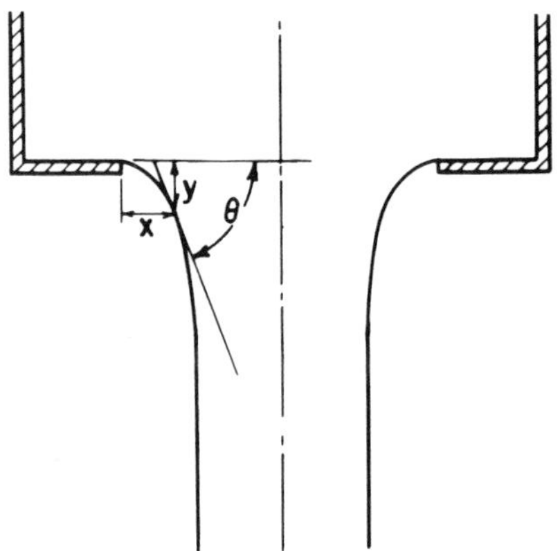

FIG. 62 IDEAL FRICTIONLESS LIQUID FLOW FROM OPENING IN BOTTOM OF TANK

with fillets having the proportions of Table I, no appreciable stress concentration effect was obtained.

To reduce the length of the streamline fillet Deutler and Harvers[146] suggested a special elliptical form based on theoretical considerations of Föppl.

Grodzinski[147] mentions fillets of parabolic form. He also gives a simple graphical method, which may be useful in making a template or a pattern for a cast part (Fig. 63). Dimensions a and b are usually dictated by space or design considerations. Divide each distance into the same number of parts and number in the order shown. Connect points having the same numbers by straight lines; this results in an envelope of gradually increasing radius, as shown in Fig. 63.

For heavy shafts or rolls, Morgenbrod[148] has suggested a tapered fillet with radii at the ends, the included angle of the tapered portion being between 15 and 20°. This is similar

TABLE I. Proportions for Streamline Fillet

y/d	d_f/d for tension	d_f/d for bending or torsion	y/d	d_f/d for tension	d_f/d for bending or torsion
0.0	1.636	1.475	0.3	1.187	1.052
0.002	1.610	1.420	0.4	1.134	1.035
0.005	1.594	1.377	0.5	1.096	1.026
0.01	1.572	1.336	0.6	1.070	1.021
0.02	1.537	1.287	0.7	1.051	1.018
0.04	1.483	1.230	0.8	1.037	1.015
0.06	1.440	1.193	0.9	1.027	1.012
0.08	1.405	1.166	1.0	1.019	1.010
0.10	1.374	1.145	1.3	1.007	1.005
0.15	1.310	1.107	1.6	1.004	1.003
0.2	1.260	1.082	∞	1.000	1.000

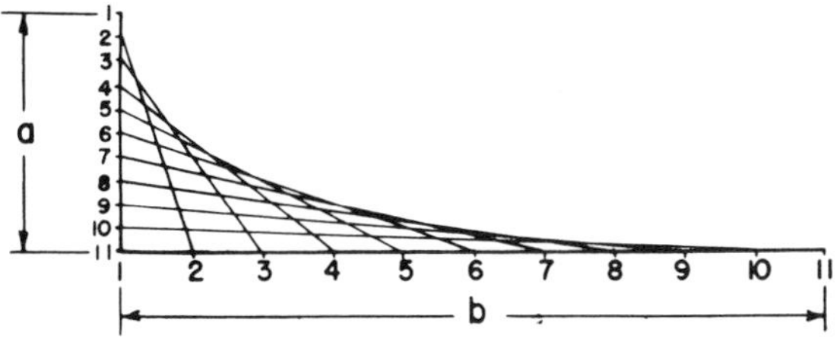

FIG. 63 CONSTRUCTION OF SPECIAL FILLET (GRODZINSKI)

to the basis of the tapered cantilever fatigue specimen of McAdam,[149] which has been shown[150] to have a stress variation of less than 1% over a length 2 in.* (The conical surface is tangent to the constant-stress cubic solid of revolution.)

Photoelastic tests have provided values for a range of useful elliptical fillets for bending (Section B.d); the degree of improvement obtained may be useful in considering a tension case. Clock[151] has approximated an elliptical fillet by using an equivalent segment of a circle and has provided corresponding K_t values.

An excellent treatment of optimum transition shapes has been provided by Heywood;[152] his discussion of the shapes found in nature (tree trunks and branches, thorns, animal bones) is most interesting.

(e) Stepped Round Bar with a Circumferential Shoulder Fillet

The K_t values for this case (Fig. 72) were obtained by ratioing the K_t values of Fig. 65 in accordance with the three- to two-dimensional notch values, as explained in Chapter 2, Section A.1; Fig. 72 is labeled "Approximate" in view of this procedure.

For d/D values considered valid for comparison (0.6, 0.7, 0.9), photoelastic results for round bars[153] are somewhat lower than the values of Fig. 72.

Recent photoelastic tests[137] give K_t values for $D/d = 1.5$ which are in good agreement with Fig. 72.

(f) Stepped Pressure Vessel Wall with a Shoulder Fillet

The K_t curve of Fig. 72*a* is based on calculated values of Griffin and Thurman.[153a] A direct comparison[153b] with a specific photoelastic test by Leven[153c] shows good agreement. The strain gage results of Heifetz and Berman[153d] are in reasonably good agreement with Fig. 72*a*; lower values have been obtained in a finite element analysis.[153e]

For comparison the model shown in.Fig. 65 may be considered to be split in half axially. The corresponding K_t curves have the same shape as in Fig. 72*a*, but they are somewhat higher; however, the cases are not strictly comparable and, furthermore, Fig. 65 is approximate.

*Nominal diameter, 1 in.

(B) BENDING

(a) Opposite Shoulder Fillets in a Flat Bar or Plate

Photoelastic values of Leven and Hartman[139] cover the r/d range from 0.03 to 0.3 whereas recent photoelastic tests[98a] cover r/d values in the 0.003 to 0.03 range. These results blend together reasonably well and form the basis of Fig. 73.

(b) Effect of Shoulder Geometry in a Flat Member

In Figs. 74, 75, and 76, K_t factors are given for various shoulder parameters.[139] For $L/D = 0$ a cusp remains: for $D/d = 1.25$ (Fig. 74) and $r/d \leqq 1/8$, $K_t = 1$ when $L/D = -1.6\,r/d$; for $D/d = 2$ (Fig. 75) and $r/d \leqq 1/2$, $K_t = 1$ when $L/D = -r/d$; for $D/d = 3$ (Fig. 76) and $r/d \leqq 1$, $K_t = 1$ when $L/D = -(2/3)\,(r/d)$. The dashed lines in Figs. 74, 75, and 76 show extrapolations to $K_t = 1$. Round bar values are not available; it is suggested that the designer obtain an adjusted value by ratioing in accordance with the corresponding Neuber three- to two-dimensional notch values (Ref. 1, p. 61).

(c) Short Beam with Opposite Shoulder Fillets

In connection with a study of gear teeth, some photoelastic tests[319] were made of short rectangular beams with opposite shoulder fillets. These results are given in Fig. 189 and are discussed in Chapter 5, Section C.

(d) Elliptical Shoulder Fillet in a Flat Member

Photoelastic tests by Berkey[154] have provided K_t factors for the flat plate in bending (Fig. 77). The corresponding factors for a round shaft should be somewhat lower; an estimate can be made by comparing the corresponding Neuber three- to two-dimensional notch factors, as discussed in Chapter 2, Section A.1.

(e) Stepped Round Bar with a Circumferential Shoulder Fillet

Photoelastic tests[139] have been made of stepped round bars in the r/d range of 0.03 to 0.3. By use of recent plane bending tests[98a] reasonable extensions of curves have been made in the r/d range below 0.03. The results are presented in Fig. 78.

In comparison with other round bar photoelastic tests,[155] for the d/D ratios considered valid for comparison, there is reasonably good agreement for $d/D = 0.6, 0.8$; but for $d/D = 0.9$ the results in Ref. 155 are lower.

In design of machinery shafts (bending and torsion are the main cases of interest), small steps (D/d near 1.0) are often used; for this region, Fig. 78 is not very suitable. Fig. 78*a* has been provided, wherein the curves go to $K_t = 1.0$ at $D/d = 1.0$.

(C) TORSION

(a) Stepped Round Bar with a Circumferential Shoulder Fillet

The design curves previously used (Ref. 1, Fig. 67) were based on the experimental data of Jacobsen[156] (electrical analog) and Weigand[157] (strain gage), the former covering the range of r/d from 0.02 to 0.12 and the latter from 0.12 to 0.25. Although the agreement at 0.12 was not as good as desired, it was possible to blend the curves in such a way as to provide a reasonable result. Jacobsen used a steel model whose thickness varied as the cube of the distance from the shaft center. To eliminate the sharp edge corresponding to a solid shaft, Jacobsen assumed a central hole and thereby could use a thin, flat edge on the inner surface of his model.

In recent years, investigations of the filleted shaft in torsion have been made by use of photoelasticity,[137,158] by use of the electrical analog,[159] and mathematically.[132] The latter work,[132] using a numerical technique based on elasticity equations and a point matching method for approximately satisfying boundary conditions, is believed to be of satisfactory accuracy; it provides K_{ts} values (Fig. 79) lower than those used previously,[1] and in the lower r/d range it provides higher values than from the electrical analog.[159]

An empirical relation,[137] based on published data including two photoelastic tests by the authors,[137] is in satisfactory agreement with the values of Fig. 79 in the area covered by the author's tests.[137]

In design of machinery shafts (bending and torsion are the main cases of interest), small steps (D/d near 1.0) are often used; for this region, Fig. 79 is not very suitable. Fig. 79*a* has been provided, wherein the curves go to $K_t = 1.0$ at $D/d = 1.0$.

(b) Stepped Round Bar with a Circumferential Shoulder Fillet and a Central Axial Hole

Central (axial) holes are used in large forgings for inspection purposes and in shafts for cooling or fluid transmission purposes.

For the hollow shaft, a reasonable design procedure is application of ratios (Fig. 80), which have been obtained from the electrical analog values,[159] to the K_t values of Fig. 79.

The strength/weight ratio of the small-diameter portion of the shaft increases with increasing hollowness ratio. However, this is usually not of substantial benefit in practical designs because of the relatively larger weight of the large-diameter portion of the shaft. An exception may occur when the diameters are close together ($D/d = 1.2$ or less).

(D) METHODS OF REDUCING STRESS CONCENTRATION AT A SHOULDER

One of the problems occurring in the design of shafting, rotors, and so forth, is the reduction of stress concentration at a shoulder fillet (Fig. 64*a*) while maintaining the positioning line A–A and dimensions D and d. This can be done in a number of ways, some of which are illustrated in Fig. 64*b*, *c*, *d*, *e*, *f*. By cutting into the shoulder a larger fillet radius can be obtained (Fig. 64*b*) without developing interference with the fitted member. A ring insert could be used as at Fig. 64*c*, but this represents an additional part. A similar result could be obtained as shown in Fig. 64*d*, except that a smooth fillet surface is more difficult to realize.

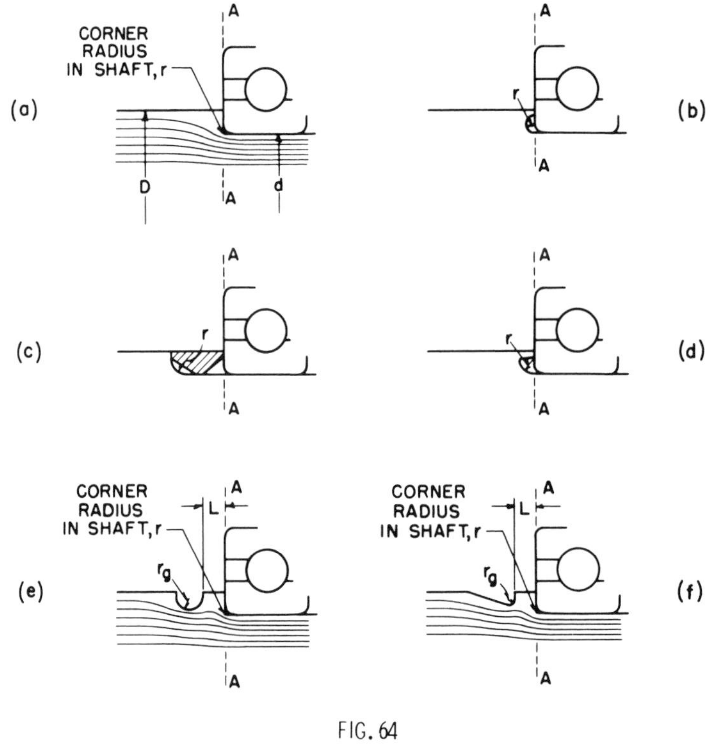

FIG. 64

RELIEF GROOVING.

Sometimes the methods of Fig. 64*b*, *c*, *d* are not helpful because the shoulder height $(D - d)/2$ is too small. A relief groove (Fig. 64*e*, *f*) may be used provided this does not conflict with location of a seal or other shaft requirements. Fatigue tests[160,161] show a considerable gain in strength due to relief grooving.

It should be mentioned that in the case at hand there is also a combined stress concentration and fretting corrosion problem at the bearing fit (see Chapter 5, Section D). The gain due to fillet improvement in this case might be limited by failure at the fitted surface. However, fatigue tests[161] showed that at least for the specific proportions tested a gain in strength was realized by the use of relief grooves.

FIG. 65

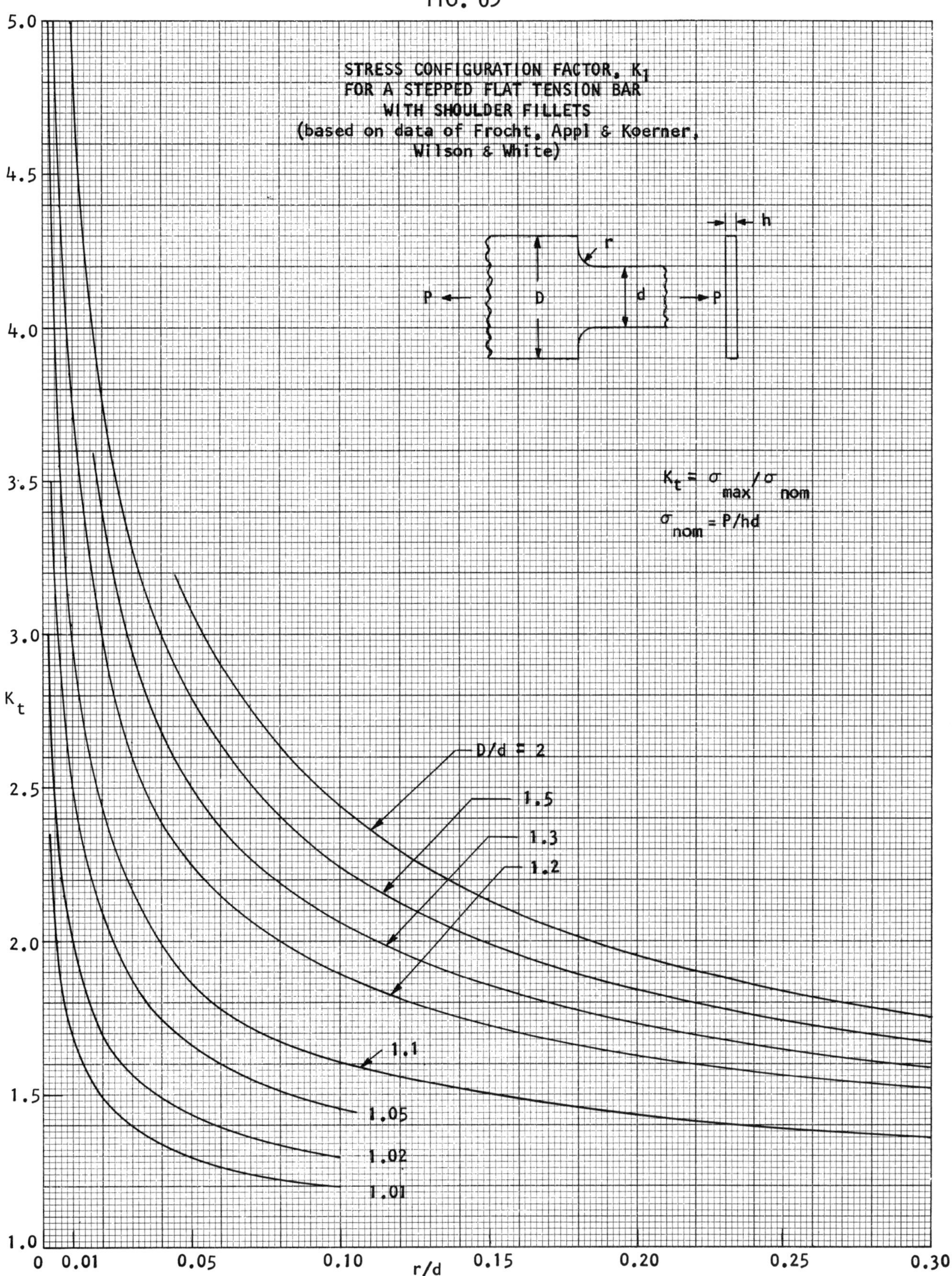

STRESS CONFIGURATION FACTOR, K_t
FOR A STEPPED FLAT TENSION BAR
WITH SHOULDER FILLETS
(based on data of Frocht, Appl & Koerner, Wilson & White)
h
r
P
D
d
P
$K_t = \sigma_{max}/\sigma_{nom}$
$\sigma_{nom} = P/hd$
D/d = 2
1.5
1.3
1.2
1.1
1.05
1.02
1.01
5.0
4.5
4.0
3.5
3.0
K_t
2.5
2.0
1.5
1.0
0
0.01
0.05
0.10
r/d
0.15
0.20
0.25
0.30

FIG. 66

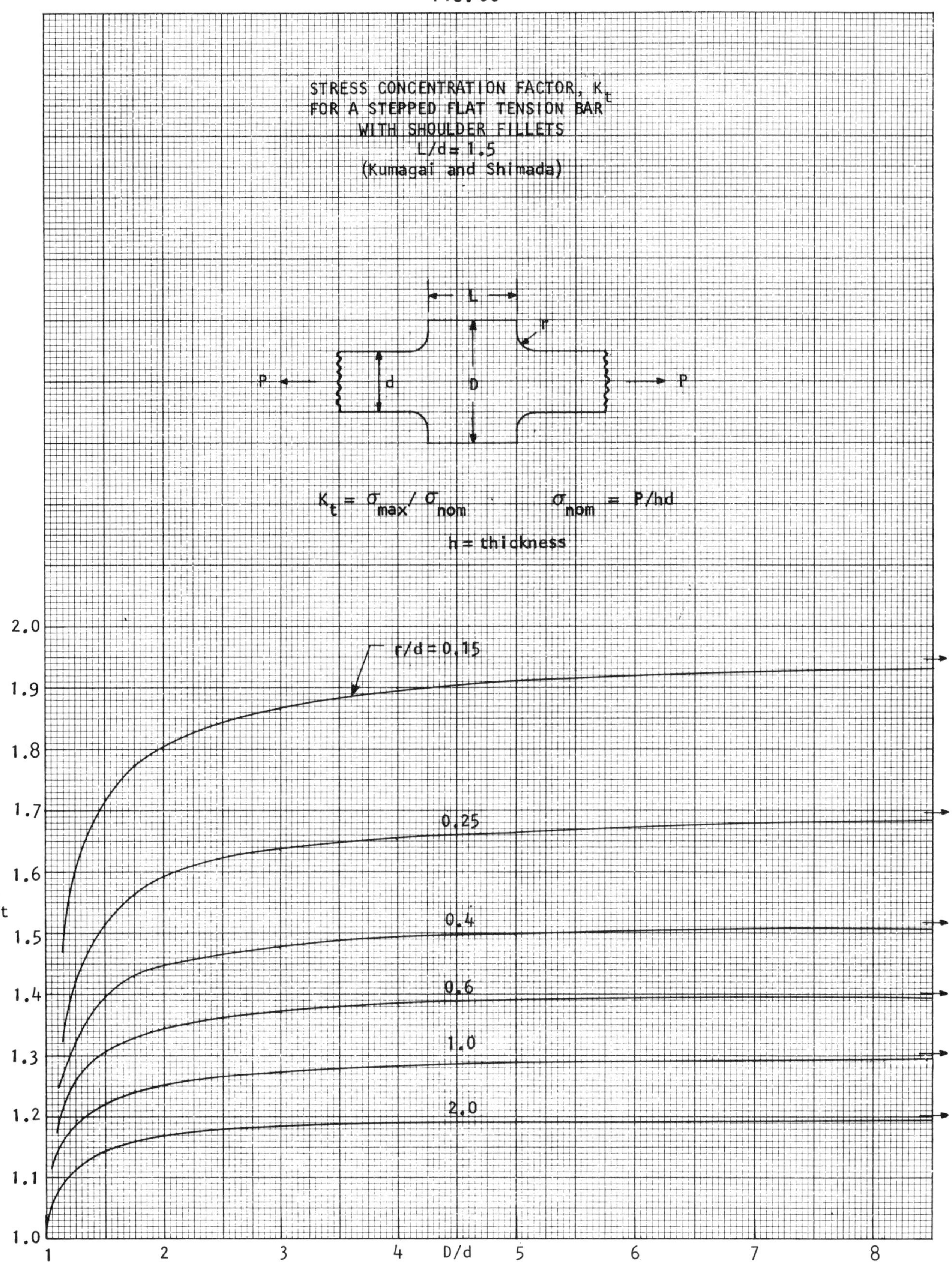
STRESS CONCENTRATION FACTOR, K_t
FOR A STEPPED FLAT TENSION BAR
WITH SHOULDER FILLETS
L/d = 1.5
(Kumagai and Shimada)
L
r
P
d
D
P
$K_t = \sigma_{max}/\sigma_{nom}$
$\sigma_{nom} = P/hd$
h = thickness
r/d = 0.15
0.25
0.4
0.6
1.0
2.0
K_t
2.0
1.9
1.8
1.7
1.6
1.5
1.4
1.3
1.2
1.1
1.0
1
2
3
4
D/d
5
6
7
8

FIG.67

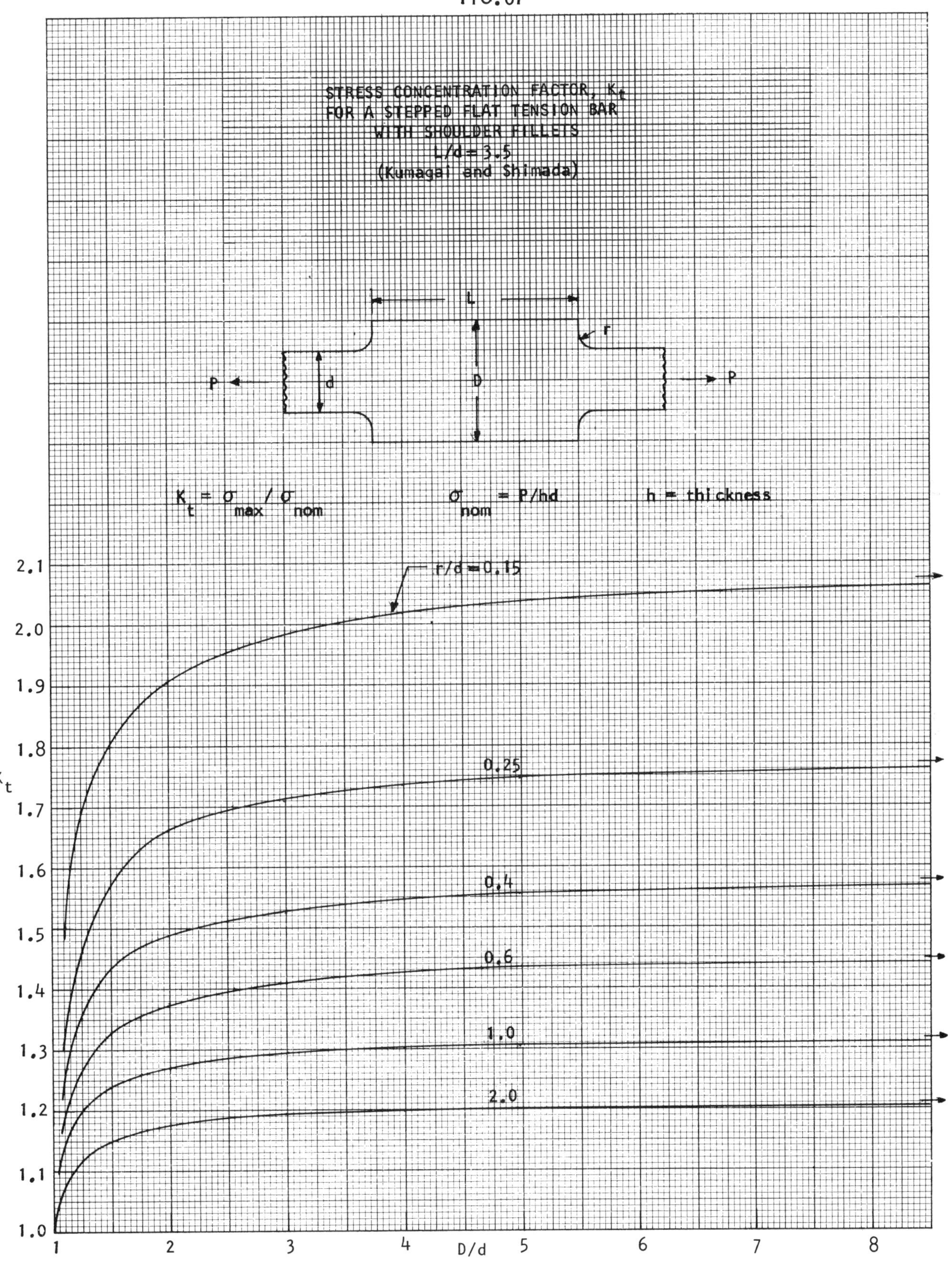
STRESS CONCENTRATION FACTOR, K_t
FOR A STEPPED FLAT TENSION BAR
WITH SHOULDER FILLETS
L/d = 3.5
(Kumagai and Shimada)
L
r
P
d
D
P
$K_t = \sigma_{max}/\sigma_{nom}$
$\sigma_{nom} = P/hd$
h = thickness
r/d = 0.15
0.25
0.4
0.6
1.0
2.0
K_t
2.1
2.0
1.9
1.8
1.7
1.6
1.5
1.4
1.3
1.2
1.1
1.0
1
2
3
4
D/d
5
6
7
8

FIG. 68

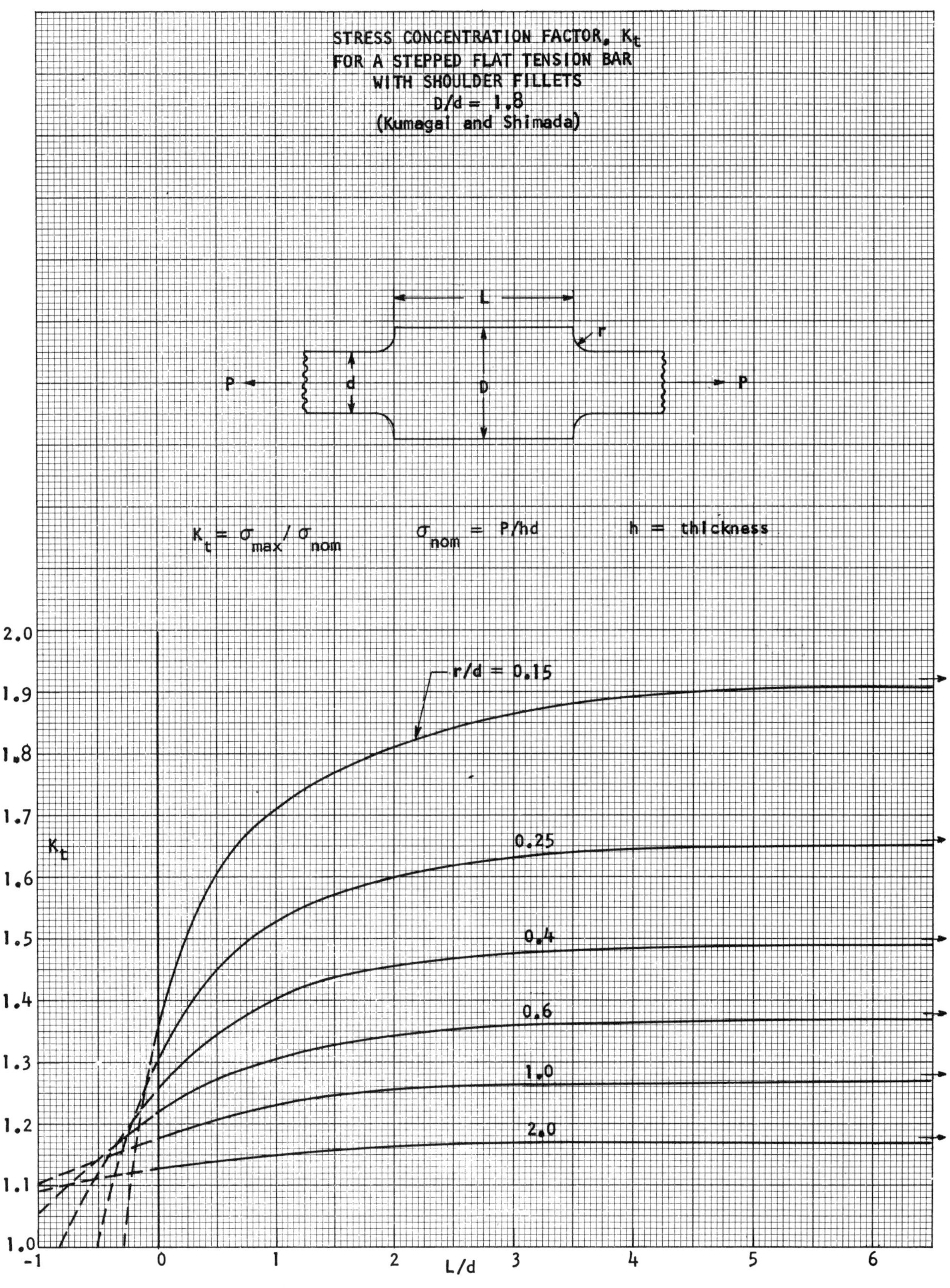

STRESS CONCENTRATION FACTOR, K_t
FOR A STEPPED FLAT TENSION BAR
WITH SHOULDER FILLETS
D/d = 1.8
(Kumagai and Shimada)
L
r
P
d
D
P
$K_t = \sigma_{max} / \sigma_{nom}$
σ_{nom} = P/hd
h = thickness
r/d = 0.15
0.25
0.4
0.6
1.0
2.0
K_t
2.0
1.9
1.8
1.7
1.6
1.5
1.4
1.3
1.2
1.1
1.0
-1
0
1
2
L/d
3
4
5
6

FIG. 69

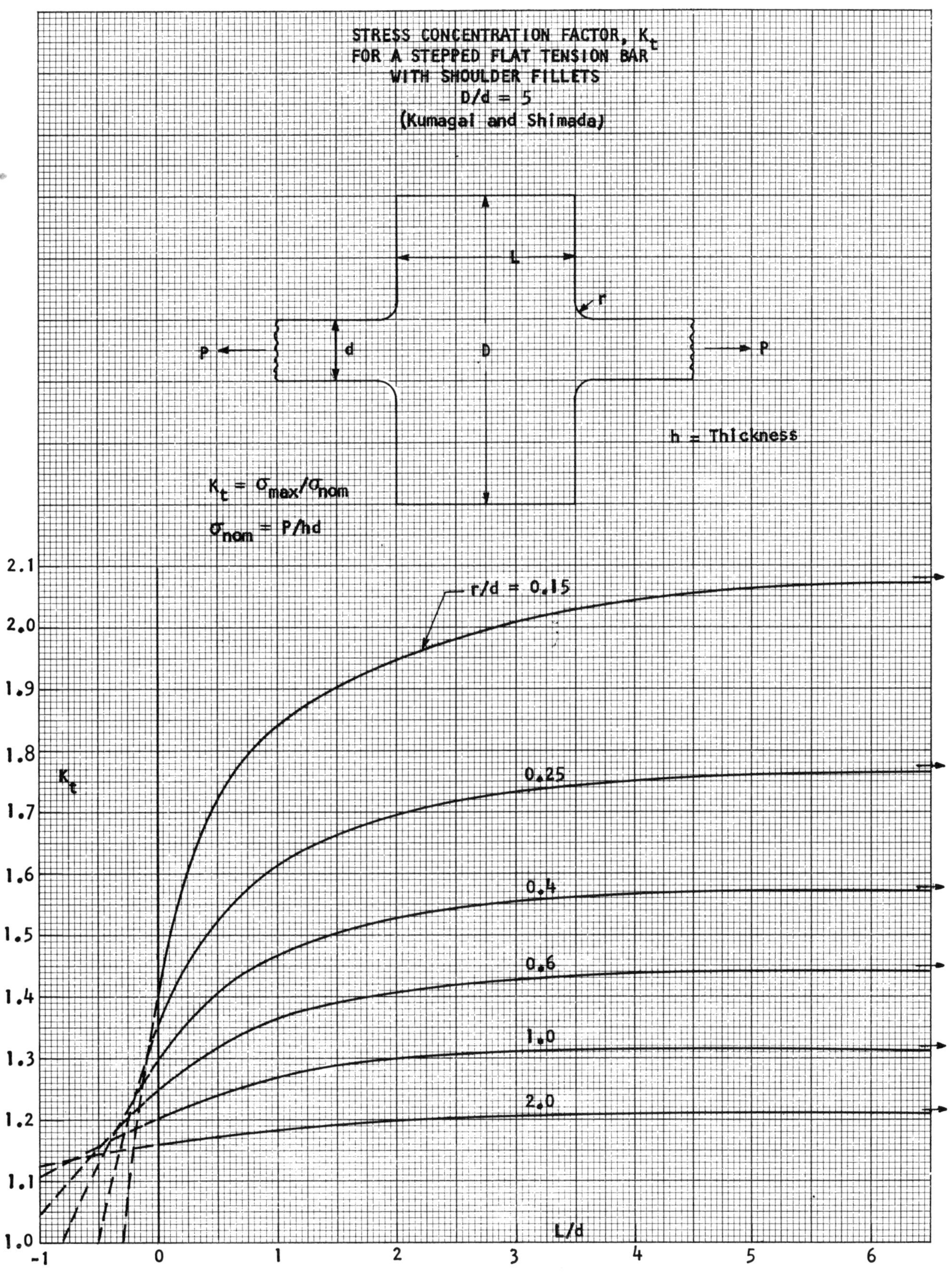
STRESS CONCENTRATION FACTOR, K_t
FOR A STEPPED FLAT TENSION BAR
WITH SHOULDER FILLETS
D/d = 5
(Kumagai and Shimada)
L
r
P
d
D
P
h = Thickness
$K_t = \sigma_{max}/\sigma_{nom}$
$\sigma_{nom} = P/hd$
r/d = 0.15
0.25
0.4
0.6
1.0
2.0
K_t
L/d
2.1
2.0
1.9
1.8
1.7
1.6
1.5
1.4
1.3
1.2
1.1
1.0
-1
0
1
2
3
4
5
6

FIG. 70

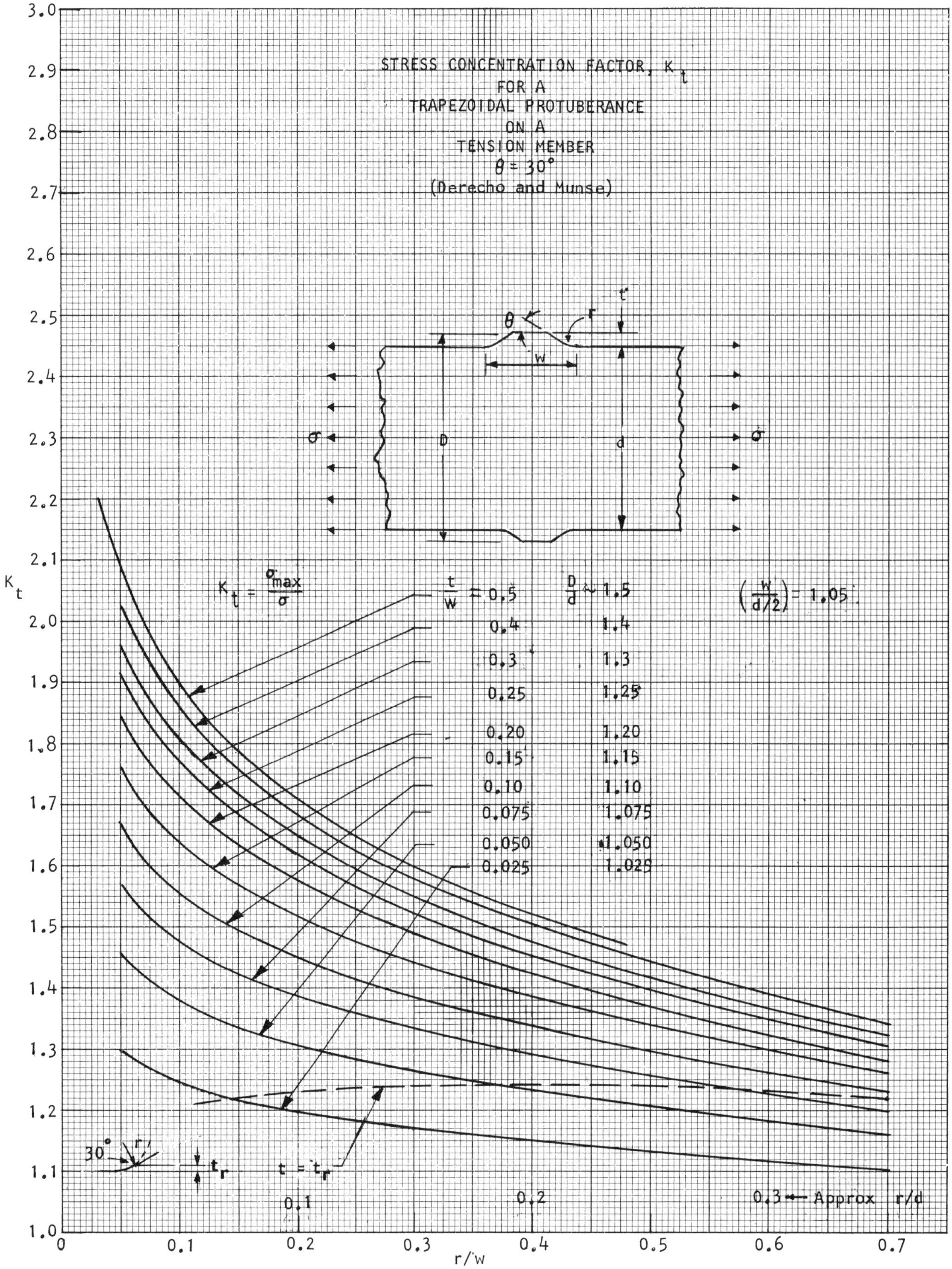

STRESS CONCENTRATION FACTOR, K_t
FOR A
TRAPEZOIDAL PROTUBERANCE
ON A
TENSION MEMBER
$\theta = 30°$
(Derecho and Munse)
$K_t = \frac{\sigma_{max}}{\sigma}$
$\frac{t}{w} = 0.5$ $\frac{D}{d} \approx 1.5$
0.4 1.4
0.3 1.3
0.25 1.25
0.20 1.20
0.15 1.15
0.10 1.10
0.075 1.075
0.050 1.050
0.025 1.025
$\left(\frac{w}{d/2}\right) = 1.05$
$t = t_r$
0.1 0.2 0.3 ← Approx r/d
K_t
r/w

FIG. 71

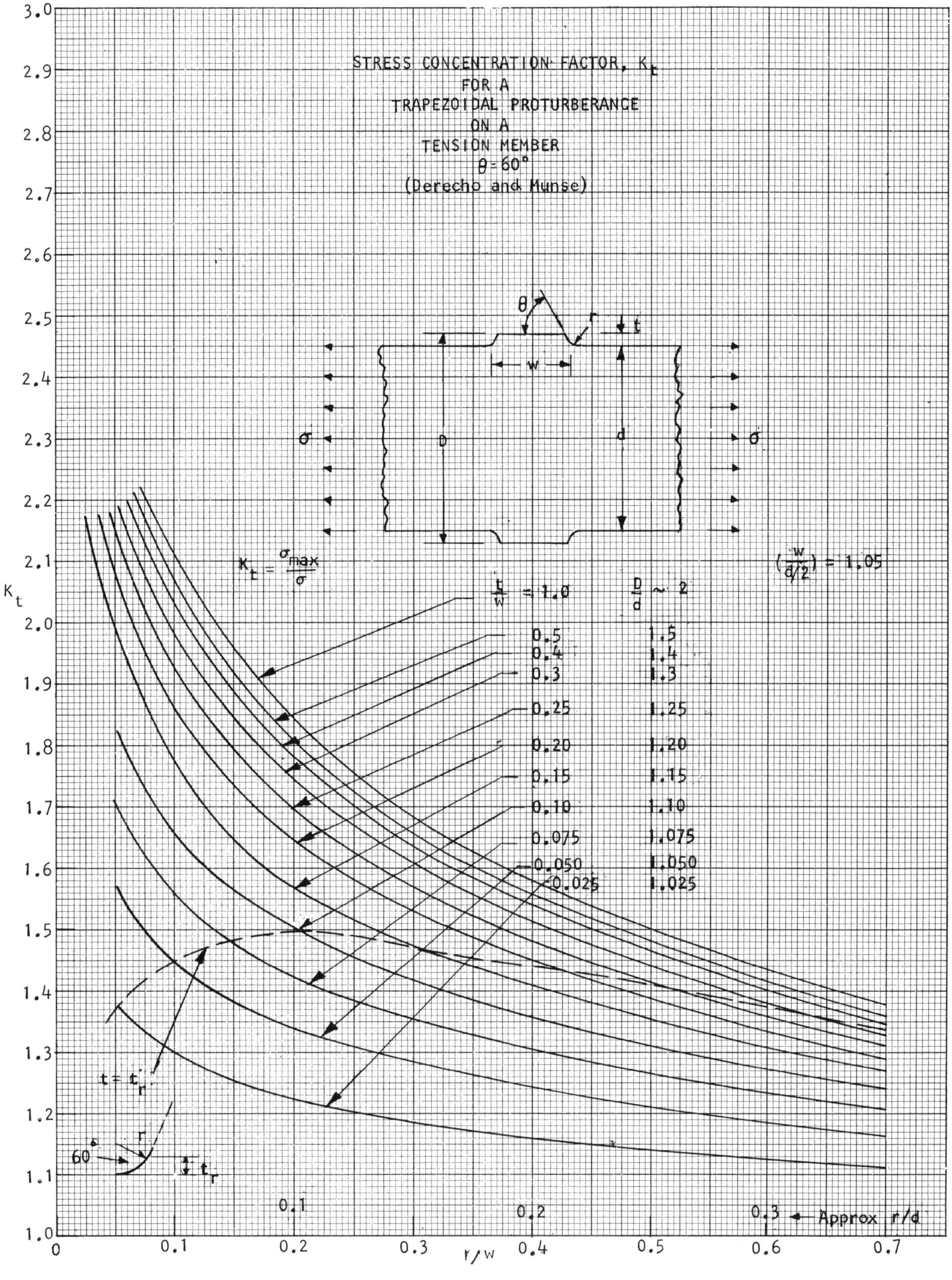

STRESS CONCENTRATION FACTOR, K_t
FOR A
TRAPEZOIDAL PROTURBERANCE
ON A
TENSION MEMBER
$\theta = 60°$
(Derecho and Munse)

θ, r, t, w, D, d, σ, σ

$K_t = \frac{\sigma_{max}}{\sigma}$

$\left(\frac{w}{d/2}\right) = 1.05$

$\frac{t}{w}$	$\frac{D}{d}$
= 1.0	~ 2
0.5	1.5
0.4	1.4
0.3	1.3
0.25	1.25
0.20	1.20
0.15	1.15
0.10	1.10
0.075	1.075
0.050	1.050
0.025	1.025

$t = t_r$

60°, r, t_r

K_t: 1.0, 1.1, 1.2, 1.3, 1.4, 1.5, 1.6, 1.7, 1.8, 1.9, 2.0, 2.1, 2.2, 2.3, 2.4, 2.5, 2.6, 2.7, 2.8, 2.9, 3.0

0.1, 0.2, 0.3 ← Approx r/d

0, 0.1, 0.2, 0.3, 0.4, 0.5, 0.6, 0.7

r/w

FIG. 72

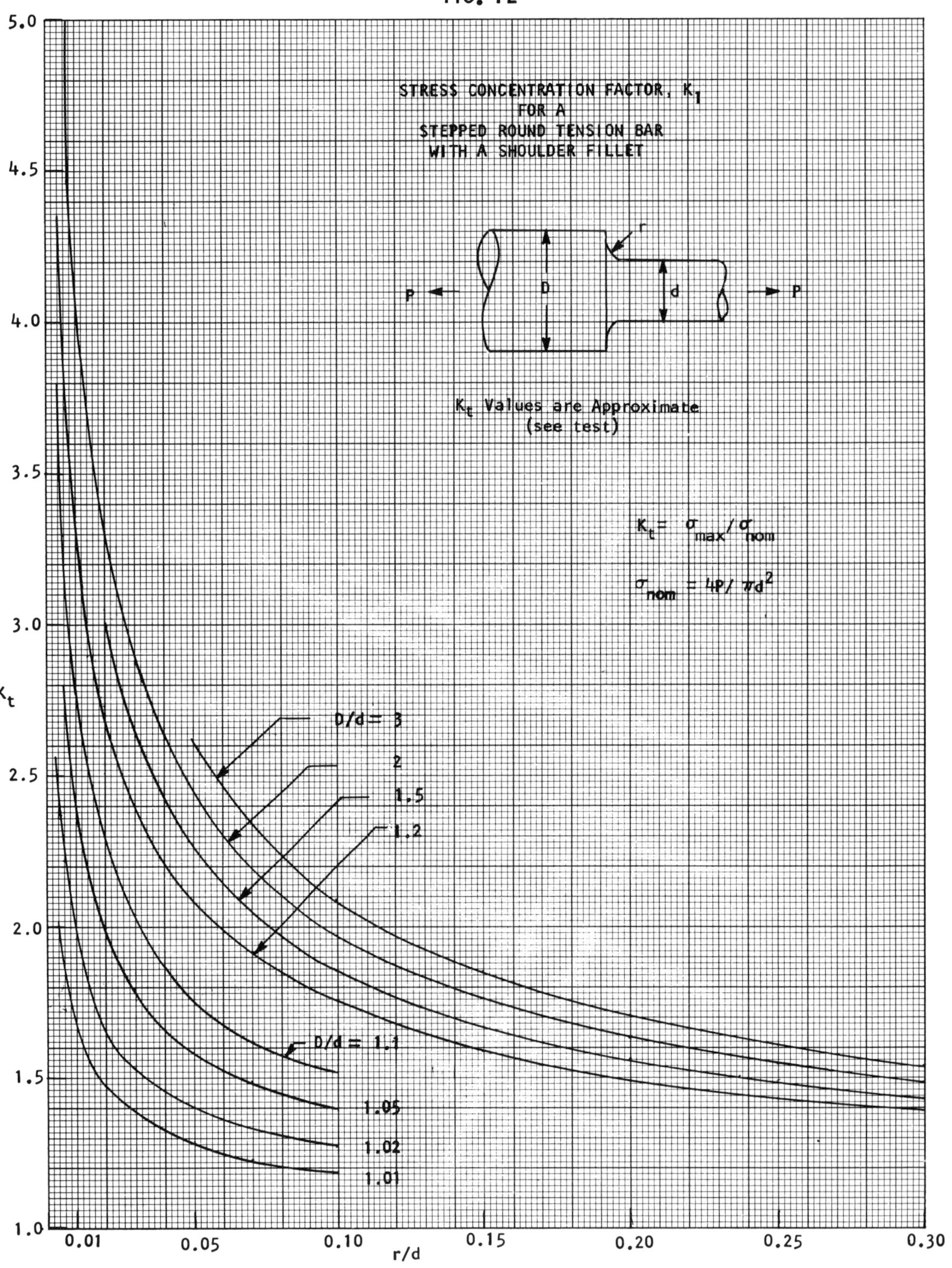

STRESS CONCENTRATION FACTOR, K_t
FOR A
STEPPED ROUND TENSION BAR
WITH A SHOULDER FILLET
r
P
D
d
P
K_t Values are Approximate
(see test)
$K_t = \sigma_{max}/\sigma_{nom}$
$\sigma_{nom} = 4P/\pi d^2$
D/d = 3
2
1.5
1.2
D/d = 1.1
1.05
1.02
1.01
5.0
4.5
4.0
3.5
3.0
2.5
2.0
1.5
1.0
K_t
0.01
0.05
0.10
0.15
0.20
0.25
0.30
r/d

FIG. 72a

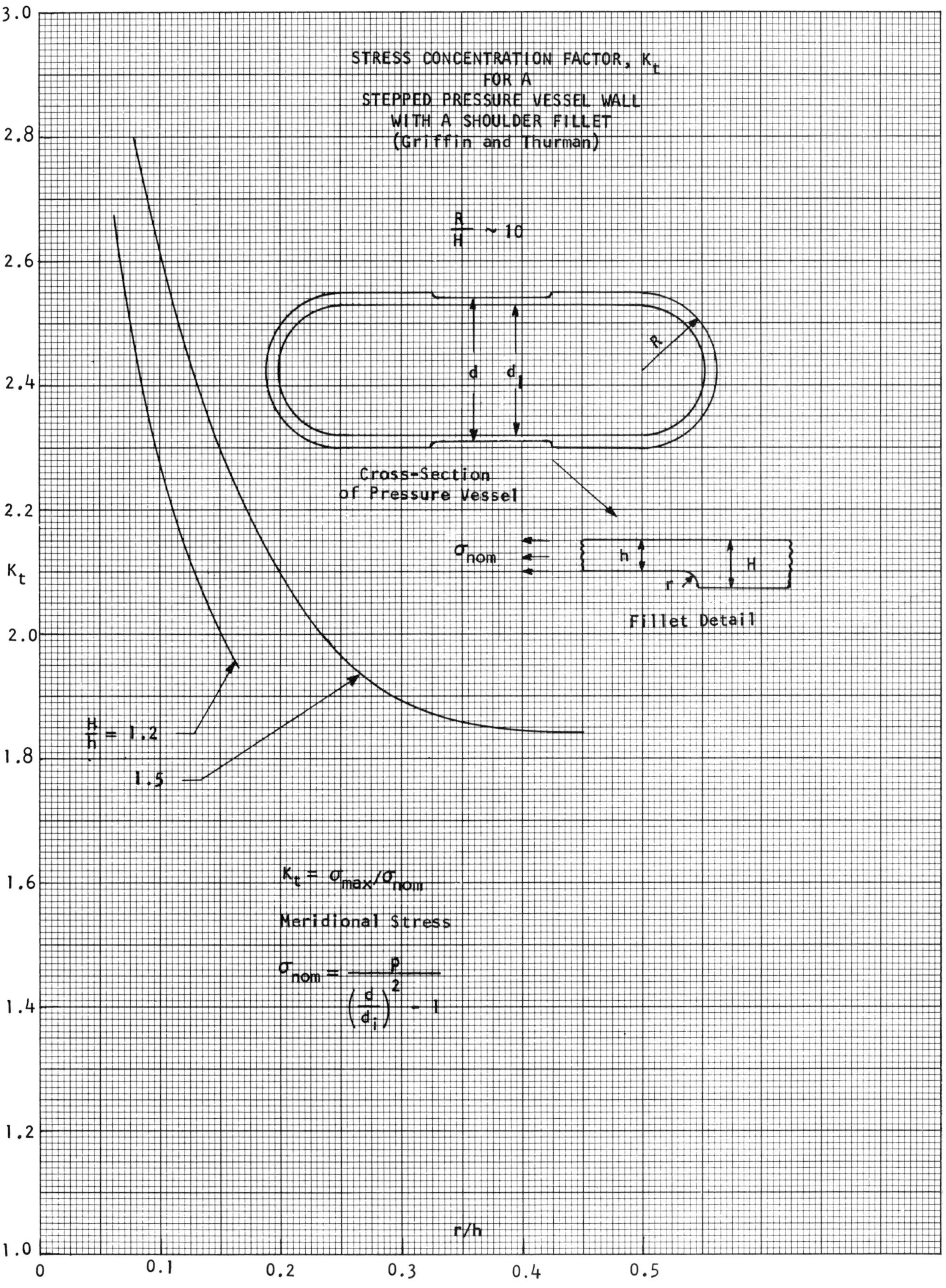
STRESS CONCENTRATION FACTOR, K_t
FOR A
STEPPED PRESSURE VESSEL WALL
WITH A SHOULDER FILLET
(Griffin and Thurman)
$\frac{R}{H} \sim 10$
d
d_i
R
Cross-Section of Pressure Vessel
σ_{nom}
h
H
r
Fillet Detail
$\frac{H}{h} = 1.2$
1.5
$K_t = \sigma_{max}/\sigma_{nom}$
Meridional Stress
$\sigma_{nom} = \frac{p}{\left(\frac{d}{d_i}\right)^2 - 1}$
K_t
r/h
3.0
2.8
2.6
2.4
2.2
2.0
1.8
1.6
1.4
1.2
1.0
0
0.1
0.2
0.3
0.4
0.5

FIG. 73

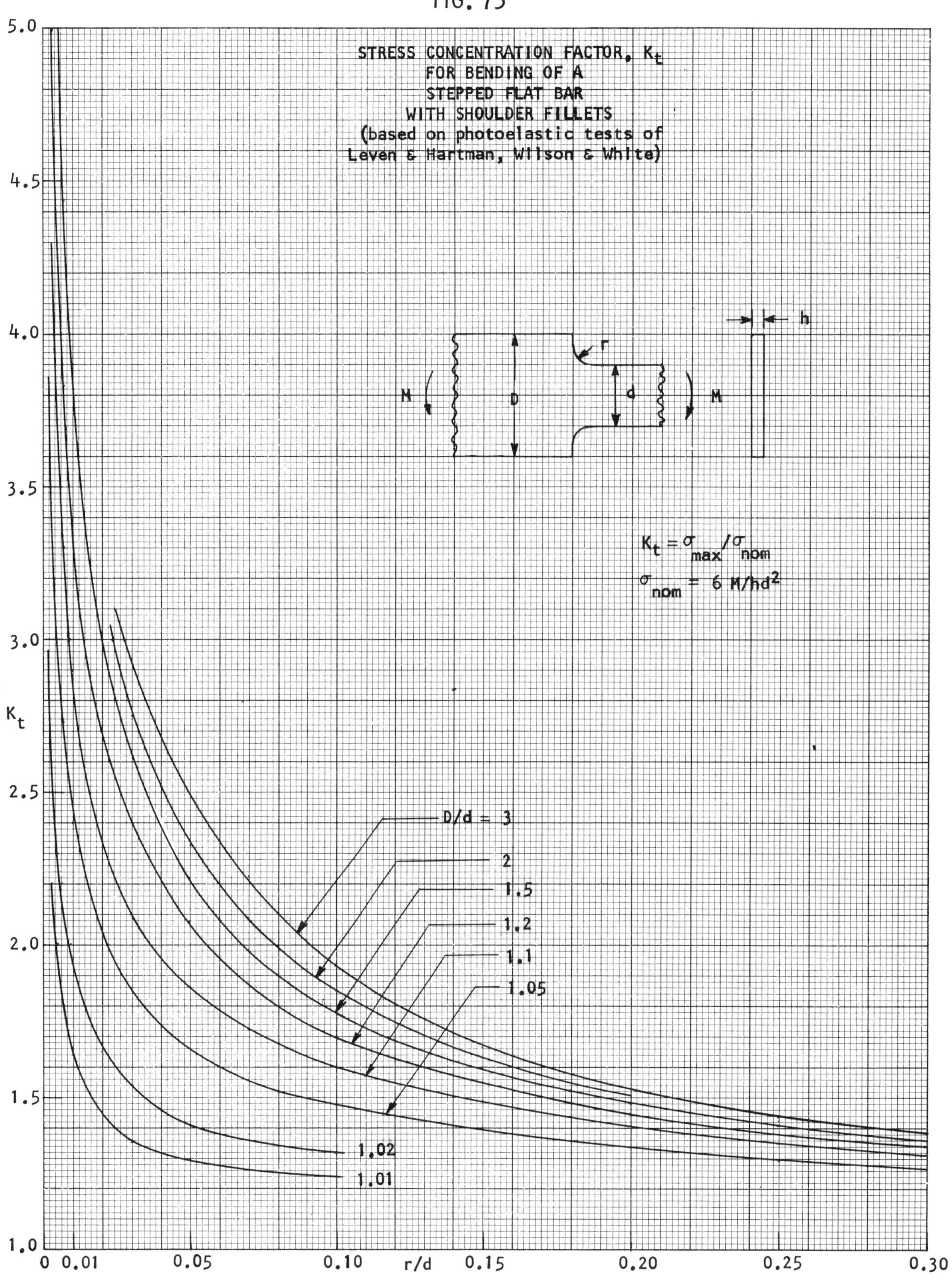

STRESS CONCENTRATION FACTOR, K_t
FOR BENDING OF A
STEPPED FLAT BAR
WITH SHOULDER FILLETS
(based on photoelastic tests of
Leven & Hartman, Wilson & White)
h
r
M
D
d
M
$K_t = \sigma_{max}/\sigma_{nom}$
$\sigma_{nom} = 6\ M/hd^2$
D/d = 3
2
1.5
1.2
1.1
1.05
1.02
1.01
K_t
5.0
4.5
4.0
3.5
3.0
2.5
2.0
1.5
1.0
0
0.01
0.05
0.10
r/d
0.15
0.20
0.25
0.30

FIG. 74

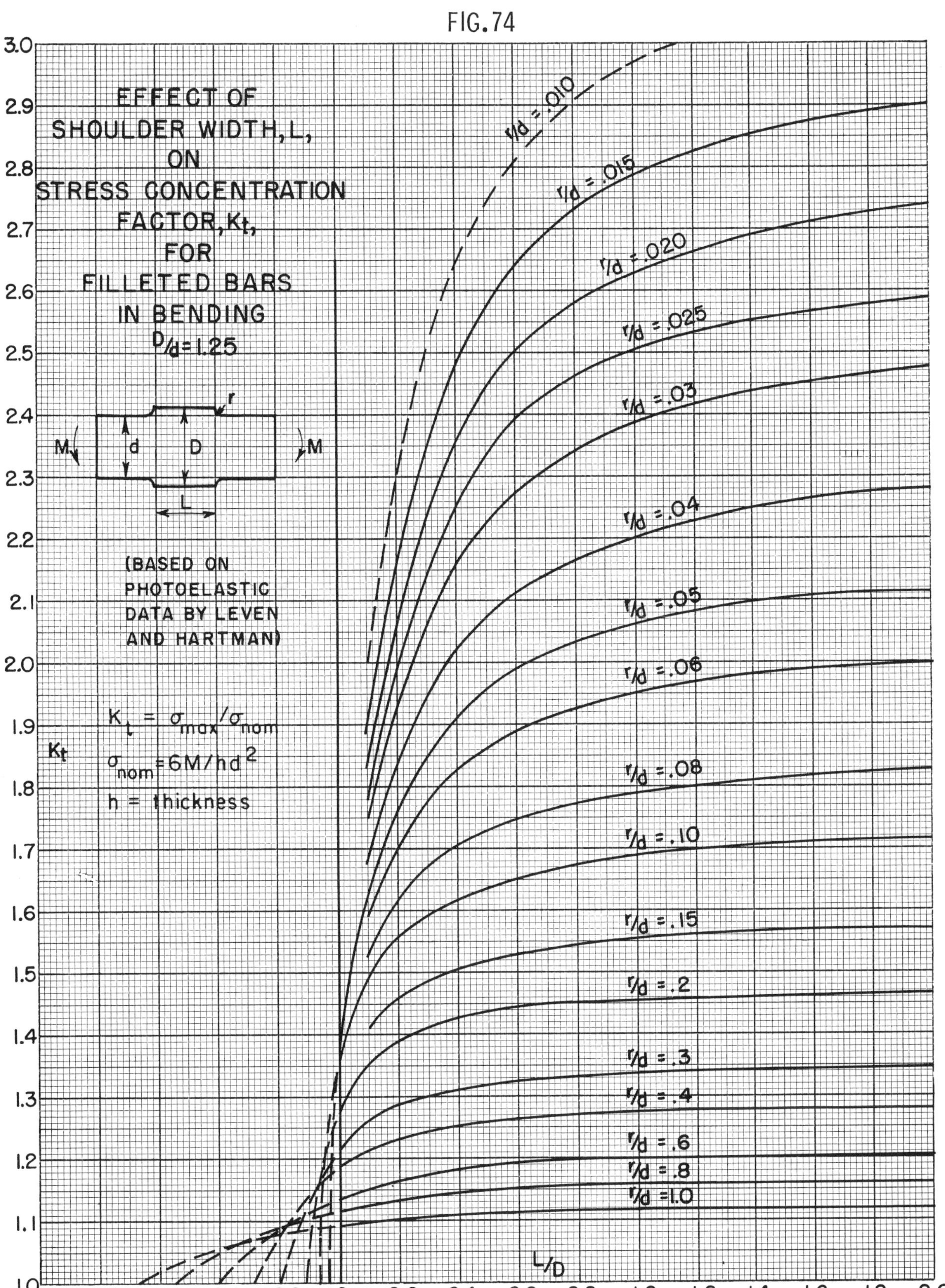

EFFECT OF SHOULDER WIDTH, L, ON STRESS CONCENTRATION FACTOR, K_t, FOR FILLETED BARS IN BENDING
D/d = 1.25
(BASED ON PHOTOELASTIC DATA BY LEVEN AND HARTMAN)
$K_t = \sigma_{max}/\sigma_{nom}$
$\sigma_{nom} = 6M/hd^2$
h = thickness
M
d
D
r
L
K_t
L/D
r/d = .010
r/d = .015
r/d = .020
r/d = .025
r/d = .03
r/d = .04
r/d = .05
r/d = .06
r/d = .08
r/d = .10
r/d = .15
r/d = .2
r/d = .3
r/d = .4
r/d = .6
r/d = .8
r/d = 1.0
3.0
2.9
2.8
2.7
2.6
2.5
2.4
2.3
2.2
2.1
2.0
1.9
1.8
1.7
1.6
1.5
1.4
1.3
1.2
1.1
1.0
−0.8
−0.6
−0.4
−0.2
0
0.2
0.4
0.6
0.8
1.0
1.2
1.4
1.6
1.8
2.0

FIG. 75

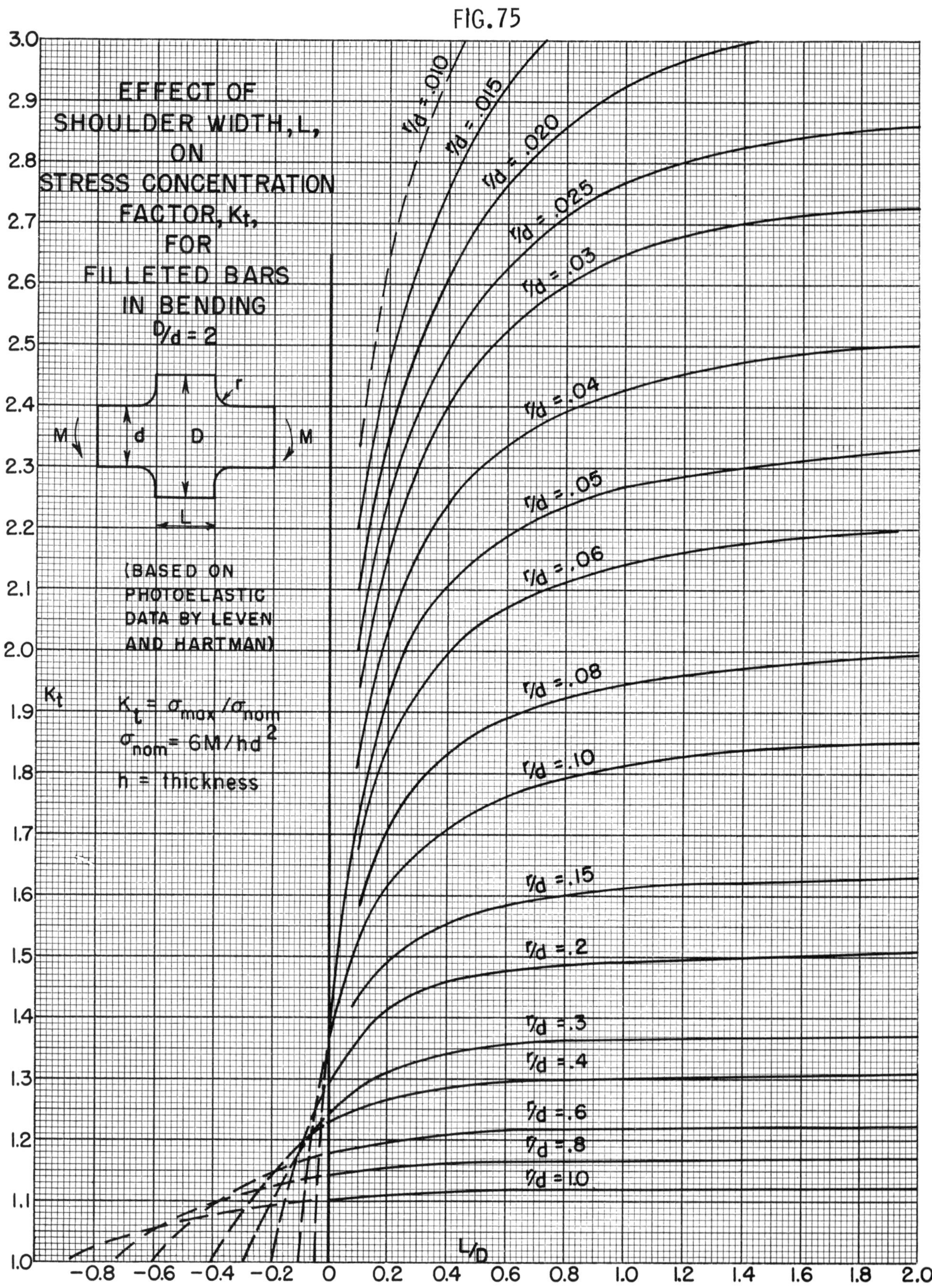

EFFECT OF
SHOULDER WIDTH, L,
ON
STRESS CONCENTRATION
FACTOR, Kt,
FOR
FILLETED BARS
IN BENDING
D/d = 2
M
d
D
r
M
L
(BASED ON
PHOTOELASTIC
DATA BY LEVEN
AND HARTMAN)
K_t
$K_t = \sigma_{max}/\sigma_{nom}$
$\sigma_{nom} = 6M/hd^2$
h = thickness
r/d = .010
r/d = .015
r/d = .020
r/d = .025
r/d = .03
r/d = .04
r/d = .05
r/d = .06
r/d = .08
r/d = .10
r/d = .15
r/d = .2
r/d = .3
r/d = .4
r/d = .6
r/d = .8
r/d = 1.0
L/D
3.0
2.9
2.8
2.7
2.6
2.5
2.4
2.3
2.2
2.1
2.0
1.9
1.8
1.7
1.6
1.5
1.4
1.3
1.2
1.1
1.0
−0.8
−0.6
−0.4
−0.2
0
0.2
0.4
0.6
0.8
1.0
1.2
1.4
1.6
1.8
2.0

FIG. 76

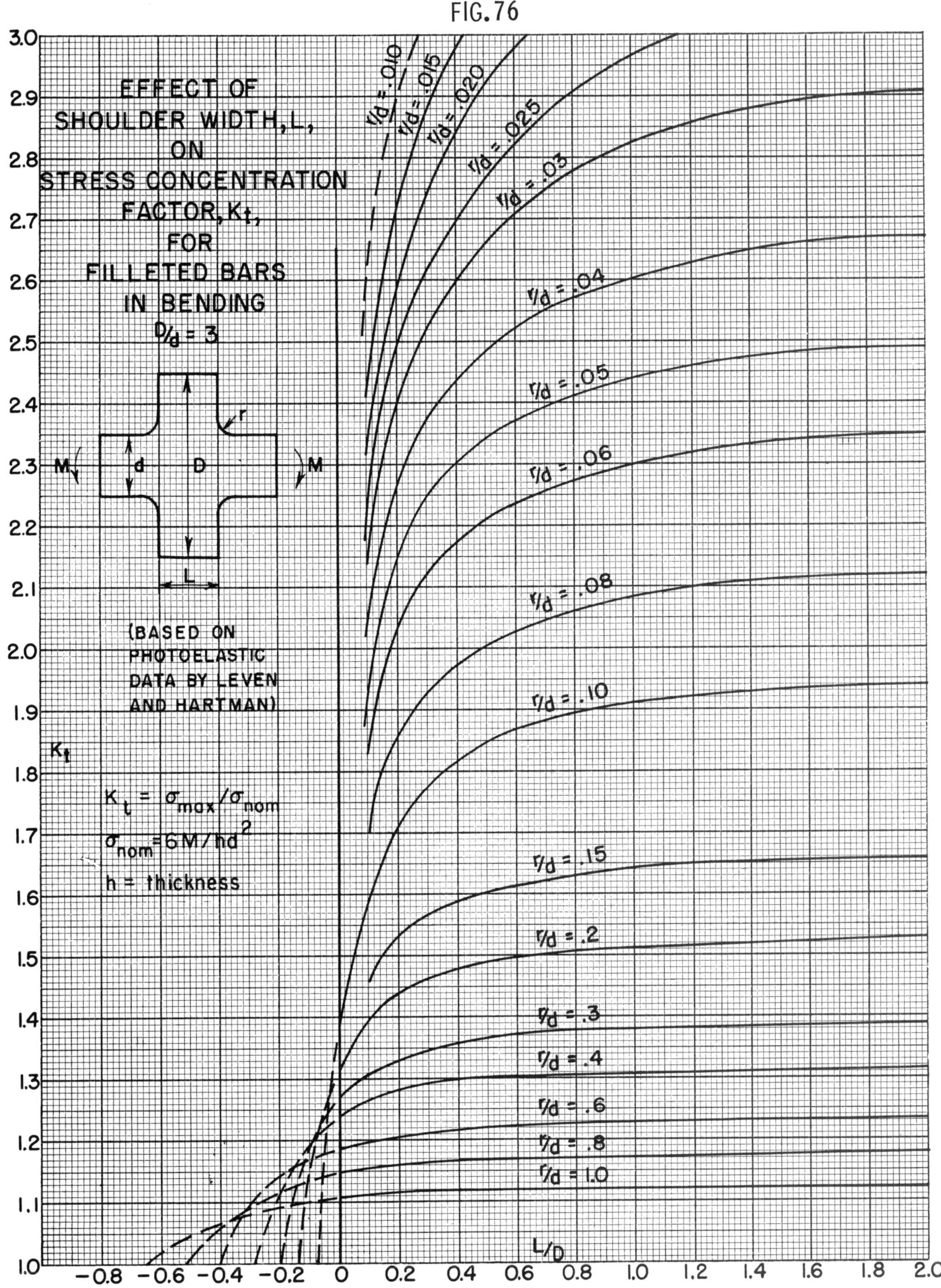
EFFECT OF
SHOULDER WIDTH, L,
ON
STRESS CONCENTRATION
FACTOR, K_t,
FOR
FILLETED BARS
IN BENDING
D/d = 3
M
d
D
r
M
L
(BASED ON
PHOTOELASTIC
DATA BY LEVEN
AND HARTMAN)
K_t
$K_t = \sigma_{max}/\sigma_{nom}$
$\sigma_{nom} = 6M/hd^2$
h = thickness
r/d = .010
r/d = .015
r/d = .020
r/d = .025
r/d = .03
r/d = .04
r/d = .05
r/d = .06
r/d = .08
r/d = .10
r/d = .15
r/d = .2
r/d = .3
r/d = .4
r/d = .6
r/d = .8
r/d = 1.0
L/D
3.0
2.9
2.8
2.7
2.6
2.5
2.4
2.3
2.2
2.1
2.0
1.9
1.8
1.7
1.6
1.5
1.4
1.3
1.2
1.1
1.0
−0.8
−0.6
−0.4
−0.2
0
0.2
0.4
0.6
0.8
1.0
1.2
1.4
1.6
1.8
2.0

FIG. 77

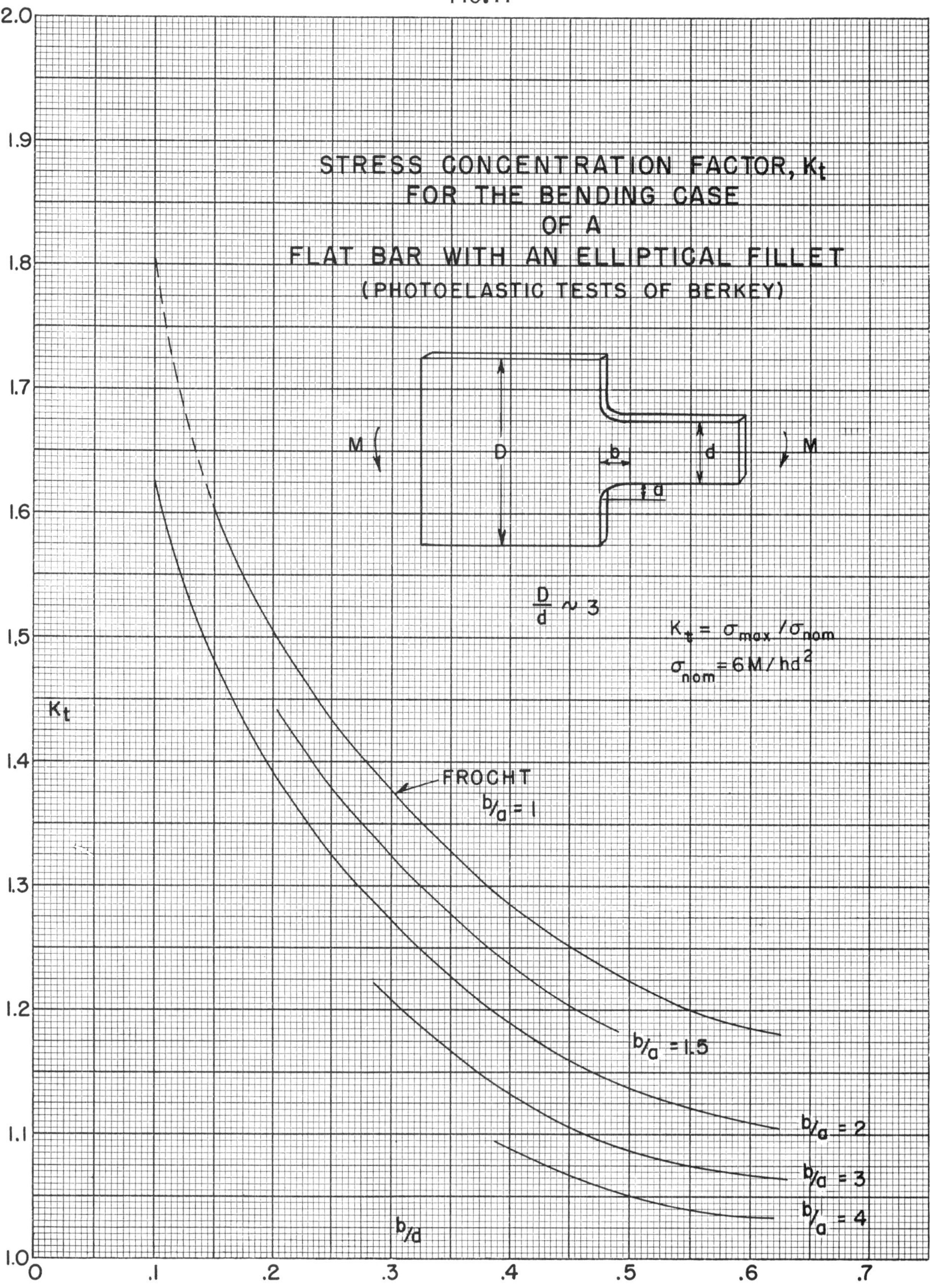

STRESS CONCENTRATION FACTOR, K_t
FOR THE BENDING CASE
OF A
FLAT BAR WITH AN ELLIPTICAL FILLET
(PHOTOELASTIC TESTS OF BERKEY)
M
D
b
d
a
M
$\frac{D}{d} \sim 3$
$K_t = \sigma_{max} / \sigma_{nom}$
$\sigma_{nom} = 6M/hd^2$
K_t
FROCHT
$b/a = 1$
$b/a = 1.5$
$b/a = 2$
$b/a = 3$
$b/a = 4$
b/d
2.0
1.9
1.8
1.7
1.6
1.5
1.4
1.3
1.2
1.1
1.0
0
.1
.2
.3
.4
.5
.6
.7

FIG. 78

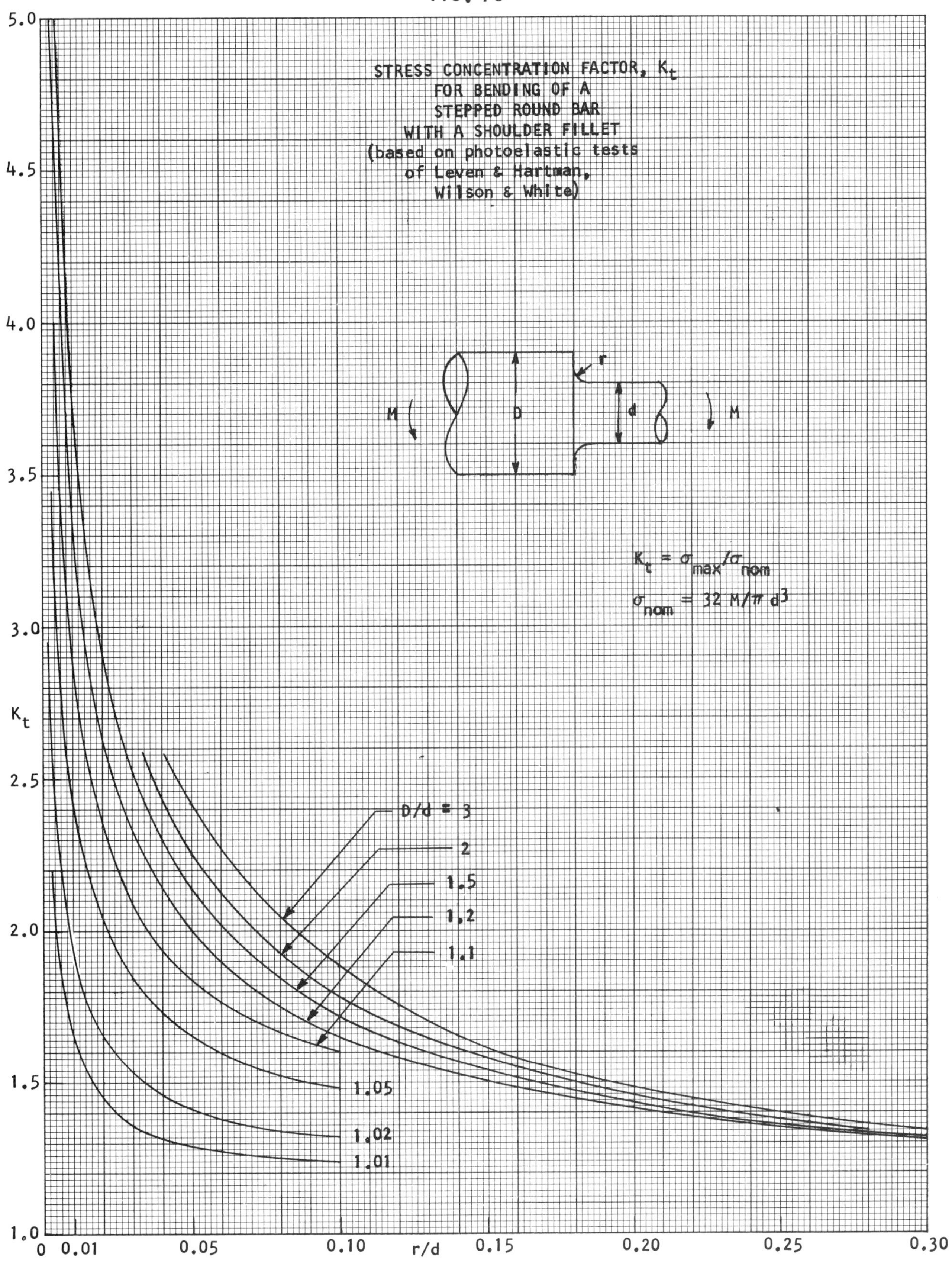
STRESS CONCENTRATION FACTOR, K_t
FOR BENDING OF A
STEPPED ROUND BAR
WITH A SHOULDER FILLET
(based on photoelastic tests
of Leven & Hartman,
Wilson & White)
$K_t = \sigma_{max}/\sigma_{nom}$
$\sigma_{nom} = 32\ M/\pi\ d^3$
M
D
d
r
M
D/d = 3
2
1.5
1.2
1.1
1.05
1.02
1.01
K_t
r/d
5.0
4.5
4.0
3.5
3.0
2.5
2.0
1.5
1.0
0
0.01
0.05
0.10
0.15
0.20
0.25
0.30

FIG. 78a

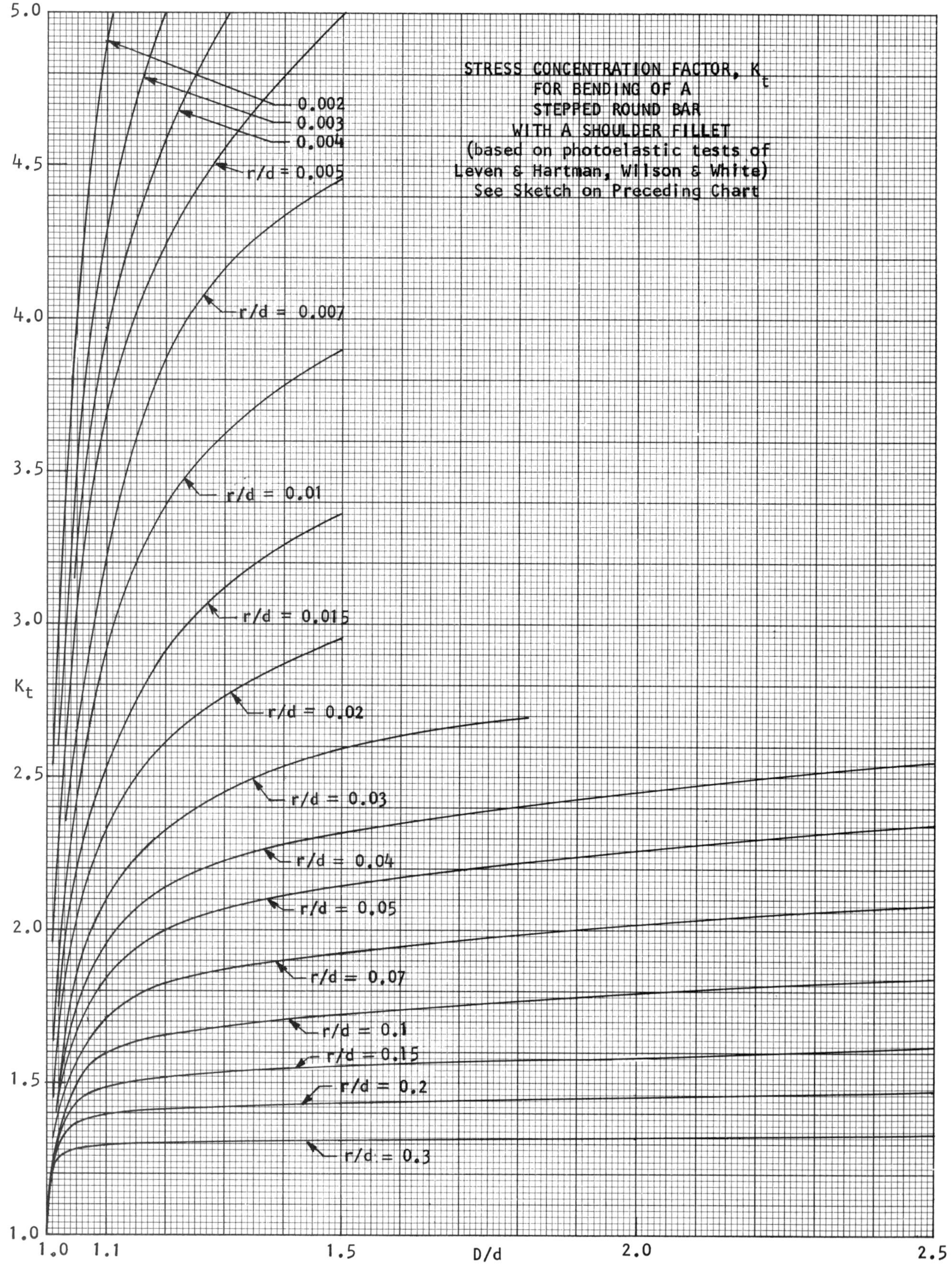

STRESS CONCENTRATION FACTOR, K_t
FOR BENDING OF A
STEPPED ROUND BAR
WITH A SHOULDER FILLET
(based on photoelastic tests of
Leven & Hartman, Wilson & White)
See Sketch on Preceding Chart
0.002
0.003
0.004
r/d = 0.005
r/d = 0.007
r/d = 0.01
r/d = 0.015
r/d = 0.02
r/d = 0.03
r/d = 0.04
r/d = 0.05
r/d = 0.07
r/d = 0.1
r/d = 0.15
r/d = 0.2
r/d = 0.3
K_t
5.0
4.5
4.0
3.5
3.0
2.5
2.0
1.5
1.0
1.0
1.1
1.5
D/d
2.0
2.5

FIG. 79

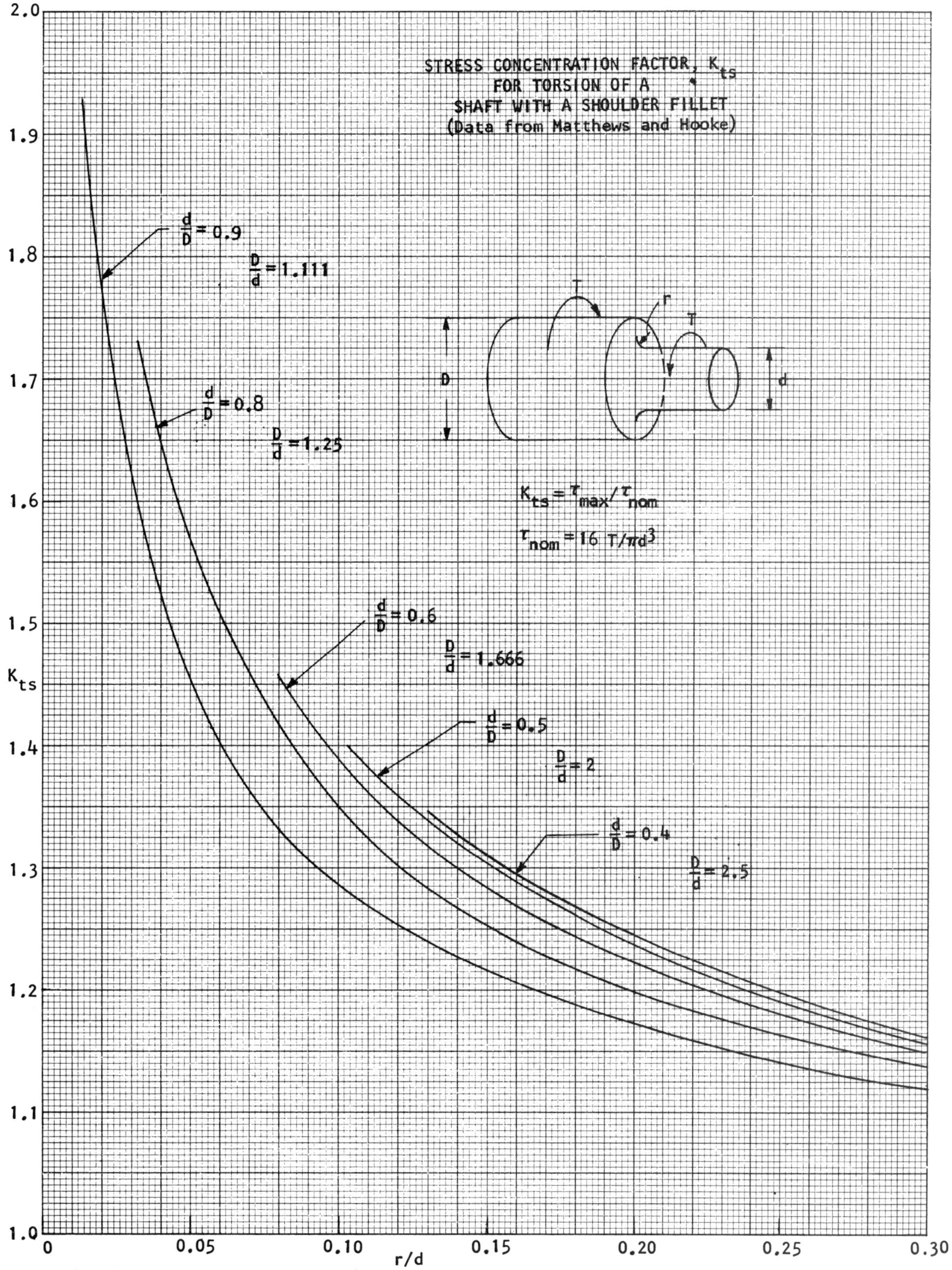

STRESS CONCENTRATION FACTOR, K_{ts}
FOR TORSION OF A
SHAFT WITH A SHOULDER FILLET
(Data from Matthews and Hooke)
$\frac{d}{D} = 0.9$
$\frac{D}{d} = 1.111$
$\frac{d}{D} = 0.8$
$\frac{D}{d} = 1.25$
$\frac{d}{D} = 0.6$
$\frac{D}{d} = 1.666$
$\frac{d}{D} = 0.5$
$\frac{D}{d} = 2$
$\frac{d}{D} = 0.4$
$\frac{D}{d} = 2.5$
T
r
T
D
d
$K_{ts} = \tau_{max}/\tau_{nom}$
$\tau_{nom} = 16\ T/\pi d^3$
K_{ts}
2.0
1.9
1.8
1.7
1.6
1.5
1.4
1.3
1.2
1.1
1.0
0
0.05
0.10
0.15
0.20
0.25
0.30
r/d

FIG. 79a

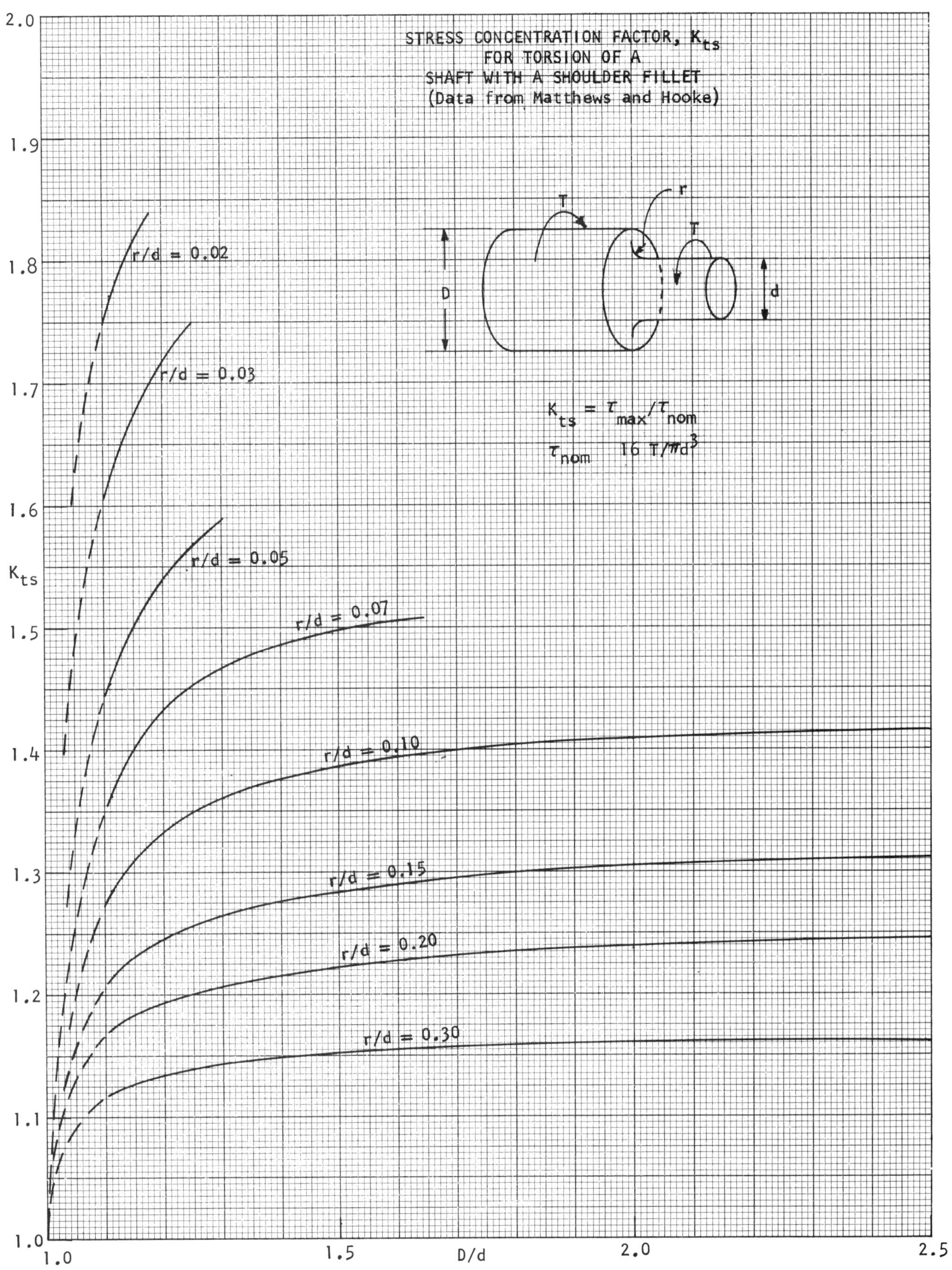

STRESS CONCENTRATION FACTOR, K_{ts}
FOR TORSION OF A
SHAFT WITH A SHOULDER FILLET
(Data from Matthews and Hooke)
T
r
T
D
d
$K_{ts} = \tau_{max}/\tau_{nom}$
τ_{nom} 16 T/πd^3
r/d = 0.02
r/d = 0.03
r/d = 0.05
r/d = 0.07
r/d = 0.10
r/d = 0.15
r/d = 0.20
r/d = 0.30
K_{ts}
2.0
1.9
1.8
1.7
1.6
1.5
1.4
1.3
1.2
1.1
1.0
1.0
1.5
D/d
2.0
2.5

FIG. 80

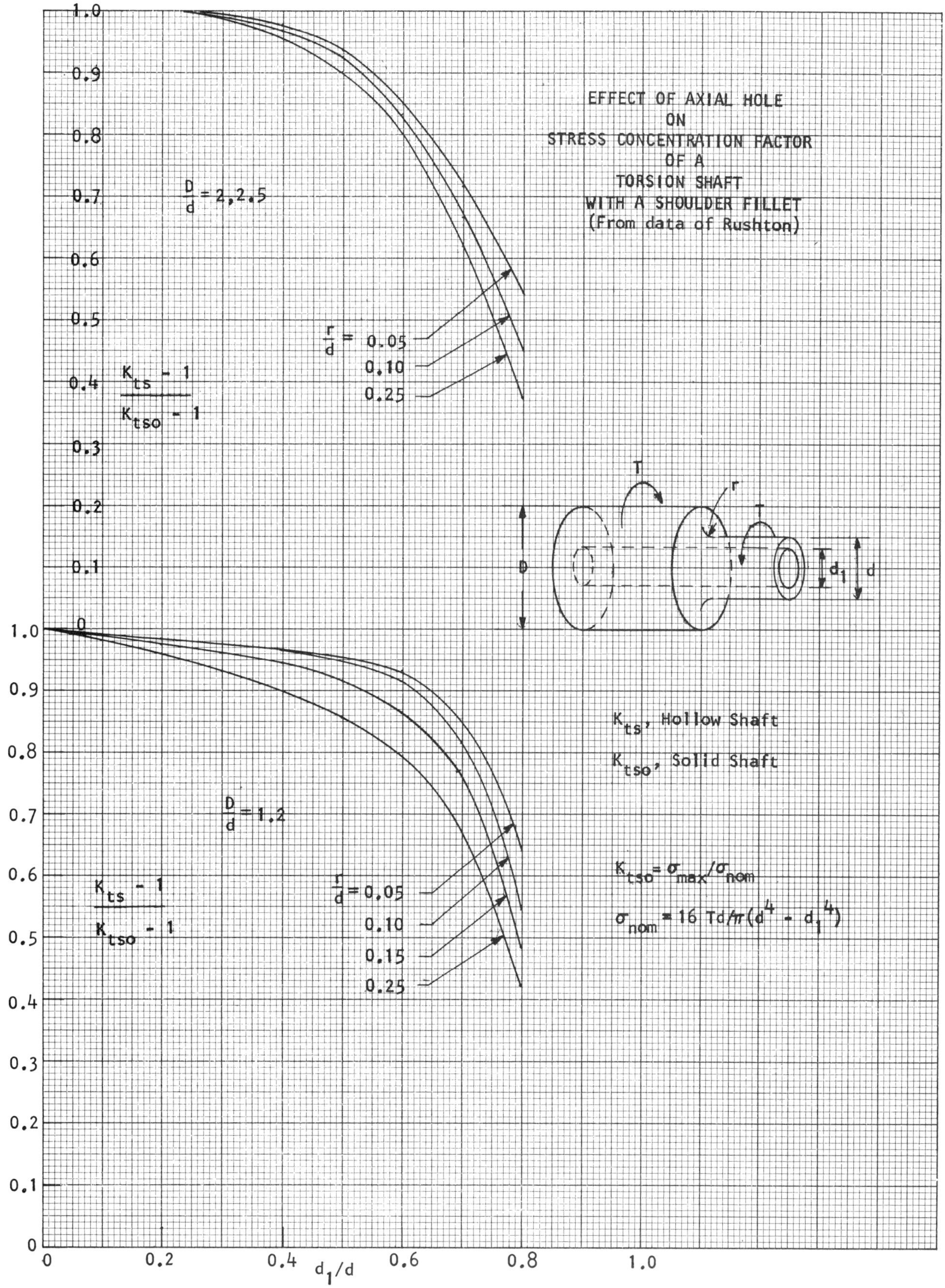

EFFECT OF AXIAL HOLE
ON
STRESS CONCENTRATION FACTOR
OF A
TORSION SHAFT
WITH A SHOULDER FILLET
(From data of Rushton)
D/d = 2, 2.5
r/d = 0.05
0.10
0.25
(Kts - 1)/(Ktso - 1)
1.0
0.9
0.8
0.7
0.6
0.5
0.4
0.3
0.2
0.1
0
T
r
T
D
d1
d
D/d = 1.2
r/d = 0.05
0.10
0.15
0.25
(Kts - 1)/(Ktso - 1)
Kts, Hollow Shaft
Ktso, Solid Shaft
Ktso = σmax/σnom
σnom = 16 Td/π(d^4 - d1^4)
0
0.2
0.4
d1/d
0.6
0.8
1.0

CHAPTER 4

HOLES

Some parts with transverse holes are shown in Fig. 81. The words "sheet," "plate," and "flat bar" are used in this chapter in the same sense in that they refer, unless otherwise specified, to the "two-dimensional" mathematical case.

A large amount of information concerning the stress distribution around holes has been published recently. This material is organized here according to stress state (tension, bending, torsion—as in preceding chapters); two- and three-dimensional cases; shape of hole (circular, elliptical, other); single and multiple holes. To fully comprehend the plan of organization and to gain proficiency in rapid use of this chapter, it may be helpful at this point to study the outline of Chapter 4 as given in Contents at the beginning of the book.

Two types of stress concentration factor are used in the literature dealing with members having holes:

$$K_{tg} = \frac{\sigma_{\max}}{\sigma} \qquad [52]$$

where K_{tg} = stress concentration factor based on gross stress
$\sigma_{\max}$ = maximum stress, at edge of hole
σ = applied stress, distant from hole

and

$$K_{tn} = \frac{\sigma_{\max}}{\sigma_{\text{nom}}} \qquad [53]$$

where K_{tn} = stress concentration factor based on net (nominal) stress
σ_{nom} = nominal (net) stress = $\sigma/(1 - a/w)$

where a = hole diameter
w = width of plate

From the foregoing,

$$K_{tn} = K_{tg}\left(1 - \frac{a}{w}\right) \qquad [54]$$

The significance of the two factors can be seen by referring to Fig. 112. Note that K_{tg} takes account of two effects: (1) increased stress due to loss of section; and (2) increased stress due to geometry. As the ligament becomes narrower, $a/b \to 1$, $K_{tg} \to \infty$. Note also that K_{tn} takes account of only one effect, increased stress due to geometry. As the ligament becomes narrower, $a/b \to 1$, it becomes in the limit, a uniform tension member, with $K_t = 1$.

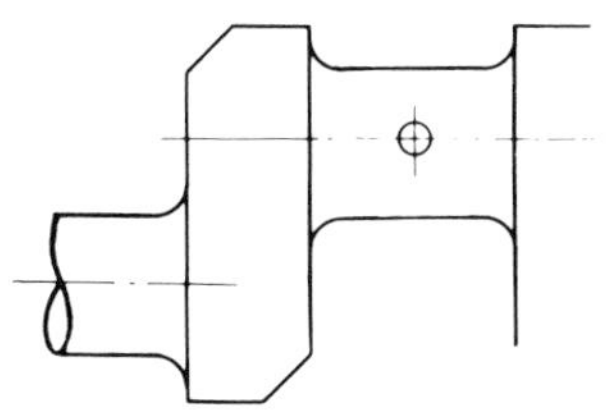

(A) OIL HOLE IN CRANKSHAFT (BENDING & TORSION)

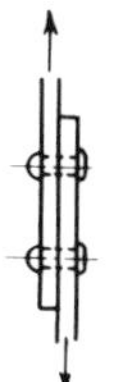

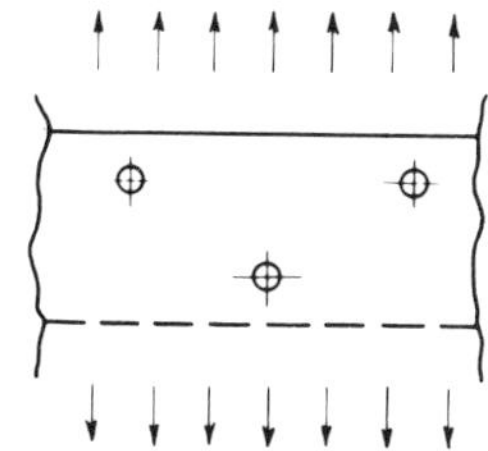

(C) RIVETED PLATES

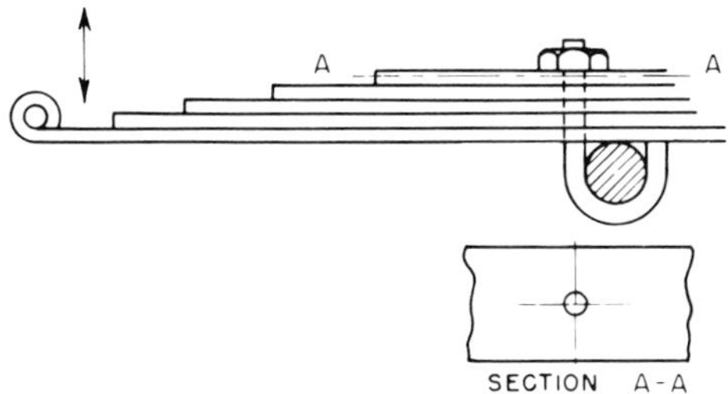

(B) CLAMPED LEAF SPRING (BENDING)

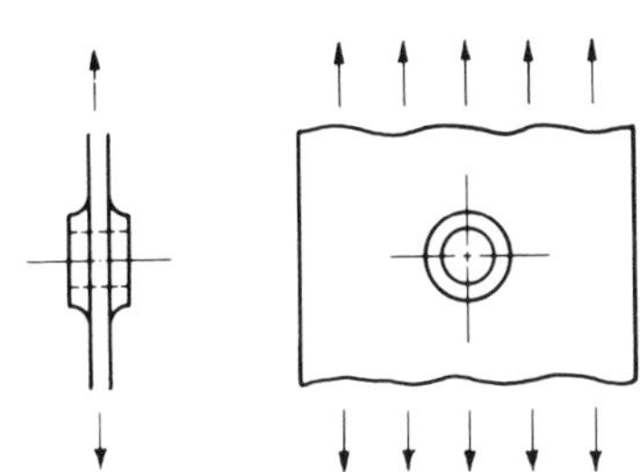

(D) HOLE WITH REINFORCING BEAD

FIG. 81

EXAMPLES OF PARTS WITH A TRANSVERSE HOLE

If one wishes only to obtain σ_{max}, the simplest procedure is to use K_{tg}. If one is concerned with stress gradient, as in certain fatigue problems, the correct factor to use is K_{tn}. This can be illustrated by the following example.

In Fig. 119, for uniaxial tension in the axial direction and $h/R = 0.1$, $K_{tg} = 10.8$ and $K_{tn} = K_{tg}\ (h/R) = 1.08$, the difference being that K_{tg} takes account of the loss of section. If the material is low carbon steel and the hole diameter is 0.048 in., we obtain $q = 0.7$ from Fig. 8. Assuming a fatigue strength (specimen without stress concentration) of 30,000 lb/in.^{2}, the estimated fatigue strength of the member with the hole (formula [12]) is

$$K_{tf} = q(K_{tn} - 1) + 1 = 0.7\ (1.08 - 1) + 1 = 1.056$$

$$\sigma_{tf} = \frac{\sigma_f}{K_{tf}} = \frac{30{,}000}{1.056} = 28{,}400 \text{ lb/in.}^2$$

This is based on nominal ligament stress and is equivalent to 2840 lb/in.^{2} based on gross stress.

If the estimate is obtained by use of K_{tg},

$$K_{tf} = q(K_{tg} - 1) = 0.7\ (10.8 - 1) + 1 = 7.86$$

$$\sigma_{tf} = \frac{\sigma_f}{K_{tf}} = \frac{30{,}000}{7.86} = 3820 \text{ lb/in.}^2$$

based on gross applied stress; or if a correction is made for loss of section,

$$\sigma_{tf} = \frac{\sigma_f}{K_{tf}(h/R)} = \frac{300{,}000}{7.86} = 38{,}200 \text{ lb/in.}^2$$

Obviously the last two results are erroneous; when q is applied it is necessary to use K_{tn}. Note that 28,400 lb/in.^{2} is close to full fatigue strength, 30,000 lb/in.^{2} This is because a narrow ligament is like a tension specimen, with small stress concentration.

In a plate with a transverse hole, the maximum tangential stress varies across the thickness of the plate, being lower at the surface and somewhat higher in the interior. For a hole of diameter D in a plate of thickness $\frac{3}{4}$ D, subjected to uniaxial tension and with $\nu = 0.3$, the maximum stress at the surface was 7% less than the thin sheet value of 3.0, whereas the stress at the midplane was less than 3% higher.[162] The authors[162] mention "the general assertion that factors of stress concentration based on two-dimensional analysis sensibly apply to plates of arbitrary thickness ratio."

In a later analysis[163] of an infinitely thick plate (semi-infinite body, mathematically) subjected to shear, and with $v = 0.3$, the maximum tangential stress at the plate surface of the hole was found to be 23% lower than the value of 4.0, and the corresponding stress at a depth of the hole radius was 3% higher.

In summarizing the foregoing discussion of stress variation in the thickness direction of a plate with a hole, it can be said that the usual two-dimensional stress concentration factors are sufficiently accurate for design application to plates of arbitrary thickness. However, this is of interest in mechanics of materials and failure analysis, since failure would be expected to start down in the hole, rather than at the surface, in the absence of other factors such as those due to processing or manufacturing.

(A) TENSION (COMPRESSION)—UNIAXIAL OR BIAXIAL; INTERNAL PRESSURE

(a) Single Circular Hole in an Infinite and in a Finite-Width Plate in Uniaxial Tension

The case of a circular hole in an infinite plate was solved by Kirsch;[164] the stress concentration factor is 3.

For the tension case of a finite-width plate with a circular hole (Fig. 86) K_t values were obtained by Howland[165] for a/w values up to 0.5. Photoelastic values[166,167] and analytical results[167a–169] are in good agreement.

For a row of holes in the axial direction with center distance/hole diameter = 3, and with $a/w = \frac{1}{2}$, Slot[95a] obtained good agreement with the Howland K_t value (Fig. 86) for the single hole with $a/w = \frac{1}{2}$.

From their photoelastic tests, Coker and Filon[169a] noted that as a/w approached unity the stress on the outside of the bar approached zero, which would correspond to $K_t = 2$. Wahl and Beeuwkes,[167] Koiter,[169b] and Heywood[170] also indicate $K_t = 2$ for $a/w \to 1$. Wahl and Beeuwkes, noting the difficulty in investigating this "thin ligament" region photoelastically, made a steel model 4.125 in. wide with a 4 in. diameter circular hole ($a/w = 0.97$); by means of strain measurements, they obtained $K_t = 1.92$, but they believed that for very small deformations the K_t value would be higher. They stated that "in case the hole diameter so closely approaches the width of the bar that the minimum section becomes an infinitely thin filament, then for any finite deformation, this filament may move inward sufficiently to allow for a uniform stress distribution, thus giving $K_t = 1$. For infinitely small deformations relative to the thickness of this filament, however, K_t may still be equal to 2." They noted that the steel model test indicated that the curve does not drop down to unity as fast as would appear from certain photoelastic tests.[170a] Since the inward movement varies with σ/E, the K_t would not drop to 1.0 as rapidly as with a plastic model;

E or σ/E for a steel model test may be of the order of 100 and 10 (respectively) times E and σ/E for a corresponding photoelastic test. Although the region $a/w \rightarrow 1$ does not have much significance from a design standpoint, the behavior in this region presents an interesting problem. A recent paper on the $a/w \rightarrow 1$ region was presented by Belie and Appl.[170b]

Heywood[170] has developed an empirical formula for K_t which covers the entire a/w range.

$$K_{tn} = 2 + \left(1 - \frac{a}{w}\right)^3 \qquad [54a]$$

The formula is in good agreement with the results of Howland[170] for $a/w < 0.3$ and is only about 1.5% lower at $a/w = \frac{1}{2}$ ($K_t = 2.125$ versus $K_t = 2.16$ for Howland). The Heywood formula is satisfactory for the usual design application, since in most cases a/w is less than $\frac{1}{3}$.

Note that the formula gives $K_{tn} = 2$ as $a/w \rightarrow 1$, which seems reasonable, owing to bending of the thin side ligaments. The Heywood formula, when expressed as K_{tg}, becomes

$$K_{tg} = \frac{2 + (1 - a/w)^3}{1 - a/w} \qquad [54b]$$

Factors for a circular hole near the edge of a semi-infinite plate in tension[171–173] are shown in Fig. 87. The load carried by the section between the hole and the edge of the plate is

$$P = \sigma ch \sqrt{1 - (r/c)^2} \qquad [55]$$

where σ = stress applied to semi-infinite plate
c = distance from center of hole to edge of plate
r = radius of hole
h = thickness of plate

In Fig. 87 the upper curve gives values of $K_{tg} = \sigma_B/\sigma$, where σ_B is the maximum stress at edge of the hole nearest the edge of the plate. Although the factor K_{tg} may be used directly in design, it was thought desirable to also compute K_{tn} based on *the load carried by the minimum net section*. The K_{tn} factor will be comparable with the stress concentration factors for other cases; this is important in analysis of experimental data as pointed out in the fourth paragraph of this chapter. Based on *actual load carried by the minimum section*, the average stress on the net section A–B is

$$\sigma_{\text{net } A-B} = \frac{\sigma ch \sqrt{1 - (r/c)^2}}{(c - r)h} = \frac{\sigma \sqrt{1 - (r/c)^2}}{1 - r/c}$$

$$K_{tn} = \frac{\sigma_B}{\sigma_{\text{net } A-B}} = \frac{\sigma_B}{\sigma} \frac{(1 - r/c)}{\sqrt{1 - (r/c)^2}} \qquad [56]$$

The case of a tension bar of finite width having an eccentrically located hole has been solved by Sjöstrom.[174] The semi-infinite plate values are in agreement with Fig. 87. Also, the special case of the centrally located hole is in agreement with the Howland solution (Fig. 86). The results of the Sjöstrom analysis are given as $K_{tg} = \sigma_{\max}/\sigma$ values in the upper part of Fig. 88. These values may be used directly in design. For reasons given in the fourth paragraph of this chapter, an attempt will be made in the following to arrive at approximate K_{tn} factors based on net section. When the hole is centrally located ($e/c = 1$, in Fig. 88), the load carried by section A–B is σch. As e/c is increased to infinity, the load carried by section A–B is, from [55], $\sigma ch\sqrt{1 - (r/c)^2}$. Assuming a linear relation between the foregoing end conditions results in the following expression for the load carried by section A–B:

$$P_{A-B} = \frac{\sigma ch \sqrt{1 - (r/c)^2}}{1 - (c/e)(1 - \sqrt{1 - (r/c)^2})} \qquad [57]$$

where the symbols are as indicated on Fig. 88. The stress on the net section A–B is

$$\sigma_{net\ A-B} = \frac{\sigma ch \sqrt{1 - (r/c)^2}}{h(c - r)\ [1 - (c/e)\ (1 - \sqrt{1 - (r/c)^2})]}$$

$$K_{tn} = \frac{\sigma_{\max}}{\sigma_{\text{net}}} = \frac{\sigma_{\max}}{\sigma} \frac{(1 - r/c)}{\sqrt{1 - (r/c)^2}} [1 - (c/e)\ (1 - \sqrt{1 - (r/c)^2})] \qquad [58]$$

It is seen from the lower part of Fig. 88 that this relation brings all the K_t curves rather closely together, so that for all practical purposes the curve for the centrally located hole ($e/c = 1$) is, under these circumstances, a reasonable approximation for all eccentricities.

(b) Single Circular Hole in an Infinite Plate in Biaxial Tension

This case is obtained by superposition of σ_1 and σ_2 uniaxial stress distributions. The K_t values are shown in Figs. 132 and 133 at $b/a = 1$:

$$K_t = 3 - \left(\frac{\sigma_2}{\sigma_1}\right) \qquad \left(\frac{\sigma_2}{\sigma_1}\right) \leqq 1 \qquad [59]$$

For σ_2 equal in magnitude to σ_1, $K_t = 2$ when both are of the same sign. When σ_1 and σ_2 are equal but of opposite sign, $K_t = 4$; this is equivalent to shear stresses $\tau = \sigma$ at 45° ($b/a = 1$ in Fig. 165).

(c) Single Hole in a Cylindrical or Spherical Shell

Circular Hole in a Cylindrical Shell (Tension or Internal Pressure)

Considerable analytical work has been done in recent years on the stresses in a cylindrical shell having a circular hole.[175–177] Stress concentration factors are given for tension in Fig. 89 and for internal pressure in Fig. 90; in both charts, factors for membrane (tension) and for total stresses (membrane plus bending) are given. (The torsion case is given in Section C, Fig. 172).

For pressure loading, the analysis assumes that the force representing the total pressure corresponding to the area of the hole is carried as a perpendicular shear force distributed around the edge of the hole; this is shown schematically in Fig. 90. Results are given as a function of a dimensionless parameter β:

$$\beta = \frac{\sqrt[4]{3(1 - \nu^2)}}{2} \left[\frac{r}{\sqrt{Rt}}\right] \qquad [60]$$

where R = mean radius of shell
t = thickness of shell
r = radius of hole
ν = Poisson's ratio

In Figs. 89 to 91, where $\nu = \frac{1}{3}$,

$$\beta = 0.639\left[\frac{r}{\sqrt{Rt}}\right] \qquad [61]$$

The analysis assumes a shallow, thin shell. Shallowness means a small curvature effect over the circumferential distance of the hole, which means a small r/R. Thinness, of course, implies a small t/R. The region of validity is shown in Fig. 91.

The physical significance of β can be evaluated by rearranging [61]:

$$\beta = 0.639\,\frac{(r/R)}{\sqrt{t/R}} \qquad [62]$$

For example, a 10-in.-diameter cylinder with a 1-in. hole would have a thickness of 0.082 in. for $\beta = \frac{1}{2}$, a thickness of 0.02 in. for $\beta = 1$, a thickness of 0.005 in. for $\beta = 2$, and a thickness of 0.0013 in. for $\beta = 4$.

Although $\beta = 4$ represents a very thin shell, Lind[178] states that in aerospace structures large values of β often occur; a formula is provided for the pressurized shell where β is large compared to unity.

The K_t factors in Figs. 89 and 90 are quite large for the larger values of β, corresponding to very thin shells. Referring to Fig. 91,

β	t/R
4	<0.003
2	<0.007
1	<0.015
$\frac{1}{2}$	<0.025

In the region of $\beta = \frac{1}{2}$, the K_t factors are not unusually large.

The theoretical results[175–177] are, with one exception, in good agreement. Experiments have been made by Houghton and Rothwell[179] and by Lekkerkerker.[175] Comparisons are made in Ref. 177; reasonably good agreement was obtained for pressure loading.[179] Poor agreement was obtained for the tension loading.[179] Referring to tests on tubular members[180] (Fig. 146), the results for $d_i/d_o = 0.9$ are in good agreement for tension loading. Photoelastic tests[180a] were made for the pressurized loading. Strain gage results[181] have been obtained for values of β up to 2 and agree reasonably well with Fig. 89.

The case of two circular holes has been analyzed;[181a,181b] the interference effect is is similar to that in an infinite plate, although the stress concentration factors are higher for the shell.[181b] The membrane and bending stresses for the single hole[181b] are in good agreement with the results[177] on which Fig. 89 and 90 are based.

Stress concentration factors have been obtained for the special case of a pressurized ribbed shell with a reinforced circular hole interrupting a rib.[181c]

Stresses around an elliptical hole in a cylindrical shell in tension have been determined.[182,182a,182b]

Circular or Elliptical Hole in a Spherical Shell (Internal Pressure)

A chart (Fig. 91*a*) based on K_t factors determined analytically[182c] covers openings varying from a circle to an ellipse with $b/a = 2$. The abscissa parameter $(b/2R)\sqrt{R/t}$ is the same as that used by the authors.[182c] Referring to Fig. 91*a*, the K_t values for the four b/a values in an infinite flat plate biaxially stressed are shown along the left-hand edge of the chart; the curves show the increase due to bending and shell curvature in relation to the flat plate values. Experimental results[182c] are in good agreement.

Application of the foregoing results to the case of an oblique nozzle is discussed in the same article.[182c]

(d) Single Symmetrically Reinforced Circular Hole in a Plate Uniaxially Stressed

The results of most interest for design application are the photoelastic test values of Seika and associates[183,184] In these tests, a plate 6 mm thick, with a hole 30 mm in diameter, had cemented symmetrically into the hole a stiffening ring of various thicknessess and containing various diameters of central hole. The width of plate was also varied. A constant in all tests was B/t = diameter of ring/thickness of plate = 5.

Figures 92 to 94 present $K_{tg} = \sigma_{\max}/\sigma$ values (where σ = gross stress) for various width ratios w/B = width of plate/diameter of ring. In all cases $\sigma_{\max}$ is located on the hole surface at 90° to the applied uniaxial tension. Only in the case of $w/B = 4$ was the effect of fillet radius investigated (Fig. 94).

For $w/B = 4$ and $B/t = 5$, Fig. 95 shows the net stress concentration factor, defined as follows:

$$P = \sigma A = \sigma_{\text{net}} A_{\text{net}}$$

$$K_{tn} = \frac{\sigma_{\max}}{\sigma_{\text{net}}} = \frac{\sigma_{\max}}{\sigma}\frac{A_{\text{net}}}{A} = \frac{K_{tg} A_{\text{net}}}{A}$$

$$K_{tn} = K_{tg}\left[\frac{(w - B)t + (B - a)h + (4 - \pi)\, r^2}{wt}\right]$$

$$K_{tn} = K_{tg}\left[\frac{[(w/B) - 1] + [1 - (a/B)]\,(h/t) + [(4 - \pi)/5]\,(r/t)^2}{w/B}\right] \qquad [63]$$

Note from Fig. 95 that the K_{tn} values are grouped closer together than the K_{tg} values. Also note that for efficient section use, the minimum h/t for $r = 0$ is about 1.5 and for the various fillet ratios the minimum h/t is about 3.

The $w/B = 4$ values are particularly useful in that they can be used without serious error for wide plate problems. This can be demonstrated by replotting the K_{tg} values in terms of a/w = diameter of hole/width of plate and extrapolating to zero (equivalent to an infinite plate) (see Fig. 96).

It will be noted from Fig. 94 that the lowest K_{tg} factor achieved by the reinforcements used in this series of tests was approximately 1.1 ,with $h/t \geqq 4$, $a/B = 0.3$, and $r/t = 0.83$. By deceasing a/B, that is, by increasing B relative to a, the K_{tg} factor can be brought to 1.0. For a wide plate, $K_{tg} = 3$ without reinforcement; to reduce this to 1, it is obvious that h/t should be 3 or somewhat greater.

An approximate solution was proposed by Timoshenko[185], based on curved bar theory. A comparison curve is shown on Fig. 94.

(e) Single Nonsymmetrically Reinforced Circular Hole in a Plate Uniaxially Stressed (Ring on One Side Only)

Photoelastic tests[186] have been made with $a/t = 1.833$ constant. Except for one series of tests, the volume of the reinforcement was made equal to the volume of the hole. In Fig. 97 the effect of varying the ring height (and the corresponding ring diameter) is shown for various a/w ratios. A minimum K_t value is reached at about $h/t = 1.45$ and $B/a = 1.8$.

As mentioned, the foregoing is for $V_R/V_H = 1$. If one wishes to lower K_t by increasing V_R/V_H, the following shape factor (see Fig. 82) should be maintained as an interim procedure:

$$C_S = \frac{B/2}{h - t} = \frac{a(B/a)}{2t\,[(h/t) - 1]} = 3.666 \qquad [64]$$

See sketch in Fig. 97 for symbols.

In Fig. 98 the abscissa scale is a/w; extrapolation is shown to $a/w = 0$, which provides intermediate values for relatively wide plates.

The curves shown are for zero fillet radius. A fillet radius 0.7 of the plate thickness reduces K_{tn} approximately 12%; for smaller radii the reduction is approximately linearly proportional to the radius (i.e., for $r/t = 0.35$, the reduction is approximately 6%).

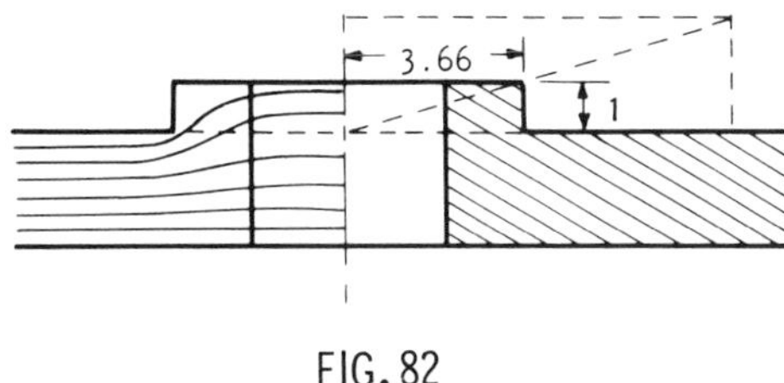

FIG. 82

(f) Single Symmetrically Reinforced Circular Hole in a Wide Plate Biaxially Stressed

Pressure vessels, turbine casings, deep sea vessels, aerospace devices, and other structures subjected to pressure require perforation of the shell by holes for introduction of working fluid, control mechanisms, windows, access for personnel, and so on. Although these designs involve complicating factors such as vessel curvature and closure details, some guidance can be obtained from the work on flat plates, especially for small openings, including those for leads and rods.

The state of stress in a pressurized spherical shell is biaxial, $\sigma_1 = \sigma_2$; for a circular hole in a biaxially stressed plate, $K_t = 2$. The state of stress in a cylindrical shell is $\sigma_2 = \sigma_1/2$; for the corresponding plate, $K_t = 2.5$. By proper reinforcement design, these factors can be reduced to 1, with a resultant large gain in strength. It has long been the practice to reinforce holes, but design information for achieving a specific K value, and in an optimum way, has not been available.

The reinforcement considered here is a ring type of rectangular cross section, symmetrically disposed on both sides of the plate (see Fig. 99). The results are for flat plates and are applicable for pressure vessels only when the diameter of the hole is small compared to the vessel diameter. The data should be useful in optimization over a fairly wide practical range.

A considerable number of theoretical analyses have been made.[187–196a] In most of the analyses it has been assumed that the edge of the hole, in an infinite sheet, is reinforced by a "compact" rim (one whose round or square cross-sectional dimensions are small compared to the diameter of the hole). Some of the analyses[187,188,196a] do not assume a compact rim. Most analyses are concerned with stresses in the sheet. Where the rim stresses are considered, they are assumed to be uniformly distributed in the thickness direction.

In pressure vessel design, the reinforcement is not "compact" (small compared to the diameter of the hole). The information of most interest in the present connection appears to be the strain gage tests made at NASA by Kaufman, Bizon, and Morgan[197] for reinforcement of circular holes of diameter eight times the thickness of the plate. A fillet radius was not used; the actual case, using a radius, would in some instances be more favorable. The authors[197] found that the degree of agreement with the theoretical results of Beskin[188] varied considerably with variation of reinforcement parameters.

Since in the strain gage tests the width of the plate is 16 times the hole diameter, it can be assumed that for practical purposes an invariant condition corresponding to an infinite plate has been attained. Since no correction has been made for the section removal by the hole, $K_{tg} = \sigma_{max}/\sigma$ is used.

The strain gage results[197] have been used to develop charts (Figs. 99 to 102) of a form more suitable for the types of problem encountered in turbine and pressure vessel design. This has involved interpolations in regions of sparce data; for this reason the charts are labeled as giving approximate stress concentration values. A more recent set of charts[197a] is based on the theoretical derivations of Gurney.[187]

In Figs. 99 to 104 the stress concentration factor $K_t = \sigma_{max}/\sigma_1$ has been used instead of $K_{te} = \sigma_{max}/\sigma_{eq}$; the former is perhaps more suitable where the designer wishes to obtain σ_{max} as simply and directly as possible. For $\sigma_1 = \sigma_2$ the two factors are the same; for $\sigma_2 = \sigma_1/2$, $K_{te} = (2/\sqrt{3})\ K_t = 1.157\ K_t$.

In drawing the charts (Figs. 99 to 102), it has been assumed that as D/d is increased an invariant condition is approached where $h/t = 2/K_t$ for $\sigma_1 = \sigma_2$; $h/t = 2.5/K_t$ for $\sigma_2 = \sigma_1/2$. It has also been assumed that for relatively small values of D/d, less than about 1.7, constant values of K_t are reached as h/t is increased; that is, the outermost part of the reinforcement in the thickness direction becomes stress-free (dead photoelastically) (see Fig. 60).

Figures 99 to 102 are plotted in terms of the two ratios defining the reinforcement proportions, D/d and h/t (see sketches in figures). When these ratios are not much greater than 1.0, the stress in the rim exceeds the stress in the plate; when the ratios are large, the reverse is true. Also note in Figs. 99 to 102 the crossover, or limit, line (dotted line denoted $K_{tg} \backsim 1$) dividing the two regions. Beyond the line (toward the upper right) the maximum stress in the reinforcement is approximately the applied stress. In the other direction (toward the lower left) the maximum stress is in the rim, with K_{tg} increasing from approximately 1 at the crossover line to a maximum (2 for $\sigma_1 = \sigma_2$ and 2.5 for $\sigma_2 = \sigma_1/2$) at the origin. It is useful to consider that the left-hand and lower straight edges of the diagrams (Figs. 99 to 102) also represent the above-maximum conditions; then one can readily interpolate an intermediate curve, as for $K_{tg} = 1.9$ in Figs, 99 and 100 or $K_{tg} = 2.3$ in Fig. 101 and 102.

The reinforcement variables D/d and h/t can be used to form two dimensionless ratios:

A/td = cross-sectional area of added reinforcement material/cross-sectional area of hole:

$$\frac{A}{td} = \frac{(D-d)\,(h-t)}{td} = \left(\frac{D}{d} - 1\right)\left(\frac{h}{t} - 1\right) \qquad [65]$$

V_R/V_H = volume of added reinforcement material/volume of hole:

$$\frac{V_R}{V_H} = \frac{(\pi/4)\,(D^2 - d^2)\,(h - t)}{(\pi/4)\,d^2t} = \left[\left(\frac{D}{d}\right)^2 - 1\right]\left[\frac{h}{t} - 1\right] \qquad [66]$$

The ratio A/td is used in pressure vessel design[198] in the form

$$A = Ftd \qquad [67]$$

where $F \leqq 1$. Although for certain specified design conditions[198] F may be less than 1, usually $F = 1$ in which case [67] may be stated

$$\frac{A}{dt} \geqq 1 \qquad [67a]$$

The ratio V_R/V_H is useful in arriving at optimum designs where weight is a consideration (aerospace devices, deep sea vehicles, etc.).

In Figs. 99 and 101 a family of A/td curves has been drawn, and in Figs. 100 and 102 a family of V_R/V_H curves has been drawn, each pair for $\sigma_1 = \sigma_2$ and $\sigma_2 = \sigma_1/2$ stress states. Note that there are locations of tangency between the A/td or V_R/V_H curves and the K_{tg} curves; these locations represent optimum design conditions, that is, for any given K_{tg} ratio, a minimum cross-sectional area or weight of reinforcement. The dot-and-dash curves provide the full range of optimum conditions. It is clear that K_{tg} does not depend solely on the reinforcement area A (as assumed in a number of analyses) but also on the shape (rectangular cross-sectional proportions) of the reinforcement.

In Figs. 103 and 104 the K_{tg} values corresponding to the dot-and-dash locus curves are presented in terms of A/td and V_R/V_H. Note that the largest gains in reducing K_{tg} are made at relatively small reinforcements and that to reduce K_{tg} from say 1.2 to 1.0 requires a relatively large volume of material.

The Pressure Vessel Code[198] formula [67a] may be compared with the values of Figs. 99 and 101, which are for symmetrical reinforcements of a circular hole in a flat plate. For $\sigma_1 = \sigma_2$ (Fig. 99), a value of K_{tg} of approximately 1 is attained at $A/td = 1.6$; for $\sigma_2 = \sigma_1/2$ (Fig. 101), a value of K_{tg} of approximately 1 is attained at A/td = approximately 3.

It must be borne in mind that the tests[197] were for d/t = diameter of hole/thickness of plate = 8. For pressure vessels d/t may be less than 8, and for aircraft windows d/t is greater than 8. If d/t is greater than 8, the stress distribution would not be expected to change markedly; furthermore, the change would be toward a more favorable distribution.[197]

However, for a markedly smaller d/t ratio, the optimum proportions corresponding to $d/t = 8$ are not satisfactory. To illustrate, Fig. 83*a* shows the approximately optimum

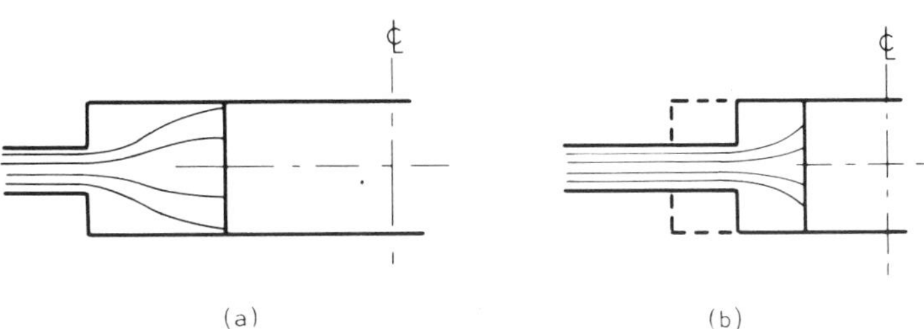

FIG. 83

proportions $h/t = 3$, $D/d = 1.8$ from Fig. 99 where $d/t = 8$. If we now consider a case where $d/t = 4$ (Fig. 83*b*) we see that the previous proportions $h/t = 3$, $D/d = 1.8$ are unsatisfactory for spreading the stress in the thickness direction. As an interim procedure, for $\sigma_1 = \sigma_2$ it is suggested that the optimum h/t value be found from Fig. 99 or 101 and D/d then be determined in such a way that the same reinforcement shape $[(D - d)/2]/[(h - t)/2]$ is maintained. For $\sigma_1 = \sigma_2$, the stress pattern is symmetrical, with the principal stresses in radial and tangential (circular) directions.

From Fig. 99, for $\sigma_1 = \sigma_2$, the optimum proportions for $K_{tg} \sim 1$ are approximately $D/d = 1.8$ and $h/t = 3$. Rearranging, the shape factor is

$$C_1 = \frac{D - d}{h - t} = \frac{[(D/d) - 1]d}{[(h/t) - 1]t} \qquad [68]$$

For $D/d = 1.8$, $h/t = 3$ and $d/t = 8$, the shape factor $C_1 = 3.2$.

For $\sigma_1 = \sigma_2$, suggested tentative reinforcement proportions for d/t values less than 8, which is the basis of Figs. 99 to 102, are found as follows:

$$\frac{h}{t} = 3 \qquad [69]$$

$$\frac{D}{d} = \frac{C_1 [(h/t) - 1]}{d/t} + 1 \qquad [70]$$

$$\frac{D}{d} = \frac{6.4}{d/t} + 1 \qquad [71]$$

For $d/t = 4$, $D/d = 2.6$ as shown by the dashed lines in Fig. 83*b*.

For $\sigma_2 = \sigma_1/2$ and $d/t < 8$, it is suggested as an interim procedure that $D/(h - t)$ for $d/t = 8$ be maintained for the smaller values of d/t (see [64], uniaxial tension):

$$C_2 = \frac{D}{h - t} = \frac{D/d}{(h/t) - 1}\left(\frac{d}{t}\right) \qquad [72]$$

For $D/d = 1.75$, $h/t = 5$, and $d/t = 8$, $C_2 = 3.5$.

For d/t less than 8:

$$\frac{h}{t} = 5 \qquad [72a]$$

$$\frac{D}{d} = \frac{14}{d/t} \qquad [73]$$

The foregoing formulas are based on $K_{tg} \sim 1$. If a higher value is used, for example, to obtain a more favorable V_R/V_H ratio (i.e., less weight), the same procedure may be followed to obtain the corresponding shape factors.

As an example of such a design "tradeoff," if, for $\sigma_2 = \sigma_1/2$, the rather high reinforcement thickness ratio of $h/t = 5$ is reduced to $h/t = 4$, we see from Fig. 101 that the K_{tg} factor increases from about 1.0 to only 1.17; also, from Fig. 104, the volume of reinforcement material is reduced 33% (V_R/V_H of 8.4 to 5.55).

The general formula for this example, based on [72], for d/t values less than 8, is

$$\frac{h}{t} = 4 \qquad [74]$$

$$\frac{D}{d} = \frac{10.5}{d/t} \qquad [75]$$

Similarly, for $\sigma_1 = \sigma_2$, if we accept $K_t = 1.1$ instead of 1.0, we see from Fig. 99 that $h/t = 2.2$ and $D/d = 1.78$; from Fig. 104, the volume of reinforcement material is reduced 41% (V_R/V_H of 4.4 to 2.6).

The general formula for this example, based on [70], for d/t values less than 8, is

$$\frac{h}{t} = 2.2 \tag{76}$$

$$\frac{D}{d} = \frac{6.25}{d/t} + 1 \tag{77}$$

The foregoing procedure may add more weight than is necessary for cases where $d/t < 8$, but from a stress standpoint, the procedure would be on the safe side.

The same procedure applied to d/t values larger than 8 would go in the direction of lighter, more "compact" reinforcements. However, owing to the planar extent of the stress distribution around the hole, it is not recommended to extend the procedure to relatively thin sheets, $d/t > 50$, such as in an airplane structure; Ref. 187 to 196 should be consulted for the applicable analysis.

Where weight is important, some further refinements may be worth considering. Due to the nature of the stress-flow lines, the outer corner region is unstressed, Fig. 84(*a*). An ideal contour would be similar to Fig. 84(*b*).

Kaufman, Bizon and Morgan[199] studied a reinforcement of triangular cross section, Fig. 84(*c*). The angular edge at *A* may not be practical, since a lid or other member often is used; a compromise shape may be considered, Fig. 84(*d*). Dhir and Brock[199a] present results for a shape like Fig. 84(*d*) and point out the large savings of weight that is attained.

Studies of a "neutral hole"[192] and of a variation of sheet thickness which results in uniform hoop stress for a circular hole in a biaxially stressed sheet[200] are worthy of further consideration for certain design applications (i.e., molded parts).

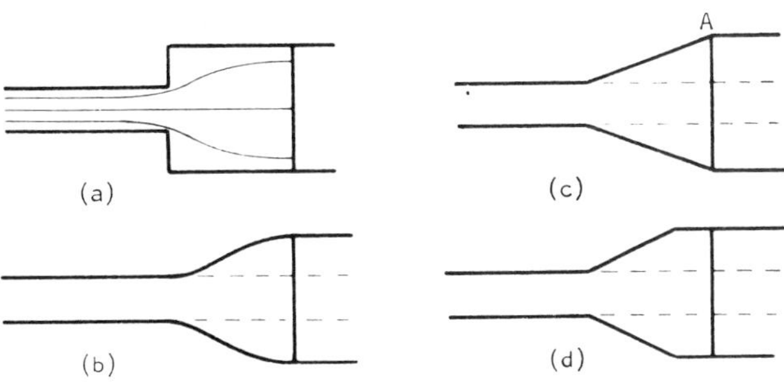

FIG. 84

(g) Circular Hole with Internal Pressure

The tensile stress at the hole edge is obtained by subtracting the pressure, p, from the biaxial tension stress distribution, $\sigma = -p$. Since $K_t = 2$ for biaxial tension in an infinite plate, $K_t = 1$ for the internal pressure case ($\sigma_{max} = -p$).

The case of a square plate with a pressurized central circular hole could be useful as a cross section of a construction conduit. The $K_t = \sigma_{max}/p$ factors[201,202] are given in Fig. 104*a*. Note that for the thinner walls ($r/a > 0.67$), the maximum stress occurs on the outside edge at the thinnest section; for the thicker walls ($r/a < 0.67$), the maximum stress occurs on the hole edge at the diagonal location (line through corners). As a matter of interest the Lamé values are shown, although for $r/a > 0.67$, these are not the maximum values. A check at r/a values of $\frac{1}{4}$ and $\frac{1}{2}$ with theoretical[203] factors shows good agreement.

A more recent analysis[201] covering a wide r/a range is in good agreement with Fig. 104*a*; by plotting $(K_t - 1)(1 - r/a)/(r/a)$ versus r/a, linear relations are obtained for small and large r/a values. Extrapolation is made to $(K_t - 1)(1 - r/a)/(r/a) = 2$ at $r/a \to 1$, as indicated by Koiter's analysis,[169b] and to 0 for $r/a \to 0$.

The upper curve (maximum) values of Fig. 104*a* are in reasonably good agreement with other recently calculated values.[95a]

For a pressurized circular hole near a corner of a large square plate,[204a] the maximum K_t values are quite close to the values for the square plate with a central hole.

For a hexagonal plate with a pressurized central circular hole,[95a] the K_t values are somewhat lower than the corresponding values of the upper curve of Fig. 104*a*, with $2a$ defined as the width across the sides of the hexagon.

For other cases involving a pressurized hole, see Sections A.n and A.r of this chapter.

For an eccentrically located hole in a circular plate, see Table 4 (Sec. A.n) and Fig. 126.

(h) Two Circular Holes of Equal Diameter in a Plate (Uniaxial or Biaxial Tension)

Stress concentration factors for two circular holes of equal diameter[205,206] are presented in Figs. 105 to 108. K_{tg} and K_{tn} are given in Figs. 106 and 108.

Referring to Fig. 106, if it is assumed that section B–B carries a load corresponding to the distance between center lines, namely, σbh, we obtain

$$K_{tnB} = (\sigma_{\max B}/\sigma)(1 - a/b) \qquad [78]$$

This corresponds to the light K_{tnB} lines of Figs. 106 and 108; it will be noted that near $b/a = 1$, the factor becomes low in value (less than 1 for the biaxial case). If the same basis is used as for [55] (i.e., actual load carried by minimum section), the heavy K_{tnB} curves of Figs. 106 and 108 are obtained. For this case

$$K_{tnB} = \frac{\sigma_{\max B}}{\sigma}\,\frac{(1 - a/b)}{\sqrt{1 - (a/b)^2}} \qquad [79]$$

Note that K_{tnB} approaches 1.0 as b/a approaches 1.0, the ligament tending to become, in effect, a uniformly stressed tension member. A photoelastic test by North[207] of a plate with two holes having $b/a = 1.055$ and uniaxially stressed transverse to the axis of the holes showed nearly uniform stress in the ligament.

In Fig. 105, $\sigma_{\max}$ is located at 90° for $b/a = 0$, 84.4° for $b/a = 1$, and approaches 90° as b/a is increased. In Fig. 107, $\sigma_{\max}$ for $\alpha = 0°$ is the same as in Fig. 105 ($\theta = 84.4°$ for $b/a = 1.055$, 89.8° for $b/a = 6$); $\sigma_{\max}$ for $\alpha = 45°$ is located at $\theta = 171.8°$ at $b/a = 1.055$ and decreases toward 135° with increasing values of b/a; $\sigma_{\max}$ for $\alpha = 90°$ is located at $\theta = 180°$.

Numerical determination[167a] of K_t for a biaxially stressed plate with two circular holes with $b/a = 2$ is in good agreement with the corresponding values of Ling[205] and Haddon.[206]

The case of two circular holes in shear is given in Section C, Fig. 167.

(i) Two Circular Holes of Unequal Diameter in a Plate (Uniaxial or Biaxial Tension)

Values for K_{tg} for uniaxial tension in an infinite plate have been obtained by Haddon;[206] his geometrical notation is used in Figs. 109 and 110 since this is more convenient in deriving the following expressions for K_{tn}.

For Fig. 109, to obtain K_{tn} exactly one must know the exact loading of the ligament in tension and bending and the relative magnitudes. (For two equal holes the loading is tensile, but its relative magnitude is not known.) In the absence of this information, two arbitrary methods were tried to determine if reasonable K_t values could be obtained.

Procedure A arbitrarily assumes (see Fig. 109) that the unit thickness load carried by s is $\sigma\ (R + r + s)$:

$$\sigma_{\text{net}}\, s = \sigma(R + r + s)$$

$$K_{tn} = \frac{\sigma_{\max}}{\sigma_{\text{net}}} = \frac{K_{tg}\, s}{R + r + s} \qquad [80]$$

Procedure B arbitrarily assumes, based on equation [55], that the unit thickness load carried by s is made up of two parts: $\sigma c_1\sqrt{1 - (R/c)^2}$ from the region of the larger hole, carried over distance $c_R = Rs/(R + r)$; $\sigma c_2\sqrt{1 - (r/c_2)^2}$ from the region of the smaller hole, carried over distance $c_r = rs/(R + r)$. In the foregoing $c_1 = R + c_R$; $c_2 = r + c_r$. For the small hole:

$$K_{tn} = \frac{K_{tg}}{\left(1 + \dfrac{(R/r) + 1}{s/r}\right)\sqrt{1 - \left(\dfrac{(R/r) + 1}{(R/r) + 1 + (s/r)}\right)^2}} \qquad [81]$$

Referring to Fig. 109, procedure A is not satisfactory in that K_{tn} for equal holes is less than 1 for values of s/r below 1. As s/r approaches 0, for two equal holes the ligament becomes essentially a tension specimen, so one would expect a condition of uniform stress ($K_{tn} = 1$) to be approached. Procedure B, which is also arbitrary, is not satisfactory below $s/r = \frac{1}{2}$, but it does provide K_{tn} values greater than 1 and for s/r greater than $\frac{1}{2}$ this curve has a reasonable shape, assuming $K_{tn} = 1$ at $s/r = 0$.

In Fig. 110, $\sigma_{\max}$ denotes the maximum tension stress; for $R/r = 5$ and for $s/r < 1$ the compression stress at $\theta = 180°$ is higher; for $R/r = 10$ and for $s/r = 3$ the compression stress at 180° is higher.

In Fig. 110, $\sigma_{\max}$ for $R/r = 5$ is located at $\theta = 77.8°$ at $s/r = 0.1$ and increases to 87.5° at $s/r = 10$; $\sigma_{\max}$ for $R/r = 10$ is located at 134.7° at $s/r = 0.1$, 90.3° at $s/r = 1$, 77.8° at $s/r = 4$, and 84.7° at $s/r = 10$. The $\sigma_{\max}$ locations for $R/r = 1$ are given in the discussion of Fig. 107; since $s/r = 2\,[(b/a) - 1]$ for $R/r = 1$, $\theta = 84.4°$ at $s/r = 0.1$, 89.8° at $s/r = 10$.

For biaxial tension, $\sigma_1 = \sigma_2$, K_{tg} values have been obtained by Salerno and Mahoney[208] (Fig. 111). The maximum stress occurs at the ligament side of the larger hole.

The foregoing results for variable spacing might also be useful in considering the approximate magnitude of the interaction effect of neighboring inclusions of different size.

(j) Single Row of Circular Holes in a Plate in Tension

For a single row of holes in an infinite plate, Howland[209] obtained $K_t = 1.62$ for $b/a = 2$ for transverse stressing. For axial stressing he obtained $K_t = 2.16$ for $b/a = 2$ and $K_t = 2.54$

for $b/a = 3.33$. Schulz[111] developed complete curves for the same cases as functions of a/b (Figs. 112 and 113). More recently calculated values of Meijers[210] are in agreement with the Schulz values. Slot[95a] found that for σ applied a distance $1.5a$ above and below the center line through the holes (Fig. 112), good agreement was found with the stress distribution for σ applied at infinity.

For the case of the plate stressed parallel to the axis of the holes (Fig. 113), when $b/a = 1$ we have an infinite row of edge notches. This portion of the curve (between $b/a = 0$ and 1) is in agreement with the work of Atsumi[109] on edge notches.

For a row of holes in the axial direction with center distance/hole diameter = 3, and with $a/w = \frac{1}{2}$, Slot[95a] obtained good agreement with the Howland K_t value (Fig. 86) for the single hole with $a/w = \frac{1}{2}$. A specific K_t value obtained by Slot[95a] for $b/a = 2$ and $a/w = \frac{1}{3}$ is in good agreement with the Schulz curves (Fig. 113).

The biaxially stressed case (Fig. 114), from the work of Hütter,[211] represents an approximation in the midregion of a/b. (Hütter's values for the uniaxial case with perpendicular stressing are inaccurate in the midregion.)

For a finite-width plate (strip), Schulz[111a] has provided K_t values for the dashed curves of Fig. 113. The K_t factors for $a/b = 0$ are the Howland[165] values; the K_t factors for the strip are in agreement with the Nisitani[241] values of Fig. 135 ($b/a = 1$).

K_t factors for a single row of holes in an infinite plate in transverse bending are given in Fig. 161, in shear in Fig. 167.

(k) Double Row of Circular Holes in a Plate in Tension

The double row of circular holes, staggered, is a configuration which is used in riveted and bolted joints. The K_{tg} values of Schulz[111] are presented in Fig. 115. Comparable values of Meijers[210] are in agreement.

In Fig. 115, as θ increases the two rows grow farther apart; at $\theta = 90°$ the K_{tg} values are the same as for a single row (Fig. 112). For $\theta = 0°$, a single row occurs with an intermediate hole in the span b; the curves for θ and 90° are basically the same, except that, as a consequence of the nomenclature of Fig. 115, b/a for $\theta = 0°$ is twice b/a for 90° for the same K_{tg}. The type of plot used in Fig. 115 makes it possible to obtain K_{tg} for intermediate values of θ by drawing θ versus b/a curves for various values of K_{tg}; in this way the important case of $\theta = 60°$, shown dashed on Fig. 115, was obtained.

In obtaining K_{tn}, based on net section, two relations are needed since for a given b/a the relation of net sections A–A and B–B depends on θ. (See Fig. 116.) For $\theta < 60°$, A–A is the minimum section and the following formula is used:

$$K_{tA} = \frac{\sigma_{\max}}{\sigma}\left[1 - 2\frac{a}{b}\cos\theta\right] \qquad [82]$$

For $\theta > 60°$, B–B is the minimum section and the formula is based on the net section in the row:

$$K_{tB} = \frac{\sigma_{\max}}{\sigma}\left[1 - \frac{a}{b}\right] \qquad [83]$$

At 60° these formulas give the same result. The K_t values in accordance with [82] and [83] are given in Fig. 116.

(l) Symmetrical Pattern of Circular Holes in a Plate (Uniaxial or Biaxial Tension)

Symmetrical triangular or square patterns of circular holes are used in heat exchanger and nuclear vessel design;[212] notations used in these fields will also be used in this section. The triangular pattern is shown in the sketch in Fig. 117; the square pattern is shown in the sketch in Fig. 119.

Ligament efficiency is defined as the minimum distance ($2h$) of solid material between two adjacent holes divided by the distance ($2R$) between the centers of the same holes, that is, ligament efficiency $= h/R$. It is assumed that the pattern repeats infinitely.

For the triangular pattern (Fig. 117), Horvay[213] obtained a solution for long and slender ligaments, taking account of tension, bending, and shear (Fig. 118). Horvay considers the results as not valid for h/R greater than 0.2. Photoelastic tests[214–216] have been made over the h/R range used in design. Computed values[210,216a,216b] are in general agreement but differ slightly in certain ranges; when this occurs the computed values[210] are used in Figs. 117 and 120. Subsequent computed values[95a] are in good agreement with Ref. 210.

For the square pattern, Bailey and Hicks[217] (with confirmation by Hulbert[218,222]) have obtained solutions for applied biaxial fields oriented in the square and diagonal directions (Figs. 119 and 120). Photoelastic tests by Nuno, Fujie, and Ohkuma[219] are in excellent agreement with mathematical results[217] for the square direction of loading but, as pointed out by O'Donnell,[220] are lower (than those of Ref. 217) for intermediate values of h/R for the diagonal direction of loading. Check tests by Leven[221] of the diagonal case resulted in agreement with the previous photoelastic tests[219] and pointed to a recheck of the mathematical solution of this case; this was done by Hulbert under PVRC sponsorship at the instigation of O'Donnell. The corrected results are given in a paper by O'Donnell[222] (Ref. 222 is essentially his Ref. 220 with the Hulbert correction). Later confirmatory results were obtained by Meijers.[210] Subsequent computed values[95a,216a] are in good agreement with Ref. 210.

The $\sigma_2 = -\sigma_1$ state of stress[220,214,217,222] shown in Fig. 120 corresponds to shear stress $\tau = \sigma$ at 45°. For shear at 0° see Section C, Fig. 168.

K_{tg} values are obtained for uniaxial tension and for various states of biaxiality of stress (Fig. 121) by superposition. Figure 121 is approximate (note that the lines are not straight, but they are so nearly straight that the curved lines drawn should not be significantly in error).

For uniaxial tension, Figs. 122, 122a, and 123 are charts for rectangular and diamond patterns.[210] Figures 169 and 170 (Section C) are corresponding charts for shear stress.

With the computer programs at hand, it would be relatively easy to compute K_t for other geometrical spacings for the cases given in this section and for other similar cases.

(m) Radially Stressed Circular Plate with a Ring of Circular Holes, with or without a Central Circular Hole

For the case of six holes in a circular plate loaded by six external radial forces, maximum K_{tg} values are given for four specific cases as shown in Table 2. K_{tg} is defined as $R_0\sigma_{max}/F$ for a plate of unit thickness. Good agreement has been obtained for the maximum K_{tg} values of Hulbert[223] and the corresponding photoelastic[224] and calculated[225] values of Buivol.

For the case of a circular plate with radial edge loading p and with a central hole and a ring of four or six holes, the maximum K_t values[226] are shown in Fig. 124 as a function of r/R_0 for two cases: all holes equal size ($r = R_i$), central hole ¼ of outside diameter of plate

Table 2
Maximum K_t for circular holes in circular plate loaded externally with concentrated radial forces

	Pattern	Spacing	Maximum K_t	Location	Reference
1	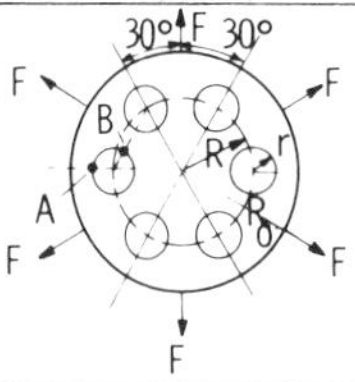	$R/R_0 = 0.65$ $r/R_0 = 0.2$	4.745	A	223, 224
		$R/R_0 = 0.7$ $r/R_0 = 0.25$	5.754	B	
2	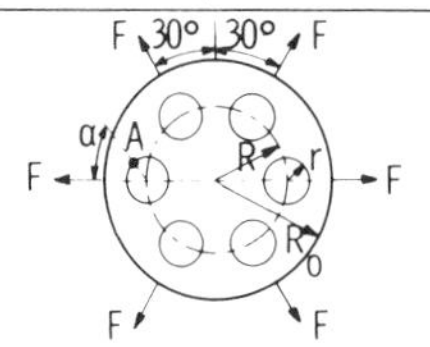	$R/R_0 = 0.65$ $r/R_0 = 0.2$	9.909	A $\alpha = 50°$	223, 224
		$R/R_0 = 0.6$ $r/R_0 = 0.2$	7.459	A $\alpha = 50°$	223, 224, 225

($R_1/R_0 = \frac{1}{4}$). Kraus[226] points out that with appropriate assumptions concerning axial stresses and strains, the results apply to both plane stress (plate) and plane strain (cylinder).

For the case of an annulus (flange) with internal pressure p and with a ring of holes located on the midradius of the flange ($R = 0.9R_0$), the maximum K_t values[227] are shown in Fig. 125 as a function of hole size and number of holes. K_t is defined as σ_{max} divided by σ_{nom}, the average tensile stress on the net radial section through a hole. In the paper[227] K_t factors are given for other values of R_i/R_0 and R/R_0.

(n) Plate with Circular Holes with Internal Pressure Only

To find the stresses for the case of internal pressure in the holes,[223,228] the pressure p is subtracted from the corresponding stresses in the biaxial stress case $\sigma_1 = \sigma_2 = -p$. For the biaxial case of a single hole in an infinite plate, $K_t = \sigma_{max}/\sigma = 2$ becomes $K_t = \sigma_{max}/p = 1$ for internal pressure only.

For two holes in an infinite plate, the $K_t = \sigma_{max}/p$ values for internal pressure only are found by subtracting 1.0 from the biaxial K_{tg} values of Fig. 108. For two specific spacings, Hulbert[223] has obtained K_t values (Table 3) which are in agreement with the results of the procedure given in the preceding sentence.

For an infinite row of circular holes, the $K_t = \sigma_{max}/p$ values for internal pressure only are found by subtracting 1.0 from the biaxial values of Fig. 114. For a specific spacing, Hulbert[223] has obtained a K_t value (Table 3) which is consistent with Fig. 114.

Maximum K_t values for specific spacings of other hole patterns are given for infinite plates in Table 3. Even for a large number of holes (19 in item 8 of Table 3), the K_t value cannot be expected to be the same as for an infinite network of holes, since the effective modulus E' of the 19-hole pattern is different from the surrounding material E. For the 19-hole pattern, $K_t = 1.652$ at A, Table 3, whereas for the infinite network, $K_t = 2.6$ (from Fig. 117, $K_t = 3.6$ at $h/R = \frac{1}{3}$, and for internal pressure $K_t = 3.6 - 1 = 2.6$).

Maximum K_t values for specific spacings of hole patterns in circular plates are given in Table 4. For the $r/R_0 = 0.5$ case of the single hole eccentrically located in a circular

Table 3
Maximum K_t for infinite plate with circular holes with internal pressure only

	Pattern	Spacing	Max. K_t	Location	Reference
1			1.0	All Around	223, 228
2	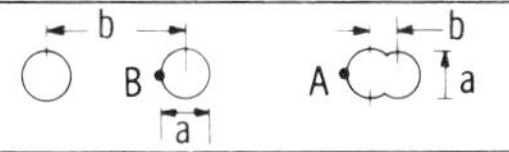	b/a = 3 b/a = ½	1.155 1.524	B A	223
3	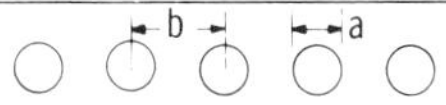	b/a = 2	1.85	A	223
4	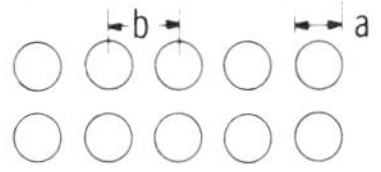	b/a = 1.5	2.83	A	223
5	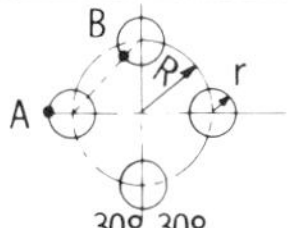	R/r = 3	1.16	A, B	223
6	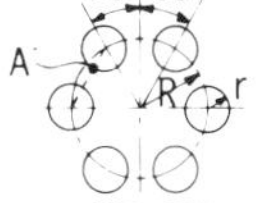	R/r = 3	1.658	A	223
7	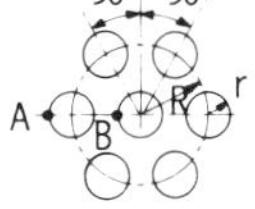	R/r = 3	1.393 1.249	A B	223
8	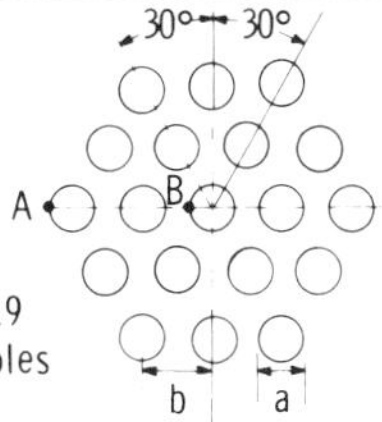 19 Holes	b/a = 1.5	1.652 1.182	A B	223
9	Infinite Triangular Pattern	b/a = 1.5	2.6	B	See Text

plate,[223,229] a sufficient number of eccentricities were calculated to provide Fig. 126. This general problem arises in evaluating piping which does not meet specifications with regard to eccentricity; unfortunately only the $r/R_0 = 0.5$ values have been calculated (Fig. 126). For zero eccentricity, the formula for an annulus[228] provides the K_t values (see Chapter 5, Section O; Fig. 210).

For a circular ring of three or four holes in a circular plate, Kraus[230] has obtained K_t for variable hole size (Fig. 127). For $r/R_0 = 0.2$, Hulbert[223] has obtained agreement using a different method. Hulbert[223] has also computed the specific four-hole case for $r/R_0 = 0.25$ (Table 4).

The case of a central hole and six holes at $R/R_0 = 0.6$ is given as Item 4 in Table 4. From Fig. 124, if we subtract 1.0 from the six-hole, $R_i = r$ case, we obtain 2.17, which is somewhat lower than the 2.278 value in Table 4, owing to the wider ligament corresponding to $R/R_0 = 0.625$.

With the computer programs available, it would be relatively easy to compute other geometrical spacings for the cases given in this section and for similar cases.

Table 4
Maximum K_t for circular plate with circular holes with internal pressure only

	Pattern	Spacing	Maximum K_t	Location	Reference
1	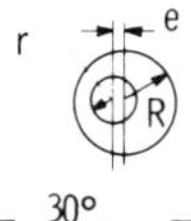	$r/R = 0.5$	See Fig. 126	See Fig. 126	223, 228, 229
2	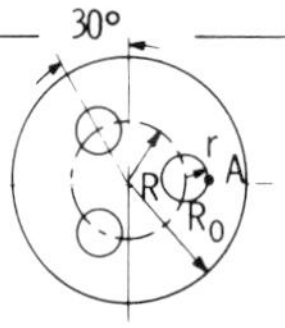	$R/R_0 = 0.5$ $r/R_0 = 0.2$	See Fig. 127	See Fig. 127	230
3	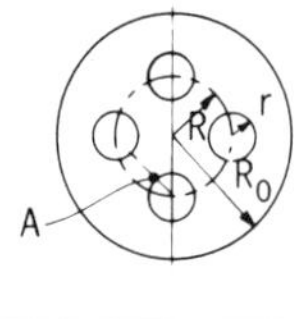	$R/R_0 = 0.5$ $r/R_0 = 0.2$	See Fig. 127	See Fig. 127	230
		$R/R_0 = 0.5$ $r/R_0 = 0.25$	2.45	A	223
4	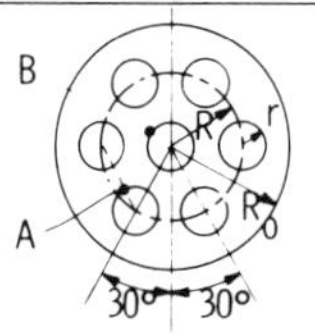	$R/R_0 = 0.6$ $r/R_0 = 0.2$	2.278 Pressure in All Holes 1.521 Pressure in Center Hole Only	A B	223

(o) Single Elliptical Hole in an Infinite Plate and in a Finite-Width Plate in Uniaxial Tension

The case of an elliptical hole in an infinite plate in tension was solved mathematically by Kolosoff[231] and by Inglis.[91] The stress concentration factor is given in Fig. 128:

$$K_t = 1 + \frac{2b}{a} \qquad [84]$$

where b = width perpendicular to applied stress (see Fig. 128)
a = width parallel to applied stress

The foregoing formula may also be written

$$K_t = 1 + 2\sqrt{t/r} \qquad [85]$$

where $t = b/2$
r = radius at intersection with axis perpendicular to applied stress

Formula [85] has been used as an approximation for the case of an elliptical notch of depth t and radius r. As can be seen from Fig. 15, K_t is higher for the notch than for the corresponding elliptical hole.

The "equivalent ellipse" concept is useful for hole geometries with the same t/r which can be enveloped by an ellipse, that is, an ovaloid, two circles connected by a slit, and so on (see Section A.t).

For an elliptical hole in a finite-width tension plate, the K_t values of Isida[169,232] are presented in Fig. 129; other analytical results[167a,233] are, with exception of one value,[233] in good agreement. Photoelastic results[233a] are lower. The remarks in Section A.a regarding extremely thin ligaments should be considered in connection with the $b/w \to 1$ condition.

Values for an elliptical hole near the edge of a semi-infinite plate[173] are given in Fig. 130. Isida[234] has also provided K_t factors for the special problem of an elliptic-sectioned tunnel as a function of the distance below the ground level.

(p) "Width Correction Factor" for a Cracklike Central Slit in a Tension Plate

For the very narrow ellipse approaching a crack (Fig. 131), a number of "finite-width correction" formulas have been proposed including the following: Dixon[235], Westergaard[100] and Irwin,[101] Brown and Srawley,[103] Fedderson,[236] and Koiter.[104] Correction factors have also been calculated by Isida.[237]

The Brown-Srawley formula[103] for $b/w < 0.6$:

$$\frac{K_{tg}}{K_{t\infty}} = 1 - 0.1\frac{b}{w} + \left(\frac{b}{w}\right)^2 \qquad [86]$$

$$\frac{K_{tn}}{K_{t\infty}} = \frac{K_{tg}}{K_{t\infty}}\left(1 - \frac{b}{w}\right)$$

The Fedderson formula:[236]

$$\frac{K_{tg}}{K_{t\infty}} = \left(\sec\frac{\pi}{2}\frac{b}{w}\right)^{1/2} \qquad [87]$$

The Koiter formula:[104]

$$\frac{K_{tg}}{K_{t\infty}} = \left[1 - 0.5\frac{b}{w} + 0.326\left(\frac{b}{w}\right)^2\right]\left[1 - \frac{b}{w}\right]^{-1/2} \qquad [88]$$

Formulas [86], [87], and [88] represent the ratios of stress-intensity factors; in the small-radius, narrow-slit limit, the ratios are valid for stress concentration.[105,106]

Formula [88] covers the entire b/w range from 0 to 1.0, (Fig. 131), with correct end conditions. Formula [86] is in good agreement for $b/w < 0.6$. Formula [87] is in good agreement[238] (generally less than 1% difference; at $b/w = 0.9$, less than 2%). Isida values are within 1% difference[238] for $b/w < 0.8$.

Photoelastic tests[235,239] of tension members with a transverse slit connecting two small holes are in reasonable agreement with the foregoing, taking into consideration the accuracy limits of the photoelastic test.

Figure 131 also provides factors for circular and elliptical holes (from Figs. 86 and 129).

Correction factors have also been developed[240] for an eccentrically located crack in a tension strip.

(q) Single Elliptical Hole in a Plate in Biaxial Tension

Referring to the sketch in Fig. 132, the stress σ_1 is perpendicular to the b dimension of the ellipse, regardless of whether b is larger or smaller than a. The abscissa scale (σ_2/σ_1) goes from -1 to $+1$; in other words, σ_2 is numerically equal to or less than σ_1.

The usual stress concentration factor, based on normal stresses, is defined as follows:

$$K_{tB} = \frac{\sigma_B}{\sigma_1} = 1 + \frac{2b}{a} - \frac{\sigma_2}{\sigma_1} \qquad [89]$$

$$K_{tA} = \frac{\sigma_A}{\sigma_1} = \frac{\sigma_2}{\sigma_1}\left[1 + \frac{2}{b/a}\right] - 1 \qquad [90]$$

These factors are shown in Fig. 132.

For $\sigma_1 = \sigma_2$:

$$K_{tB} = \frac{2b}{a} \qquad [91]$$

$$K_{tA} = \frac{2}{b/a} \qquad [92]$$

Setting [89] = [90], we find that the stresses at A and B are equal when

$$\frac{\sigma_2}{\sigma_1} = \frac{b}{a} \qquad [93]$$

The tangential stress is uniform around the ellipse for condition [93]. Relation [93] is shown by the dot-dash curve on Fig. 132; this condition occurs only for σ_2/σ_1 between 0 and 1, with the minor axis perpendicular to the major stress σ_1. Relation [93] provides a means of design optimization for elliptical openings. For example, for $\sigma_2 = \sigma_1/2$, $\sigma_A = \sigma_B$ for $b/a = \frac{1}{2}$, with $K_t = 1.5$. Keeping $\sigma_2 = \sigma_1/2$ constant, note that if b/a is *decreased*, K_{tB} becomes less than 1.5 but K_{tA} becomes greater than 1.5; for example, for $b/a = \frac{1}{4}$, $K_{tB} = 1$, $K_{tA} = 3.5$. If b/a is *increased*, K_{tA} becomes less than 1.5, but K_{tB} becomes greater than 1.5; for example, for $b/a = 1$ (circular opening), $K_{tA} = 0.5$, $K_{tB} = 2.5$.

One usually thinks of a circular hole as having the lowest stress concentration, but this depends on the stress system; we see that for $\sigma_2 = \sigma_1/2$ the maximum stress for a circular hole greatly exceeds that for the optimum ellipse ($b/a = \frac{1}{2}$) by a factor of $2.5/1.5 = 1.666$.

An airplane cabin is basically a cylinder with $\sigma_2 = \sigma_1/2$ where σ_1 = hoop stress, σ_2 = axial stress. This indicates that a favorable shape for a window would be an ellipse of height 2 and width 1, similar to that on the Viscount. The 2 to 1 factor is for a single hole in an infinite sheet; it should be added that there are other modifying factors, the proximity of adjacent windows, the stiffness of the structures, and so on. A round opening, which is often used, does not seem the most favorable design from a stress standpoint, although other considerations may enter.

It is sometimes said that what has a pleasing appearance often turns out to be technically correct. That this is not always so can be illustrated by the following. In the foregoing consideration of airplane windows, a stylist would no doubt wish to orient elliptical windows with long axis in the horizontal direction to give a "streamline" effect, as was done with decorative "portholes" in the hood of one of the automobiles of the past. The horizontal arrangement would be most unfavorable from a stress standpoint, where $K_{tB} = 4.5$ as against 1.5 oriented vertically.

The stress concentration factor based on maximum shear stress (Fig. 132) is defined as follows:

$$K_{ts} = \frac{\sigma_{\max}/2}{\tau_{\max}}$$

$$\tau_{\max} = \frac{\sigma_1 - \sigma_2}{2}, \quad \frac{\sigma_1 - \sigma_3}{2}, \quad \frac{\sigma_2 - \sigma_3}{2}$$

In a sheet $\sigma_3 = 0$:

$$\tau_{\max} = \frac{\sigma_1 - \sigma_2}{2}, \quad \frac{\sigma_1}{2}, \quad \frac{\sigma_2}{2}$$

For $0 \gtreqless (\sigma_2/\sigma_1) \lesseqgtr 1$:

$$K_{ts} = \frac{\sigma_{\max}/2}{\sigma_1/2} = K_t \qquad [94]$$

For $-1 \gtreqless (\sigma_2/\sigma_1) \lesseqgtr 0$:

$$K_{ts} = \frac{\sigma_{\max}/2}{(\sigma_1 - \sigma_2)/2} = \frac{K_t}{1 - (\sigma_2/\sigma_1)} \qquad [95]$$

Since σ_2 is negative, the denominator is greater than σ_1, resulting in a lower numerical value of K_{ts} as compared to K_t, as seen in Fig. 132. For $\sigma_2 = -\sigma_1$, $K_{ts} = K_t/2$.

The stress concentration factor based on equivalent stress is defined as follows:

$$K_{te} = \frac{\sigma_{\max}}{\sigma_{\text{eff}}}$$

$$\sigma_{\text{eff}} = \frac{1}{\sqrt{2}} \sqrt{(\sigma_1 - \sigma_2)^2 + (\sigma_1 - \sigma_3)^2 + (\sigma_2 - \sigma_3)^2}$$

For $\sigma_3 = 0$:

$$\sigma_{\text{eff}} = \frac{1}{\sqrt{2}} \sqrt{(\sigma_1 - \sigma_2)^2 + \sigma_1^2 + \sigma_2^2}$$

$$= \sigma_1 \sqrt{1 - (\sigma_2/\sigma_1) + (\sigma_2/\sigma_1)^2}$$

$$K_{te} = \frac{K_t}{\sqrt{1 - (\sigma_2/\sigma_1) + (\sigma_2/\sigma_1)^2}} \qquad [96]$$

K_{te} values are shown in Fig. 133.

For obtaining $\sigma_{\max}$, the simplest factor K_t is adequate. For mechanics of materials problems, the latter two factors, which are associated with failure theory are useful.

The condition $\sigma_2/\sigma_1 = -1$ is equivalent to pure shear oriented 45° to the ellipse axes; this case and the case where the shear stresses are parallel to the ellipse axes are discussed in Section C, Fig. 165.

Some values for biaxial stressing of a finite plate with an elliptical hole are noted in Ref. 233.

Stresses around an elliptical hole in a cylindrical shell in tension have been determined.[182,182a,182b]

Values for an elliptical hole in a pressurized spherical shell are presented in Fig. 91*a*.

(r) Elliptical Hole with Internal Pressure

Subtracting 1.0 from [91]:

$$K_t = 2\frac{b}{a} - 1 \qquad [97]$$

(s) Infinite Row of Elliptical Holes in an Infinite Plate and in a Finite-Width Plate

The results of Nisitani[241] are given in Fig. 134; the ordinate values are plotted as K_t/K_{to}, where $K_{to} = K_t$ for the single hole (Fig. 128). The results are in agreement with Schulz[111] for circular holes and are consistent with the results of Atsumi[109] for the semicircular notches. The effect of finite width[241] is shown in Fig. 135 as a ratio of K_{to} for the single hole (Fig. 129);

these results are in agreement with Schulz[111a] for circular holes (Fig. 113). The author[241] concludes that the interference effect is proportional to the square of the ratio of the major semiwidth of the ellipse over the distance between the centers of the holes.

Photoelastic tests[233] have been made of a 20- × 20-in. plate with two elliptical holes of major axis 1.5 in. and minor axis 0.75 in. The plate was subjected to various states of biaxial stress; the distance between the elliptical holes was varied to establish the degree of interference effect.

K_t factors for a row of elliptical holes in an infinite plate in transverse bending are shown in Fig. 163.

(t) Tension Plate with an Ovaloid (Slot Having Semicircular Ends); Two Holes Connected by a Slit; Equivalent Ellipse

The "equivalent ellipse" concept[242–244] is useful for the ovaloid and other openings such as two holes connected by a slit; if such a shape is enveloped (Fig. 84*a*) by an ellipse (same major width b and minimum radius r), the K_t values for the ovaloid and the ellipse will be very nearly the same (within 2%). K_t for the ellipse is obtained from [85].

For two tangential circular holes (Fig. 106), $K_t = 3.869$, as compared to the "equivalent ellipse" value of $K_t = 3.828$. The cusps resulting from the enveloping ellipse are, in effect, stress-free ("dead" photoelastically); a similar situation exists for two holes connected by a slit.

The round-cornered square hole oriented 45° to the applied uniaxial stress,[245] although not completely enveloped by the ellipse, is approximately represented by the "equivalent ellipse."

Previously published values for a slot with semicircular ends[246] are low compared with the K_t values for the elliptical hole (Fig. 129) and for a circular hole (Fig. 86). It is suggested that the values for the equivalent ellipse be used.

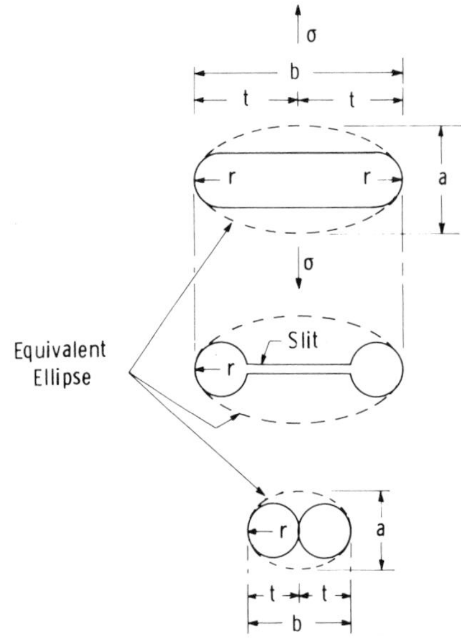

Fig. 84a – Equivalent ellipse

It has been shown[242] that the equivalent ellipse applies for tension but is not applicable for shear.

A photoelastic investigation[247] of a slot of constant $b/a = 3.24$ found the optimum elliptical slot end as a function of b/w, where w = plate width. The optimum shape was an ellipse of a/c about 3 (see Fig. 136) and this resulted in a reduction of K_t, from the value for the semicircular end, of about 22% at $b/w = 0.6$ to about 30% for $b/w = 0.2$, with an average reduction of about 26%. The authors state that the results may prove useful in the design of solid propellant grains. Although the numerical conclusions apply only to $b/a = 3.24$, it is clear that the same method of optimization may be useful in other design configurations with the possibility of significant stress reductions.

(u) Tension Plate with a Circular Hole with Opposite Semicircular Lobes

This particular geometry has been used for fatigue tests of sheet materials since the stress concentrator can be readily produced, with minimum variation from piece to piece.[248,249] Mathematical results[250] for an infinitely wide plate are shown in Fig. 137 and are compared with an ellipse of the same overall width and minimum radius (equivalent ellipse).

For a finite-width plate (Fig. 138), representative of a test piece, the following empirical formula was developed by Mitchell:[250]

$$K_t = K_{t\infty}\left[1 - \frac{b}{w} + \left(\frac{6}{K_{t\infty}} - 1\right)\left(\frac{b}{w}\right)^2 + \left(1 - \frac{4}{K_{t\infty}}\right)\left(\frac{b}{w}\right)^3\right] \qquad [98]$$

where $K_{t\infty}$ = K_t for infinitely wide plate (see Fig. 137)
b = width of hole plus lobes
w = width of plate

For $w = \infty$, $b/w = 0$, $K_t = K_{t\infty}$.

For $r/R \to 0$, $K_{t\infty}$ is obtained by multiplying K_t for the hole, 3.0, by K_t for the semicircular notch,[90] 3.065, resulting in $K_{t\infty} = 9.195$; the Mitchell[250] value is $3(3.08) = 9.24$.

For r/R greater than about 1.5, the middle hole is entirely swallowed up by the lobes; the resulting geometry, with middle opposite cusps, is the same as in Fig. 106 ($b/a < 2/3$).

For $r/R \to \infty$, a circle is obtained, $K_{t\infty} = 3$; relation [98] reduces to the Heywood formula:[170]

$$K_t = 2 + \left(1 - \frac{b}{w}\right)^3 \qquad [99]$$

Photoelastic tests by Cheng[251] confirm the accuracy of the Mitchell formula.

Miyao[251a] has solved the case for one lobe. The K_t values are lower, varying from 0% at $r/R \to 0$ to about 10% at $r/R = 1.0$ (ovaloid, see Fig. 139). Miyao[251a] also gives K_t values for biaxial tension.

(v) Infinite Plate with a Rectangular Hole with Rounded Corners (Uniaxial, Biaxial Stress)

The rectangular opening with rounded corners is often found in structures, such as ship hatch openings and airplane windows.

Referring to the sketch in Fig. 139, the stress σ_1 is perpendicular to the b dimension, regardless of whether b is larger or smaller than a.

Mathematical results, with specific data obtained by computer, have been published.[243,252–254] The top dashed curve of Fig. 139 is for the ovaloid (slot with semicircular ends). In Section A.t it was noted that for uniaxial tension and for the same t/r, the ovaloid and the equivalent ellipse are the same for all practical purposes. In the published results[243,253] for the rectangular hole, the ovaloid values are close to the elliptical values; the latter are used in Fig. 139 to give the ovaloid curve a smoother shape. The method of plotting[252,254] in Figs. 139 to 142 shows clearly the minimum K_t as a function of r/b.

A photoelastic test[255] in tension with $b/a = \frac{1}{4}$, $r/b = 0.1$, and $b/w = \frac{1}{4}$ gave $K_{tg} = 3.25$, which compares favorably with 3.38 in Fig. 139, considering that the latter corresponds to $w/b = \infty$.

For the $\sigma_2 = \sigma_1/2$ case (Fig. 140), the "equivalent stress" concentration factor (Mises) is $K_{te} = (2/\sqrt{3})K_t = 1.157K_t$; for $\sigma_2 = \sigma_1$ and $\sigma_2 = 0$, $K_{te} = K_t$.

Note that for $\sigma_1 = \sigma_2$ (Fig. 141), the square opening has the highest K_t for a given r/b.

The $\sigma_2 = -\sigma_1$ case (Fig. 142) corresponds to shear $\tau = \sigma$ at 45°. The (Mises) stress concentration factor is $K_{te} = (1/\sqrt{3})K_t = 0.577K_t$. The curve for the ovaloid, $r = a/2$, is not in agreement with elliptical hole values; as pointed out by Cox,[242] the equivalent ellipse concept is not valid in shear. For the case of shear stress in line with the elliptical hole axes see Section C, Fig. 166.

Commenting on Figs. 139 to 142 in general, it will be noted that for $b/a > 1$, either the ovaloid (slot with semicircular ends) represents the minimum K_t (see Fig. 141), or the rectangular hole with an intermediate optimum radius (r/b between 0 and $\frac{1}{2}$) represents the minimum K_t (Figs. 139, 140, 142).

A possible design problem is to select a shape of opening having a minimum K_t within rectangular limits a and b. In Fig. 143 the following shapes are compared: ellipse; ovaloid; rectangle with rounded corners (for radius giving minimum K_t). The data are from Figs. 132, 139, 140, and 141.

For the uniaxial case (top three dashed curves of Fig. 143) the ovaloid has a lower K_t than the ellipse for $b/a > 1$ and a higher value for $b/a < 1$. The K_t for the optimum rectangle is lower than (or equal to) the K_t for the ovaloid; it is lower than the K_t for the ellipse for $b/a > 0.85$, higher for $b/a < 0.85$.

One might think that a circular opening in a tension plate would have a lower maximum stress than a round-cornered square opening having a width equal to the circle diameter. From Figs. 139 and 143 it can be seen that *a square opening with corner radii of about a third* of the width has a lower maximum stress than a circular opening of the same width.* If one studies photoelastic results for notches and shoulder fillets, it is seen that there is a similar effect.

In the 1930s Richmond[255a] made photoelastic tests of round-cornered square holes in a tension plate and found a minimum K_t value less than the $K_t = 3$ value for the circular hole. The tests were made in connection with the tunnel problem; Mindlin[255b] in the discussion closure stated that Richmond's finding was a result of some importance that might be applied usefully in other fields of stress analysis and design.

The preceding remarks apply to the uniaxial tension case; they do not apply for $\sigma_1 = \sigma_2$. For $\sigma_2 = \sigma_1/2$, the optimum opening has only a slightly lower K_t.

The full line curves of Fig. 143, representing $\sigma_2 = \sigma_1/2$, the stress state of a cylindrical shell under pressure, show that the ovaloid and optimum rectangle are fairly comparable

*Minimum at $r/b = 0.37$; K_t is 6% lower than for the circular opening.

and that their K_t values are lower than the ellipse for $b/a > 1$ and $b/a < 0.38$, greater for $b/a > 0.38$ and $b/a < 1$.

Note that for $b/a = \frac{1}{2}$, K_t reaches the *low value of 1.5* for the ellipse. This was discussed in Section A.q in comparison to the circle and its relation to airplane windows. It is to be noted here that the ellipse is in this case superior to the ovaloid, $K_t = 1.5$ as compared to $K_t = 2.08$.

For the equal biaxial state, found in a pressurized spherical shell, the dot-dash curves of Fig. 143 show that the ovaloid is the optimum opening in this case, and gives a lower K_t than the ellipse (except, of course, at $b/a = 1$, where both become circles).

Referring to Fig. 143, it may be possible in some instances to obtain a lower K_t by use of a more complex shape (see Fig. 136); no systematic data of this kind are available. There are certain conditions where the minimum has been attained (see dot-dash curve, $K_{tA} = K_{tB}$, of Fig. 132 and relations [93]).

For a round-cornered square hole oriented 45° to the applied tension, Hirano[244] has shown that the "equivalent ellipse" concept (see Section A.t) is applicable.

(w) Finite-Width Tension Plate with Round-Cornered Square Hole

In comparing the K_t values of Isida[245] (for the finite-width strip) with Fig. 139, it appears that satisfactory agreement is obtained only for small values of b/w. As b/w, the hole width/plate width, increases, K_t increases in approximately the following way: $K_{tg}/K_{t\infty} = \backsim 1.01$, 1.03, 1.05, 1.09, 1.13, for $b/w = 0.1, 0.2, 0.3, 0.4$, and 0.5 respectively.

(x) Round-Cornered Equilateral Triangular Hole in an Infinite Plate under Various States of Tension

The triangular hole with rounded corners has been used as an airplane window shape and has also been used in certain architectural designs.

The stress distribution around a triangular hole with rounded corners has been studied by Savin.[229]

The K_t values for $\sigma_2 = 0$ (σ_1 only), $\sigma_2 = \sigma_1/2$, and $\sigma_2 = \sigma_1$ in Fig. 144 were determined by Wittrick[256] by a complex variable method using a polynomial transformation function for mapping the contour. The corner radius is not constant; r is the minimum radius, positioned symmetrically at the corners of the triangle. For $\sigma_2 = \sigma_1/2$, the equivalent stress concentration factor (Mises), $K_{te} = 2/\sqrt{3}\ K_t = 1.157\ K_t$; for $\sigma_2 = \sigma_1$ and $\sigma_2 = 0$, $K_{te} = K_t$. In Fig. 145, the K_t factors of Fig. 144 are replotted as a function of σ_2/σ_1.

(y) Slotted Rocket Grain under External Pressure

The results of photoelastic tests[256a] of models of solid propellant rocket grains under external pressure are shown in Fig. 145*a*. The effect of the number of "star points," n, (six in the sketch in Fig. 145*a*) on the maximum stress varies as $\sqrt[3]{n}$; this factor is included in the ordinate scale. K_t is arbitrarily defined as σ_{max}/p. For example, for eight star points, the K_t values are half of the ordinate values shown.

The foregoing results are in excellent agreement with an analytical solution by Wilson[256b] of the four-star configuration. Extensions[256c] of Fig. 145*a* have been made for arbitrary internal pressures and for the effects of an external case. Photoelastic tests[256d] have been made of a six-star model subjected to pressure or temperature changes.

The possibility of lowering stress concentration by use of a flat ellipse instead of the constant radius, r, at the slot end is indicated by Figs. 41 and 136 (see Chapter 2, Section B.f).

(z) Reactor Hole Pattern (Tension)

A photoelastic test of a tension plate with a nuclear reactor hole pattern[256e] (Fig. 84*b*) has been made; because of edge effects, the K_t values are unreliable.

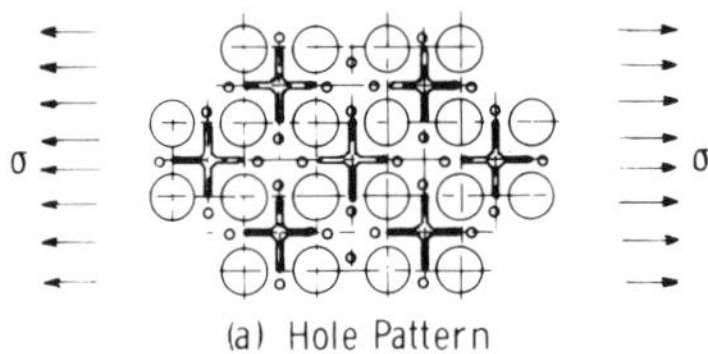

(a) Hole Pattern

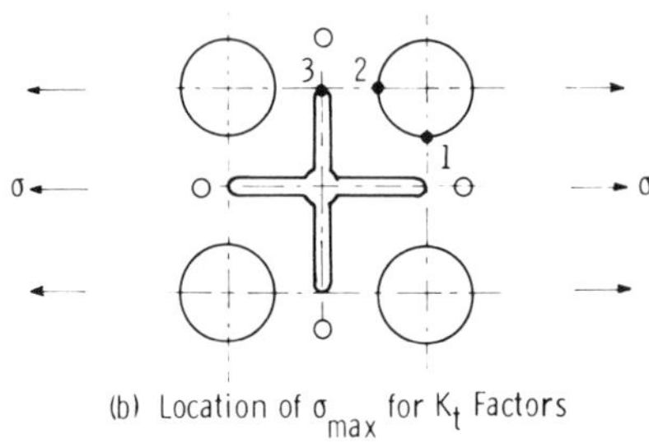

(b) Location of σ_{max} for K_t Factors

Fig. 84b—Hole pattern for nuclear reactor core plate

(z-1) Uniaxially Stressed Round Bar or Tube with a Transverse Circular Hole

The transverse hole in a round bar or tube occurs in engineering practice in lubricant and coolant ducts in shafts, in connectors for control or transmission rods, and in various types of tubular framework.

Stress concentration factors K_{tg} and K_{tn} are shown in Fig. 146. The results of Leven[257] and of Thum and Kirmser[258] for the solid shaft are in close agreement; the curve represents both sets of data. The results for the tubes are from British data.[259,260] The factors are defined as follows:

$$K_{tg} = \frac{\sigma_{\text{max}}}{\sigma_{\text{gross}}} = \frac{\sigma_{\text{max}}}{P/A_{\text{tube}}} = \frac{\sigma_{\text{max}}}{P/[(\pi/4)(d^2 - d_i^2)]} \qquad [100]$$

$$K_{tn} = \frac{\sigma_{\text{max}}}{\sigma_{\text{net}}} = \frac{\sigma_{\text{max}}}{P/A_{\text{net}}} = K_{tg}\frac{A_{\text{net}}}{A_{\text{tube}}} \qquad [100a]$$

Symbols are defined in Fig. 146.

The ratio A_{net}/A_{tube} has been determined mathematically;[261] the formulas and charts will not be repeated here (specific values can be obtained by dividing the Fig. 146 values of K_{tn} by K_{tg}). If the hole is sufficiently small relative to the shaft diameter, the hole may be considered to be of rectangular cross section:

$$\frac{A_{net}}{A_{tube}} = 1 - \frac{4\pi(a/d)\,[1 - (d_i/d)]}{1 - (d_i/d)^2} \qquad [101]$$

It can be seen from the bottom curves of Fig. 146 that the error due to this approximation is small below $a/d = 0.3$.

Thum and Kirmser[258] found that the maximum stress did not occur on the surface of the shaft but at a small distance inside the hole on the surface of the hole; this was later corroborated by other investigators.[257,259] The σ_{max} value used in developing Fig. 146 is the maximum stress inside the hole. No factors are given for the somewhat lower stress at the shaft surface; if these factors are of interest, they can be found in the references.[257–259]

(z-2) Round Pin Joint in Tension

The case of a pinned joint in an infinite plate has been solved mathematically by Bickley.[262] The finite-width case has been solved by Knight[263] (plate width equal twice hole diameter a) and by Theocaris[263a] for $a/w = 0.2$ to 0.5. Experimental results (strain gage or photoelastic) have been obtained by Coker and Filon,[264] Stoltenberg,[265] Frocht and Hill,[266] Jessop, Snell, and Holister,[266a] and Cox and Brown.[266b]

Two methods have been used in defining K_{tn}:

Nominal stress based on net section:

$$\sigma_{na} = \frac{P}{(w - a)h}$$

$$K_{tna} = \frac{\sigma_{max}}{\sigma_{na}} = \sigma_{max}\frac{(w - a)h}{P} \qquad [102]$$

Nominal stress based on bearing area:

$$\sigma_{nb} = \frac{P}{ah}$$

$$K_{tnb} = \frac{\sigma_{max}}{\sigma_{nb}} = \frac{\sigma_{max}\,ah}{P} \qquad [103]$$

Note that

$$\frac{K_{tna}}{K_{tnb}} = \frac{1}{a/w} - 1 \qquad [103a]$$

In Fig. 147 the K_{tnb} curve corresponds to the Theocaris[263a] values from $a/w = 0.2$ to 0.5; the values of Frocht and Hill[266] and Cox and Brown[266b] are in good agreement with Fig. 147 (slightly lower). From a/w 0.5 to 0.75 the foregoing 0.2 – 0.5 curve is extended to be consistent with the Frocht and Hill values. The resulting curve is for joints where m/w is 1.0 or greater; for $m/w = 0.5$ the K_{tn} values are somewhat higher.[266b,266c]

From [103a], $K_{tna} = K_{tnb}$ at $a/w = \frac{1}{2}$. It would seem more logical to use the lower (full line) branches of the curves on Fig. 147; since in practice a/w is usually less than $\frac{1}{4}$ this means that [103], based on bearing area, generally is used.

Figure 147 is for closely fitting pins. The K_t factors are increased by clearance in the pin fit. For example, at $a/w = 0.15$, K_{tnb} values[266b] of approximately 1.1, 1.3, and 1.8 were obtained for clearances of 0.7, 1.3, and 2.7%, respectively.

The effect of interference fits is to reduce the stress concentration factor.[266a,266b]

A joint having an infinite row of pins has been analyzed.[266d] It is assumed that the plate is thin (two-dimensional case), that there are no friction effects, and that the pressure on the hole wall is distributed as a cosine function over half of the hole. The stress concentration factors (Fig. 147a) have been recalculated from Ref. 266d to be related to $\sigma_{nom} = F/a$ rather than the mean peripheral pressure, in order to be defined in the same way as in Fig. 147. It is seen from Fig. 147*a* that decreasing e/a from a value of 1.0 results in a progressively increasing stress concentration factor. Also, as in Fig. 147, increasing a/b or a/w results in a progressively increasing stress concentration factor.

The end pins in a row carry a relatively greater share of the load; the exact distribution depends on the elastic constants and the joint geometry.[267]

(z-3) Inclined Round Hole in an Infinite Plate Subjected to Various States of Tension

The inclined round hole is found in oblique nozzles and control rod holes in nuclear and other pressure vessels.

The curve for uniaxial stressing and $\nu = 0.5$, second curve from top of Fig. 148 (which is for an inclination of 45°), is based on the photoelastic tests of Leven,[268] Daniel,[269] and McKenzie and White[270] and the strain gage tests of Ellyin.[271] The K_t factors,[268,270,271] adjusted to the same K_t definition (to be explained in the next paragraph) for $t/a \backsim 0.5$ are in good agreement (K_t in Ref. 269 is for $t/a = 2.4$). Theoretical K_t factors[272] are considerably higher than the experimental factors as the angle of inclination increases; however, the theoretical curves are used in estimating the effect of Poisson's ratio and in estimating the effect of state of stress. At $t/a \to 0$, the $K_{t\infty}$ values are for the corresponding ellipse (Fig. 128). For t/a large, the $K_{t\infty}$ values at the midsection are for a circular hole ($b/a = 1$ in Fig. 132); this result is a consequence of the flow lines in the middle region of a thick plate taking a direction perpendicular to the axis of the hole. For uniaxial stress σ_2, the midsection $K_{t\infty}$ is the maximum value; for uniaxial stress σ_1, the surface $K_{t\infty}$ is the maximum value.

For design use, it is desirable to start with a factor corresponding to infinite width and then have a method of correcting this to the b/w ratio involved in any particular design (b = major width of surface hole; w = width of plate). This can be done, for design purposes, in the following way. For any inclination, the surface ellipse has a corresponding b/a ratio; in Fig. 129, we obtain K_{tn}, K_{tg}, and $K_{t\infty}$ for the b/w ratio of interest ($K_{t\infty}$ is the value at $b/w \to 0$). Ratios of these values were used to adjust the experimental values to $K_{t\infty}$ in Figs. 148 and 149. In design the same ratio method is used in going from $K_{t\infty}$ to the K_t corresponding to the actual b/w ratio.

In Fig. 149 the effect of inclination angle θ is given. The $K_{t\infty}$ curve is based on the photoelastic K_{tg} values of McKenzie and White[270] adjusted to $K_{t\infty}$ as described above. The curve is for $t/a = 0.533$, corresponding to the flat peak region of Fig. 148. The effect of Poisson's ratio is estimated by use of the ratio of values given in Fig. 11 (for $t/a = 0.6$) of Ref. 272.

For uniaxial stress σ_1 in plates, the maximum stress is located at A, Fig. 148. An attempt to reduce this stress by rounding the edge of the hole with a contour radius $a/2$ produced the surprising result[269] of increasing the maximum stress (for $t/a = 2.4$, 30% higher for $\theta = 45°$, 50% higher for $\theta = 60°$). The maximum stress was located at the point of tangency of the contour radius with a line perpendicular to the plate surfaces. The stress increase has been explained[269] by the stress concentration due to the egg-shaped cross section in the horizontal plane. For $\theta = 75.5°$ and $t/a = 0.533$, it was found[270] that for $r/a < 1/9$, a small decrease in stress was obtained by rounding the corner, but above $r/a = 1/9$, the stress increased rapidly, which is consistent with the $a/2$ result.[269]

Strain gage tests were made by Ellyin and Izmiroglu[272a] on 45° and 60° oblique holes in 1-in. thick steel plates subjected to tension. The effects of rounding the corner A (Fig. 148) and of blunting the corner with a cut perpendicular to the plate surface were evaluated. In most of the tests $t/a = \sim 0.4$. For $\theta = 45°$ a small decrease in maximum stress was obtained in the region $r/t < 0.2$, but for $r/t > 0.2$ the maximum stress was increased by rounding.

It is difficult to compare the various investigations of plates with an oblique hole having a rounded corner because of large variations in t/a; also the effect depends on r/t and r/a. Leven[268] has obtained a 25% maximum stress reduction in a 45° oblique nozzle in a pressure vessel model by blunting the acute nozzle corner with a cut perpendicular to the vessel axis. From a consideration of flow lines it appears that the stress lines would not be as concentrated for the vertical cut as for the "equivalent" radius.

(z-4) Pressure Vessel Nozzle (Reinforced Cylindrical Opening)

A nozzle in pressure vessel and nuclear reactor technology denotes an integral tubular opening in the pressure vessel wall (see Fig. 85). Extensive strain gage[272b] and photoelastic tests[273,273a] have been made of various geometric reinforcement contours aimed at reducing stress concentration; Fig. 85 is an example of a resulting "balanced" design.[274] Stress concentration factors for oblique nozzles (nonperpendicular intersection) are also available.[275]

(z-5) Spherical or Ellipsoidal Cavity

Stress concentration factors for cavities are useful in evaluating the effects of porosity in materials.[276]

The stress distribution around a cavity having the shape of an ellipsoid of revolution has been obtained by Neuber[277] for various types of stressing. The case of tension in the direction of the axis of revolution is shown in Fig. 150; note that the effect of Poisson's ratio, ν, is relatively small. It is seen from Fig. 150 that high K_t factors are obtained as the ellipsoid becomes thinner and approaches the condition of a disk-shaped transverse crack.

The case of stressing perpendicular to the axis has been solved for an internal cavity having the shape of an elongated ellipsoid of revolution.[278] From Fig. 151 it is seen that for a circularly cylindrical hole ($b = \infty$, $a/b \to 0$) the value of $K_t = 3$ is obtained and that this reduces to $K_t = 2.05$ for the spherical cavity ($a/b = 1$). If we now consider an elliptical

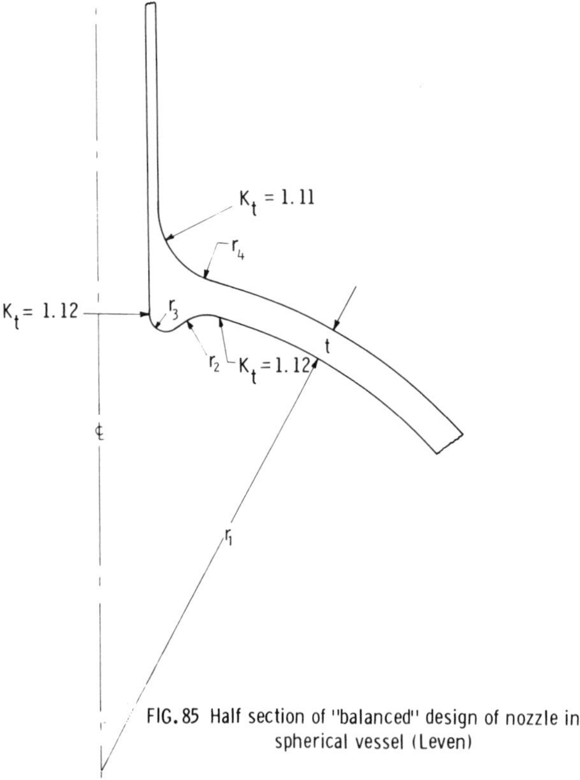

FIG. 85 Half section of "balanced" design of nozzle in spherical vessel (Leven)

shape, $b/a = 3$, $(t/r = 9)$, from [84], [85] (Figs. 128 and 150) we find that for a cylindrical hole of elliptical cross section $K_t = 7$; for a circular cavity of elliptical cross section (Fig. 150), $K_t = 4.6$; and for an ellipsoid of revolution (Fig. 151), $K_t = 2.69$. The order of the factors quoted above seems reasonable if one considers the course streamlines must take in going around the shapes under consideration.

Sternberg and Sadowsky[279] studied the "interference" effect of two spherical cavities in an infinite body subjected to hydrostatic tension. With a space of one diameter between the cavities, the factor was increased less than 5%, $K_t = 1.57$ as compared to infinite spacing (single cavity), $K_t = 1.50$. This compares with approximately 20% for the analogous plane problem (circular holes in biaxially stressed plate) of Fig. 108.

In Fig. 152, stress concentration factors K_{tg} and K_{tn} are given for tension of a circular cylinder with a central spherical cavity.[280] The value for the infinite body is [281]

$$K_t = \frac{27 - 15\nu}{14 - 10\nu} \qquad [104]$$

where ν = Poisson's ratio. For $\nu = 0.3$, $K_t = 2.045$.

For a large spherical cavity in a round tension bar, Ling shows $K_t = 1$ for $a/w \to 1$; Koiter[169b] obtains the following for $a/w \to 1$:

$$K_t = (6 - 4\nu)\,\frac{1 + \nu}{5 - 4\nu^2} \qquad [104a]$$

The remarks in Section A.a regarding extremely thin ligaments should be considered in connection with the $a/w \to 1$ condition.

In Fig. 152, a curve for K_{tg} is given for a biaxially stressed plate with a central spherical cavity.[282] For infinite thickness[281]

$$K_{tg} = \frac{12}{7 - 5\nu} \quad [105]$$

This value corresponds to the pole position on the spherical surface perpendicular to the plane of the applied stress.*

The curve for the flat plate of Fig. 152 was calculated for $\nu = \frac{1}{4}$. The value for $a/w = 0$ and $\nu = 0.3$ is also shown.

For the corresponding case of a semi-infinite body,[282a] with Poisson's ratio of $\frac{1}{4}$, r = radius of circular cavity, and c = distance from center of cavity to surface: $K_t = 2.21$ for $r/c = 0.5$, 2.32 for $r/c = 0.6$, 2.51 for $r/c = 0.7$, 2.76 for $r/c = 0.8$, and 3.3 for $r/c = 0.9$. Uniaxial photoelastic results[282b] are close to the foregoing results.

Calculations have also been made for the case of the torsion cylinder with a central spherical cavity.[283]

The effect of spacing for a row of "disk-shaped" ellipsoidal cavities[241] is shown in Fig. 153 in terms of K_t/K_{t0}, where $K_{t0} = K_t$ for the single cavity (Fig. 150). These results are for Poisson's ratio = 0.3. Nisitani[241] concludes that the interference effect is proportional to the cube of the ratio of the major semiwidth of the cavity over the distance between the centers of the cavities. In the case of holes in plates (Section A.s) the proportionality was as the square of the ratio.

The results for two cavities[284] are also shown (for Poisson's ratio $\nu = \frac{1}{4}$) in Fig. 153.

Some results for a row of spherical cavities in a round bar in tension[284a] are shown in Fig. 152.

For an infinite body containing a circular cylindrical cavity of length equal to diameter, with a corner radius equal to $\backsim\frac{1}{4}$ diameter and with the cylinder axis in line with the applied tension,[284b] $K_t = \backsim 2$, with σ_{max} occurring at an angle of $\backsim 13°$ measured from the beginning of the corner radius on the cylinder.

(z-6) Spherical or Ellipsoidal Inclusion

The evaluation of the effect of inclusions on the strength of materials, especially in fatigue and brittle fracture, is an important consideration in engineering technology.

The stresses around an inclusion have been analyzed by considering that the hole or cavity is filled with a material having a different modulus of elasticity, E', and that adhesion between the two materials is perfect. Donnell[285] has obtained relations for cylindrical inclusions of elliptical cross section in a plate for E'/E varying from 0 (hole) to ∞ (rigid inclusion). Donnell found that for Poisson's ratio $\nu = 0.25$ to 0.3, the plane stress and plane strain values were sufficiently close that he could use a formulation giving a value between the two cases (approximation differs from exact values 1.5% or less). Edwards[286] extended the work of Goodier and Donnell to cover the case of the inclusion having the shape of an ellipsoid of revolution.

Curves for E'/E for $\frac{1}{4}$, $\frac{1}{3}$, and $\frac{1}{2}$ are shown in Figs. 128 and 151; these ratios are in the range of interest in considering the effect of silicate inclusions in steel. It is seen that the hole or cavity represents a more critical condition than a corresponding inclusion of the type mentioned.

*In Ref. 276, the K_t value is for the plane of stressing and is not the maximum.

For a rigid spherical inclusion, $E'/E = \infty$, in an infinite member, Goodier[287] obtained the following relations for uniaxial tension.

For the maximum adhesion (radial) stress at the axial (pole) position:

$$K_t = \frac{2}{1+\nu} + \frac{1}{4-5\nu} \qquad [106]$$

For $\nu = 0.3$, $K_t = 1.94$.

For the tangential stress at the equator (position perpendicular to the applied stress):

$$K_t = \frac{\nu}{1+\nu} - \frac{5\nu}{8-10\nu} \qquad [106a]$$

For $\nu = 0.3$, $K_t = -.069$.

For $\nu = 0.2$, $K_t = 0$; for $\nu > 0.2$, K_t is negative, that is, the tangential stress is compressive. The same results have been obtained[284b] by using a different method.

The case of a rigid circular cylindrical inclusion may be useful in the design of plastic members and concrete structures reinforced with steel wires or rods. Goodier[287] has obtained the following plane strain relation for a circular cylindrical inclusion, with $E'/E = \infty$:

$$K_t = \frac{1}{2}\left(3 - 2\nu + \frac{1}{3-4\nu}\right) \qquad [106b]$$

For $\nu = 0.3$, $K_t = 1.478$.

Studies have been made of the stresses in an infinite body containing a circular cylindrical inclusion of length one and two times the diameter, with a corner radius and with the cylinder axis in line with the applied tension.[284b] The results may provide some guidance for a design condition where a reinforcing rod ends within a concrete member. For a length/diameter ratio of 2 and a corner radius/diameter ratio of ¼, the following K_{ta} values were obtained ($K_{ta} = \sigma_a/\sigma$ = max normal stress/applied stress): $K_{ta} = 2.33$ for $E'/E = \infty$; $K_{ta} = 1.85$ for $E'/E = 8$; $K_{ta} = 1.63$ for $E'/E = 6$. For a length/diameter ratio $= 1$, K_{ta} does not vary greatly with corner radius/diameter ratio varying from 0.1 to 0.5 (spherical, $K_t = 1.94$); below $r/d = 0.1$, K_{ta} rises rapidly ($K_{ta} = 2.85$ at $r/d = 0.05$). Defining $K_{tb} = \tau_{max}/\sigma$ for the bond shear stress, the following values were obtained: $K_{tb} = 2.35$ at $r/d = 0.05$; $K_{tb} = 1.3$ at $r/d = ¼$; $K_{tb} = 1.05$ at $r/d = ½$ (spherical).

Donnell[285] obtained the following relations for a rigid elliptical cylindrical inclusion:

Pole position B, Fig. 154:

$$K_{tB} = \frac{\sigma_{\max B}}{\sigma} = \frac{3}{16}\left(1 - \frac{b}{a}\right) \qquad [107]$$

Midposition A:

$$K_{tA} = \frac{\sigma_{\max A}}{\sigma} = \frac{3}{16}\left(5 + 3\frac{a}{b}\right) \qquad [108]$$

These stresses are radial (normal to the ellipse), adhesive tension at B and compression at A. The tangential stresses are one-third of the foregoing values.

It would seem that for the elliptical inclusion with its major axis in the tension direction, failure would start at the pole by rupture of the bond, with the crack progressing perpendicular to the applied tensile stress.

For the inclusion with its major axis perpendicular to the applied tensile stress, it would seem that for a/b less than about 0.15, the compressive stress at the end of the ellipse would cause plastic deformation, but that cracking would eventually occur at position B by rupture of the bond, followed by progressive cracking perpendicular to the applied tensile stress.

Nisitani[241] has obtained exact values for the plane stress and plane strain radial stresses for the pole position B, Fig. 154, of the rigid elliptical cylindrical inclusion:

$$K_t = \frac{(\gamma + 1)\,[(\gamma + 1)\,(a/b) + (\gamma + 3)]}{8\gamma} \qquad]109]$$

where $\gamma = 3 - 4\nu$ for plane strain
$\gamma = (3 - \nu)/(1 + \nu)$ for plane stress
a = ellipse width parallel to applied stress
b = ellipse width perpendicular to applied stress
ν = Poisson's ratio

For plane strain

$$K_t = \frac{(1 - \nu)\,[2\,(1 - \nu)\,(a/b) + 3 - 2\nu]}{3 - 4\nu} \qquad [110]$$

Formula [110] reduces to [106] for the circular cylindrical inclusion. As stated, [109] and [110] are sufficiently close to [107] so that a single curve can be used in Fig. 154.

A related case of a plate having a circular hole with a bonded cylindrical insert (r_i/r_o = 0.8) having a modulus of elasticity 11.5 times the modulus of elasticity of the plate has been studied by a combined photoelasticity and Moiré analysis.[287a]

The effect of spacing on a row of rigid elliptical inclusions[241] is shown in Fig. 155 as a ratio of the K_t for the row to the K_{t0} for the single inclusion (Fig. 154).

Shioya[287b] has obtained K_t factors for an infinite tension plate with two circular inclusions.

(z-7) Cylindrical Tunnel

Mindlin[287c] has solved the following cases of an infinitely long cylindrical tunnel: (I) hydrostatic pressure, $-cw$, at the tunnel location before tunnel is formed (c = distance from surface to center of tunnel, w = weight per unit volume of material); (II) material restrained from lateral displacement; (III) no lateral stress.

Results for case I are shown in Fig. 155*a* in dimensionless form, $\sigma_{max}/2wr$ versus c/r (see Fig. 155*a* for notation). It is seen that the minimum value of peripheral stress, σ_{max}, is reached at values of c/r = 1.2, 1.25, and 1.35 for ν = 0, ¼, and ½, respectively. For smaller values of c/r, the increased stress is due to the thinness of the "arch" over the hole, whereas for larger values of c/r, the increased stress is due to the increased pressure created by the material above.

An arbitrary stress concentration factor may be defined as $K_t = \sigma_{max}/p = \sigma_{max}/(-cw)$, where p = hydrostatic pressure, equal to $-cw$. Figure 155*a* may be converted to K_t as shown in Fig. 155*b* by dividing the $\sigma_{max}/2wr$ ordinates of Fig. 155*a* by $c/2r$, half of the abscissa values of Fig. 155*a*. It is seen from Fig. 155*b* that for large values of c/r, K_t approaches 2, the well-known K_t for a hole in a hydrostatic, or biaxial, stress field.

For a deep tunnel, c/r large[287c]

$$\sigma_{\max} = -2cw - rw\left[\frac{3 - 4\nu}{2(1-\nu)}\right] \qquad [110a]$$

By writing (rw) as (r/c) (cw), we can factor out $(-cw)$ to obtain K_t:

$$K_t = \frac{\sigma_{\max}}{-cw} = 2 + \frac{1}{c/r}\left[\frac{3 - 4\nu}{2(1-\nu)}\right] \qquad [110b]$$

The second term arises from the weight of the material removed from the hole; as c/r becomes large this term becomes negligible and K_t approaches 2, as indicated in Fig. 155*b*.

Solutions for various tunnel shapes (circular, elliptical, rounded square), at depths not influenced by the surface, have been obtained, with and without a rigid liner.[287d]

(z-8) Intersecting Cylindrical Holes

The intersecting cylindrical holes[287e] are in the form of a cross (+), a tee (T) or a round-cornered ell (L) with the plane containing the hole axes perpendicular to the applied uni-axial stress. This case is of interest in tunnel design[287e] and in various geometrical arrangements of fluid ducting in machinery.

Three-dimensional photoelastic tests[287e] were made of an axially compressed cylinder with these intersecting cylindrical hole forms located with the hole axes in a midplane perpendicular to the applied uniaxial stress. The cylinder was 8-in. diameter and all holes were 1.5-in. diameter. The maximum nominal stress concentration factor K_{tn} for the three intersection forms was found to be 3.6, corresponding to the maximum tangential stress at the intersection of the holes at the plane containing the hole axes.

The K_{tn} value of 3.6, based on nominal stress, applies only for the cylinder tested. A more useful value is an estimate of $K_{t\infty}$ in an infinite body; we next attempt to obtain this value.

First, it is observed that K_{tn} for the cylindrical hole away from the intersection is 2.3. The gross (applied) stress concentration factor is $K_{tg} = K_{tn}/(A_{\text{net}}/A) = 2.3/(0.665) = 3.46$ for the T intersection (A = cross-sectional area of cylinder, A_{net} = cross-sectional area in plane of hole axes). Referring to Fig. 86, it is seen that for $a/w = 1 - 0.665 = 0.335$, the same values of $K_{tn} = 2.3$ and $K_{tg} = 3.46$ are obtained and that the $K_{t\infty}$ value for the infinite width, $a/w \rightarrow 0$, is 3. The agreement is not as close for the cross and L geometries, about 6% deviation.

Next we start with $K_{tn} = 3.6$ and make the assumption that $K_{t\infty}/K_{tn}$ is the same as in Fig. 86 for the same a/w. $K_{t\infty} = 3.6\ (3/2.3) = 4.7$. This estimate is more useful generally than the specific test geometry value, $K_{tn} = 3.6$.

The author[287e] points out that stresses are highly localized at the intersection, decreasing to the value of the cylindrical hole within an axial distance equal to the hole diameter; also noted is the small value of the axial stresses.

The experimental determination of maximum stress at the very steep stress gradient at the sharp corner is difficult; it may be that the value just given is too low. For example, $K_{t\infty}$ for the intersection of a small hole into a large one would theoretically* be 9.

*The situation with respect to multiplying of stress concentration factors is somewhat similar to the case discussed in Section A.u and illustrated in Fig. 137.

It would seem that a rounded corner at the intersection (in the plane of the hole axes) would be beneficial in reducing K_t. This would be a practical expedient in the case of a tunnel or a cast metal part, but it does not seem to be practically attainable in the case where the holes have been drilled. An investigation of three-dimensional photoelastic models with the corner radius varied would be of interest.

For pressurized thick-walled cylinders with crossholes and sideholes, see Chapter 5, Section Q.

(B) BENDING

For thin plates or beams, two kinds of bending are presented: in-plane bending, Sections B.a, B.b, and B.c; transverse bending, Sections B.d, B.e, B.f, and B.g. For transverse bending, two cases are considered: simple bending ($M_1 = 1$, $M_2 = 0$) and cylindrical bending ($M_1 = 1$, $M_2 = \nu$). The cylindrical bending case removes the anticlastic bending resulting from the Poisson's ratio effect. At the beginning of application of bending, the simple condition occurs. As the deflection increases, the anticlastic effect is not realized, except for a slight curling at the edges. In the region of the hole, it is reasonable to assume that the cylindrical bending condition exists. For design problems, the cylindrical bending case is generally more applicable than the simple bending case.

It would seem that for transverse bending, rounding or chamfering of the hole edge would result in reducing the stress concentration factor.

For $M_1 = M_2$, isotropic transverse bending, K_t is independent of a/h, diameter of hole over thickness of plate; the case corresponds to biaxial tension of a plate with a hole.

(a) In-Plane Bending of a Beam with a Central Circular Hole

An effective method of weight reduction for a beam in bending is to remove material near the neutral axis, often in the form of a circular hole, or a row of circular holes.

Howland and Stevenson[288] have obtained mathematically the K_{tg} values represented by the curve of Fig. 156:

$$K_{tg} = \frac{\sigma_{\max}}{6M/w^2t} \qquad [111]$$

Symbols are defined in Fig. 156. K_{tg} is the ratio of $\sigma_{\max}$ to σ at the beam edge distant axially from the hole.

Photoelastic tests by Ryan and Fischer[289] and by Frocht and Leven[246] are in good agreement with the mathematical results.[288]

K_{tn} is based on the section modulus of the net section; the distance from the neutral axis is taken as $a/2$, so that σ_{nom} is at the edge of the hole:

$$K_{tn} = \frac{\sigma_{\max}}{6Ma/(w^3 - a^3)t} \qquad [112]$$

Another form of K_{tn} has been used where σ_{nom} is at the edge of the beam:

$$K'_{tn} = \frac{\sigma_{\max}}{6Mw/(w^3 - a^3)t} \qquad [113]$$

Udoguti[290] and Heywood[290a] noted that K'_{tn} versus a/w is a linear relation, $K'_{tn} = 2a/w$. Heywood[290a] further noted that $K_{tn} = 2$, commenting that this provides the "curious result that the stress concentration factor is independent of the relative size of the hole, and forms the only known case of a notch showing such independency."

Note from Fig. 156 that the hole does not weaken the beam for $a/w < \backsim 0.45$ (for design, $K_{tg} = 1$ for $a/w < \backsim 0.45$).

On the outer edge, the stress has peaks at F, F, but this stress is less than at E, except at and to the left of a transition zone in the region of C where $K_t = 1$ is approached. Angle $\alpha = 30°$ was found to be independent of $a/(w - a)$ over the range investigated.

(b) In-Plane Bending of a Beam with a Circular Hole Displaced from the Center Line

The K_t factor, as defined by [111], has been obtained by Isida[291] and is shown in Fig. 157. At line A—A, $K_{tgB} = K_{tgC}$, corresponding to maximum stress at B and C, respectively (see sketch in Fig. 157). Above A—A, K_{tgB} is the greater of the two stresses; below A—A, K_{tgC} is the greater, approaching $K_t = 1$, or no effect of the hole.

At $c/e = 1$, the hole is central, with factors as given in the preceding section. For $r/c \to 0$, K_{tg} is 3 multiplied by the ratio of the distance from the center line to the edge; in terms of c/e:

$$K_{tg} = 3\,\frac{1 - c/e}{1 + c/e} \qquad [114]$$

Photoelastic results of Nisida[292] are in agreement with the calculated values of Isida.[291]

(c) In-Plane Bending of a Beam with an Elliptical Hole; Slot with Semicircular Ends (Ovaloid); or Round-Cornered Square Hole

K'_{tn} factors, as defined by relation [113], were obtained by Isida;[169] these factors have been recalculated for K_{tg}, relation [111], and for K_{tn}, relation [112], and are presented in Fig. 158. The photoelastic values of Frocht and Leven[246] for a slot with semicircular ends are in reasonably good agreement when compared with an ellipse having the same a/r.

Note in Fig. 158 that the hole does not weaken the beam for a/w values less than at points A, B, and C for $a/r = 8$, 4, and 2, respectively (for design, $K_t = 1$ to the left of the intersection points).

On the outer edge, the stress has peaks at F,F, but this stress is less than at E, except at and to the left of a transition zone in the region of A, B, or C, where $K_t = 1$ is approached. In the photoelastic tests,[246] $\alpha = 35°$, $32.5°$ and $30°$ for $a/r = 8$, 4, and 2, respectively, independent of $a/(w - a)$ over the range investigated.

For shapes approximating ovaloids and round-cornered square holes (parallel and at 45°), K'_{tg} factors have been obtained[293] for central holes, small compared to the beam depth:

$$K'_{tg} = \frac{\sigma_{\max}}{6Ma/w^3t} \qquad [115]$$

(d) Transverse Bending of an Infinite and of a Finite-Width Plate with a Single Circular Hole

For simple bending ($M_1 = 1$, $M_2 = 0$), Reissner[294] obtained K_t as a function of a/h, as shown in Fig. 159. For $a/h \to 0$, $K_t = 3$. For $a/h \to \infty$:

$$K_t = \frac{5 + 3\nu}{3 + \nu} \qquad [116]$$

For $\nu = 0.3$, $K_t = 1.788$.

For cylindrical bending ($M = 1$, $M_2 = \nu$), $K_t = 2.7$ for $a/h \to 0$. For $a/h \to \infty$, Goodier[295] obtained

$$K_t = (5 - \nu)\frac{1 + \nu}{3 + \nu} \qquad [117]$$

For $\nu = 0.3$, $K_t = 1.852$.

For design problems, the cylindrical bending case is usually more applicable; see introductory remarks to section B.

For $M_1 = M_2$, isotropic bending, K_t is independent of a/h, and the case corresponds to biaxial tension of a plate with a hole.

For finite plate width and various a/h values, K_t is given in Fig. 160, based on Figs. 86 and 159, with the K_{tn} gradient at $a/w = 0$ equal to

$$\frac{dK_{tn}}{d(a/w)} = - K_{tn} \qquad [118]$$

Photoelastic tests[296,297] and strain gage measurements[298] are in reasonably good agreement with Fig. 160.

(e) Transverse Bending of an Infinite Plate with a Row of Circular Holes

Tamate[299] has obtained K_t values for simple bending ($M_1 = 1$, $M_2 = 0$) and for cylindrical bending ($M_1 = 1$, $M_2 = \nu$) with M_1 bending in the x and y directions (Fig. 161). For design problems, the cylindrical bending case is usually more applicable; see introductory remarks to Section B. The K_t value for $a/b \to 0$ corresponds to the single hole (Fig. 159). The dashed curve is for two holes[300] in a plate subjected to simple bending ($M_1 = 1$, $M_2 = 0$).

For bending about the x direction nominal stresses are used, resulting in curves that decrease as a/b increases; K_{tg} values, σ_{max}/σ, increase as a/b increases. $K_{tg} = K_{tn}/(1 - a/b)$.

(f) Transverse Bending of an Infinite Plate with a Single Elliptical Hole

Stress concentration factors for the elliptical hole[241,301] are given in Fig. 162.

For simple bending ($M_1 = 1$, $M_2 = 0$),

$$K_t = 1 + \frac{2(1 + \nu)\,(b/a)}{3 + \nu} \qquad [119]$$

where b = width of ellipse perpendicular to M_1 bending direction (Fig. 162)
a = width of ellipse perpendicular to width b
ν = Poisson's ratio

For cylindrical bending[241,295] ($M_1 = 1$, $M_2 = \nu$),

$$K_t = \frac{(1 + \nu)\,[2(b/a) + 3 - \nu]}{3 + \nu} \qquad [120]$$

For design problems, the cylindrical bending case is usually more applicable; see introductory remarks to Section B.

For $M_1 = M_2$, isotropic bending, K_t is independent of a/h, and the case corresponds to biaxial tension of a plate with a hole.

(g) Transverse Bending of an Infinite Plate with a Row of Elliptical Holes

Figure 163 presents the effect of spacing[241] for a row of elliptical holes. K_t values are given as a ratio of the single hole value (Fig. 162). The ratios are so close for simple and cylindrical bending that these cases can be represented by a single set of curves (Fig. 163). For bending about the y axis, a row of edge notches is obtained for $b/c \geqq 1.0$. For bending about the x axis, nominal stress is used, resulting in curves that decrease as b/c increases; K_{tg} values, σ_{max}/σ, increase as b/c increases. $K_{tg} = K_{tn}/(1 - b/c)$.

(h) Round Bar or Tube with a Transverse Circular Hole Having its Axis in the Plane of Bending

The K_t relations are presented in Fig. 164. The curve for the solid shaft is based on blending the data of Thum and Kirmser[258] and the British data.[259,260] There is some uncertainty regarding the exact position of the light dashed portion of the curve.

A photoelastic test by Fessler and Roberts[301a] is in good agreement with Fig. 164.

The factors are defined as follows:

$$K_{tg} = \frac{\sigma_{max}}{\sigma_{gross}} = \frac{\sigma_{max}}{M/Z_{tube}} = \frac{\sigma_{max}}{Md/2I_{tube}}$$

$$= \frac{\sigma_{max}}{32\ Md/\pi\ (d^4 - d_i^4)} \qquad [121]$$

$$K_{tn} = \frac{\sigma_{max}}{\sigma_{net}} = \frac{\sigma_{max}}{M/Z_{net}} = \frac{\sigma_{max}}{Mc/I_{net}} \qquad [122]$$

where $c = \sqrt{r^2 - (a/2)^2}$, and

$$K_{tn} = K_{tg} \frac{Z_{net}}{Z_{tube}}$$

Symbols are defined in Fig. 164.

Thum and Kirmser[258] found that the maximum stress did not occur on the surface of the shaft but at a small distance inside the hole on the surface of the hole. The σ_{max} value used in developing Fig. 164 is the maximum stress inside the hole. No factors are given for the somewhat lower stress at the shaft surface; if these factors are of interest, Ref. 258 should be examined.

The ratio Z_{net}/Z_{tube} has been determined mathematically;[261] the formulas will not be repeated here (specific values can be obtained by dividing the Fig. 164 values of K_{tn} by K_{tg}). If the hole is sufficiently small relative to the shaft diameter, the hole may be considered to be of rectangular cross section:

$$\frac{Z_{net}}{Z_{tube}} = 1 - \frac{(16/3\pi)\ (a/d)\ [1 - (d_i/d)^3]}{1 - (d_i/d)^4} \qquad [123]$$

It can be seen from the bottom curves of Fig. 164 that the error due to this approximation is small below $a/d = 0.2$.

(C) SHEAR, TORSION

(a) Shear Stressing of Infinite Plate with Circular or Elliptical Hole

By superposition of σ_1 and $\sigma_2 = -\sigma_1$ uniaxial stress distributions, the shear case $\tau = \sigma_1$ is obtained. For the circular hole, $K_t = \sigma_{max}/\tau = 4$, as shown in Fig. 132, $b/a = 1$, $\sigma_2/\sigma_1 = -1$; this is also found in Fig. 165 at $b/a = 1$. Further, $K_{ts} = \tau_{max}/\tau = (\sigma_{max}/2)/\tau = 2$.

For the elliptical hole, Fig. 165 shows K_t for shear stress orientations in line with the ellipse axes[301b] and at 45° to the axes. The 45° case corresponds to $\sigma_2 = -\sigma_1 = \tau$, as shown in Figs. 132 and 133.

The case of shearing forces parallel to the major axis of the elliptical hole, with the shearing force couple counterbalanced by a symmetrical remotely located opposite couple, has been solved by Neuber.[301c] The K_t factors are higher than the parallel shear factors in Fig. 165; for example, for a circle the "shearing force" K_t factor[301c] is 6 as compared to 4 in Fig. 165.

(b) Shear Stressing of an Infinite Plate with a Round-Cornered Rectangular Hole

In Fig. 166, $K_t = \sigma_{max}/\tau$ is given for shear stressing in line with the hole axes.[243,253] In Fig. 142, $\sigma_2 = -\sigma_1$ is equivalent to shear stress τ at 45° to the hole axes.[252,254]

(c) Shear Stressing of an Infinite Plate with Two Circular Holes or a Row of Circular Holes

Figure 167 presents $K_t = \sigma_{max}/\tau$ for shear stressing in line with the hole axis.[210,302] The location of σ_{max} varies from $\theta = 0°$ for $b/a \to 1$ to $\theta = 45°$ for $b/a \to \infty$.

(d) Shear Stressing of an Infinite Plate with an Infinite Pattern of Circular Holes

In Fig. 168, $K_t = \sigma_{max}/\tau$ is given for square and equilateral triangular patterns of circular holes for shear stressing in line with the pattern axis.[210,214,217,222] Subsequent computed values[95a,216a] are in good agreement with Ref. 210. Note that $\sigma_2 = -\sigma_1$ is equivalent to shear stressing τ at 45° to the pattern axis in Fig. 120.

In Figs. 169 and 170, $K_t = \sigma_{max}/\tau$ is given for rectangular and diamond (triangular, not limited to equilateral) patterns,[210] respectively.

(e) Twisted Infinite Plate with a Circular Hole

In Fig. 171, $M_x = 1$, $M_y = -1$ corresponds to a twisted plate.[294] $K_t = \sigma_{max}/\sigma$ due to bending moment M. For $h/a \to \infty$, $(a/h \to 0)$, $K_t = 4$. For $a/h \to \infty$,

$$K_t = 1 + \frac{1 + 3\nu}{3 + \nu} \qquad [124]$$

For $\nu = 0.3$, $K_t = 1.575$.

(f) Torsion of a Cylindrical Shell with a Circular Hole

For a discussion of the parameters, see Section A.c. The K_t factors[177] of Fig. 172 have been compared with experimental results,[175,179] with reasonably good agreement. The case of an elliptical hole has also been investigated.[303]

(g) Torsion of a Round Bar or a Tube with a Transverse Circular Hole

Stress concentration factors, (Fig. 173) are based on photoelastic tests[259,260] and strain gage tests.[258] The factors are defined as follows:

$$K_{tg} = \frac{\sigma_{\max}}{\tau_{\text{gross}}} = \frac{\sigma_{\max}}{Td/2J_{\text{tube}}}$$

$$= \frac{\sigma_{\max}}{16Td/\pi\,(d^4 - d_i^4)} \qquad [125]$$

$$K_{tn} = \frac{\sigma_{\max}}{\tau_{\text{net}}} = \frac{\sigma_{\max}}{Td/2J_{\text{net}}} = K_{tg}\frac{J_{\text{net}}}{J_{\text{tube}}} \qquad [126]$$

Symbols are defined in Fig. 173.

Thum and Kirmser[258] found that the maximum stress did not occur on the surface of the shaft but at a small distance inside the hole on the surface of the hole; this has been corroborated by later investigators.[257,259] In this report, $\sigma_{\max}$ denotes the maximum stress inside the hole. No factors are given for the somewhat lower stress at the shaft surface; if they are of interest, they can be found in the references.[257,259]

The $J_{\text{net}}/J_{\text{tube}}$ ratios have been determined mathematically;[261] the formulas and charts will not be repeated here (specific values can be obtained by dividing the Fig. 173 values of K_{tn} by K_{tg}). If the hole is sufficiently small relative to the shaft diameter, the hole may be considered to be of rectangular cross section:

$$\frac{J_{\text{net}}}{J_{\text{tube}}} = 1 - \frac{(8/3\pi)\,(a/d)\,\{[1 - (d_i/d)^3] + (a/d)^2\,[1 - (d_i/d)]\}}{1 - (d_i/d)^4} \qquad [127]$$

The bottom curves of Fig. 173 show that the error due to the foregoing approximation is small below $a/d = 0.2$.

The maximum stress, $\sigma_{\max}$, is uniaxial; the maximum shear stress $\tau_{\max} = \sigma_{\max}/2$ at 45° from the tangential direction of $\sigma_{\max}$ Maximum shear stress concentration factors can be defined:

$$K_{tsg} = \frac{K_{tg}}{2} \qquad [128]$$

$$K_{tsn} = \frac{K_{tn}}{2} \qquad [129]$$

If in Fig. 173, the ordinate values are divided by 2, maximum shear stress concentration factors will be represented.

Another stress concentration factor can be defined, based on equivalent stress of the applied system. The applied shear stress τ corresponds to principal stresses σ and $-\sigma$, 45°

from the shear stress directions. The equivalent stress $\sigma_{eq} = \sqrt{3}\,\sigma$. The equivalent stress concentration factors are

$$K_{teg} = \frac{\sigma_{max}}{\sigma_{eq}} = \frac{K_{tg}}{\sqrt{3}} \qquad [130]$$

$$K_{ten} = \frac{K_{tn}}{\sqrt{3}} \qquad [131]$$

Referring to Fig. 173, the ordinate values divided by $\sqrt{3}$ give the corresponding K_{te} factors.

Factors from formulas [128] to [131] are useful in mechanics of materials problems where one wishes to determine the initial plastic condition.

The case of a torsion cylinder with a central spherical cavity has been analyzed.[283]

FIG. 86

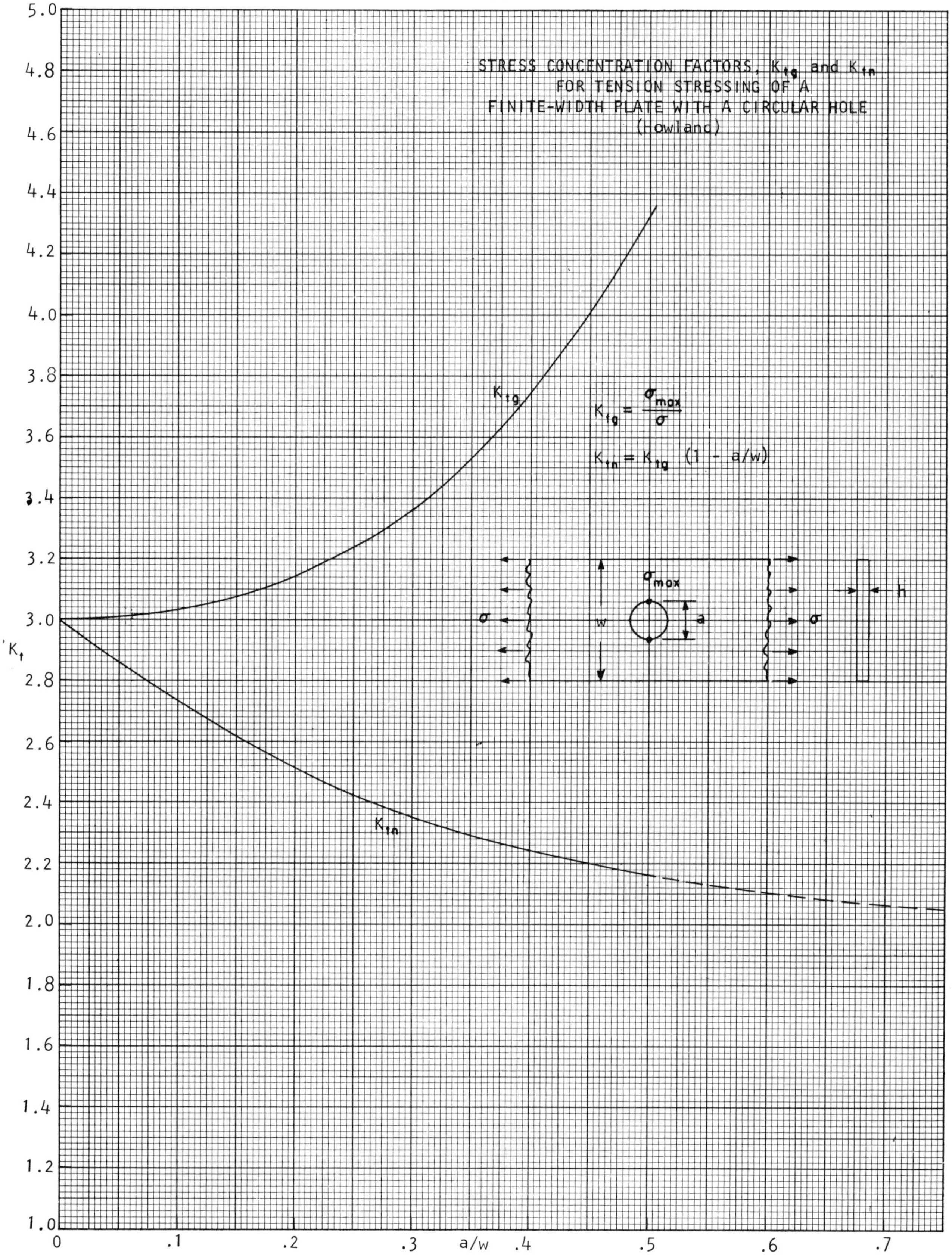

STRESS CONCENTRATION FACTORS, K_{tg} and K_{tn}
FOR TENSION STRESSING OF A
FINITE-WIDTH PLATE WITH A CIRCULAR HOLE
(Howland)

$$K_{tg} = \frac{\sigma_{max}}{\sigma}$$

$$K_{tn} = K_{tg}\ (1 - a/w)$$

K_{tg}

K_{tn}

σ_{max}

σ w   a σ h

K_t: 5.0, 4.8, 4.6, 4.4, 4.2, 4.0, 3.8, 3.6, 3.4, 3.2, 3.0, 2.8, 2.6, 2.4, 2.2, 2.0, 1.8, 1.6, 1.4, 1.2, 1.0

a/w: 0, .1, .2, .3, .4, .5, .6, .7

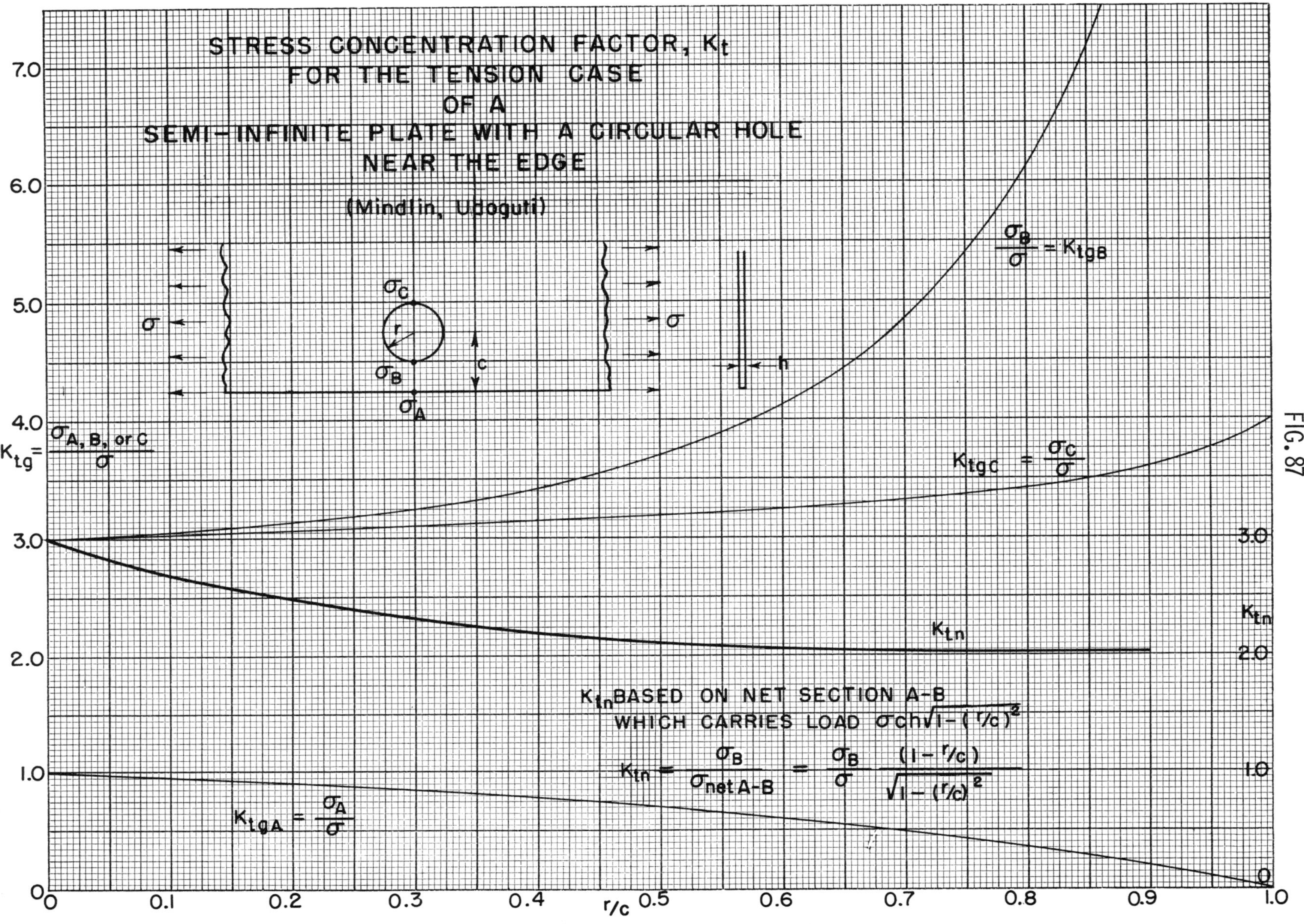

STRESS CONCENTRATION FACTOR, K_t
FOR THE TENSION CASE
OF A
SEMI-INFINITE PLATE WITH A CIRCULAR HOLE
NEAR THE EDGE
(Mindlin, Udoguti)
$\frac{\sigma_B}{\sigma} = K_{tgB}$
$K_{tgC} = \frac{\sigma_C}{\sigma}$
K_{tn}
K_{tn} BASED ON NET SECTION A-B
WHICH CARRIES LOAD $\sigma ch\sqrt{1-(r/c)^2}$
$K_{tn} = \frac{\sigma_B}{\sigma_{net\,A-B}} = \frac{\sigma_B}{\sigma} \frac{(1-r/c)}{\sqrt{1-(r/c)^2}}$
$K_{tgA} = \frac{\sigma_A}{\sigma}$
$K_{tg} = \frac{\sigma_{A,\,B,\,or\,C}}{\sigma}$
σ_C
σ_B
σ_A
σ
r
c
h
r/c
0
0.1
0.2
0.3
0.4
0.5
0.6
0.7
0.8
0.9
1.0
7.0
6.0
5.0
4.0
3.0
2.0
1.0

FIG. 87

FIG. 88

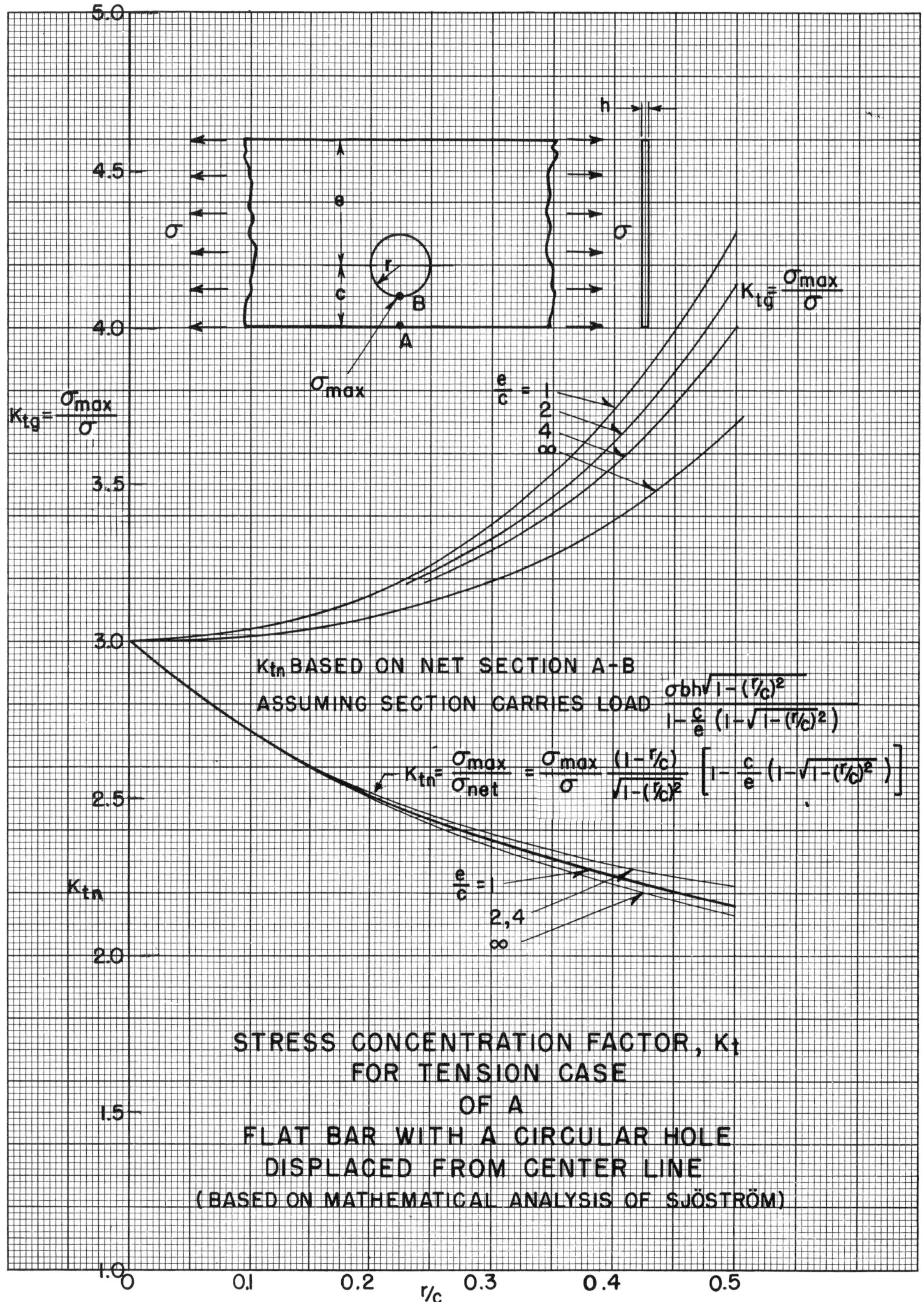

STRESS CONCENTRATION FACTOR, K_t
FOR TENSION CASE
OF A
FLAT BAR WITH A CIRCULAR HOLE
DISPLACED FROM CENTER LINE
(BASED ON MATHEMATICAL ANALYSIS OF SJÖSTRÖM)

FIG. 89

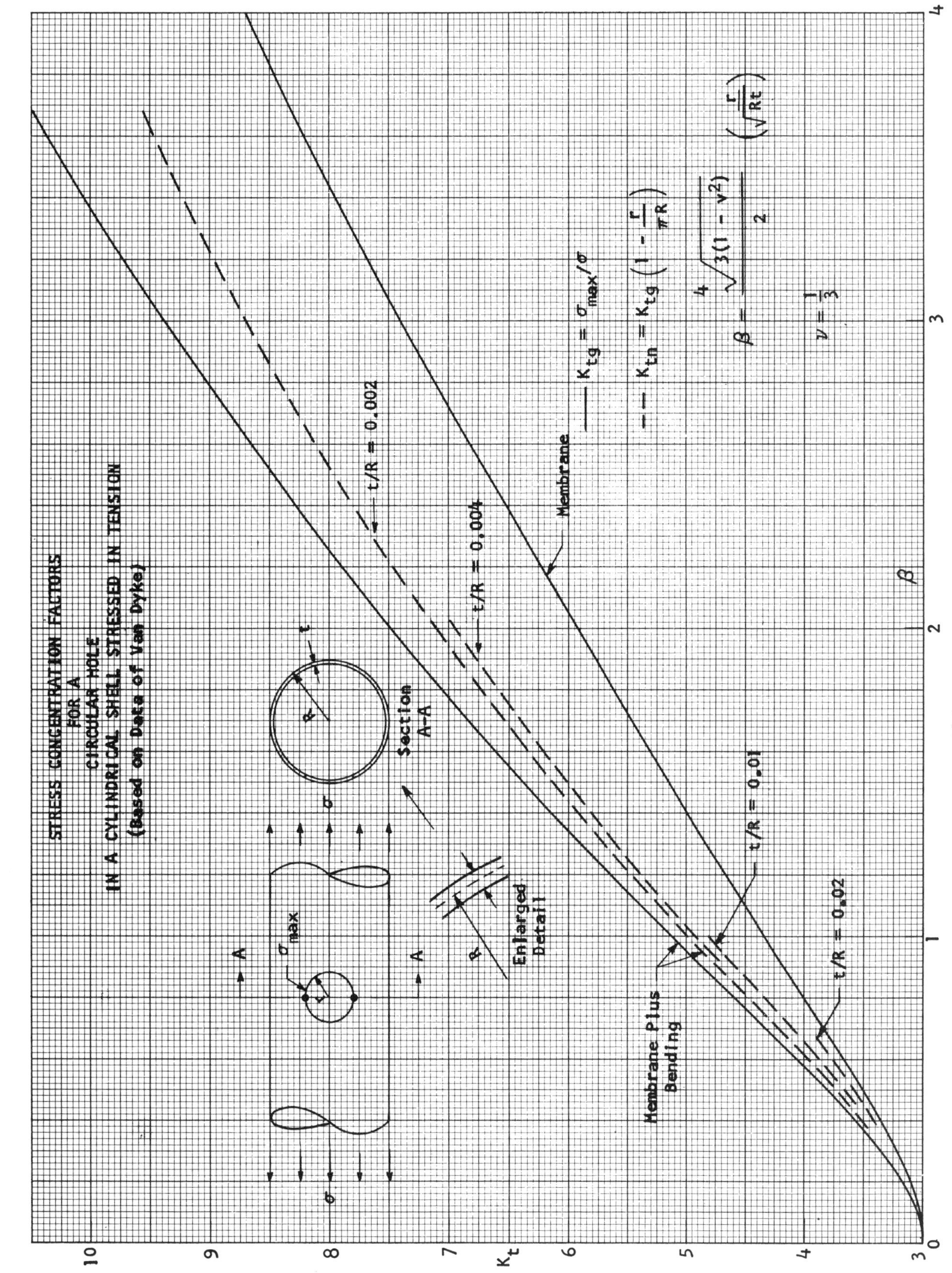

STRESS CONCENTRATION FACTORS
FOR A
CIRCULAR HOLE
IN A CYLINDRICAL SHELL STRESSED IN TENSION
(Based on Data of Van Dyke)
A
A
σ_{max}
r
σ
σ
R
t
Section
A-A
R
Enlarged
Detail
t/R = 0.002
t/R = 0.004
t/R = 0.01
t/R = 0.02
Membrane Plus
Bending
Membrane
$K_{tg} = \sigma_{max}/\sigma$
$K_{tn} = K_{tg}\left(1 - \frac{r}{\pi R}\right)$
$\beta = \frac{\sqrt[4]{3(1 - \nu^2)}}{2}\left(\frac{r}{\sqrt{Rt}}\right)$
$\nu = \frac{1}{3}$
K_t
3
4
5
6
7
8
9
10
β
0
1
2
3
4

FIG. 90

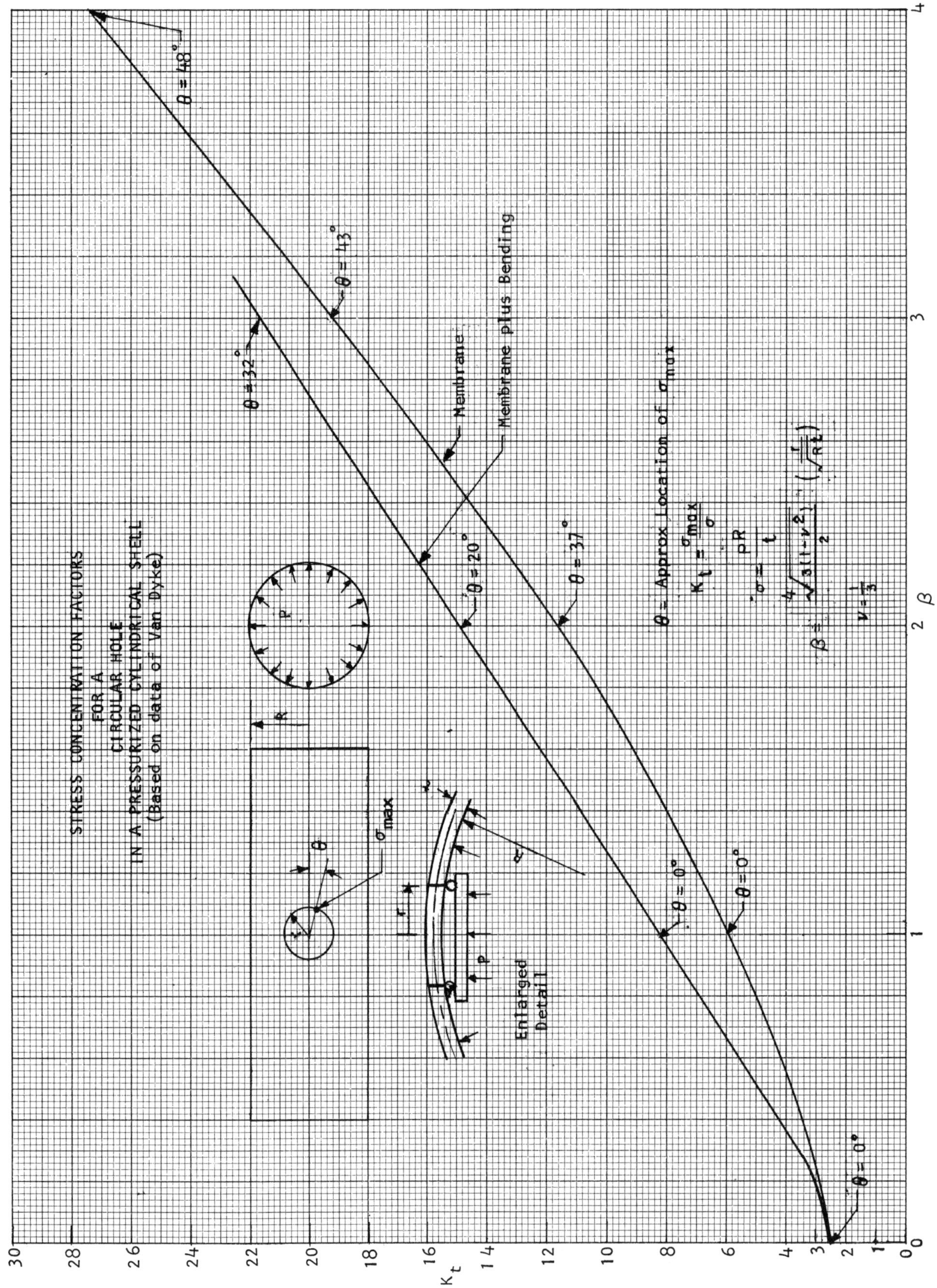

STRESS CONCENTRATION FACTORS
FOR A
CIRCULAR HOLE
IN A PRESSURIZED CYLINDRICAL SHELL
(Based on data of Van Dyke)

θ = Approx Location of σ_{max}

$$K_t = \frac{\sigma_{max}}{\sigma}$$

$$\sigma = \frac{pR}{t}$$

$$\beta = \frac{\sqrt[4]{3(1-\nu^2)}}{2}\left(\frac{r}{\sqrt{Rt}}\right)$$

$$\nu = \frac{1}{3}$$

Membrane

Membrane plus Bending

$\theta = 0°$, $\theta = 20°$, $\theta = 32°$, $\theta = 37°$, $\theta = 43°$, $\theta = 48°$

σ_{max}, θ, R, p, r, t

Enlarged Detail

K_t: 0, 1, 2, 3, 4, 6, 8, 10, 12, 14, 16, 18, 20, 22, 24, 26, 28, 30

β: 0, 1, 2, 3, 4

FIG. 91

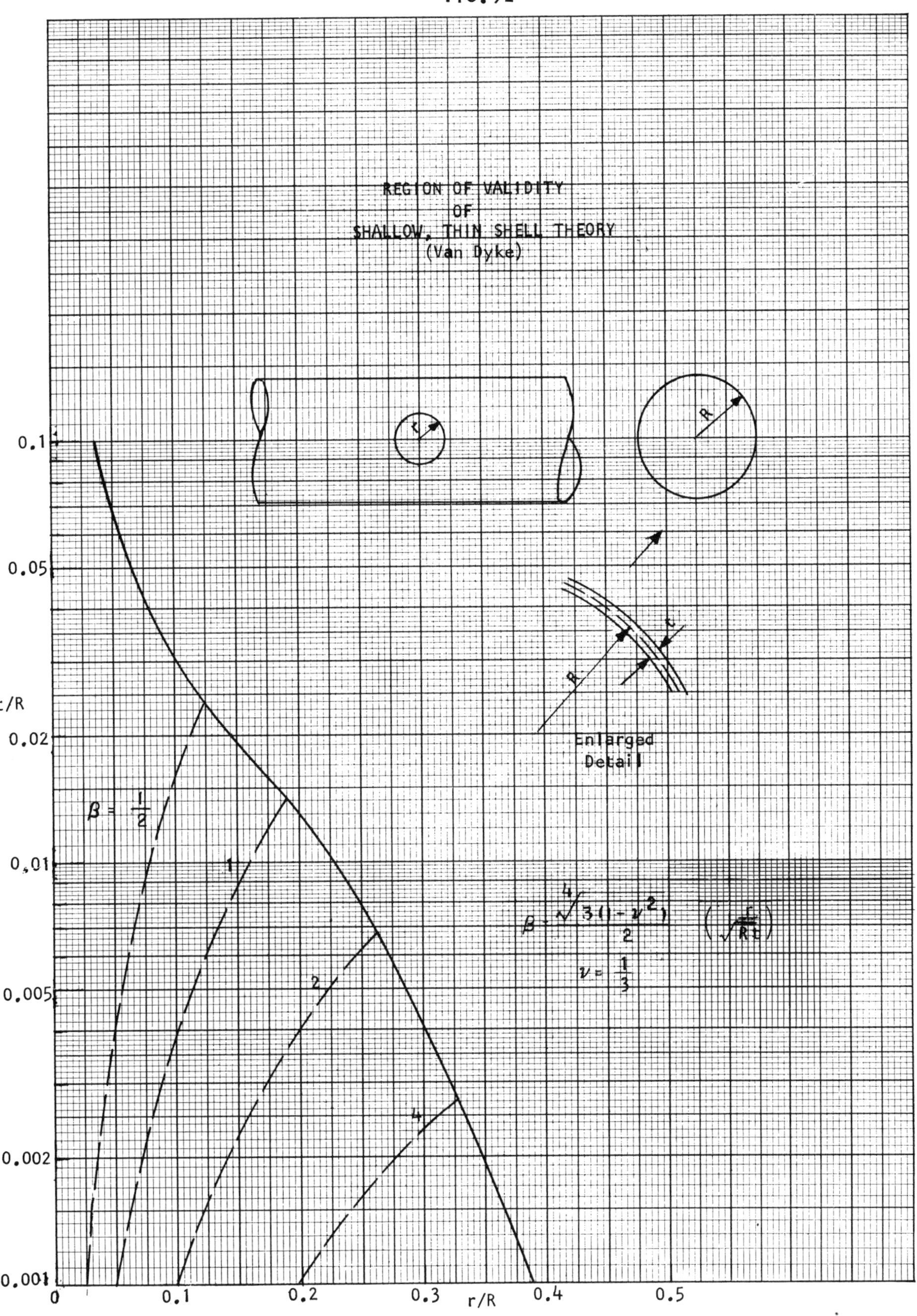

REGION OF VALIDITY
OF
SHALLOW, THIN SHELL THEORY
(Van Dyke)
r
R
R
t
Enlarged
Detail
$\beta = \frac{1}{2}$
1
2
4
$\beta = \frac{\sqrt[4]{3(1-\nu^2)}}{2} \left(\frac{r}{\sqrt{Rt}}\right)$
$\nu = \frac{1}{3}$
t/R
0.1
0.05
0.02
0.01
0.005
0.002
0.001
0
0.1
0.2
0.3
0.4
0.5
r/R

FIG. 91a

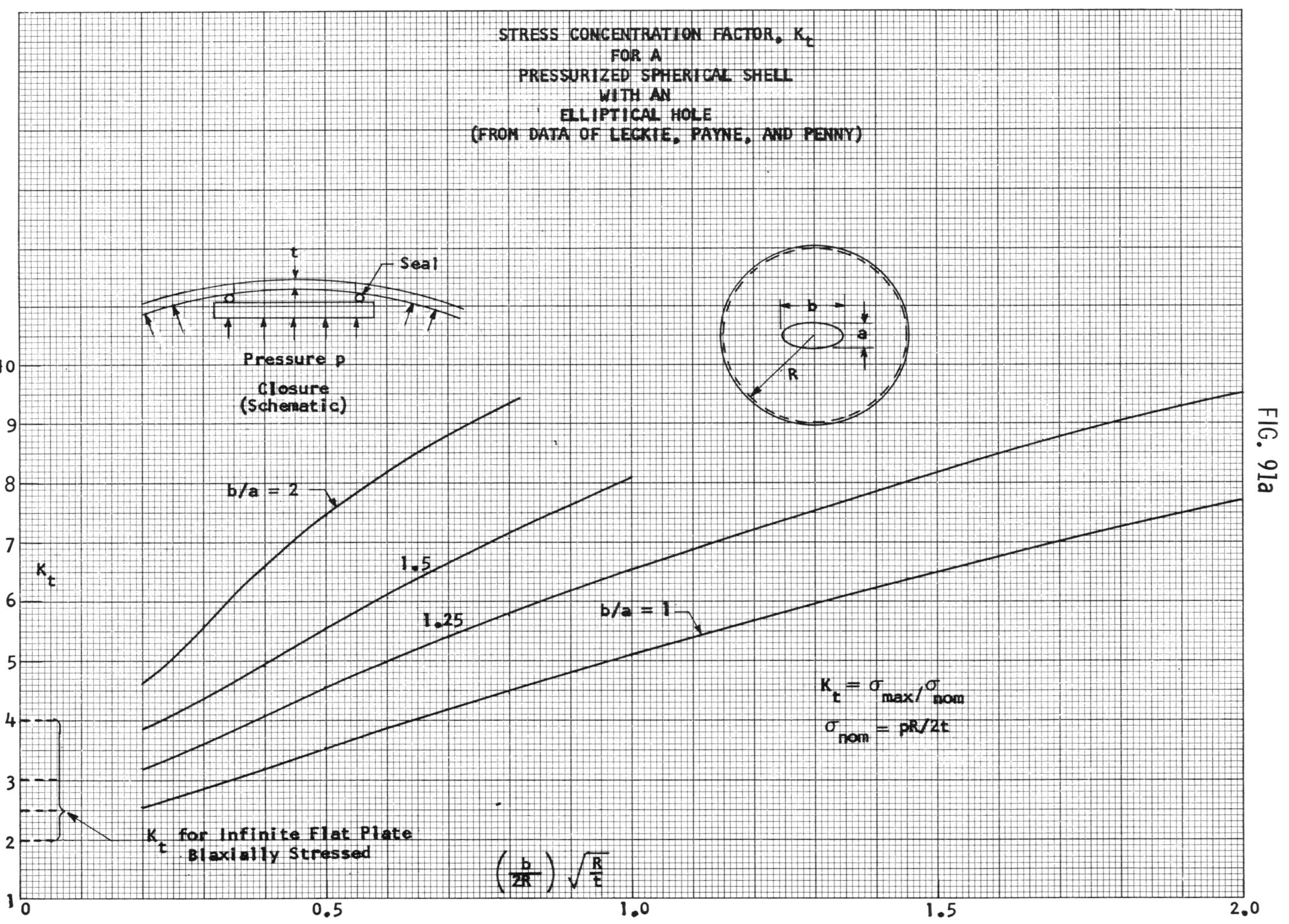

STRESS CONCENTRATION FACTOR, K_t
FOR A
PRESSURIZED SPHERICAL SHELL
WITH AN
ELLIPTICAL HOLE
(FROM DATA OF LECKIE, PAYNE, AND PENNY)
t
Seal
Pressure p
Closure
(Schematic)
b
a
R
b/a = 2
1.5
1.25
b/a = 1
$K_t = \sigma_{max}/\sigma_{nom}$
$\sigma_{nom} = pR/2t$
K_t
K_t for Infinite Flat Plate
Biaxially Stressed
1
2
3
4
5
6
7
8
9
10
0
0.5
1.0
1.5
2.0
$\left(\frac{b}{2R}\right)\sqrt{\frac{R}{t}}$

FIG.92

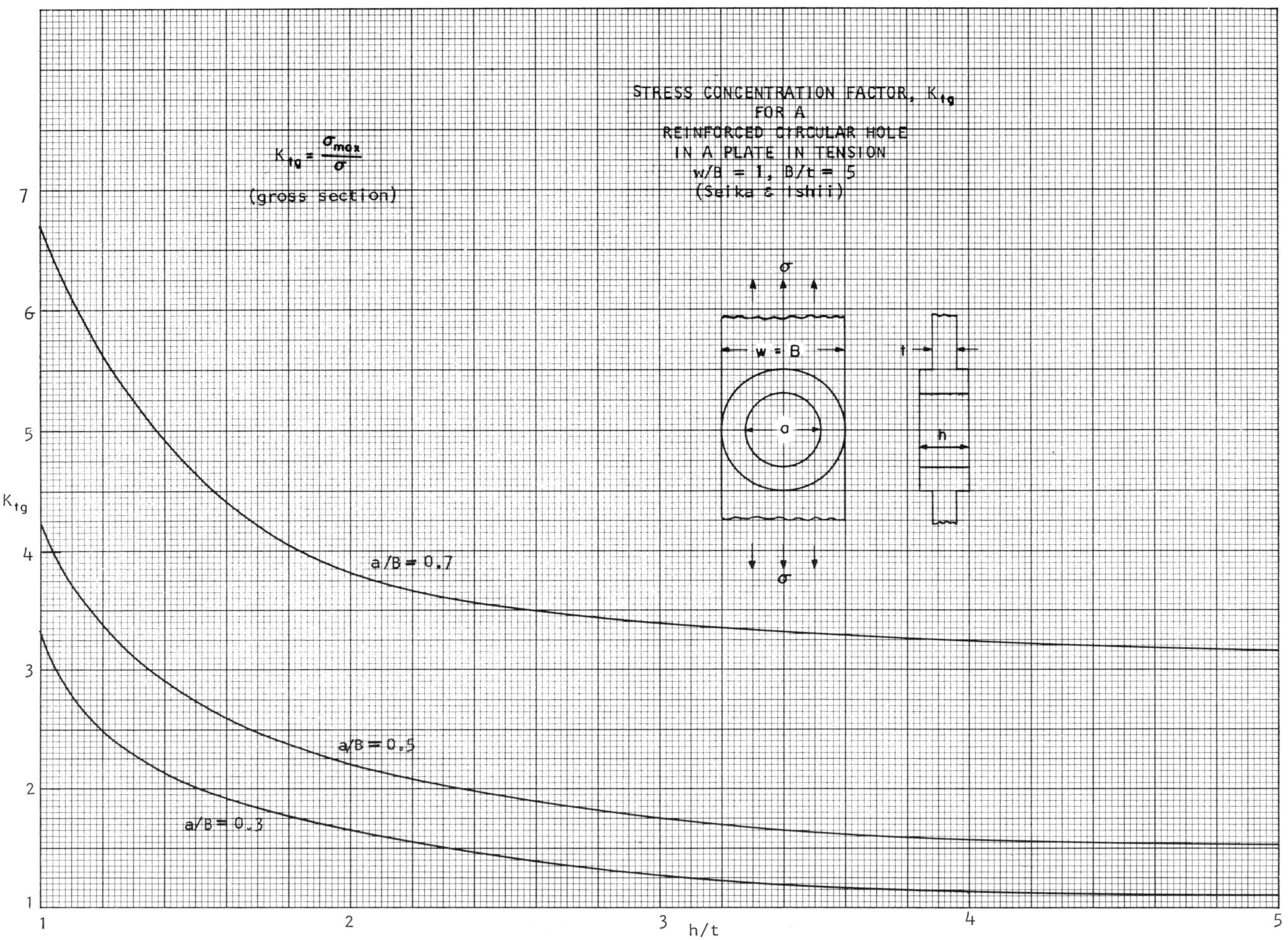

STRESS CONCENTRATION FACTOR, K_{tg}
FOR A
REINFORCED CIRCULAR HOLE
IN A PLATE IN TENSION
w/B = 1, B/t = 5
(Seika & Ishii)
$K_{tg} = \frac{\sigma_{max}}{\sigma}$
(gross section)
σ
w = B
a
t
h
a/B = 0.7
a/B = 0.5
a/B = 0.3
K_{tg}
1
2
3
4
5
6
7
h/t
1
2
3
4
5

FIG. 93

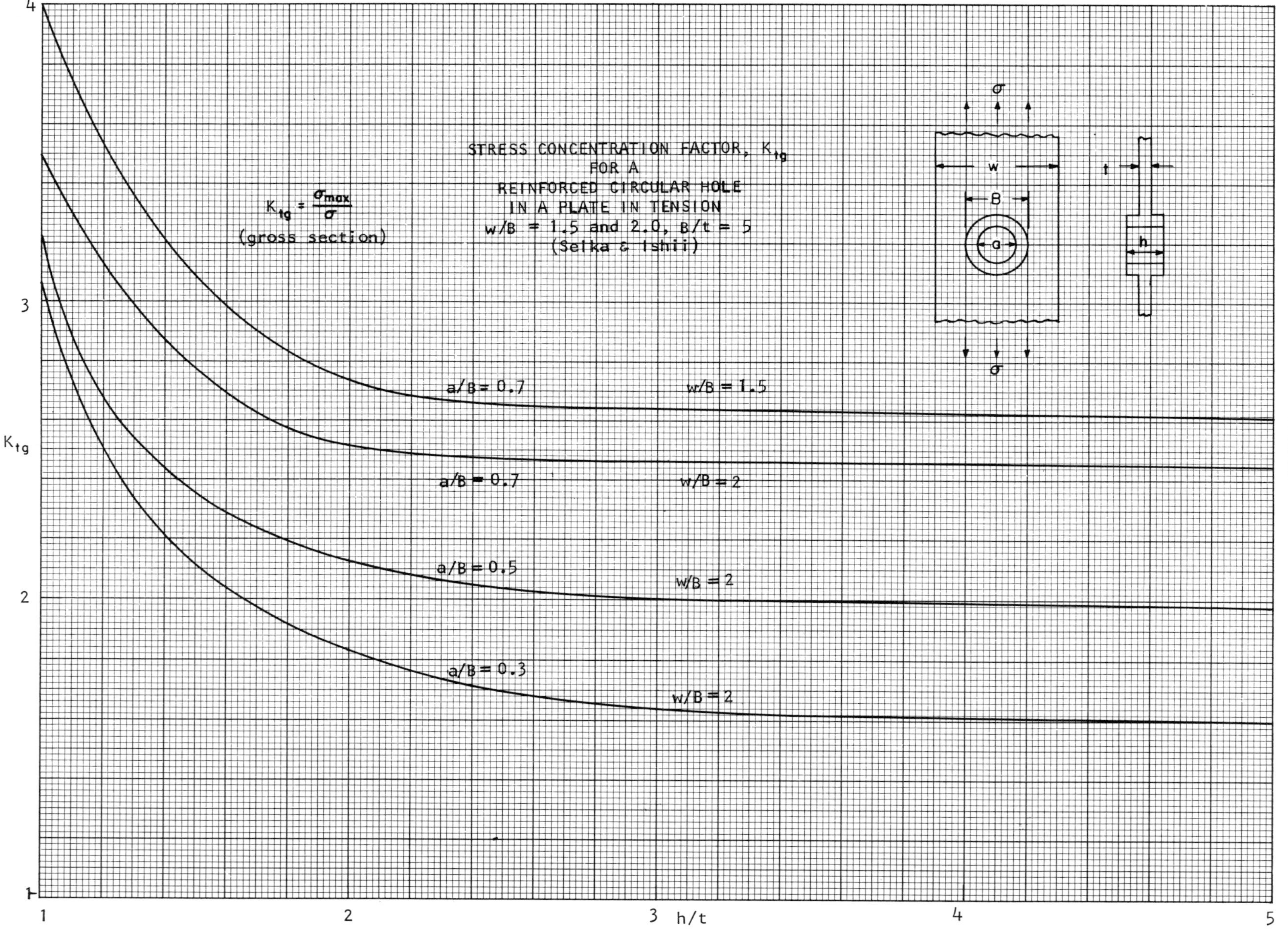

STRESS CONCENTRATION FACTOR, K_{tg}
FOR A
REINFORCED CIRCULAR HOLE
IN A PLATE IN TENSION
w/B = 1.5 and 2.0, B/t = 5
(Seika & Ishii)
$K_{tg} = \frac{\sigma_{max}}{\sigma}$
(gross section)
σ
w
B
a
t
h
a/B = 0.7
w/B = 1.5
a/B = 0.7
w/B = 2
a/B = 0.5
w/B = 2
a/B = 0.3
w/B = 2
K_{tg}
4
3
2
1
h/t
1
2
3
4
5

FIG. 94

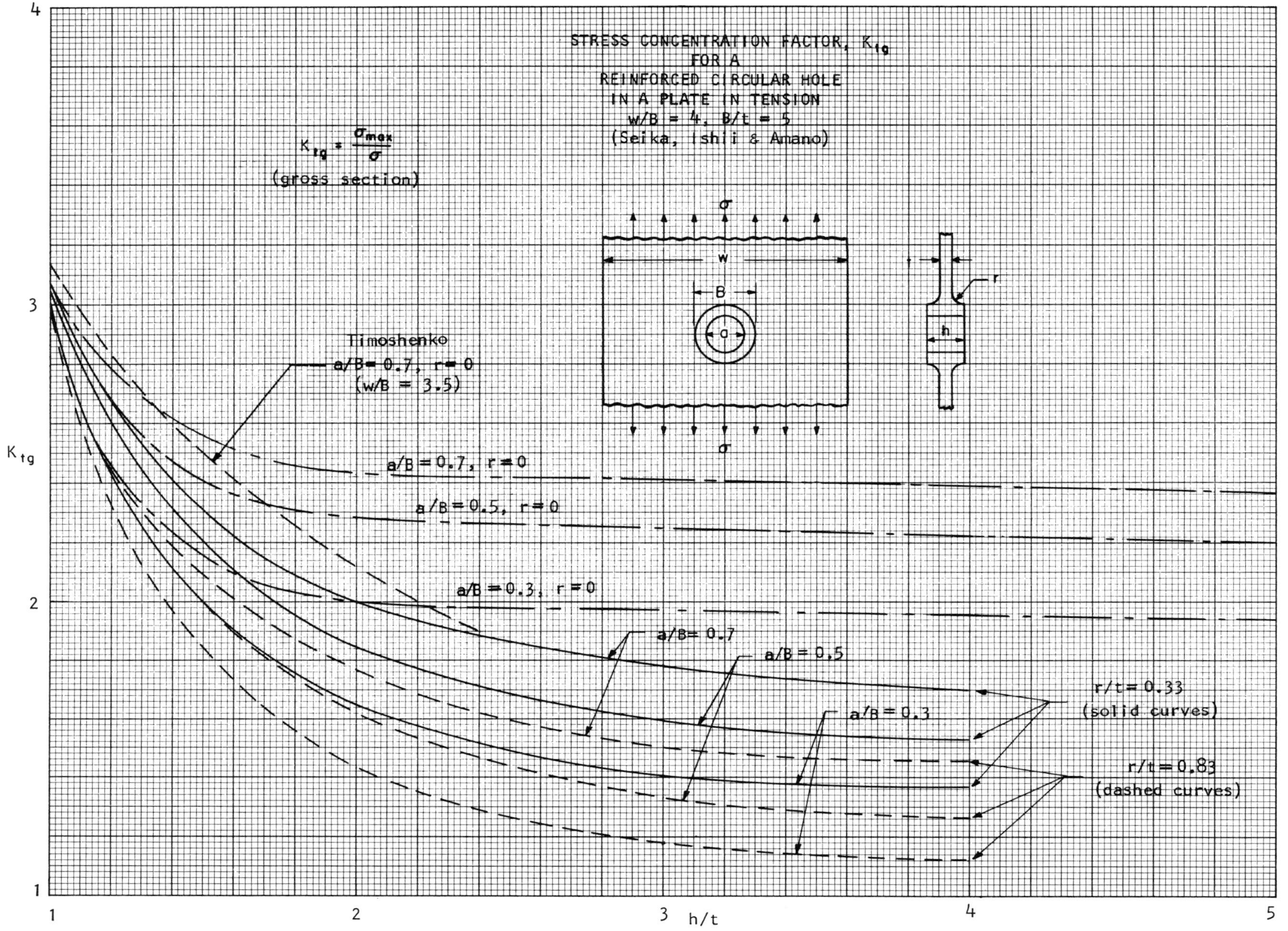

STRESS CONCENTRATION FACTOR, K_tg
FOR A
REINFORCED CIRCULAR HOLE
IN A PLATE IN TENSION
w/B = 4, B/t = 5
(Seika, Ishii & Amano)
K_tg = σ_max / σ
(gross section)
Timoshenko
a/B = 0.7, r = 0
(w/B = 3.5)
a/B = 0.7, r = 0
a/B = 0.5, r = 0
a/B = 0.3, r = 0
a/B = 0.7
a/B = 0.5
a/B = 0.3
r/t = 0.33
(solid curves)
r/t = 0.83
(dashed curves)
σ
w
B
a
h
r
t
K_tg
h/t
1
2
3
4
5

FIG. 95

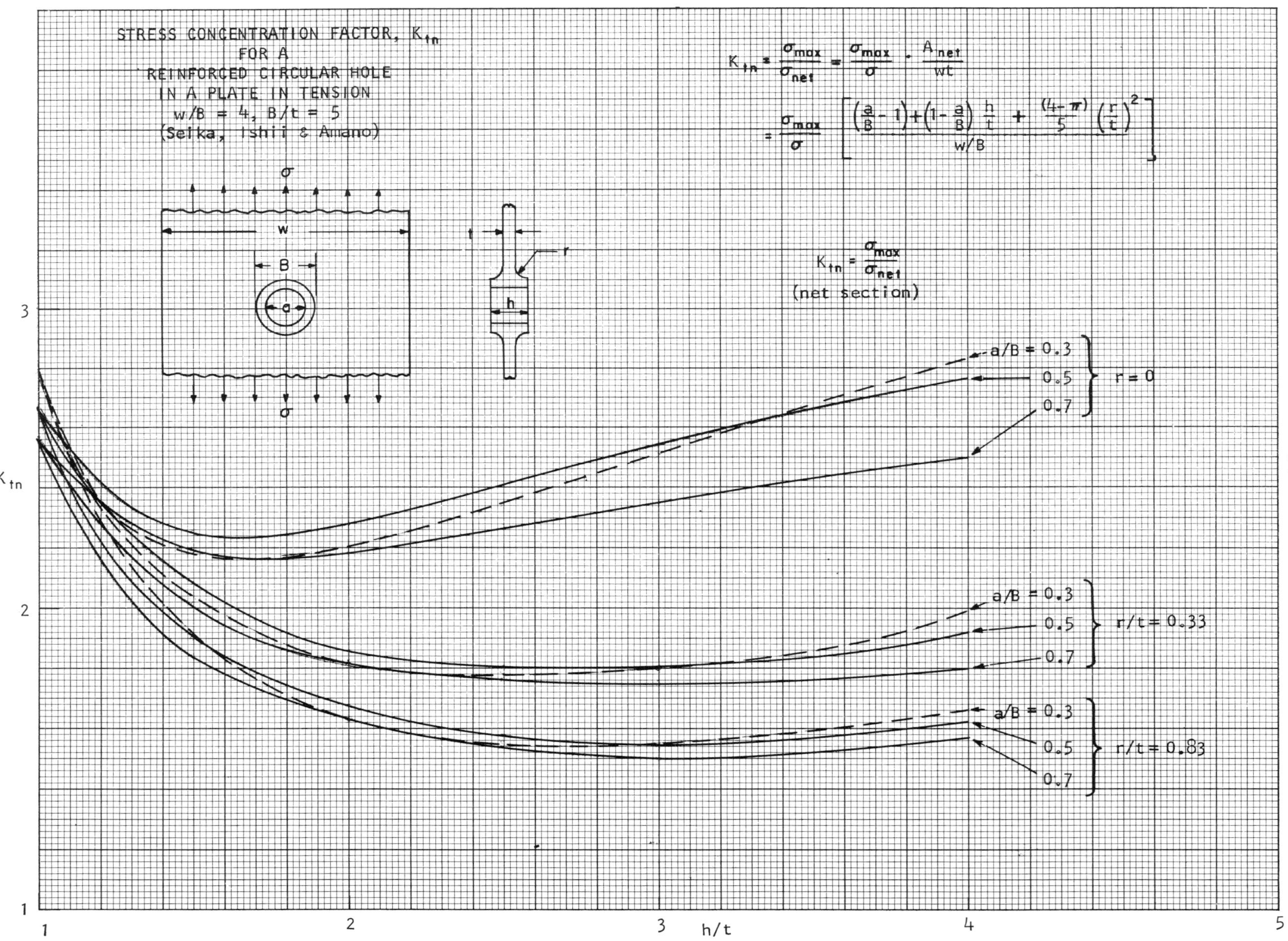

STRESS CONCENTRATION FACTOR, K_{tn}
FOR A
REINFORCED CIRCULAR HOLE
IN A PLATE IN TENSION
$w/B = 4$, $B/t = 5$
(Seika, Ishii & Amano)
$K_{tn} = \frac{\sigma_{max}}{\sigma_{net}} = \frac{\sigma_{max}}{\sigma} \cdot \frac{A_{net}}{wt}$
$= \frac{\sigma_{max}}{\sigma} \left[\frac{\left(\frac{a}{B} - 1\right) + \left(1 - \frac{a}{B}\right)\frac{h}{t} + \frac{(4-\pi)}{5}\left(\frac{r}{t}\right)^2}{w/B}\right]$
$K_{tn} = \frac{\sigma_{max}}{\sigma_{net}}$
(net section)
σ
w
B
a
t
r
h
a/B = 0.3
0.5
0.7
r = 0
a/B = 0.3
0.5
0.7
r/t = 0.33
a/B = 0.3
0.5
0.7
r/t = 0.83
K_{tn}
1
2
3
h/t
1
2
3
4
5

FIG.96

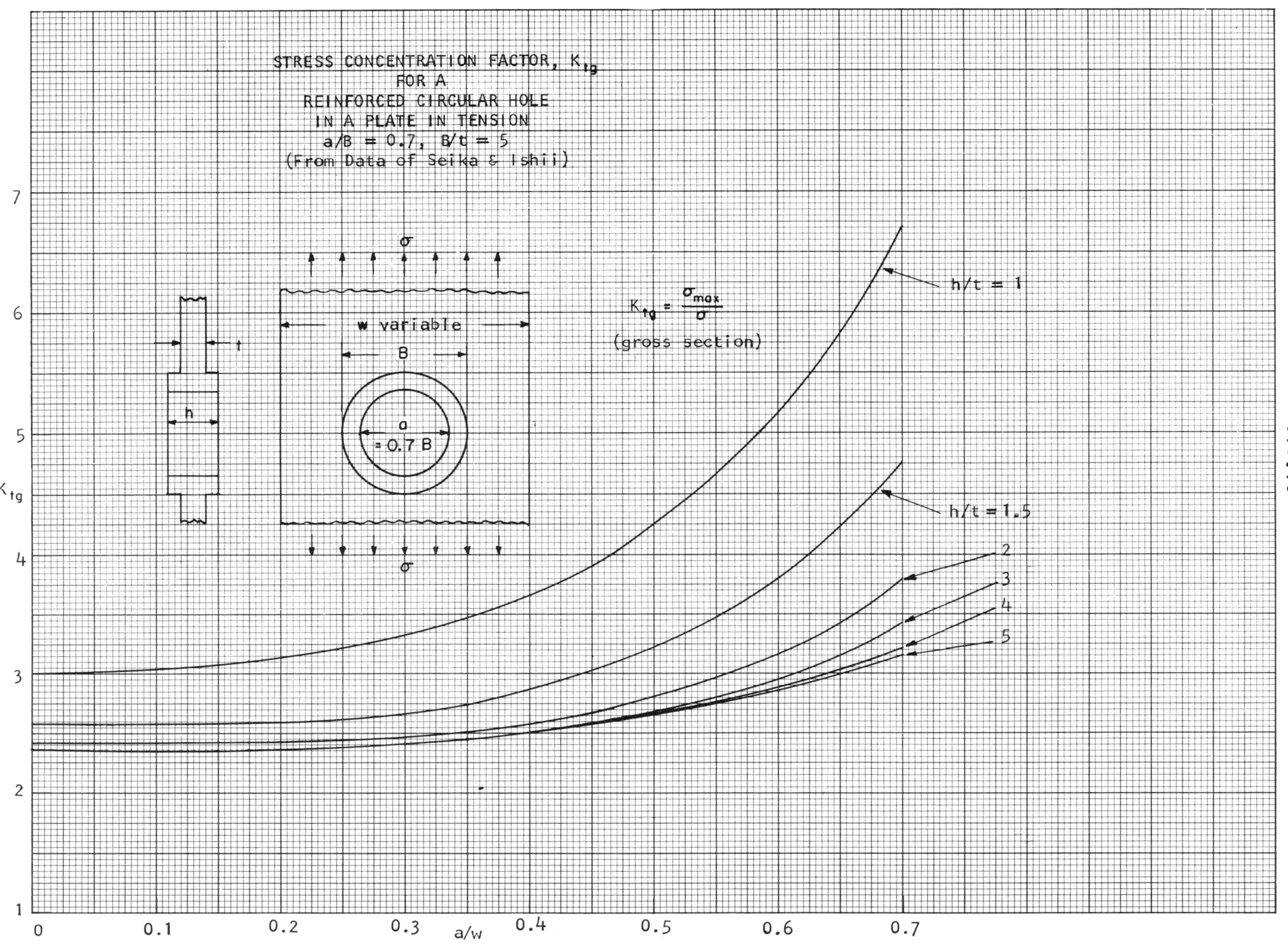

STRESS CONCENTRATION FACTOR, K_{tg}
FOR A
REINFORCED CIRCULAR HOLE
IN A PLATE IN TENSION
a/B = 0.7, B/t = 5
(From Data of Seika & Ishii)
$K_{tg} = \frac{\sigma_{max}}{\sigma}$
(gross section)
h/t = 1
h/t = 1.5
2
3
4
5
σ
w variable
B
a = 0.7 B
h
t
K_{tg}
a/w
0
0.1
0.2
0.3
0.4
0.5
0.6
0.7
1
2
3
4
5
6
7

FIG. 97
STRESS CONCENTRATION FACTORS 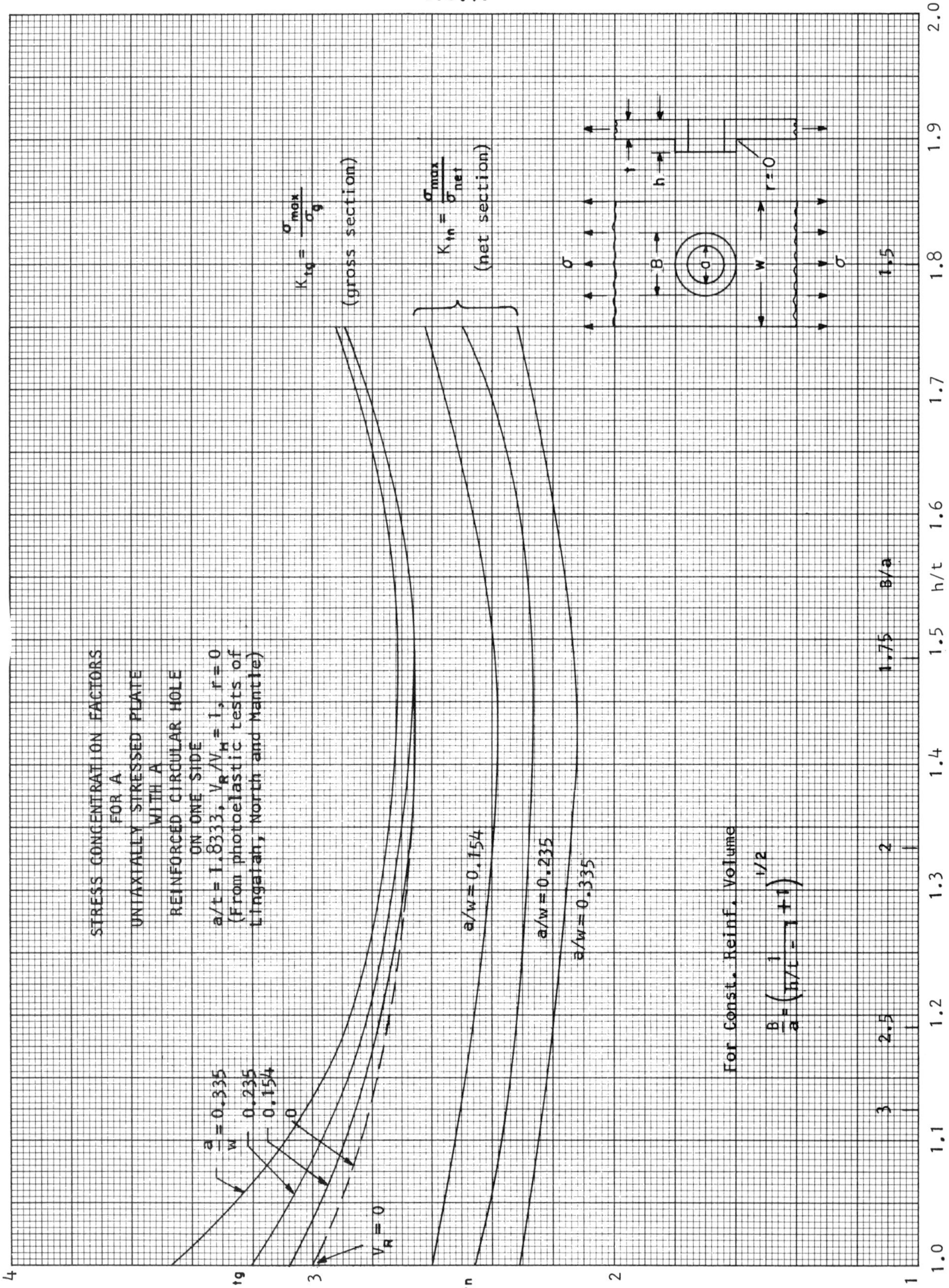
FOR A
UNIAXIALLY STRESSED PLATE
WITH A
REINFORCED CIRCULAR HOLE
ON ONE SIDE
$a/t = 1.8333,\ V_R/V_H = 1,\ r = 0$
(From photoelastic tests of
Lingaiah, North and Mantle)
$K_{tg} = \frac{\sigma_{max}}{\sigma_g}$
(gross section)
$K_{tn} = \frac{\sigma_{max}}{\sigma_{net}}$
(net section)
$\frac{a}{w} = 0.335$
0.235
0.154
0
$V_R = 0$
$a/w = 0.154$
$a/w = 0.235$
$a/w = 0.335$
For Const. Reinf. Volume
$\frac{B}{a} = \left(\frac{1}{h/t - 1} + 1\right)^{1/2}$
K_{tg}
K_{tn}
4
3
2
1
1.0
1.1
1.2
1.3
1.4
1.5
h/t
1.6
1.7
1.8
1.9
2.0
3
2.5
2
1.75
B/a
1.5
σ
B
a
w
t
h
r = 0

FIG.98

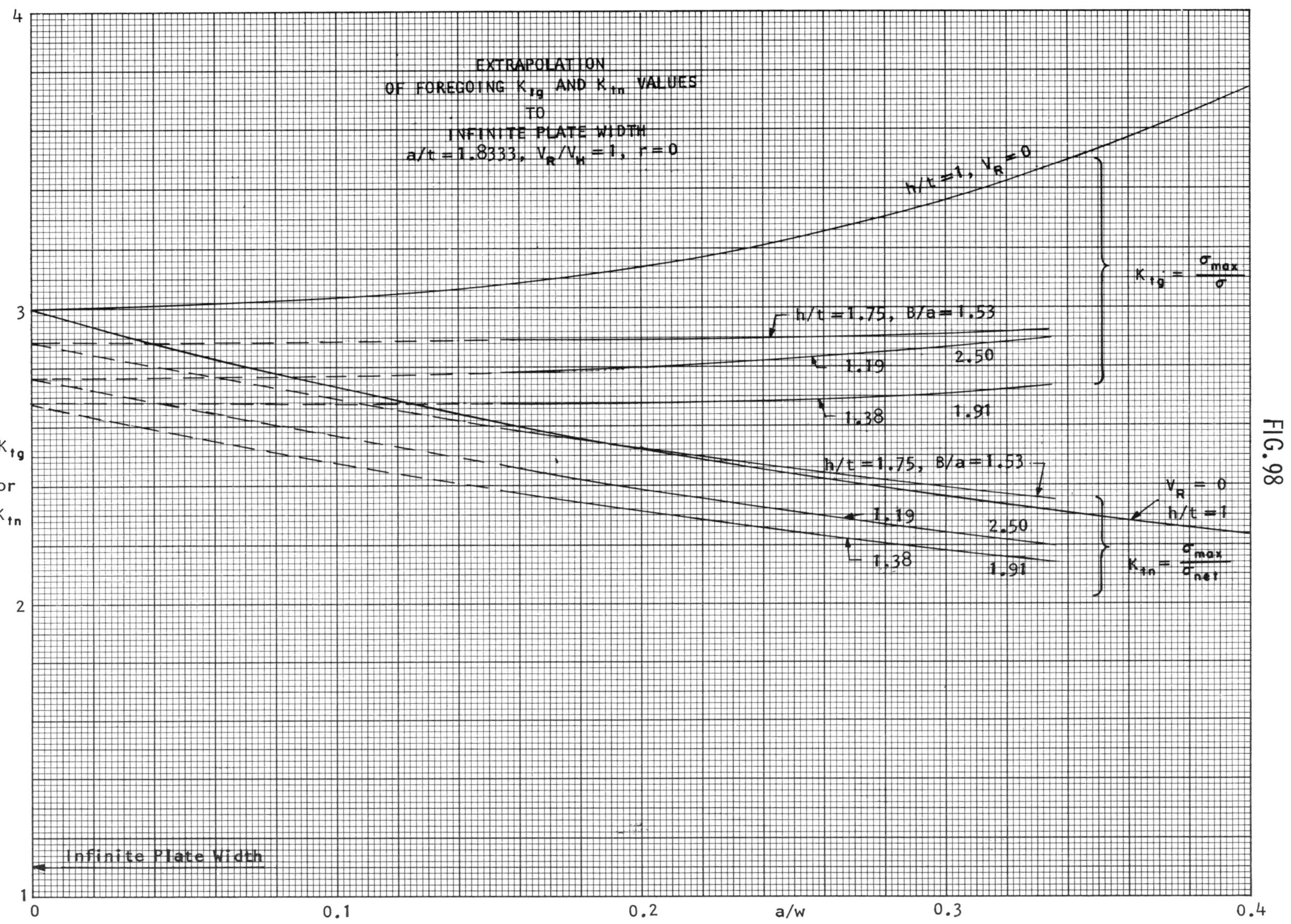

EXTRAPOLATION
OF FOREGOING K_{tg} AND K_{tn} VALUES
TO
INFINITE PLATE WIDTH
$a/t = 1.8333,\ V_R/V_W = 1,\ r = 0$
$h/t = 1,\ V_R = 0$
$K_{tg} = \frac{\sigma_{max}}{\sigma}$
$h/t = 1.75,\ B/a = 1.53$
1.19
2.50
1.38
1.91
$h/t = 1.75,\ B/a = 1.53$
$V_R = 0$
$h/t = 1$
1.19
2.50
1.38
1.91
$K_{tn} = \frac{\sigma_{max}}{\sigma_{net}}$
K_{tg}
or
K_{tn}
4
3
2
1
0
0.1
0.2
a/w
0.3
0.4
Infinite Plate Width

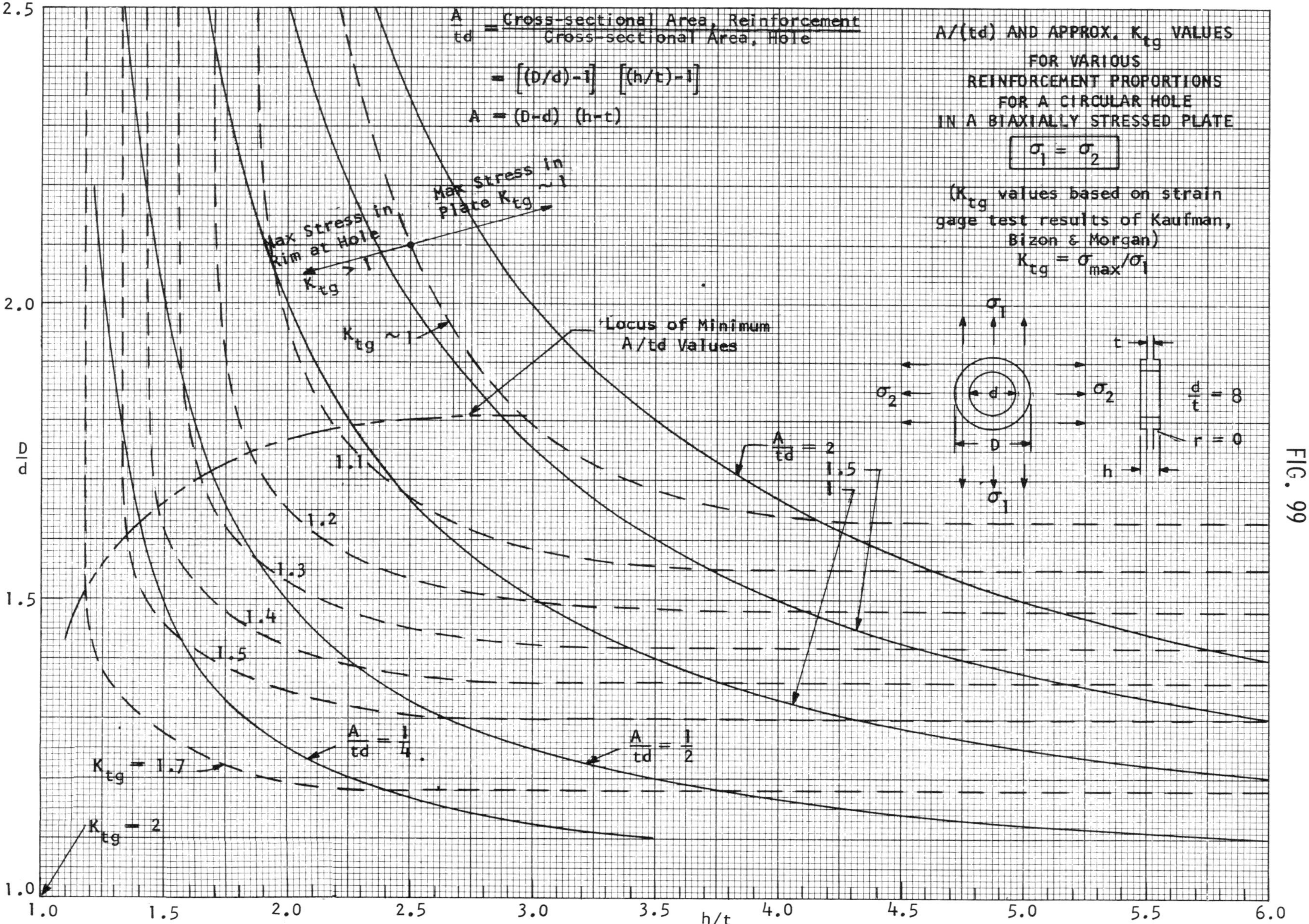

$\frac{A}{td} = \frac{\text{Cross-sectional Area, Reinforcement}}{\text{Cross-sectional Area, Hole}}$
$= \left[(D/d)-1\right]\left[(h/t)-1\right]$
$A = (D-d)(h-t)$
A/(td) AND APPROX. K_{tg} VALUES
FOR VARIOUS
REINFORCEMENT PROPORTIONS
FOR A CIRCULAR HOLE
IN A BIAXIALLY STRESSED PLATE
$\sigma_1 = \sigma_2$
(K_{tg} values based on strain gage test results of Kaufman, Bizon & Morgan)
$K_{tg} = \sigma_{max}/\sigma_1$
Max Stress in Plate $K_{tg} \sim 1$
Max Stress in Rim at Hole $K_{tg} > 1$
$K_{tg} \sim 1$
Locus of Minimum A/td Values
σ_1
σ_2
d
D
t
h
$\frac{d}{t} = 8$
$r = 0$
$\frac{A}{td} = 2$
1.5
1
1.1
1.2
1.3
1.4
1.5
$\frac{A}{td} = \frac{1}{4}$
$\frac{A}{td} = \frac{1}{2}$
$K_{tg} = 1.7$
$K_{tg} = 2$
$\frac{D}{d}$
2.5
2.0
1.5
1.0
h/t
1.0
1.5
2.0
2.5
3.0
3.5
4.0
4.5
5.0
5.5
6.0

FIG. 99

FIG. 100

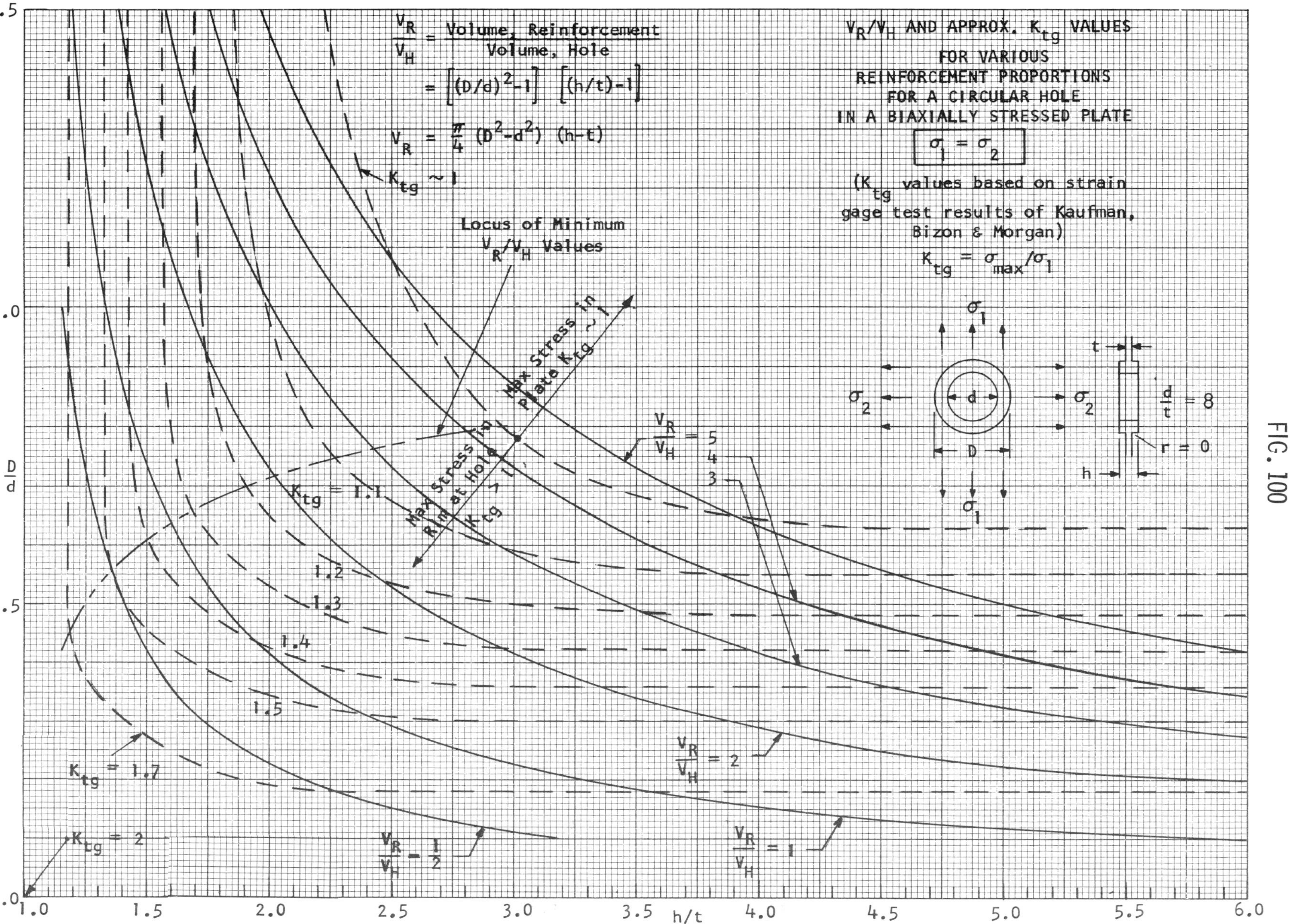

$\frac{V_R}{V_H} = \frac{\text{Volume, Reinforcement}}{\text{Volume, Hole}}$
$= \left[(D/d)^2-1\right] \left[(h/t)-1\right]$
$V_R = \frac{\pi}{4} (D^2-d^2) (h-t)$
V_R/V_H AND APPROX. K_{tg} VALUES
FOR VARIOUS
REINFORCEMENT PROPORTIONS
FOR A CIRCULAR HOLE
IN A BIAXIALLY STRESSED PLATE
$\sigma_1 = \sigma_2$
(K_{tg} values based on strain gage test results of Kaufman, Bizon & Morgan)
$K_{tg} = \sigma_{max}/\sigma_1$
$\frac{d}{t} = 8$
$r = 0$
Locus of Minimum V_R/V_H Values
Max Stress in Plate $K_{tg} \sim 1$
Max Stress in Rim at Hole $K_{tg} > 1$
$K_{tg} \sim 1$
$K_{tg} = 1.1$
1.2
1.3
1.4
1.5
$K_{tg} = 1.7$
$K_{tg} = 2$
$\frac{V_R}{V_H} = 5$
4
3
$\frac{V_R}{V_H} = 2$
$\frac{V_R}{V_H} = 1$
$\frac{V_R}{V_H} = \frac{1}{2}$
$\frac{D}{d}$
h/t

FIG. 101

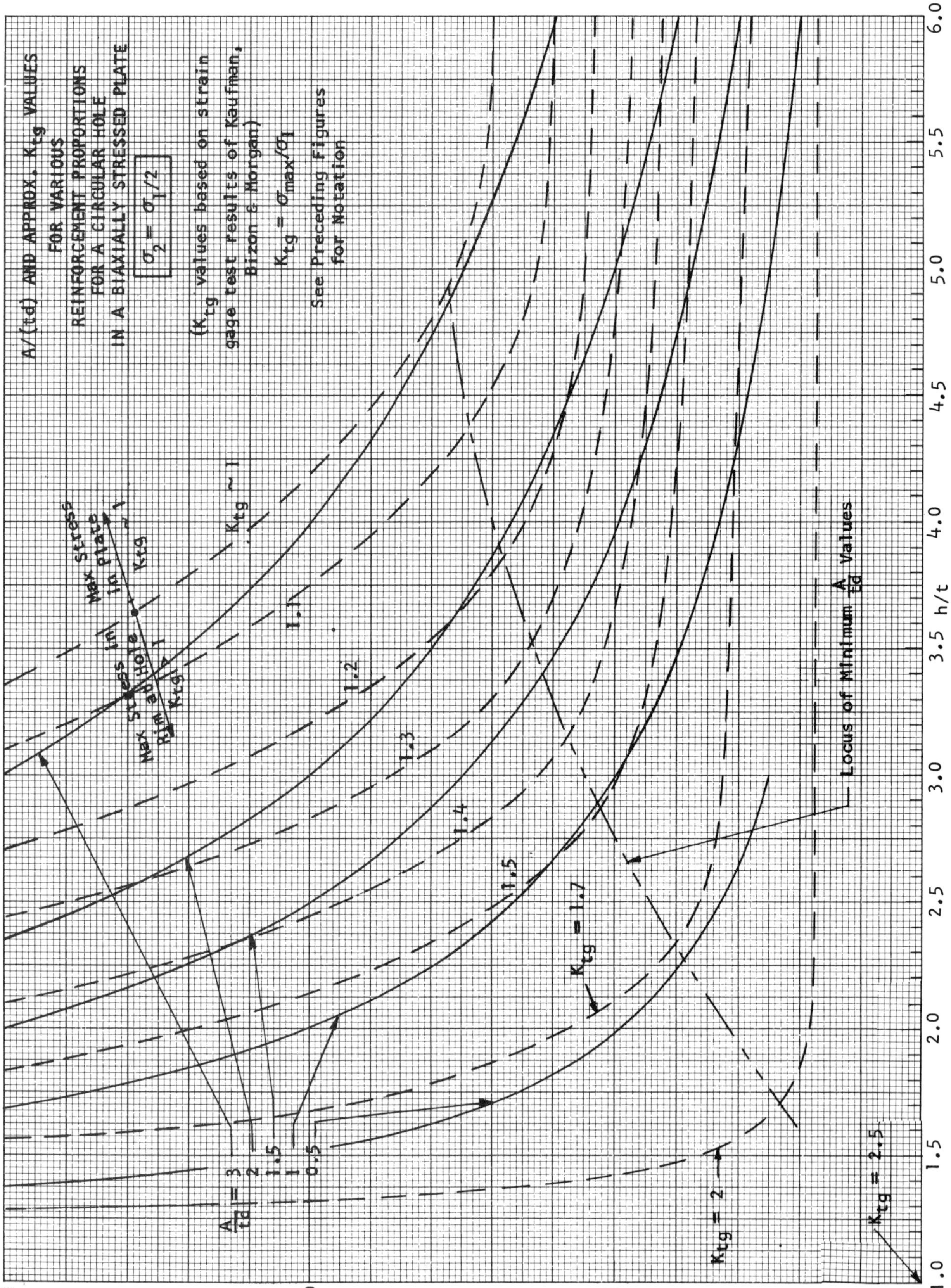

A/(td) AND APPROX. K_{tg} VALUES FOR VARIOUS REINFORCEMENT PROPORTIONS FOR A CIRCULAR HOLE IN A BIAXIALLY STRESSED PLATE

$\sigma_2 = \sigma_1/2$

(K_{tg} values based on strain gage test results of Kaufman, Bizon & Morgan)

$K_{tg} = \sigma_{max}/\sigma_1$

See Preceding Figures for Notation

Max Stress in Plate K_{tg}

Max Stress in Rim at Hole K_{tg}

$K_{tg} \sim 1$

1.1

1.2

1.3

1.4

1.5

$K_{tg} = 1.7$

$K_{tg} = 2$

$K_{tg} = 2.5$

$\frac{A}{td} = 3$, 2, 1.5, 1, 0.5

Locus of Minimum $\frac{A}{td}$ Values

$\frac{D}{d}$: 1.0, 1.5, 2.0, 2.5

h/t: 1.0, 1.5, 2.0, 2.5, 3.0, 3.5, 4.0, 4.5, 5.0, 5.5, 6.0

FIG. 102

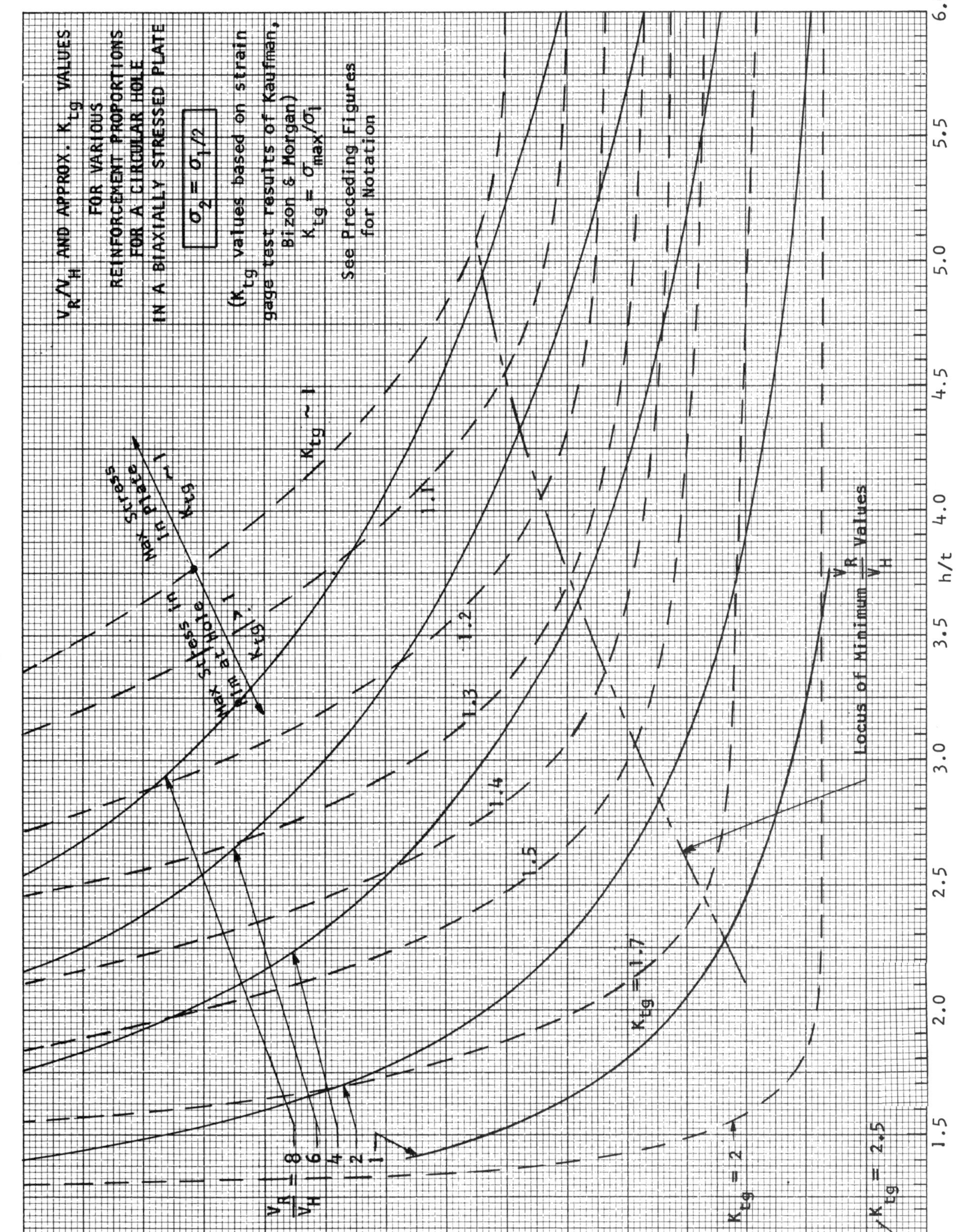

V_R/V_H AND APPROX. K_{tg} VALUES
FOR VARIOUS
REINFORCEMENT PROPORTIONS
FOR A CIRCULAR HOLE
IN A BIAXIALLY STRESSED PLATE

$\sigma_2 = \sigma_1/2$

(K_{tg} values based on strain gage test results of Kaufman, Bizon & Morgan)
$K_{tg} = \sigma_{max}/\sigma_1$

See Preceding Figures for Notation

Max Stress in Rim at Hole $K_{tg} > 1$
Max Stress in Plate $K_{tg} \sim 1$
$K_{tg} \sim 1$
1.1
1.2
1.3
1.4
1.5
$K_{tg} = 1.7$
$K_{tg} = 2$
$K_{tg} = 2.5$

$\frac{V_R}{V_H} = 8$, 6, 4, 2, 1

Locus of Minimum $\frac{V_R}{V_H}$ Values

$\frac{D}{d}$: 1.0, 1.5, 2.0, 2.5

h/t: 1.0, 1.5, 2.0, 2.5, 3.0, 3.5, 4.0, 4.5, 5.0, 5.5, 6.0

FIG. 103

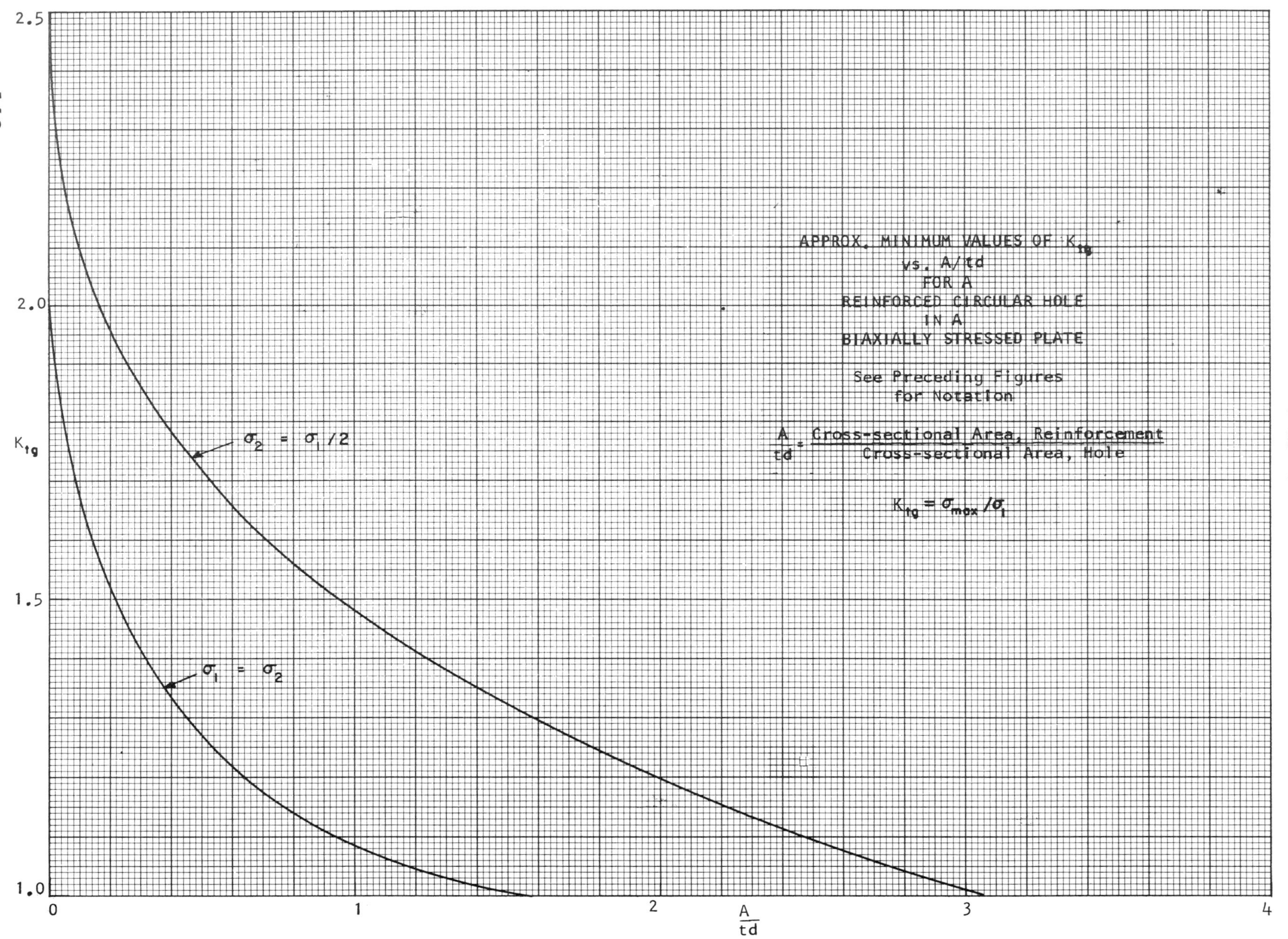

APPROX. MINIMUM VALUES OF K_{tg}
vs. A/td
FOR A
REINFORCED CIRCULAR HOLE
IN A
BIAXIALLY STRESSED PLATE
See Preceding Figures
for Notation
$\frac{A}{td} = \frac{\text{Cross-sectional Area, Reinforcement}}{\text{Cross-sectional Area, Hole}}$
$K_{tg} = \sigma_{max}/\sigma_1$
$\sigma_2 = \sigma_1/2$
$\sigma_1 = \sigma_2$
K_{tg}
2.5
2.0
1.5
1.0
$\frac{A}{td}$
0
1
2
3
4

FIG. 104

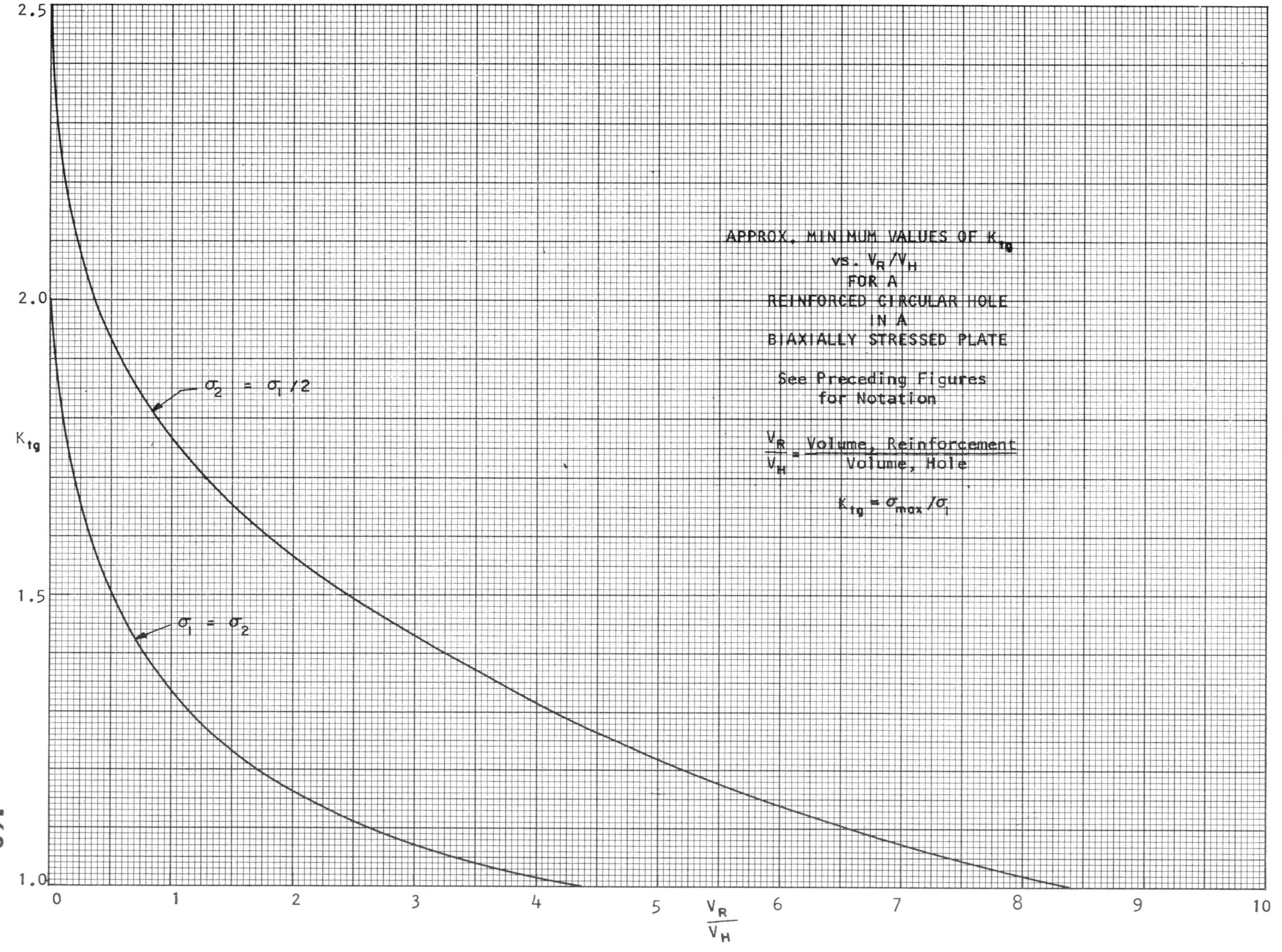

APPROX. MINIMUM VALUES OF K_{tg}
vs. V_R/V_H
FOR A
REINFORCED CIRCULAR HOLE
IN A
BIAXIALLY STRESSED PLATE
See Preceding Figures for Notation
$\frac{V_R}{V_H} = \frac{\text{Volume, Reinforcement}}{\text{Volume, Hole}}$
$K_{tg} = \sigma_{max}/\sigma_1$
$\sigma_2 = \sigma_1/2$
$\sigma_1 = \sigma_2$
K_{tg}
2.5
2.0
1.5
1.0
$\frac{V_R}{V_H}$
0
1
2
3
4
5
6
7
8
9
10

FIG. 104a

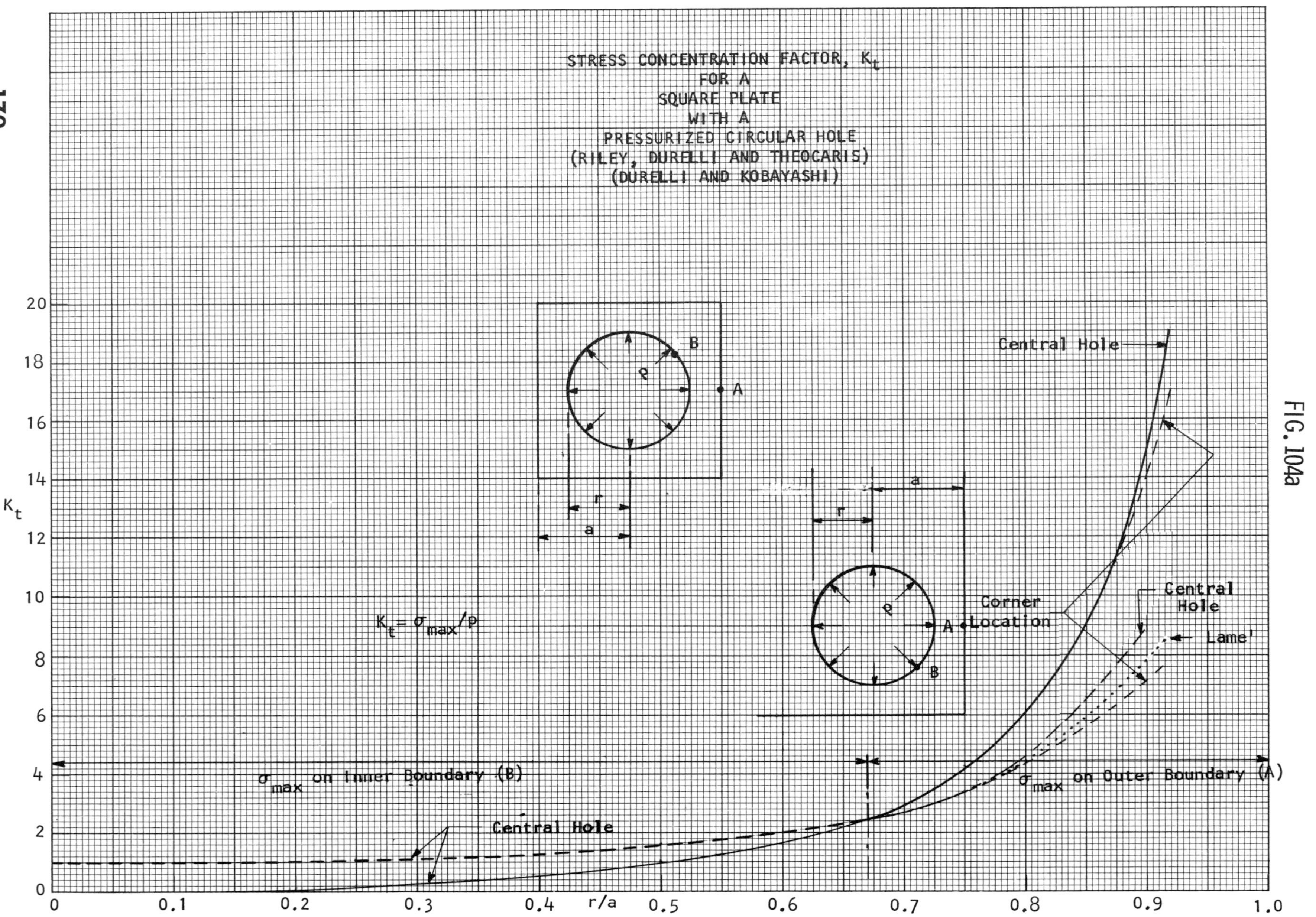

STRESS CONCENTRATION FACTOR, K_t
FOR A
SQUARE PLATE
WITH A
PRESSURIZED CIRCULAR HOLE
(RILEY, DURELLI AND THEOCARIS)
(DURELLI AND KOBAYASHI)
$K_t = \sigma_{max}/p$
Central Hole
Corner Location
Central Hole
Lame'
σ_{max} on Inner Boundary (B)
σ_{max} on Outer Boundary (A)
Central Hole
A
B
p
r
a
K_t
r/a
0
2
4
6
8
10
12
14
16
18
20
0
0.1
0.2
0.3
0.4
0.5
0.6
0.7
0.8
0.9
1.0

FIG. 105

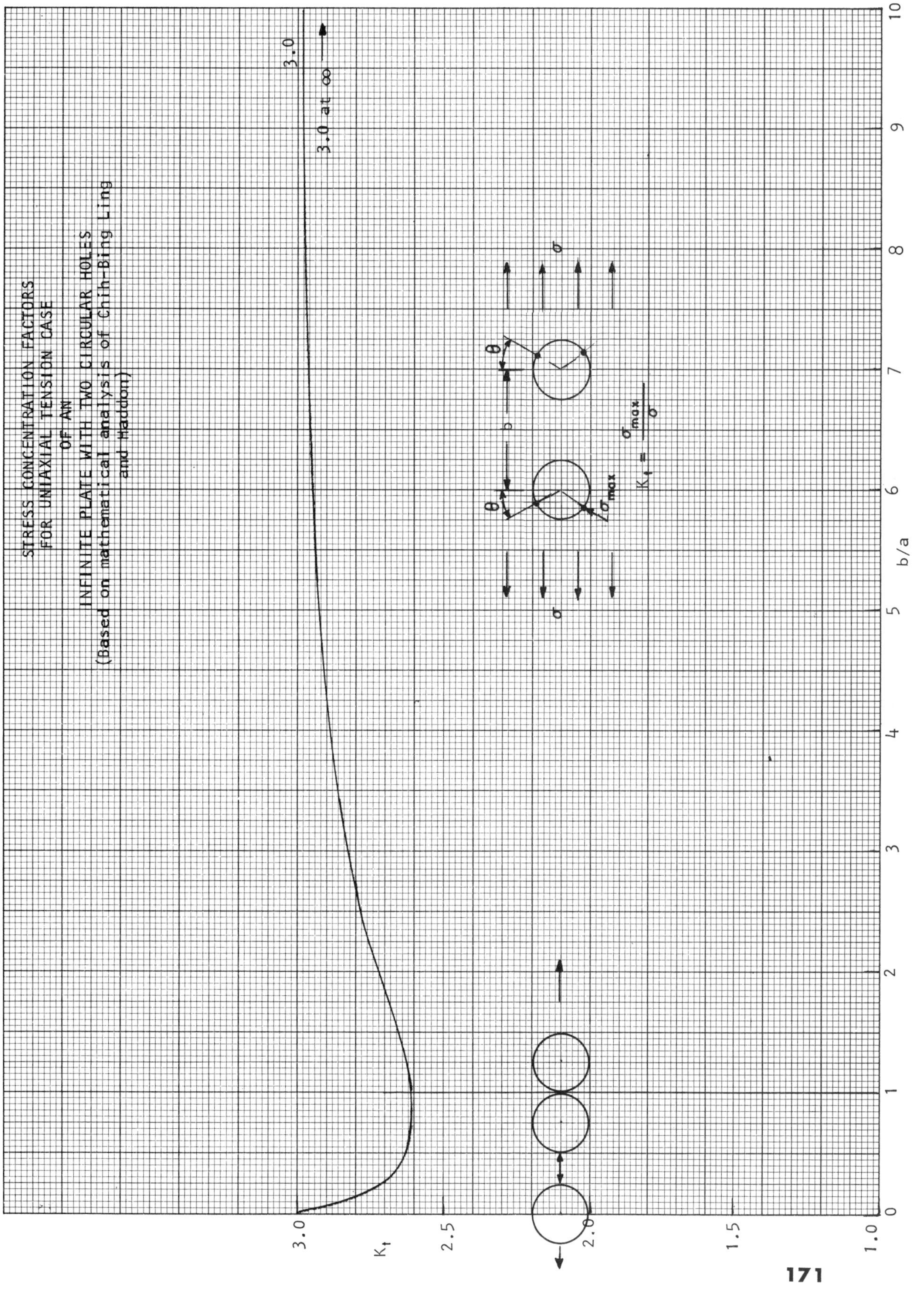

STRESS CONCENTRATION FACTORS
FOR UNIAXIAL TENSION CASE
OF AN
INFINITE PLATE WITH TWO CIRCULAR HOLES
(Based on mathematical analysis of Chih-Bing Ling
and Haddon)

K_t

3.0
2.5
2.0
1.5
1.0

0 1 2 3 4 5 6 7 8 9 10

b/a

3.0

3.0 at ∞

σ

θ

b

σ_{max}

$K_t = \frac{\sigma_{max}}{\sigma}$

FIG. 106

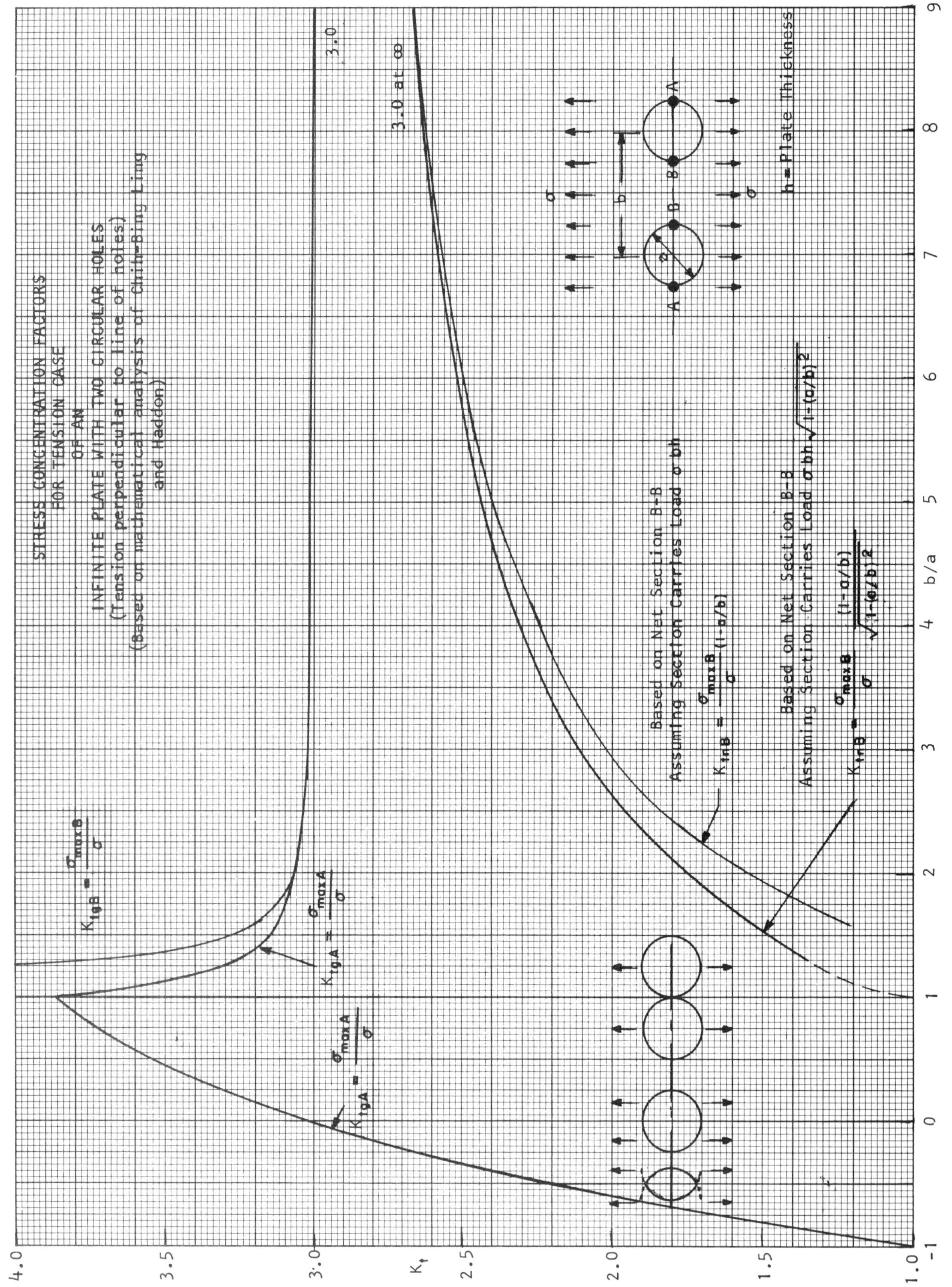

STRESS CONCENTRATION FACTORS
FOR TENSION CASE
OF AN
INFINITE PLATE WITH TWO CIRCULAR HOLES
(Tension perpendicular to line of holes)
(Based on mathematical analysis of Chih-Bing Ling and Haddon)

$K_{tgB} = \frac{\sigma_{max B}}{\sigma}$

$K_{tgA} = \frac{\sigma_{max A}}{\sigma}$

$K_{tgA} = \frac{\sigma_{max A}}{\sigma}$

3.0

3.0 at ∞

Based on Net Section B-B
Assuming Section Carries Load σbh

$K_{tnB} = \frac{\sigma_{max B}}{\sigma}(1-a/b)$

Based on Net Section B-B
Assuming Section Carries Load $\sigma bh\sqrt{1-(a/b)^2}$

$K_{tnB} = \frac{\sigma_{max B}}{\sigma} \frac{(1-a/b)}{\sqrt{1-(a/b)^2}}$

h = Plate Thickness

σ, A, B, B, A, a, b

K_t: 1.0, 1.5, 2.0, 2.5, 3.0, 3.5, 4.0

b/a: -1, 0, 1, 2, 3, 4, 5, 6, 7, 8, 9

FIG. 107

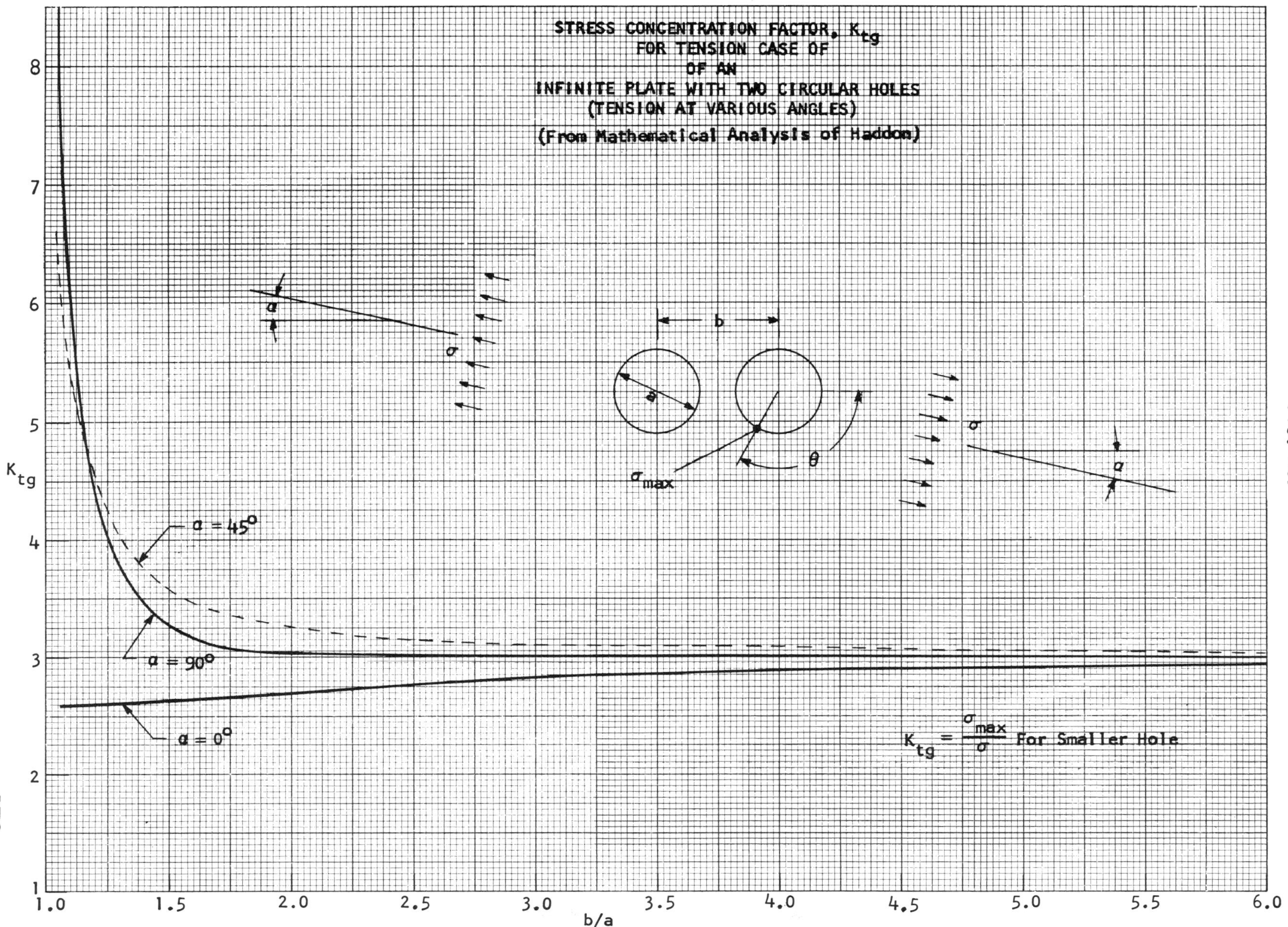
STRESS CONCENTRATION FACTOR, K_{tg}
FOR TENSION CASE OF
OF AN
INFINITE PLATE WITH TWO CIRCULAR HOLES
(TENSION AT VARIOUS ANGLES)
(From Mathematical Analysis of Haddon)
$\alpha = 45^o$
$\alpha = 90^o$
$\alpha = 0^o$
$K_{tg} = \frac{\sigma_{max}}{\sigma}$ For Smaller Hole
σ
σ_{max}
θ
a
b
K_{tg}
b/a
1.0
1.5
2.0
2.5
3.0
3.5
4.0
4.5
5.0
5.5
6.0
1
2
3
4
5
6
7
8

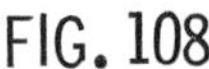

FIG. 108

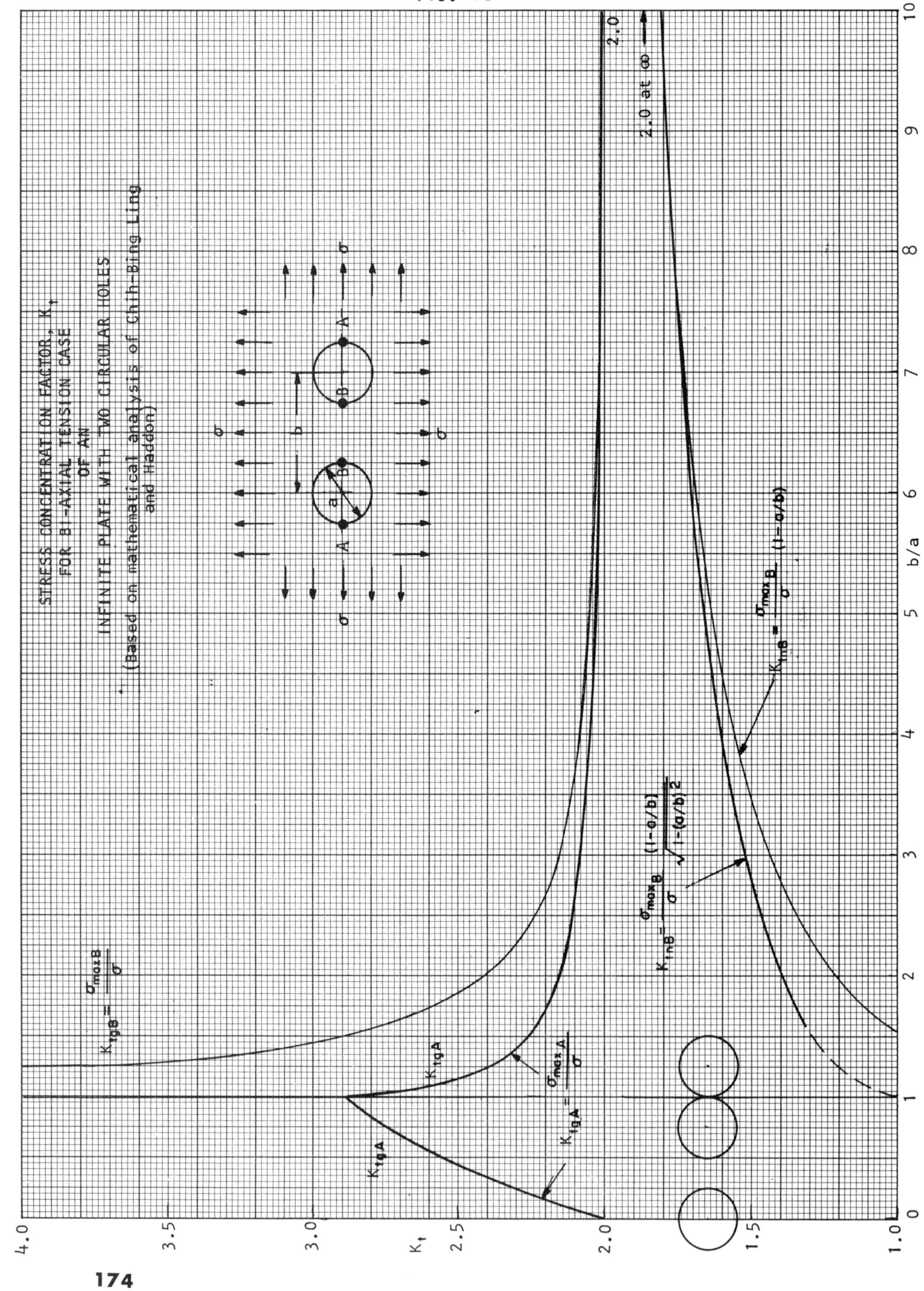

STRESS CONCENTRATION FACTOR, K_t
FOR BI-AXIAL TENSION CASE
OF AN
INFINITE PLATE WITH TWO CIRCULAR HOLES
(Based on mathematical analysis of Chih-Bing Ling and Haddon)
$K_{tgB} = \frac{\sigma_{max\,B}}{\sigma}$
K_{tgA}
$K_{tgA} = \frac{\sigma_{max\,A}}{\sigma}$
$K_{tnB} = \frac{\sigma_{max\,B}}{\sigma} \frac{(1-a/b)}{\sqrt{1-(a/b)^2}}$
$K_{tnB} = \frac{\sigma_{max\,B}}{\sigma}(1-a/b)$
2.0 at ∞
2.0
K_t
b/a
4.0
3.5
3.0
2.5
2.0
1.5
1.0
0
1
2
3
4
5
6
7
8
9
10

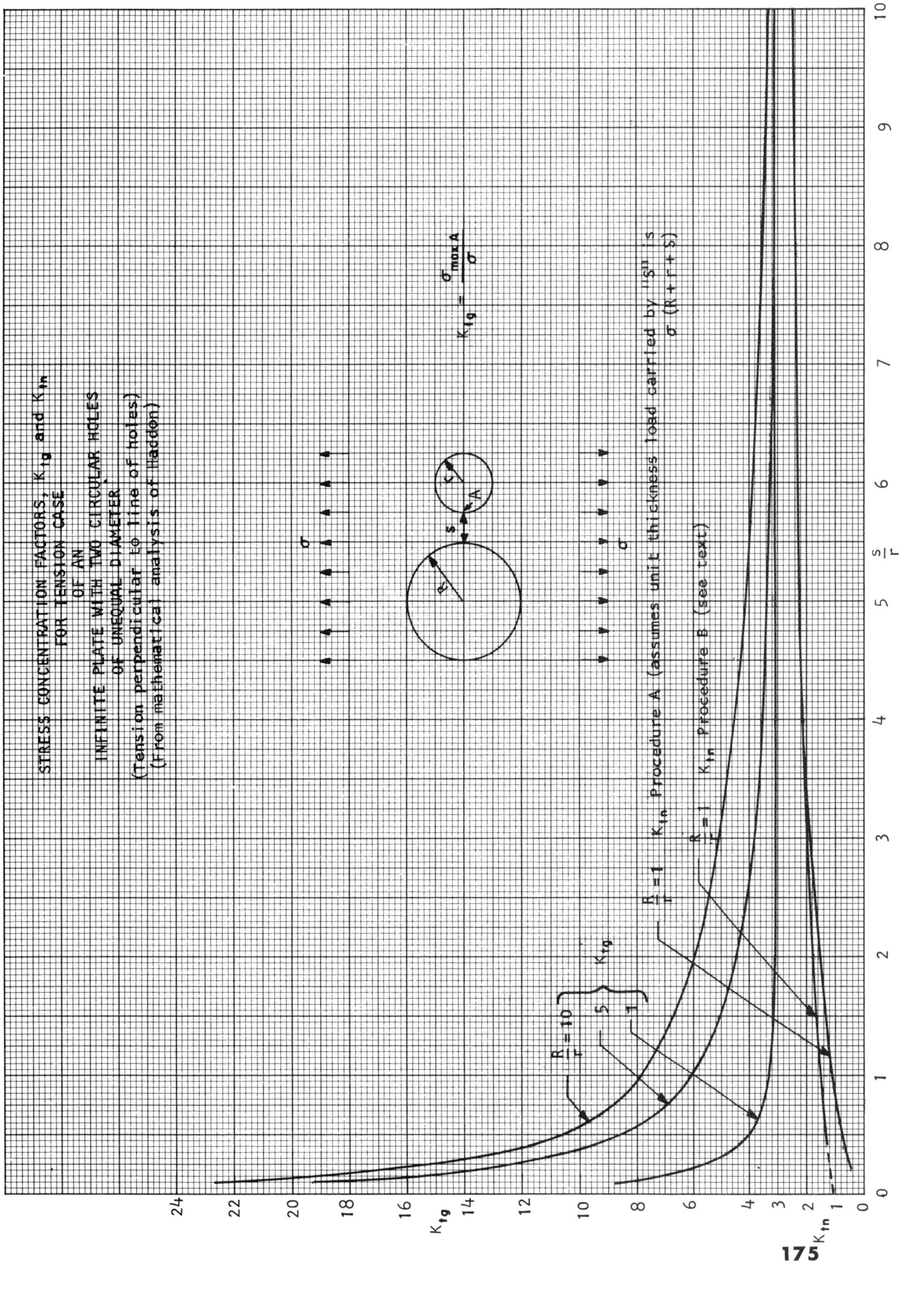

STRESS CONCENTRATION FACTORS, K_{tg} and K_{tn}
FOR TENSION CASE
OF AN
INFINITE PLATE WITH TWO CIRCULAR HOLES
OF UNEQUAL DIAMETER
(Tension perpendicular to line of holes)
(From mathematical analysis of Haddon)

σ

R s r A

σ

$$K_{tg} = \frac{\sigma_{max\,A}}{\sigma}$$

$\frac{R}{r} = 10$, 5, 1 K_{tg}

$\frac{R}{r} = 1$ K_{tn} Procedure A (assumes unit thickness load carried by "S" is $\sigma\,(R + r + S)$

$\frac{R}{r} = 1$ K_{tn} Procedure B (see text)

K_{tg}: 24, 22, 20, 18, 16, 14, 12, 10, 8, 6, 4, 3, 2, 1, 0

K_{tn}

$\frac{s}{r}$: 0, 1, 2, 3, 4, 5, 6, 7, 8, 9, 10

FIG. 110

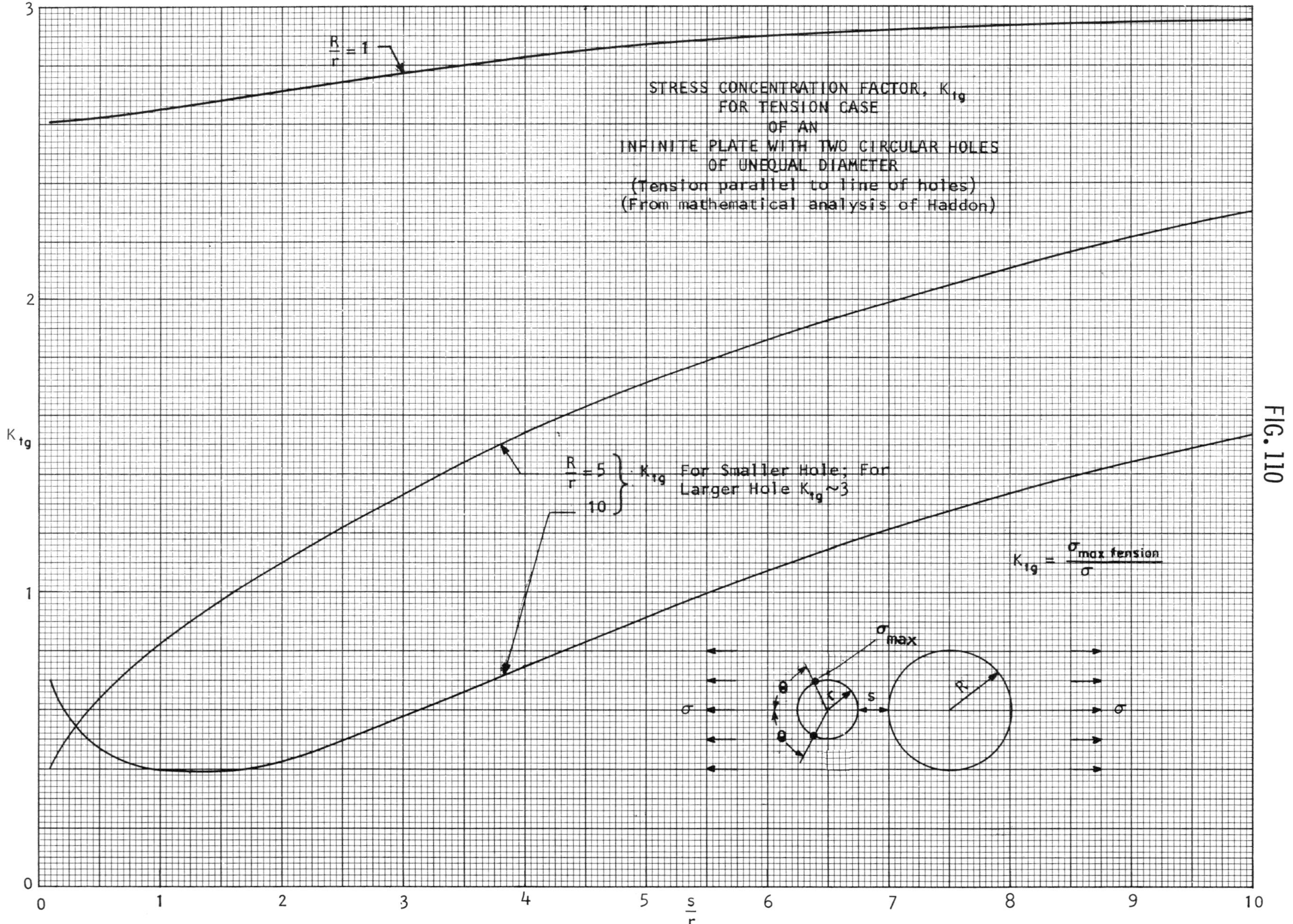

STRESS CONCENTRATION FACTOR, K_{tg}
FOR TENSION CASE
OF AN
INFINITE PLATE WITH TWO CIRCULAR HOLES
OF UNEQUAL DIAMETER
(Tension parallel to line of holes)
(From mathematical analysis of Haddon)
$\frac{R}{r} = 1$
$\frac{R}{r} = 5$ } K_{tg} For Smaller Hole; For
10 } Larger Hole $K_{tg} \sim 3$
$K_{tg} = \frac{\sigma_{max\ tension}}{\sigma}$
σ_{max}
θ
θ
r
s
R
σ
σ
K_{tg}
0
1
2
3
$\frac{s}{r}$
0
1
2
3
4
5
6
7
8
9
10

FIG. 111

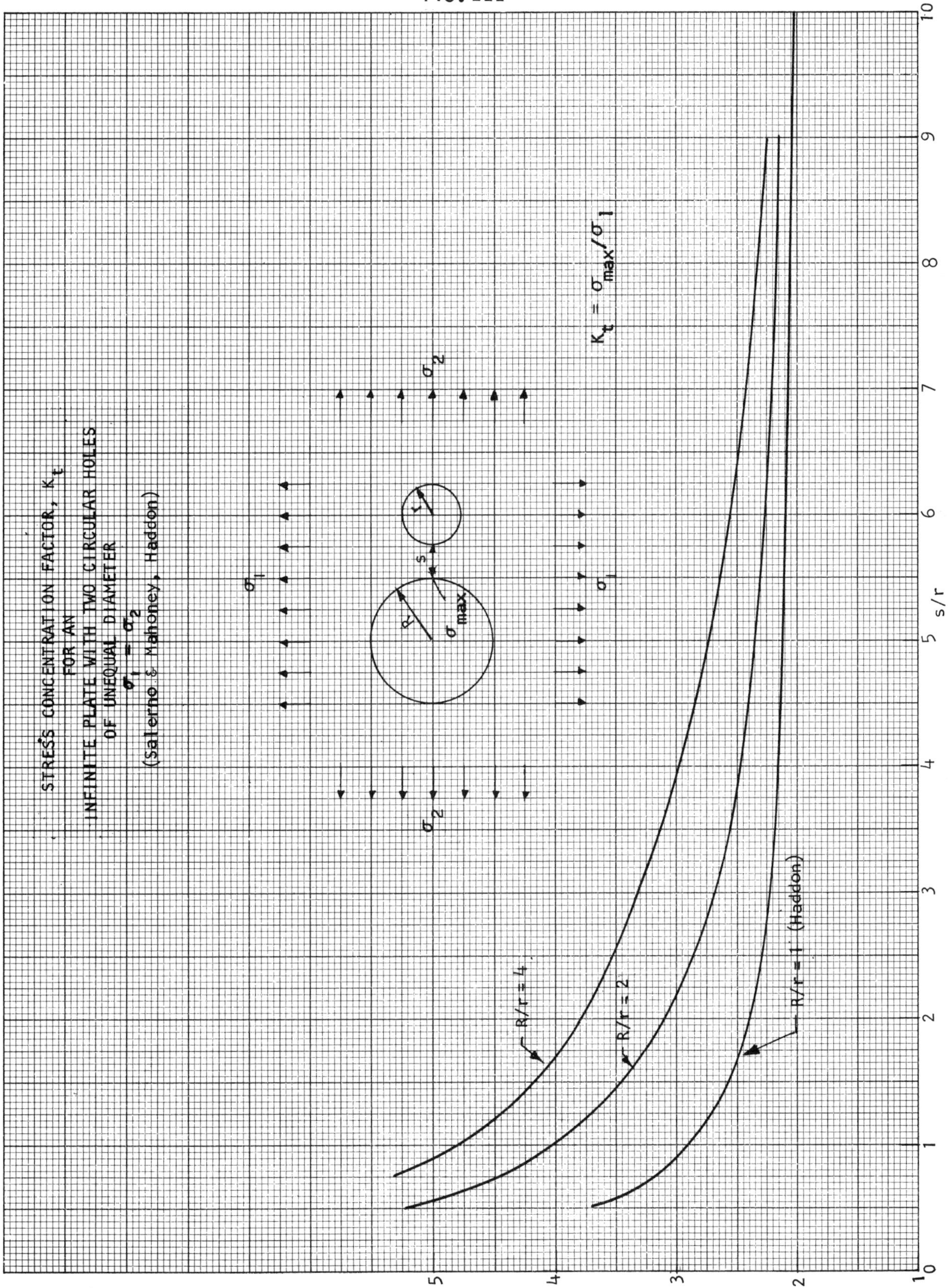

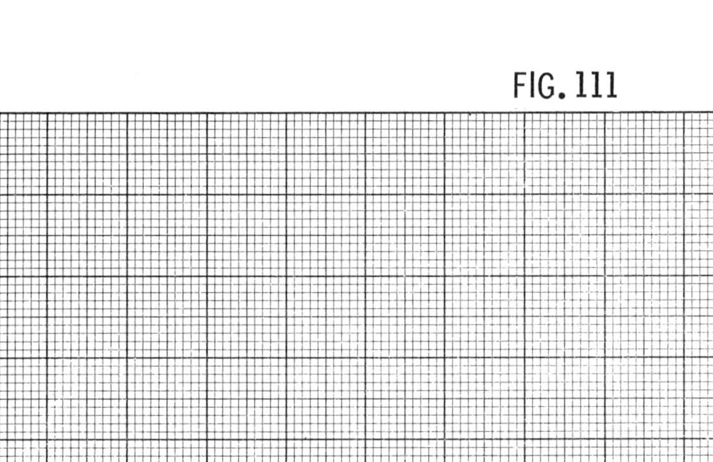

STRESS CONCENTRATION FACTOR, K_t
FOR AN
INFINITE PLATE WITH TWO CIRCULAR HOLES
OF UNEQUAL DIAMETER
$\sigma_1 = \sigma_2$
(Salerno & Mahoney, Haddon)
$K_t = \sigma_{max}/\sigma_1$
σ_1
σ_2
R
r
s
σ_{max}
R/r = 4
R/r = 2
R/r = 1 (Haddon)
s/r
K_t
0
1
2
3
4
5
6
7
8
9
10

FIG. 112

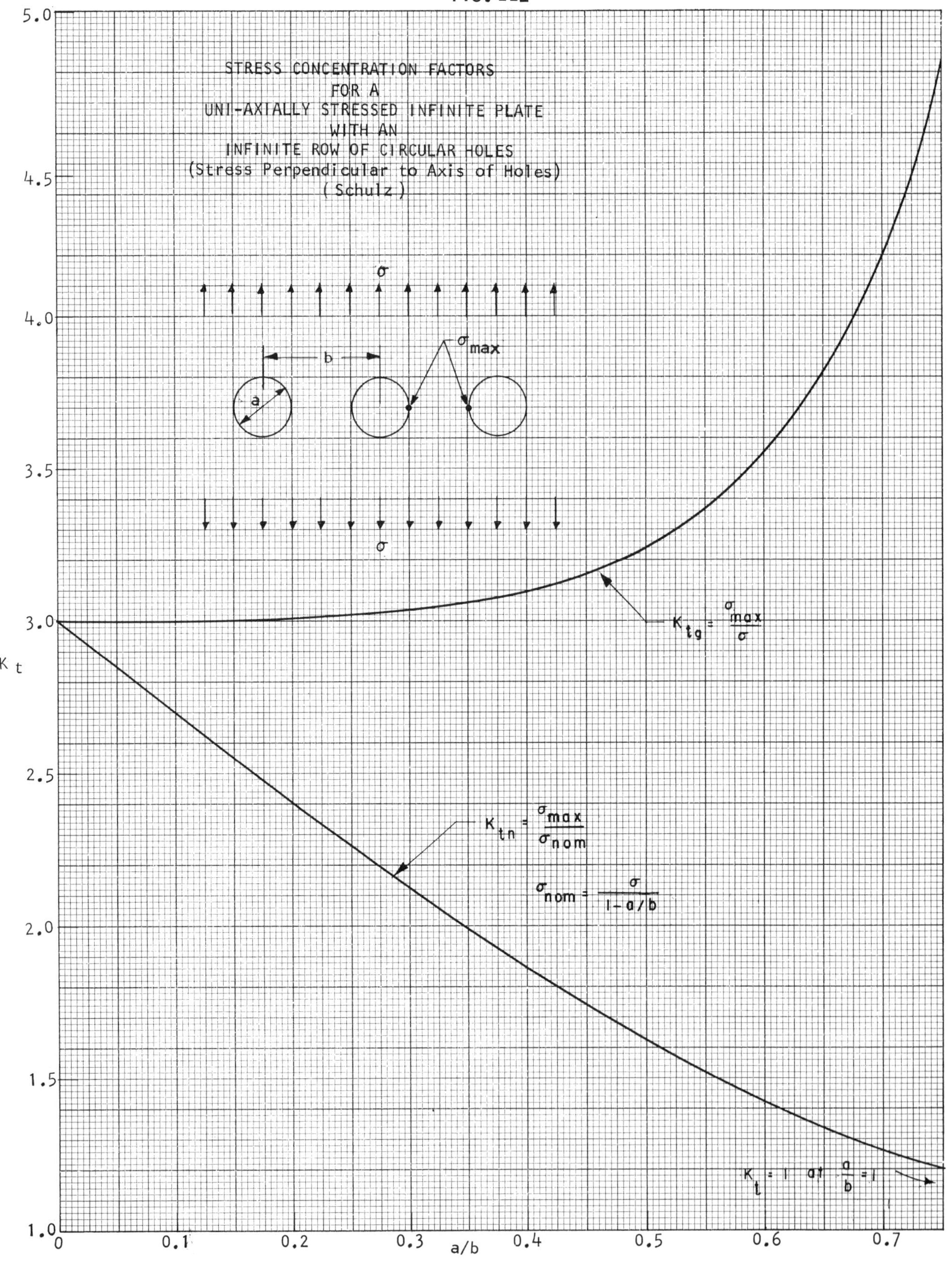
STRESS CONCENTRATION FACTORS
FOR A
UNI-AXIALLY STRESSED INFINITE PLATE
WITH AN
INFINITE ROW OF CIRCULAR HOLES
(Stress Perpendicular to Axis of Holes)
(Schulz)
σ
b
a
σ max
σ
$K_{tg} = \frac{\sigma_{max}}{\sigma}$
$K_{tn} = \frac{\sigma_{max}}{\sigma_{nom}}$
$\sigma_{nom} = \frac{\sigma}{1 - a/b}$
$K_t = 1$ at $\frac{a}{b} = 1$
K_t
a/b
5.0
4.5
4.0
3.5
3.0
2.5
2.0
1.5
1.0
0
0.1
0.2
0.3
0.4
0.5
0.6
0.7

FIG. 113

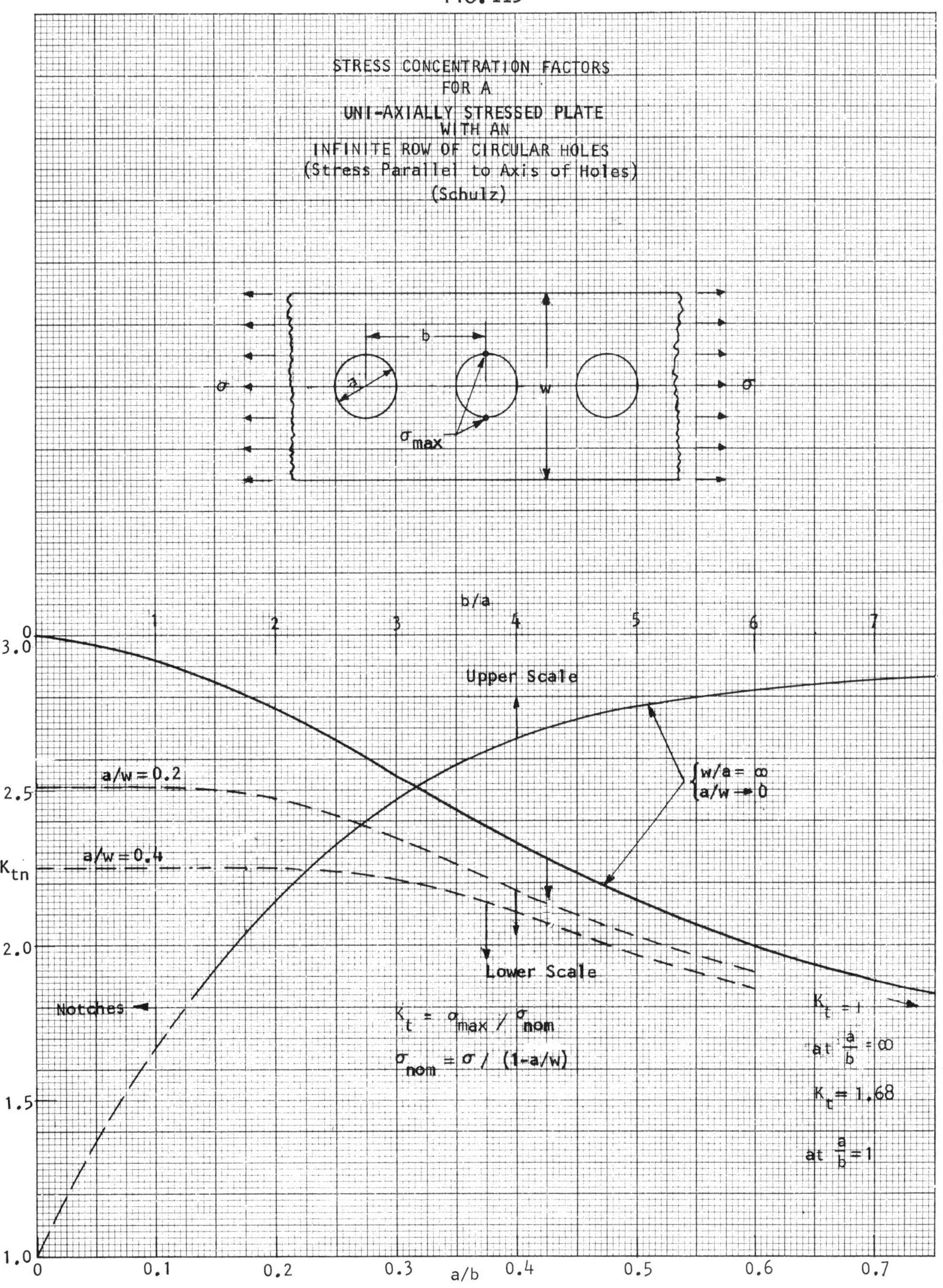
STRESS CONCENTRATION FACTORS
FOR A
UNI-AXIALLY STRESSED PLATE
WITH AN
INFINITE ROW OF CIRCULAR HOLES
(Stress Parallel to Axis of Holes)
(Schulz)
b
a
σ
w
σ
σmax
b/a
0
1
2
3
4
5
6
7
3.0
2.5
2.0
1.5
1.0
Ktn
Upper Scale
a/w = 0.2
a/w = 0.4
w/a = ∞
a/w → 0
Lower Scale
Notches
Kt = σmax / σnom
σnom = σ / (1-a/w)
Kt = 1
at a/b = ∞
Kt = 1.68
at a/b = 1
0
0.1
0.2
0.3
0.4
0.5
0.6
0.7
a/b

FIG. 114

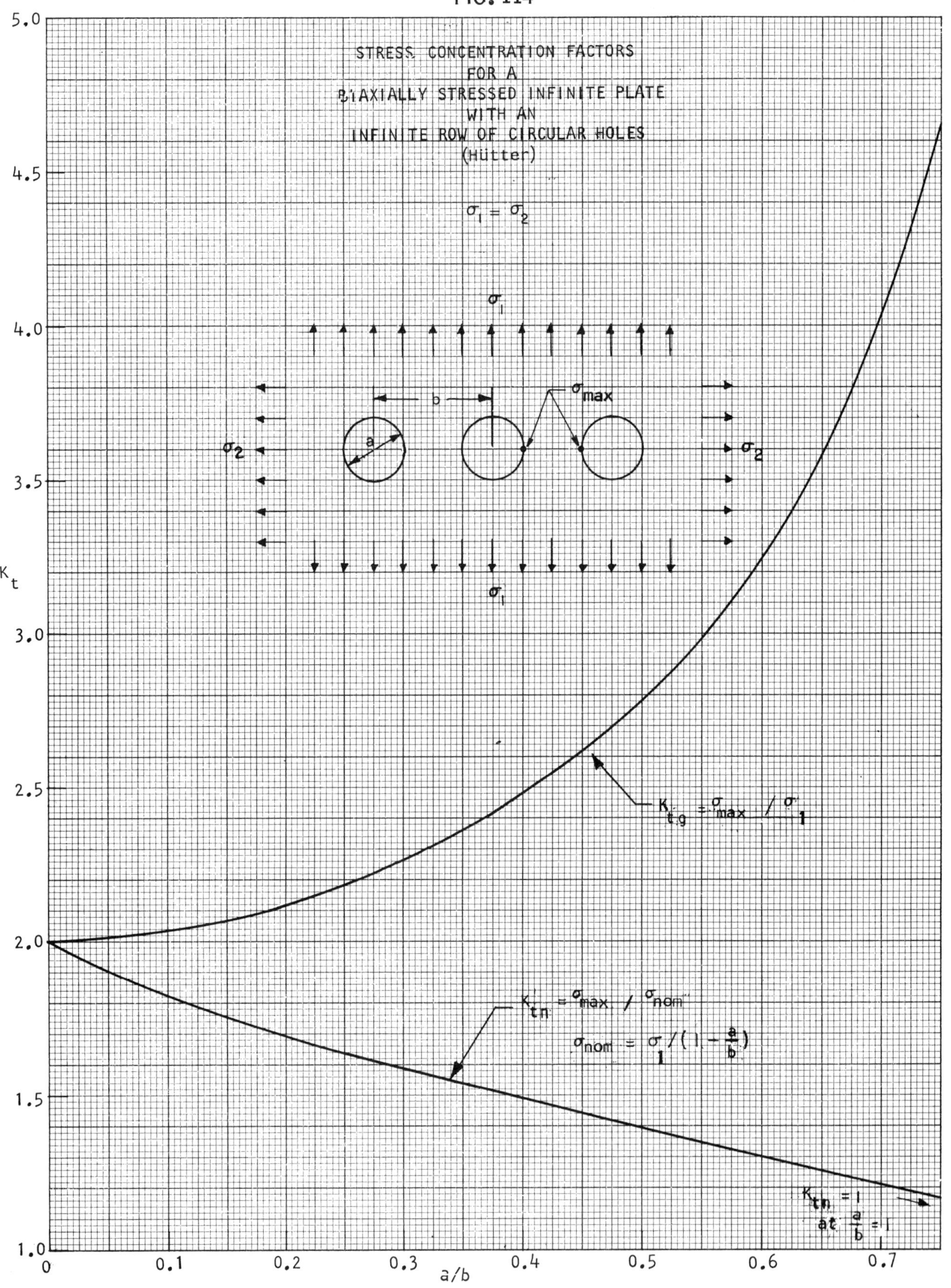

STRESS CONCENTRATION FACTORS
FOR A
BIAXIALLY STRESSED INFINITE PLATE
WITH AN
INFINITE ROW OF CIRCULAR HOLES
(Hütter)
$\sigma_1 = \sigma_2$
σ_1
σ_2
b
a
σ_{max}
$K_{tg} = \sigma_{max} / \sigma_1$
$K_{tn} = \sigma_{max} / \sigma_{nom}$
$\sigma_{nom} = \sigma_1 / (1 - \frac{a}{b})$
$K_{tn} = 1$
at $\frac{a}{b} = 1$
K_t
a/b
5.0
4.5
4.0
3.5
3.0
2.5
2.0
1.5
1.0
0
0.1
0.2
0.3
0.4
0.5
0.6
0.7

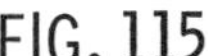

FIG. 115

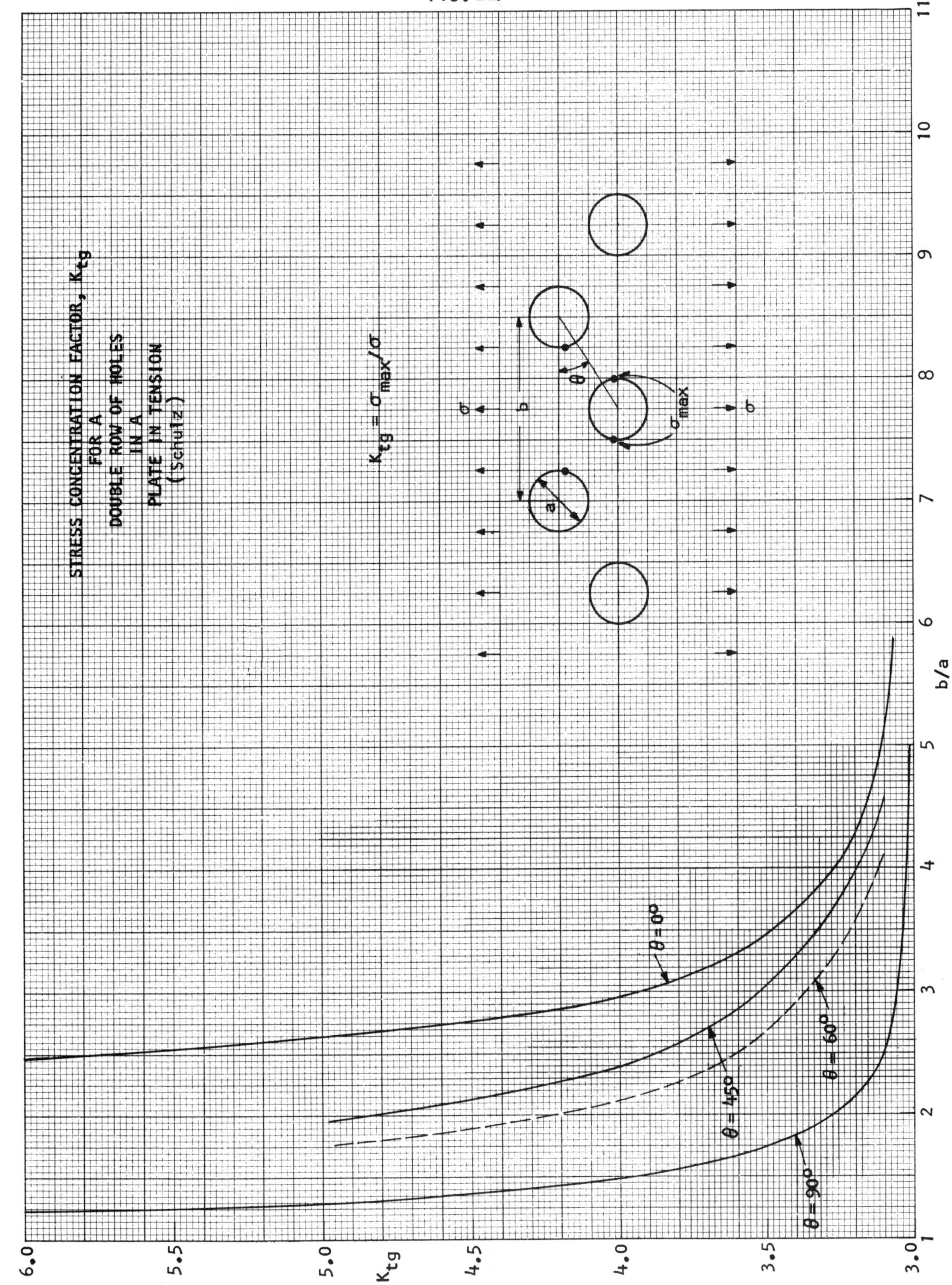

STRESS CONCENTRATION FACTOR, K_{tg}
FOR A
DOUBLE ROW OF HOLES
IN A
PLATE IN TENSION
(Schulz)
$K_{tg} = \sigma_{max}/\sigma$
σ
b
a
θ
σ_{max}
θ = 0°
θ = 45°
θ = 60°
θ = 90°
K_{tg}
b/a
6.0
5.5
5.0
4.5
4.0
3.5
3.0
1
2
3
4
5
6
7
8
9
10
11

FIG. 116

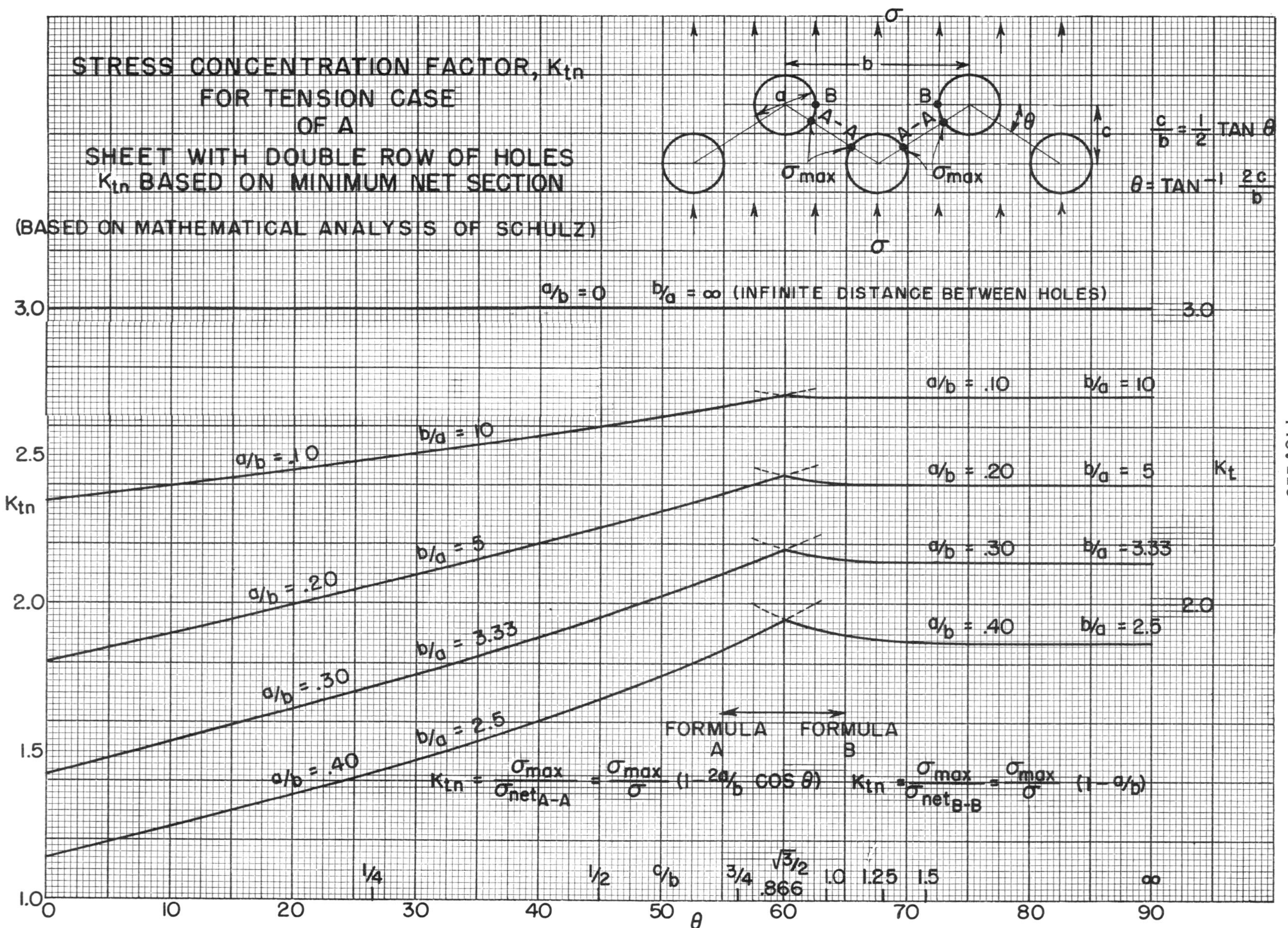
STRESS CONCENTRATION FACTOR, K_{tn}
FOR TENSION CASE
OF A
SHEET WITH DOUBLE ROW OF HOLES
K_{tn} BASED ON MINIMUM NET SECTION
(BASED ON MATHEMATICAL ANALYSIS OF SCHULZ)
$\frac{c}{b} = \frac{1}{2} \tan\theta$
$\theta = \tan^{-1} \frac{2c}{b}$
a/b = 0   b/a = ∞ (INFINITE DISTANCE BETWEEN HOLES)
a/b = .10   b/a = 10
a/b = .20   b/a = 5
a/b = .30   b/a = 3.33
a/b = .40   b/a = 2.5
K_t
K_{tn}
FORMULA A
FORMULA B
$K_{tn} = \frac{\sigma_{max}}{\sigma_{net_{A-A}}} = \frac{\sigma_{max}}{\sigma}(1 - 2a/b \cos\theta)$
$K_{tn} = \frac{\sigma_{max}}{\sigma_{net_{B-B}}} = \frac{\sigma_{max}}{\sigma}(1 - a/b)$
c/b
1/4 1/2 3/4 $\sqrt{3}/2$ .866 1.0 1.25 1.5 ∞
θ
0 10 20 30 40 50 60 70 80 90
1.0 1.5 2.0 2.5 3.0

FIG. 117

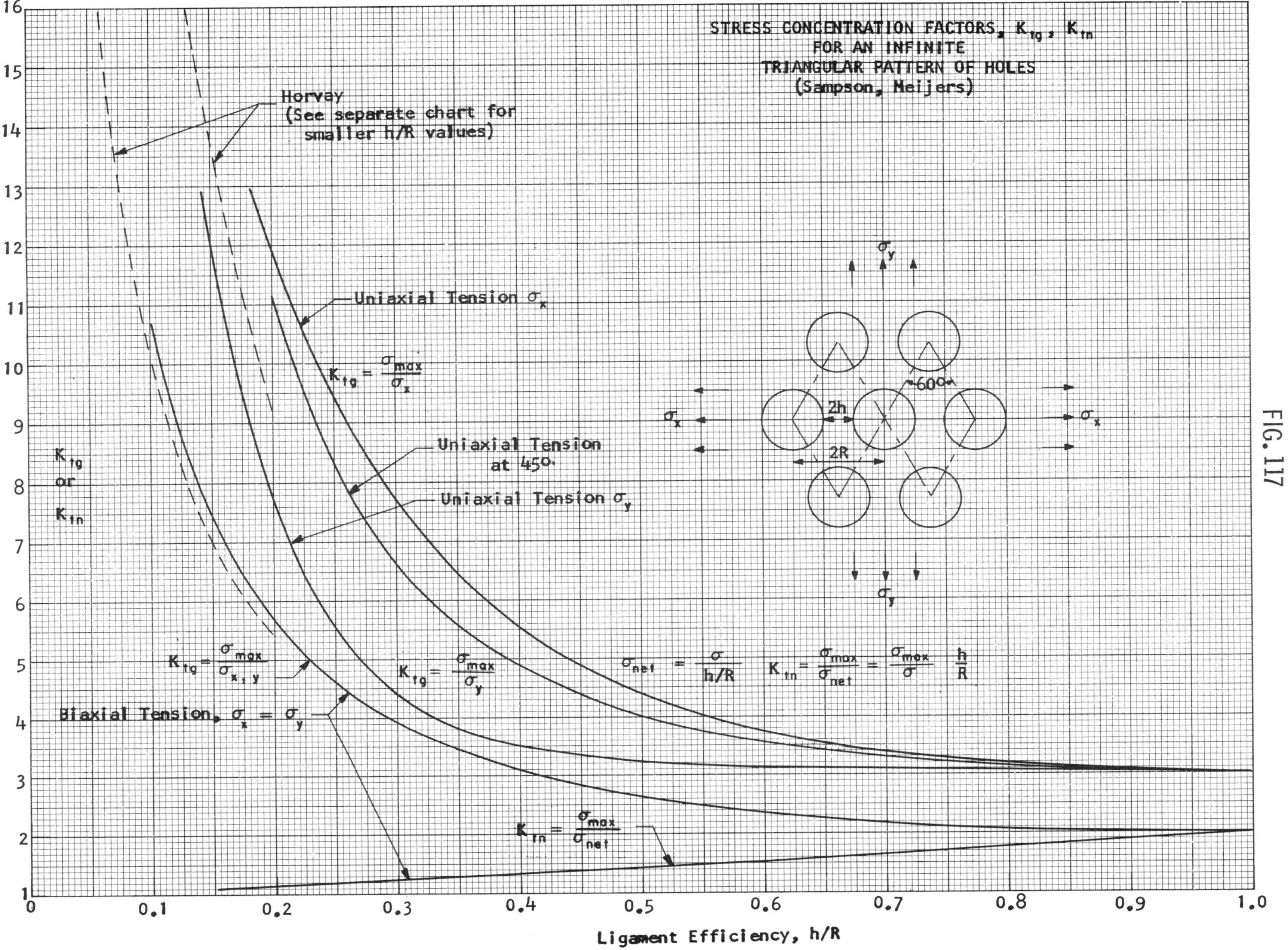
STRESS CONCENTRATION FACTORS, K_{tg}, K_{tn}
FOR AN INFINITE
TRIANGULAR PATTERN OF HOLES
(Sampson, Meijers)
Horvay
(See separate chart for
smaller h/R values)
Uniaxial Tension σ_x
$K_{tg} = \frac{\sigma_{max}}{\sigma_x}$
Uniaxial Tension
at 45°
Uniaxial Tension σ_y
$K_{tg} = \frac{\sigma_{max}}{\sigma_{x,y}}$
$K_{tg} = \frac{\sigma_{max}}{\sigma_y}$
Biaxial Tension, $\sigma_x = \sigma_y$
$\sigma_{net} = \frac{\sigma}{h/R}$
$K_{tn} = \frac{\sigma_{max}}{\sigma_{net}} = \frac{\sigma_{max}}{\sigma} \frac{h}{R}$
$K_{tn} = \frac{\sigma_{max}}{\sigma_{net}}$
σ_x
σ_y
60°
2h
2R
K_{tg}
or
K_{tn}
Ligament Efficiency, h/R
0
0.1
0.2
0.3
0.4
0.5
0.6
0.7
0.8
0.9
1.0
1
2
3
4
5
6
7
8
9
10
11
12
13
14
15
16

FIG. 118

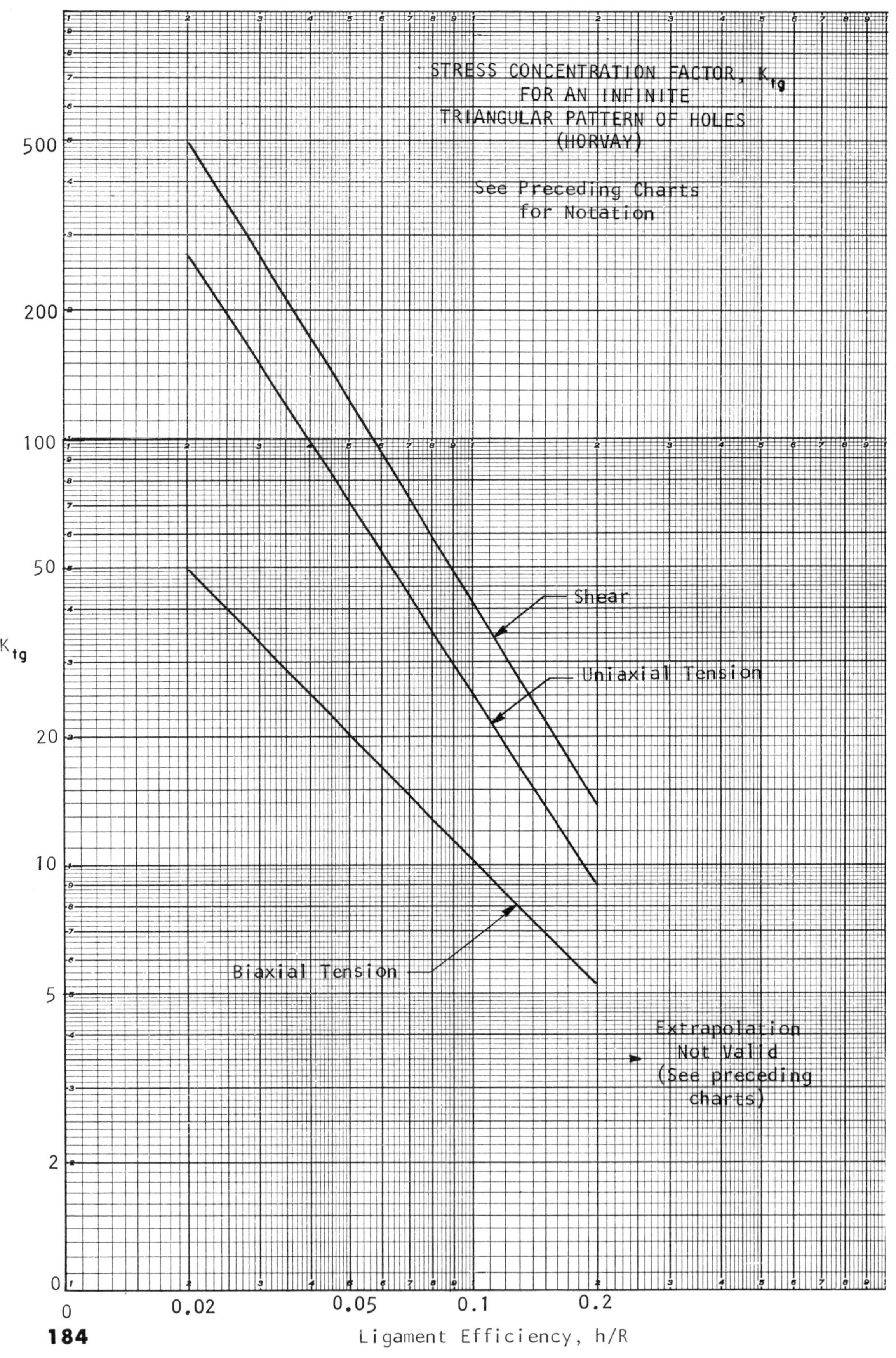

STRESS CONCENTRATION FACTOR, K_{tg}
FOR AN INFINITE
TRIANGULAR PATTERN OF HOLES
(HORVAY)
See Preceding Charts
for Notation
Shear
Uniaxial Tension
Biaxial Tension
Extrapolation
Not Valid
(See preceding
charts)
K_{tg}
500
200
100
50
20
10
5
2
0
0
0.02
0.05
0.1
0.2
Ligament Efficiency, h/R

FIG. 119

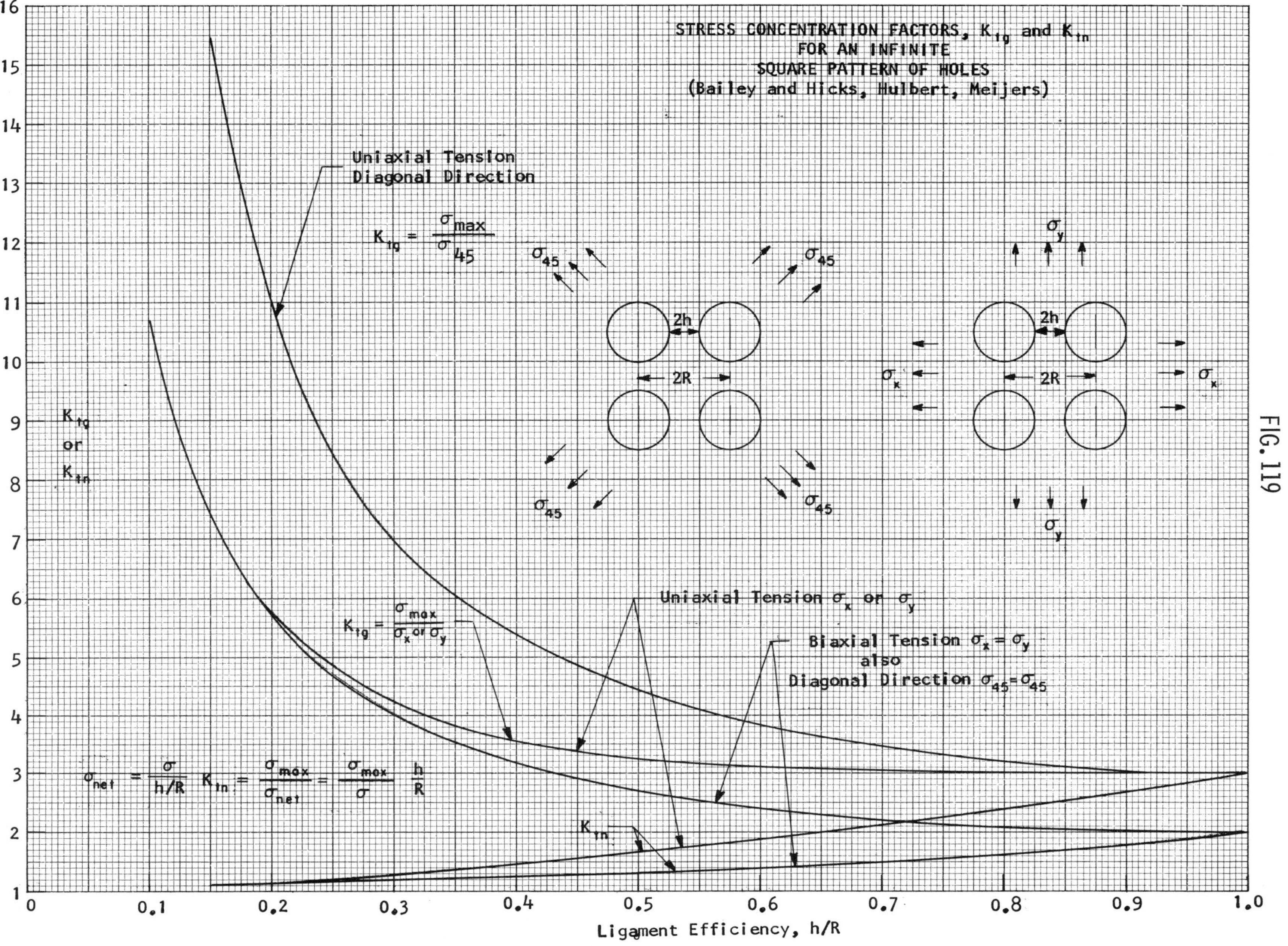

STRESS CONCENTRATION FACTORS, K_{tg} and K_{tn}
FOR AN INFINITE
SQUARE PATTERN OF HOLES
(Bailey and Hicks, Hulbert, Meijers)
Uniaxial Tension
Diagonal Direction
$K_{tg} = \frac{\sigma_{max}}{\sigma_{45}}$
σ_{45}
σ_x
σ_y
2h
2R
K_{tg}
or
K_{tn}
$K_{tg} = \frac{\sigma_{max}}{\sigma_x \text{ or } \sigma_y}$
Uniaxial Tension σ_x or σ_y
Biaxial Tension $\sigma_x = \sigma_y$
also
Diagonal Direction $\sigma_{45} = \sigma_{45}$
$\sigma_{net} = \frac{\sigma}{h/R}$ $K_{tn} = \frac{\sigma_{max}}{\sigma_{net}} = \frac{\sigma_{max}}{\sigma} \frac{h}{R}$
K_{tn}
Ligament Efficiency, h/R
0
0.1
0.2
0.3
0.4
0.5
0.6
0.7
0.8
0.9
1.0
1
2
3
4
5
6
7
8
9
10
11
12
13
14
15
16

FIG. 120

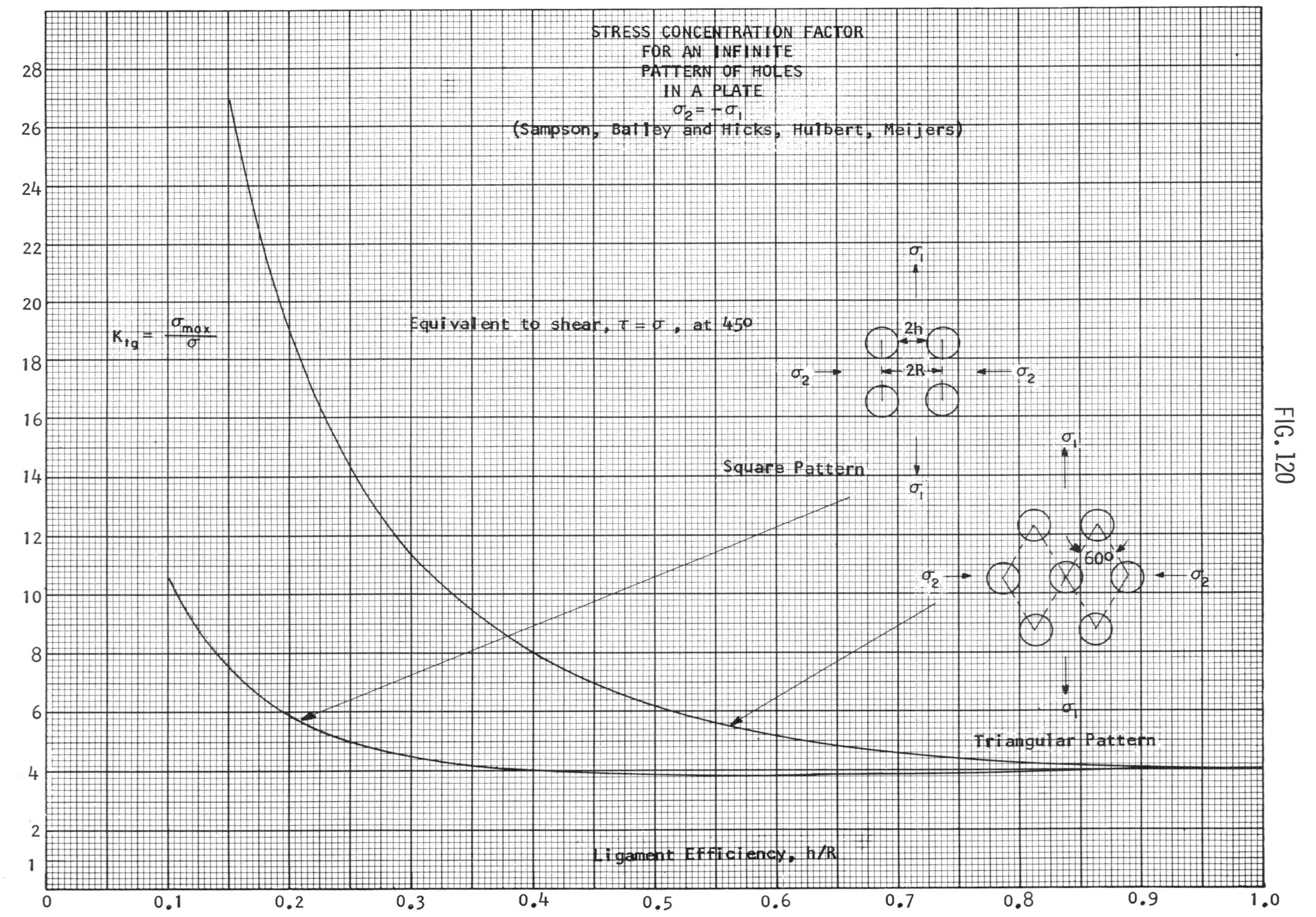
STRESS CONCENTRATION FACTOR
FOR AN INFINITE
PATTERN OF HOLES
IN A PLATE
$\sigma_2 = -\sigma_1$
(Sampson, Bailey and Hicks, Hulbert, Meijers)
$K_{tg} = \frac{\sigma_{max}}{\sigma}$
Equivalent to shear, $\tau = \sigma$, at 45°
2h
2R
σ_1
σ_2
Square Pattern
60°
Triangular Pattern
Ligament Efficiency, h/R
0
0.1
0.2
0.3
0.4
0.5
0.6
0.7
0.8
0.9
1.0
1
2
4
6
8
10
12
14
16
18
20
22
24
26
28

FIG. 121

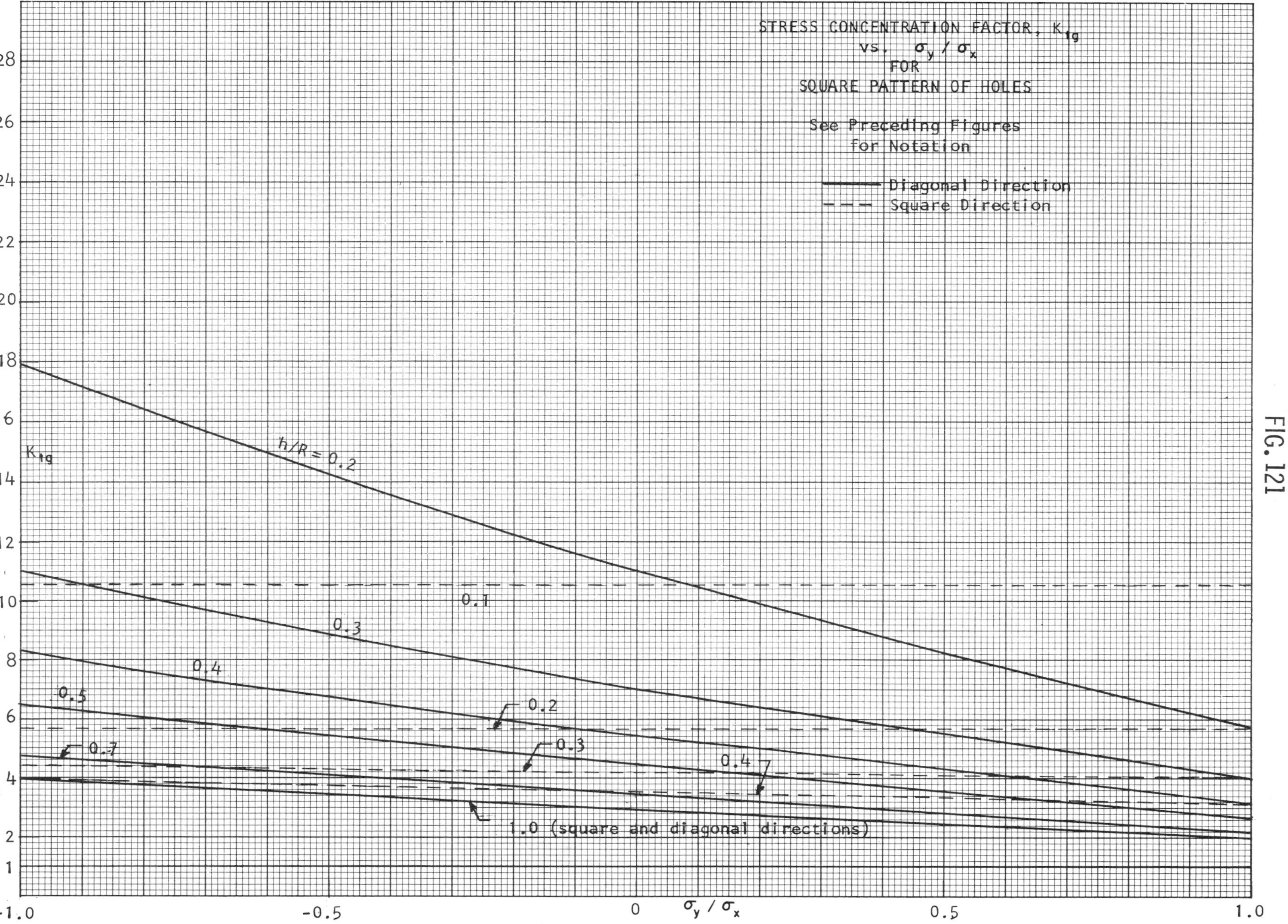
STRESS CONCENTRATION FACTOR, K_{tg}
vs. σ_y / σ_x
FOR
SQUARE PATTERN OF HOLES
See Preceding Figures
for Notation
Diagonal Direction
Square Direction
$h/R = 0.2$
0.1
0.3
0.4
0.5
0.2
0.7
0.3
0.4
1.0 (square and diagonal directions)
K_{tg}
28
26
24
22
20
18
16
14
12
10
8
6
4
2
1
-1.0
-0.5
0
0.5
1.0
σ_y / σ_x

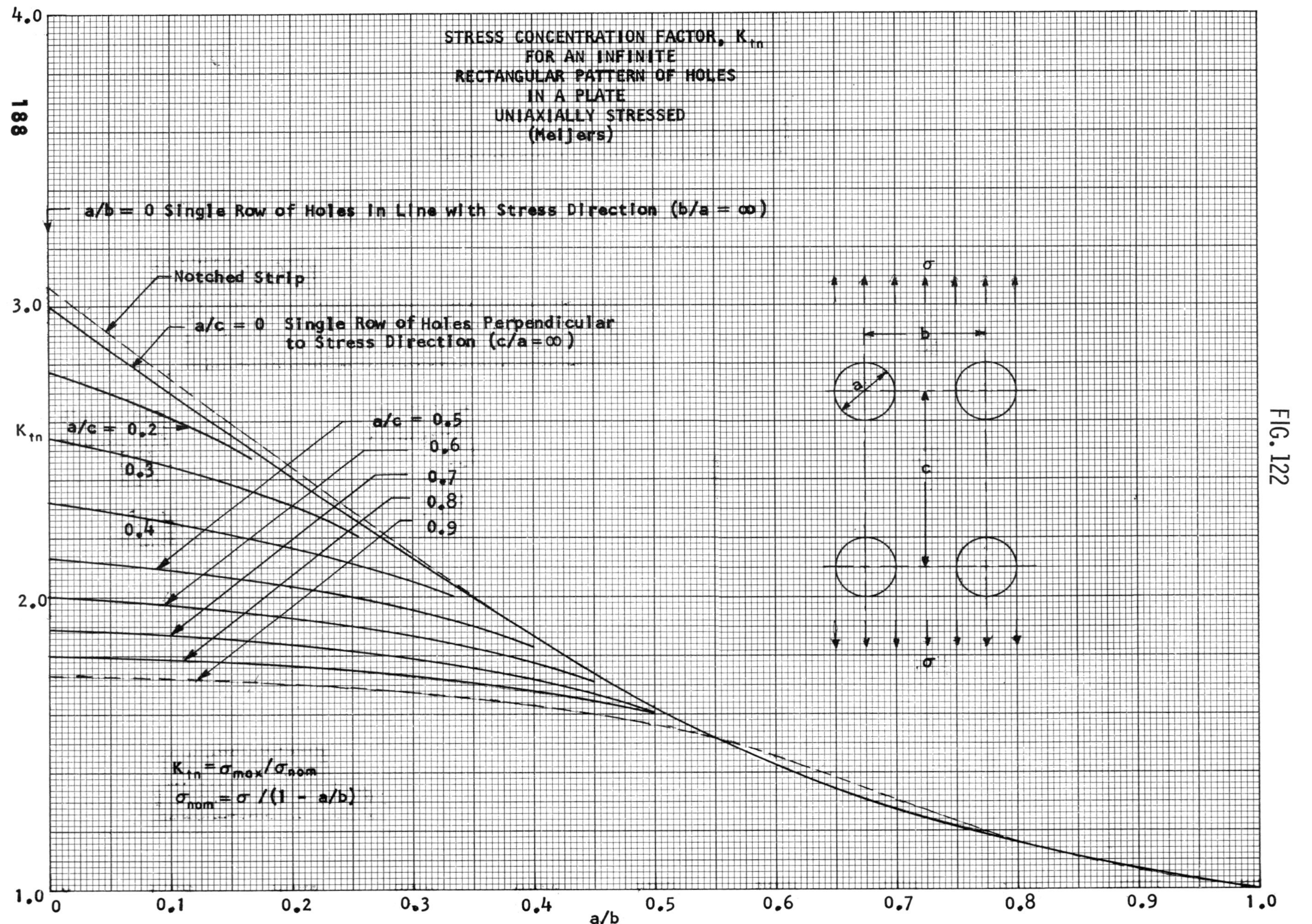

STRESS CONCENTRATION FACTOR, K_{tn}
FOR AN INFINITE
RECTANGULAR PATTERN OF HOLES
IN A PLATE
UNIAXIALLY STRESSED
(Meijers)
a/b = 0 Single Row of Holes in Line with Stress Direction (b/a = ∞)
Notched Strip
a/c = 0 Single Row of Holes Perpendicular to Stress Direction (c/a = ∞)
a/c = 0.2
0.3
0.4
a/c = 0.5
0.6
0.7
0.8
0.9
$K_{tn} = \sigma_{max}/\sigma_{nom}$
$\sigma_{nom} = \sigma/(1 - a/b)$
σ
b
a
c
σ
K_{tn}
4.0
3.0
2.0
1.0
0
0.1
0.2
0.3
0.4
0.5
0.6
0.7
0.8
0.9
1.0
a/b

FIG. 122

FIG. 122a

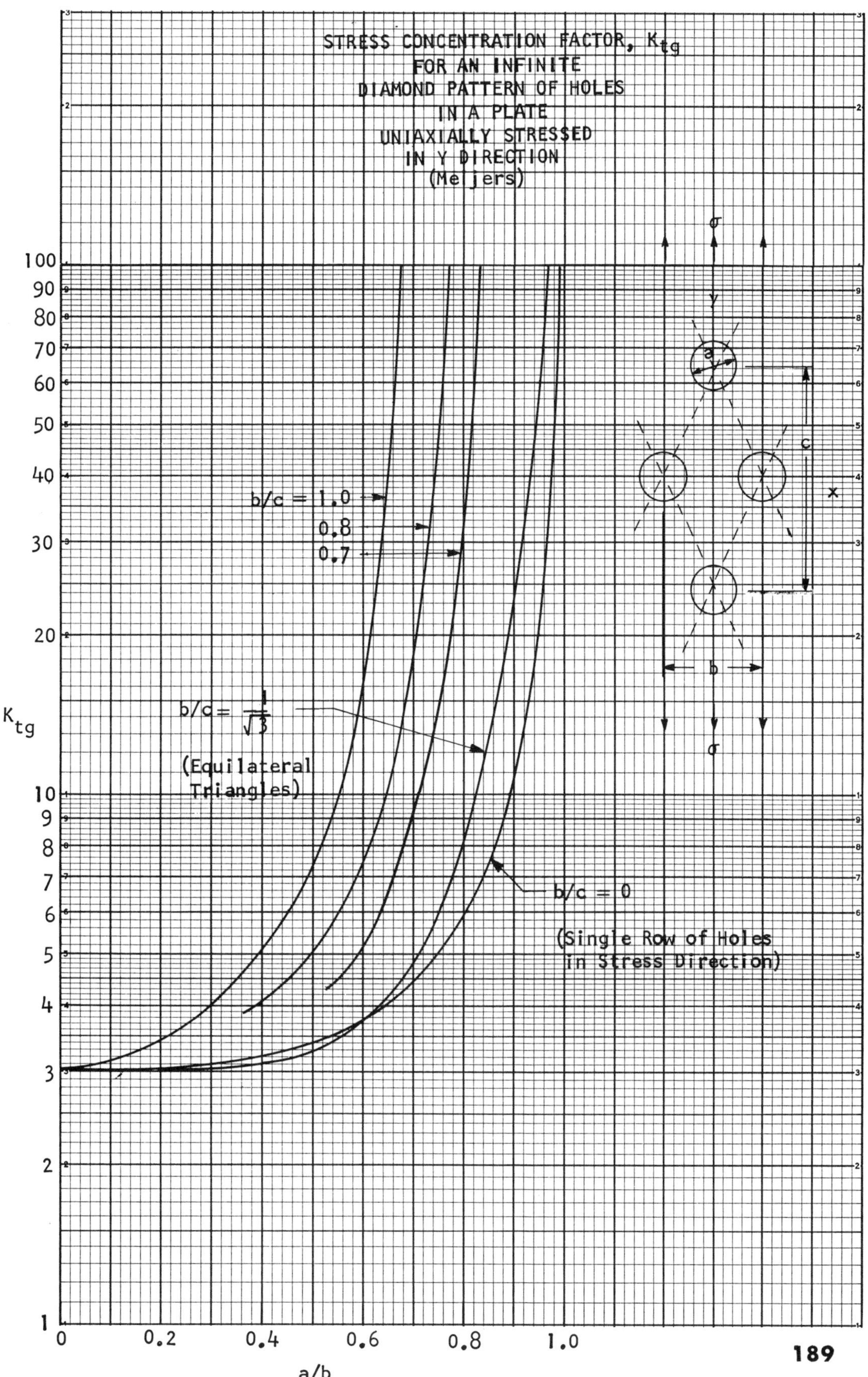
STRESS CONCENTRATION FACTOR, K_{tg}
FOR AN INFINITE
DIAMOND PATTERN OF HOLES
IN A PLATE
UNIAXIALLY STRESSED
IN Y DIRECTION
(Meijers)
b/c = 1.0
0.8
0.7
b/c = $\frac{1}{\sqrt{3}}$
(Equilateral
Triangles)
b/c = 0
(Single Row of Holes
in Stress Direction)
σ
y
a
c
x
b
K_{tg}
a/b
0
0.2
0.4
0.6
0.8
1.0
1
2
3
4
5
6
7
8
9
10
20
30
40
50
60
70
80
90
100

FIG. 123

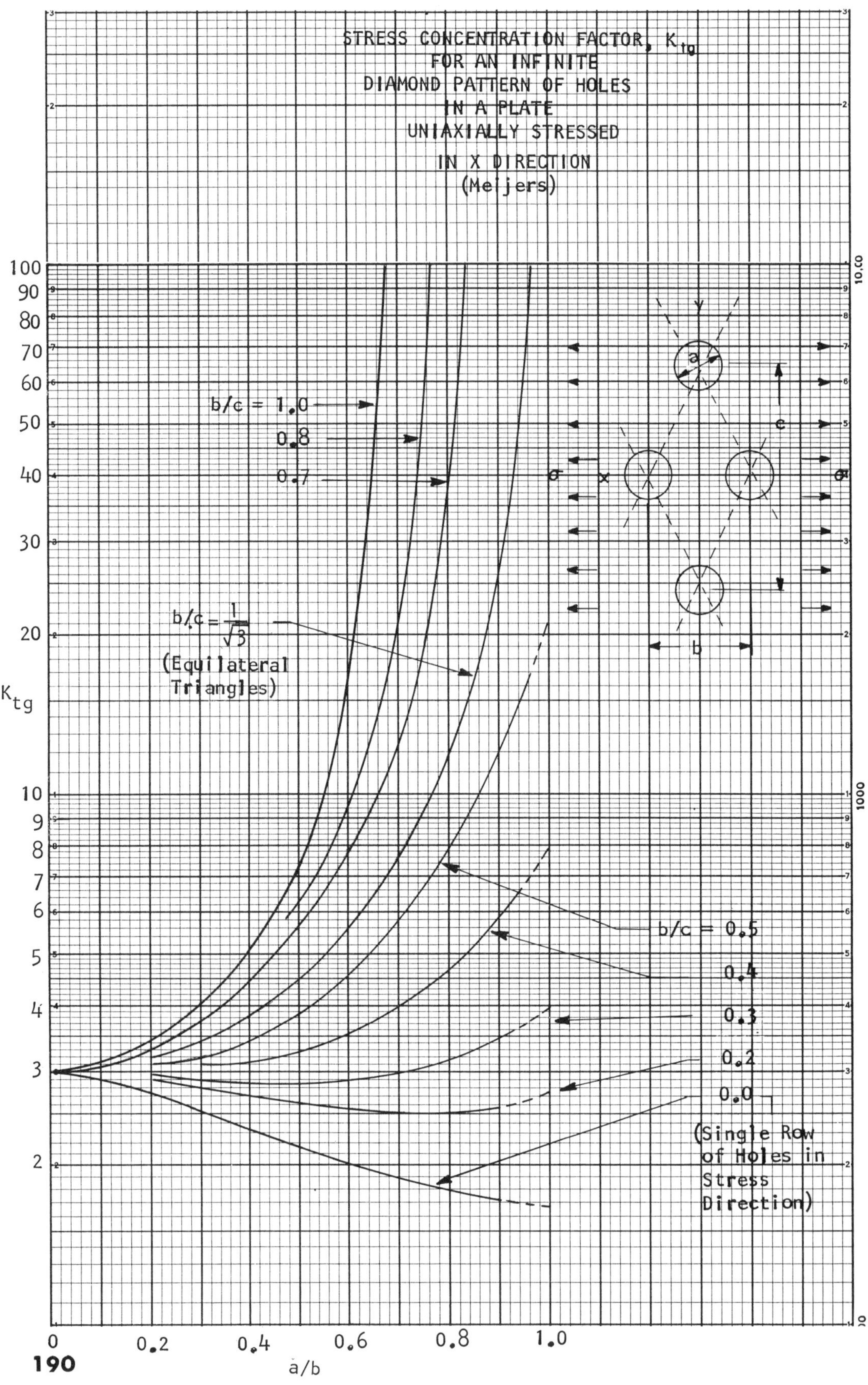

STRESS CONCENTRATION FACTOR, K_{tg}
FOR AN INFINITE
DIAMOND PATTERN OF HOLES
IN A PLATE
UNIAXIALLY STRESSED
IN X DIRECTION
(Meijers)
b/c = 1.0
0.8
0.7
b/c = $\frac{1}{\sqrt{3}}$
(Equilateral Triangles)
b/c = 0.5
0.4
0.3
0.2
0.0
(Single Row of Holes in Stress Direction)
K_{tg}
a/b
a
b
c
σ
x
100
90
80
70
60
50
40
30
20
10
9
8
7
6
5
4
3
2
0
0.2
0.4
0.6
0.8
1.0

FIG. 124

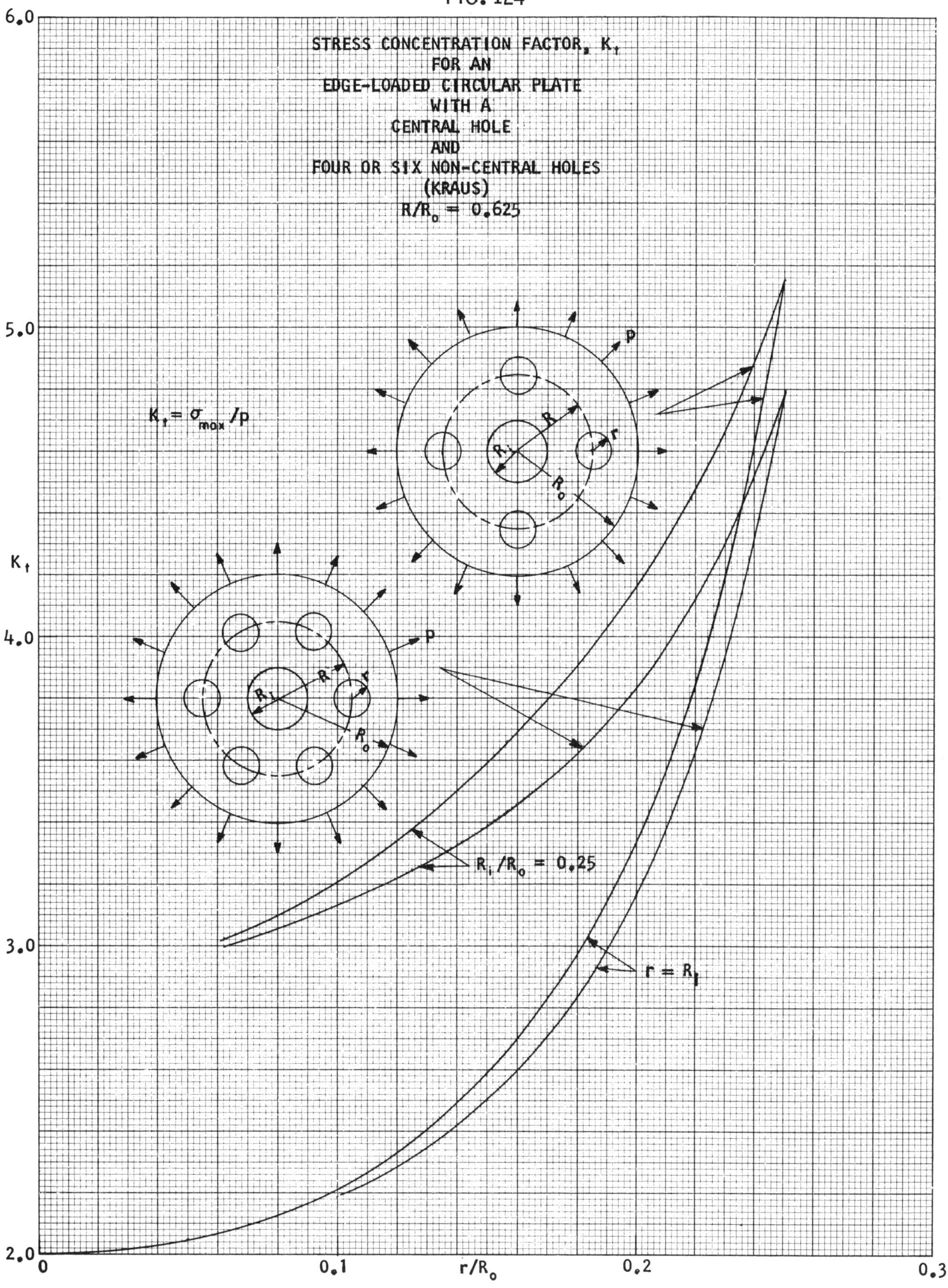

STRESS CONCENTRATION FACTOR, K_t
FOR AN
EDGE-LOADED CIRCULAR PLATE
WITH A
CENTRAL HOLE
AND
FOUR OR SIX NON-CENTRAL HOLES
(KRAUS)
$R/R_o = 0.625$
$K_t = \sigma_{max}/p$
K_t
p
R
r
R_i
R_o
$R_i/R_o = 0.25$
$r = R_i$
6.0
5.0
4.0
3.0
2.0
0
0.1
r/R_o
0.2
0.3

FIG. 125

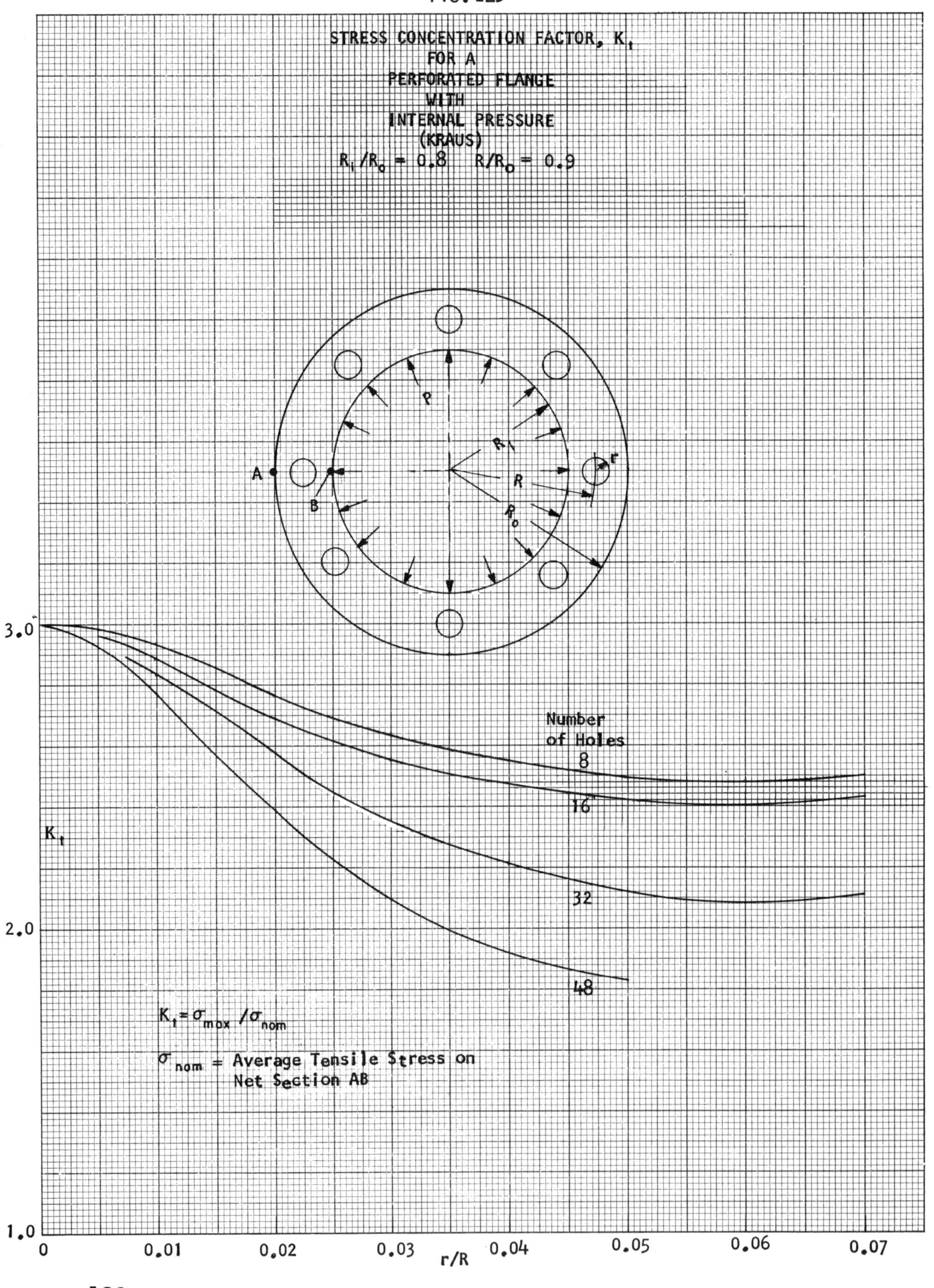

STRESS CONCENTRATION FACTOR, K_t
FOR A
PERFORATED FLANGE
WITH
INTERNAL PRESSURE
(KRAUS)
$R_i/R_o = 0.8$ $R/R_o = 0.9$
A
B
P
R_i
R
R_o
r
Number of Holes
8
16
32
48
K_t
3.0
2.0
1.0
0
0.01
0.02
0.03
0.04
0.05
0.06
0.07
r/R
$K_t = \sigma_{max} / \sigma_{nom}$
σ_{nom} = Average Tensile Stress on Net Section AB

FIG. 126

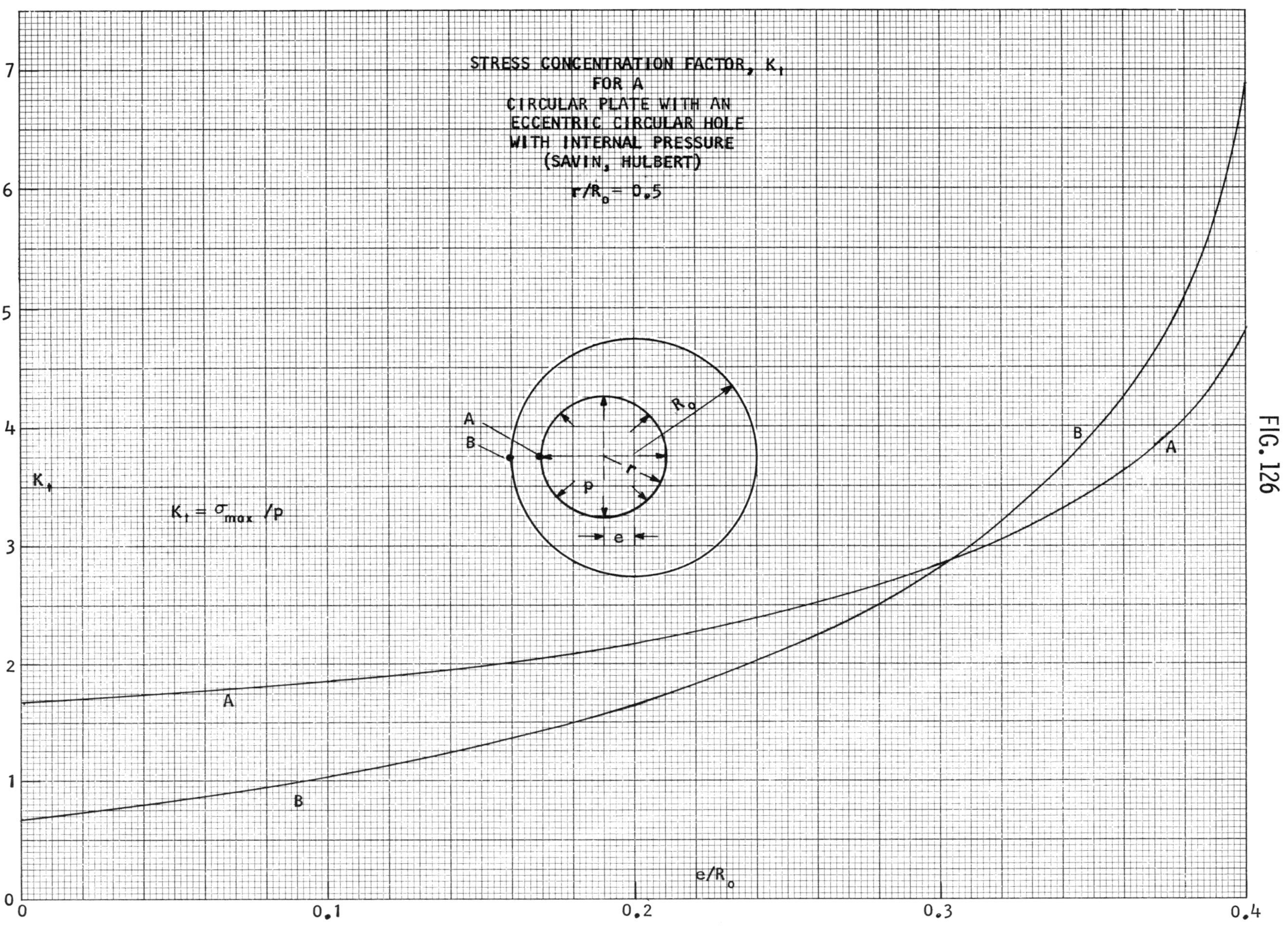

STRESS CONCENTRATION FACTOR, K_t
FOR A
CIRCULAR PLATE WITH AN
ECCENTRIC CIRCULAR HOLE
WITH INTERNAL PRESSURE
(SAVIN, HULBERT)
$r/R_o = 0.5$
$K_t = \sigma_{max}/p$
K_t
e/R_o
R_o
r
p
e
A
B
0
1
2
3
4
5
6
7
0
0.1
0.2
0.3
0.4

FIG. 127

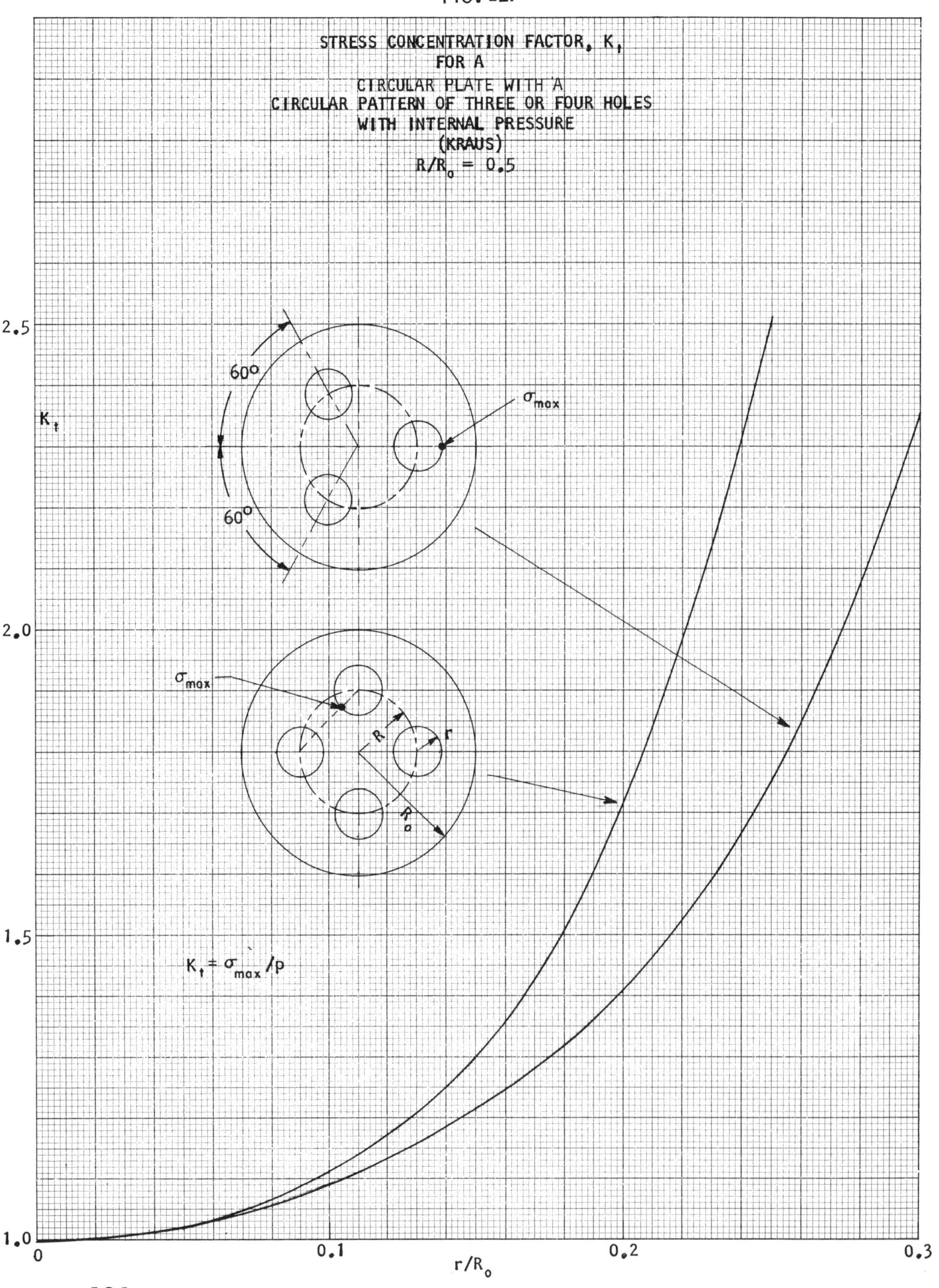
STRESS CONCENTRATION FACTOR, K_t
FOR A
CIRCULAR PLATE WITH A
CIRCULAR PATTERN OF THREE OR FOUR HOLES
WITH INTERNAL PRESSURE
(KRAUS)
$R/R_o = 0.5$
2.5
2.0
1.5
1.0
K_t
60°
60°
σ_{max}
σ_{max}
R
r
R_o
$K_t = \sigma_{max}/p$
0
0.1
0.2
0.3
r/R_o

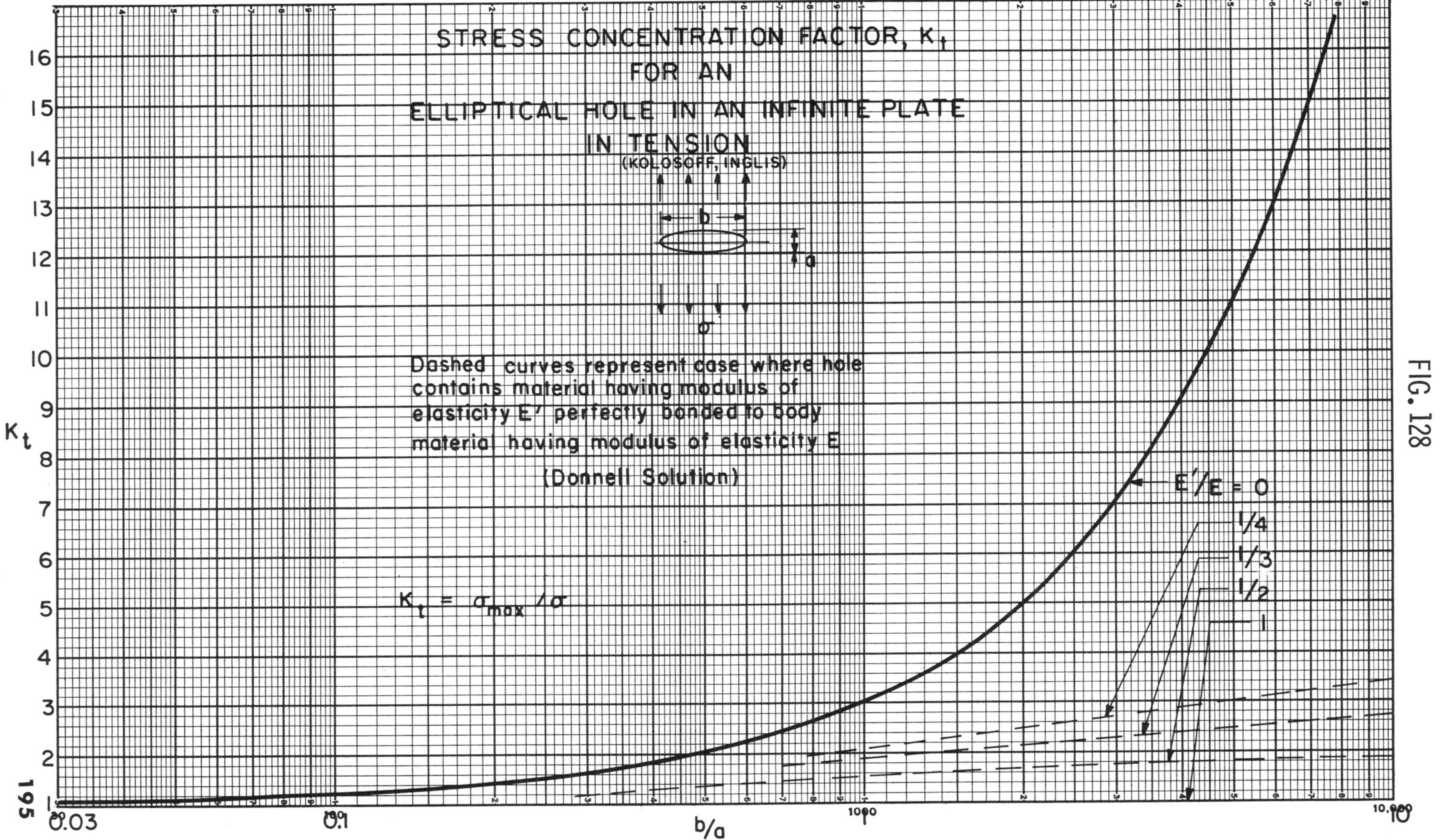

STRESS CONCENTRATION FACTOR, K_t
FOR AN
ELLIPTICAL HOLE IN AN INFINITE PLATE
IN TENSION
(KOLOSOFF, INGLIS)
b
a
σ
Dashed curves represent case where hole
contains material having modulus of
elasticity E' perfectly bonded to body
material having modulus of elasticity E
(Donnell Solution)
$K_t = \sigma_{max} / \sigma$
E'/E = 0
1/4
1/3
1/2
1
K_t
16
15
14
13
12
11
10
9
8
7
6
5
4
3
2
1
b/a
0.03
0.1
1
10

FIG. 128

FIG. 129

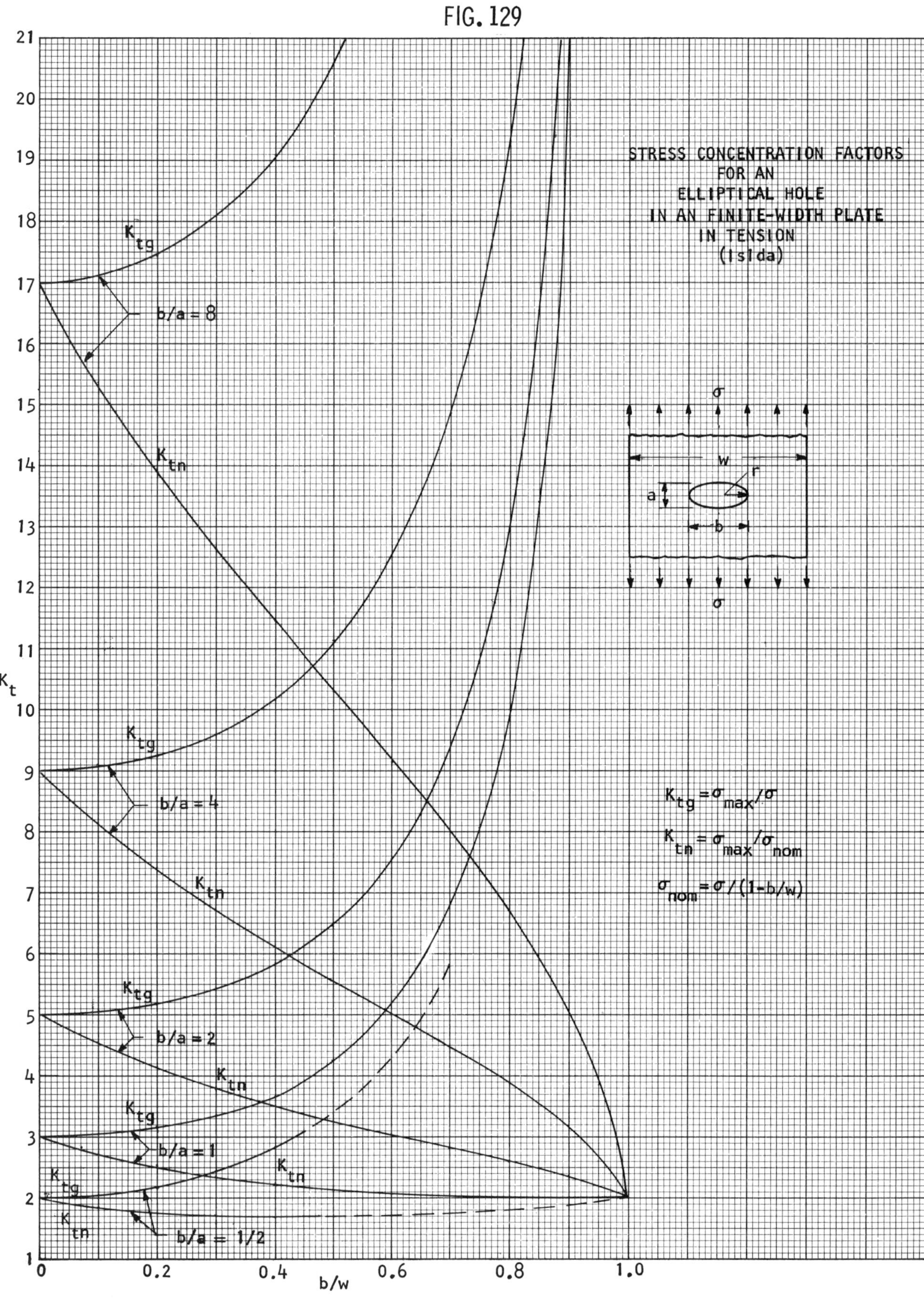
STRESS CONCENTRATION FACTORS
FOR AN
ELLIPTICAL HOLE
IN AN FINITE-WIDTH PLATE
IN TENSION
(Isida)
K_{tg}
K_{tn}
b/a = 8
b/a = 4
b/a = 2
b/a = 1
b/a = 1/2
K_t
b/w
σ
w
r
a
b
$K_{tg} = \sigma_{max}/\sigma$
$K_{tn} = \sigma_{max}/\sigma_{nom}$
$\sigma_{nom} = \sigma/(1-b/w)$

FIG. 130

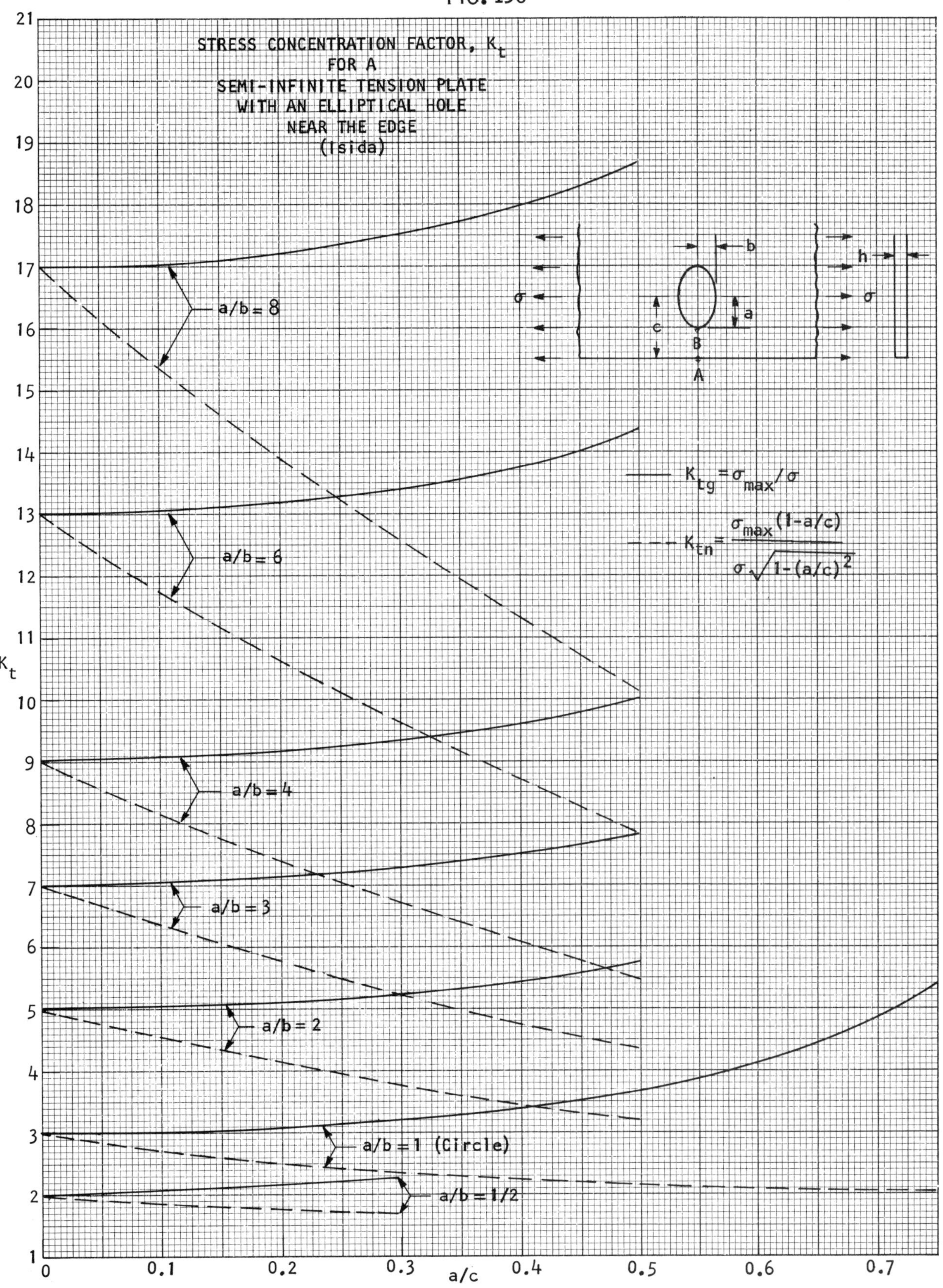

STRESS CONCENTRATION FACTOR, K_t
FOR A
SEMI-INFINITE TENSION PLATE
WITH AN ELLIPTICAL HOLE
NEAR THE EDGE
(Isida)
a/b = 8
a/b = 6
a/b = 4
a/b = 3
a/b = 2
a/b = 1 (Circle)
a/b = 1/2
$K_{tg} = \sigma_{max}/\sigma$
$K_{tn} = \frac{\sigma_{max}(1-a/c)}{\sigma\sqrt{1-(a/c)^2}}$
σ
b
h
a
c
B
A
K_t
a/c

FIG. 131

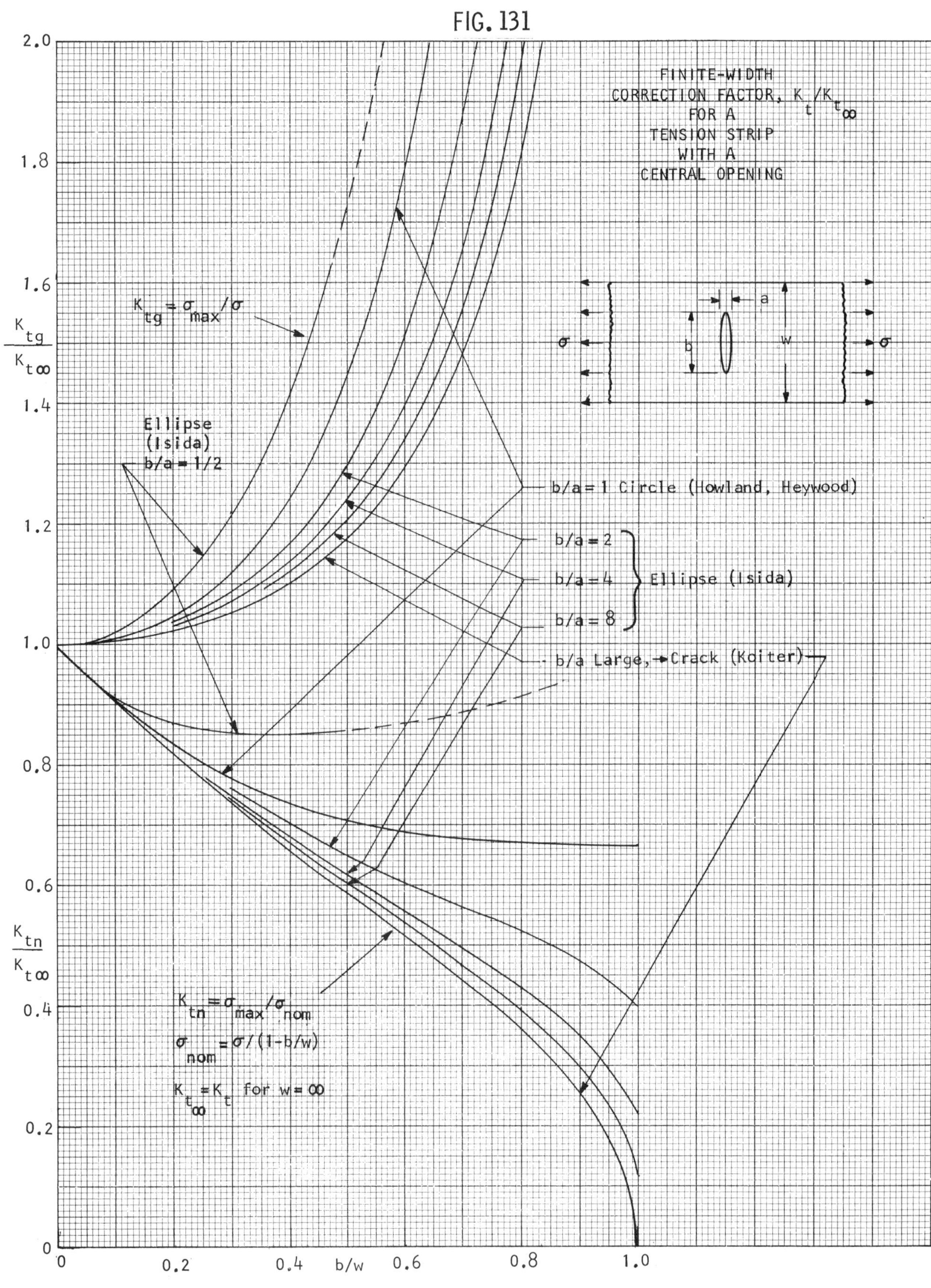

FINITE-WIDTH CORRECTION FACTOR, $K_t/K_{t\infty}$ FOR A TENSION STRIP WITH A CENTRAL OPENING
$K_{tg} = \sigma_{max}/\sigma$
$\frac{K_{tg}}{K_{t\infty}}$
Ellipse (Isida) b/a = 1/2
b/a = 1 Circle (Howland, Heywood)
b/a = 2
b/a = 4
b/a = 8
Ellipse (Isida)
b/a Large, → Crack (Koiter)
$\frac{K_{tn}}{K_{t\infty}}$
$K_{tn} = \sigma_{max}/\sigma_{nom}$
$\sigma_{nom} = \sigma/(1-b/w)$
$K_{t\infty} = K_t$ for $w = \infty$
σ
a
b
w
2.0
1.8
1.6
1.4
1.2
1.0
0.8
0.6
0.4
0.2
0
0.2
0.4
b/w
0.6
0.8
1.0

FIG. 132

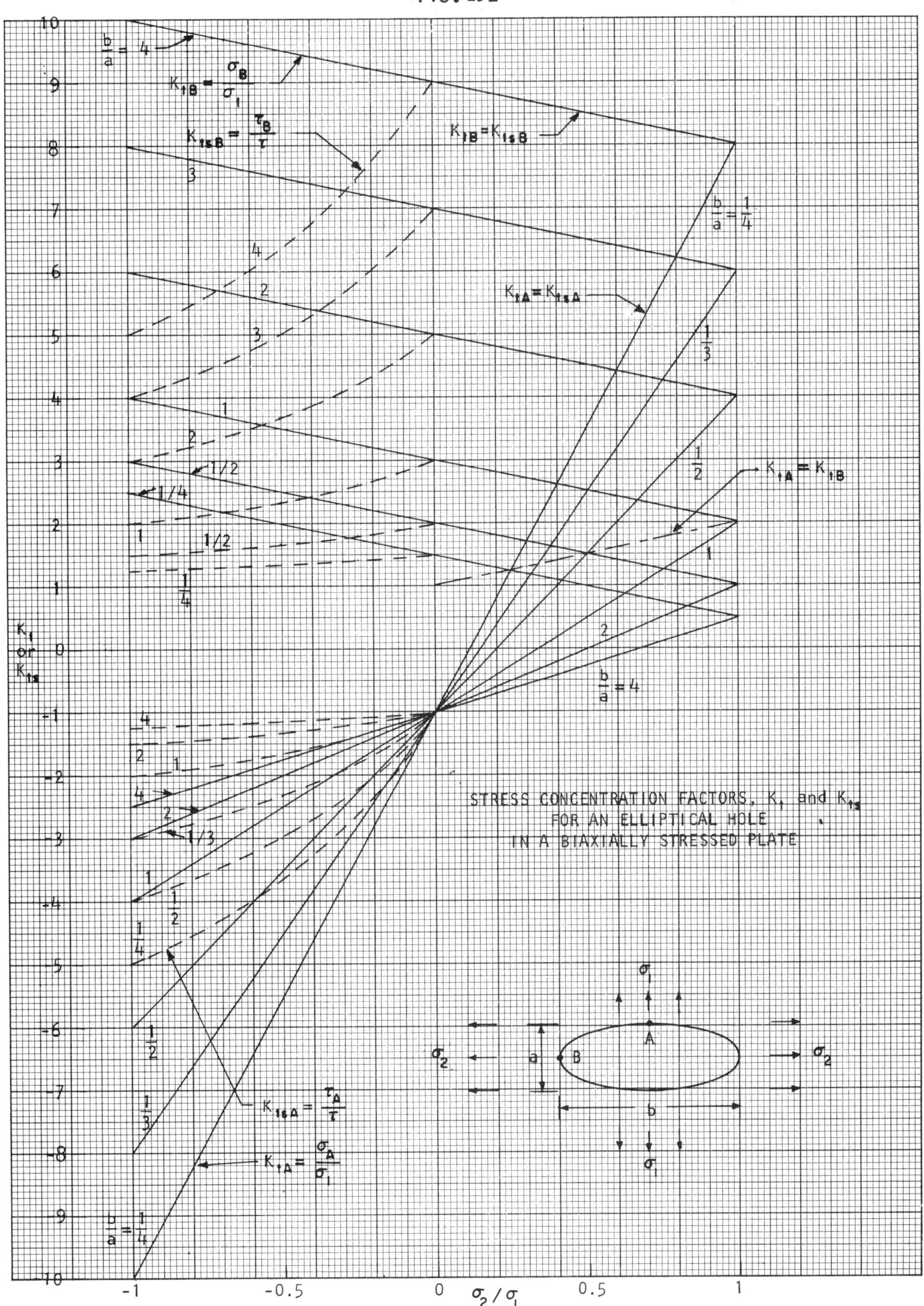

STRESS CONCENTRATION FACTORS, K_t and K_{ts}
FOR AN ELLIPTICAL HOLE
IN A BIAXIALLY STRESSED PLATE
K_t or K_{ts}
σ_2 / σ_1
$K_{tB} = \frac{\sigma_B}{\sigma_1}$
$K_{tsB} = \frac{\tau_B}{\tau}$
$K_{tB} = K_{tsB}$
$K_{tA} = K_{tsA}$
$K_{tA} = K_{tB}$
$K_{tsA} = \frac{\tau_A}{\tau}$
$K_{tA} = \frac{\sigma_A}{\sigma_1}$
$\frac{b}{a} = 4$
$\frac{b}{a} = \frac{1}{4}$
σ_1
σ_2
A
B
a
b
-1
-0.5
0
0.5
1

FIG. 133

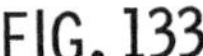

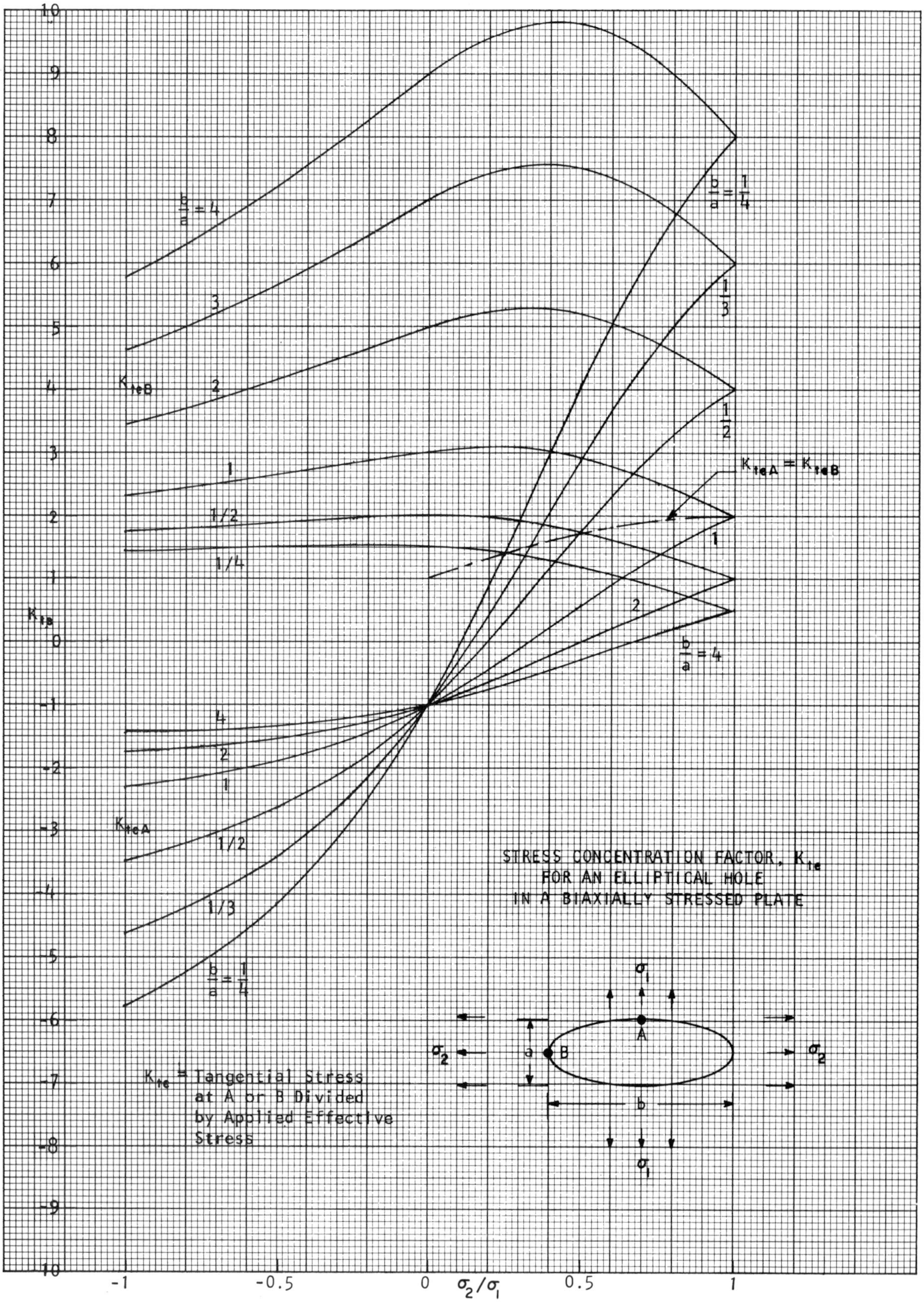

STRESS CONCENTRATION FACTOR, K_{te}
FOR AN ELLIPTICAL HOLE
IN A BIAXIALLY STRESSED PLATE
K_{te} = Tangential Stress at A or B Divided by Applied Effective Stress
$K_{teA} = K_{teB}$
K_{teB}
K_{teA}
K_{te}
$\frac{b}{a} = 4$
3
2
1
1/2
1/4
4
2
1
1/2
1/3
$\frac{b}{a} = \frac{1}{4}$
$\frac{b}{a} = \frac{1}{4}$
$\frac{1}{3}$
$\frac{1}{2}$
1
2
$\frac{b}{a} = 4$
σ_1
σ_2
A
B
a
b
σ_2/σ_1
-1
-0.5
0
0.5
1
10
9
8
7
6
5
4
3
2
1
0
-1
-2
-3
-4
-5
-6
-7
-8
-9
-10

FIG. 134

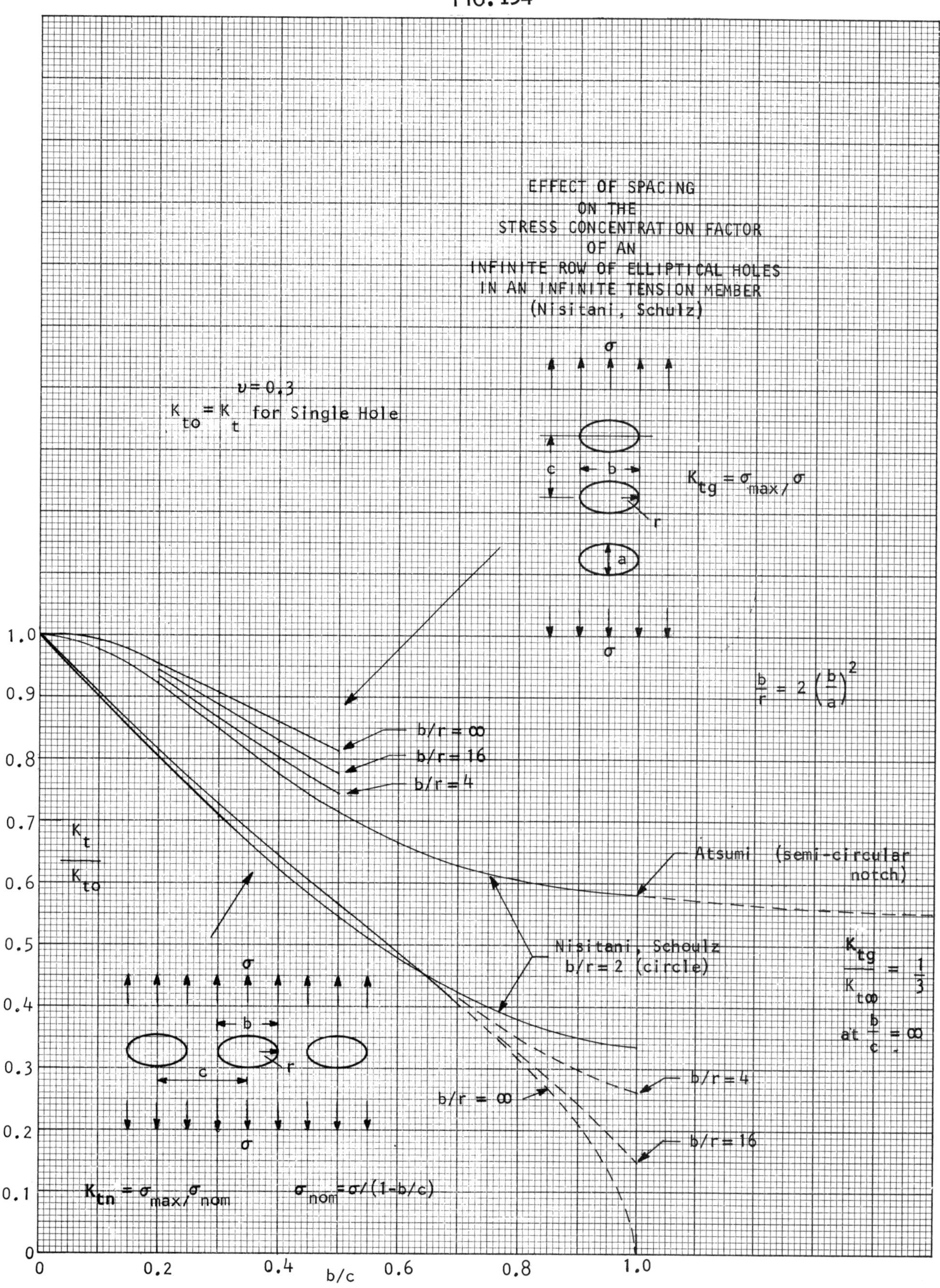

EFFECT OF SPACING
ON THE
STRESS CONCENTRATION FACTOR
OF AN
INFINITE ROW OF ELLIPTICAL HOLES
IN AN INFINITE TENSION MEMBER
(Nisitani, Schulz)
$\nu = 0.3$
$K_{to} = K_t$ for Single Hole
$K_{tg} = \sigma_{max}/\sigma$
$\frac{b}{r} = 2\left(\frac{b}{a}\right)^2$
b/r = ∞
b/r = 16
b/r = 4
Atsumi (semi-circular notch)
Nisitani, Schoulz
b/r = 2 (circle)
$\frac{K_{tg}}{K_{t\infty}} = \frac{1}{3}$ at $\frac{b}{c} = \infty$
b/r = 4
b/r = ∞
b/r = 16
$\frac{K_t}{K_{to}}$
$K_{tn} = \sigma_{max}/\sigma_{nom}$
$\sigma_{nom} = \sigma/(1-b/c)$
σ
b
c
r
a
b/c
0
0.1
0.2
0.3
0.4
0.5
0.6
0.7
0.8
0.9
1.0
0.2
0.4
0.6
0.8
1.0

FIG. 135

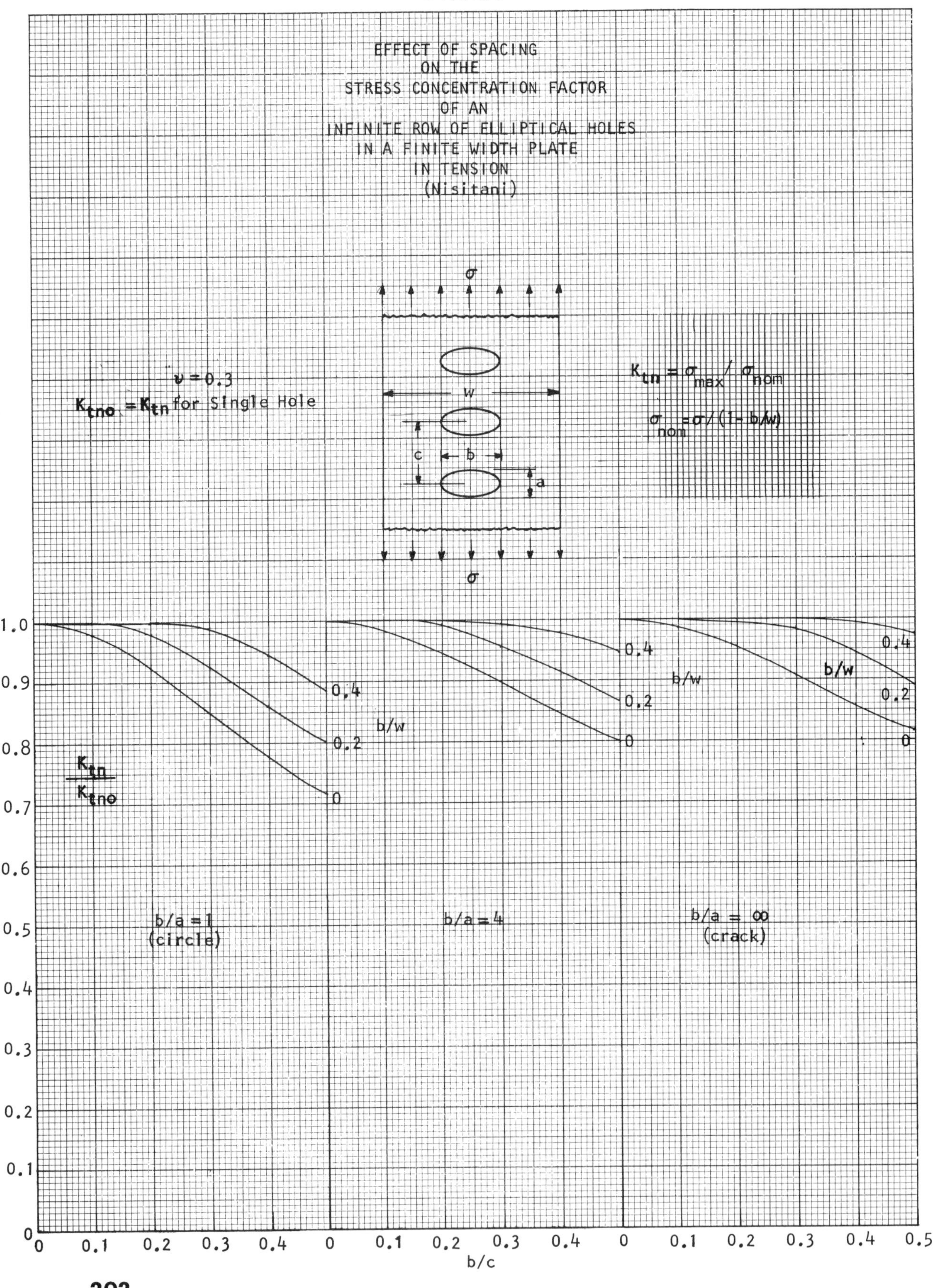

EFFECT OF SPACING
ON THE
STRESS CONCENTRATION FACTOR
OF AN
INFINITE ROW OF ELLIPTICAL HOLES
IN A FINITE WIDTH PLATE
IN TENSION
(Nisitani)
σ
w
c
b
a
ν = 0.3
Ktno = Ktn for Single Hole
Ktn = σmax / σnom
σnom = σ/(1 - b/w)
Ktn / Ktno
b/w
0.4
0.2
0
b/a = 1
(circle)
b/a = 4
b/a = ∞
(crack)
b/c
1.0
0.9
0.8
0.7
0.6
0.5
0.4
0.3
0.2
0.1
0
0.1
0.2
0.3
0.4
0.5

FIG. 136

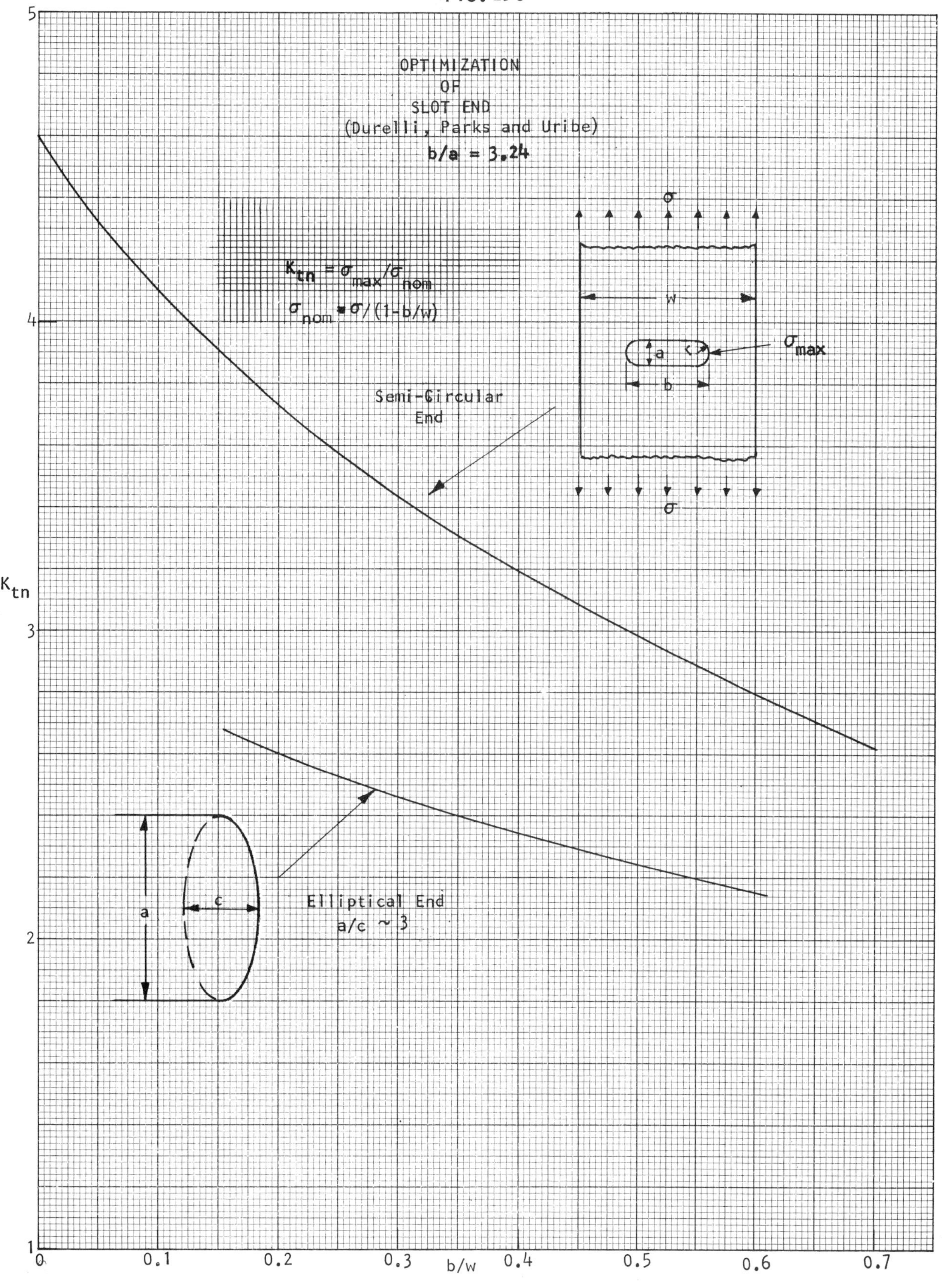
OPTIMIZATION
OF
SLOT END
(Durelli, Parks and Uribe)
b/a = 3.24
$K_{tn} = \sigma_{max}/\sigma_{nom}$
$\sigma_{nom} = \sigma/(1-b/W)$
σ
W
a
c
b
σ_{max}
Semi-Circular
End
Elliptical End
a/c ~ 3
a
c
K_{tn}
b/W
5
4
3
2
1
0
0.1
0.2
0.3
0.4
0.5
0.6
0.7

FIG. 137

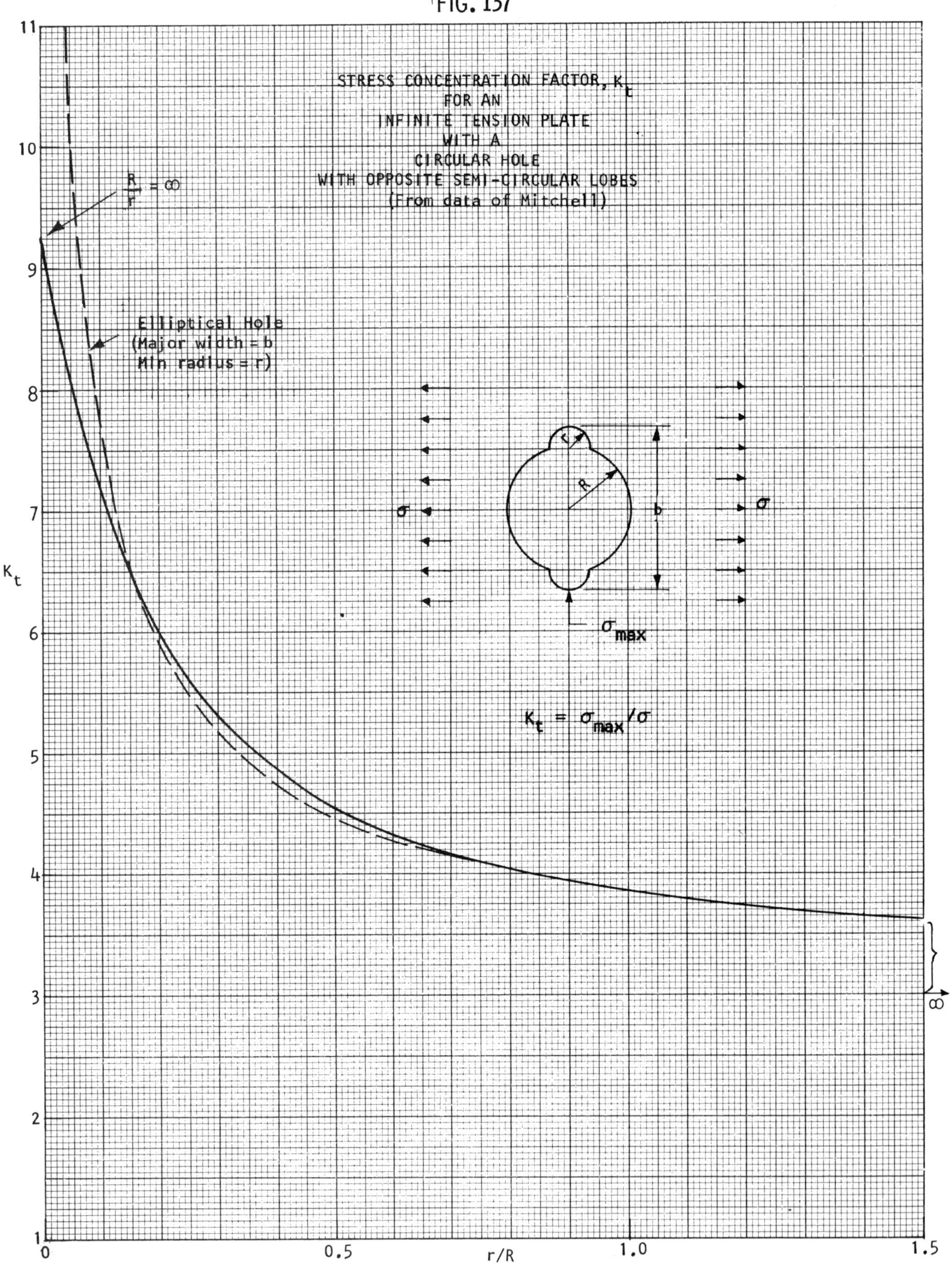
STRESS CONCENTRATION FACTOR, K_t
FOR AN
INFINITE TENSION PLATE
WITH A
CIRCULAR HOLE
WITH OPPOSITE SEMI-CIRCULAR LOBES
(From data of Mitchell)
$\frac{R}{r} = \infty$
Elliptical Hole
(Major width = b
Min radius = r)
σ
r
R
b
σ
σ_{max}
$K_t = \sigma_{max}/\sigma$
K_t
11
10
9
8
7
6
5
4
3
2
1
0
0.5
1.0
1.5
r/R
∞

FIG. 138

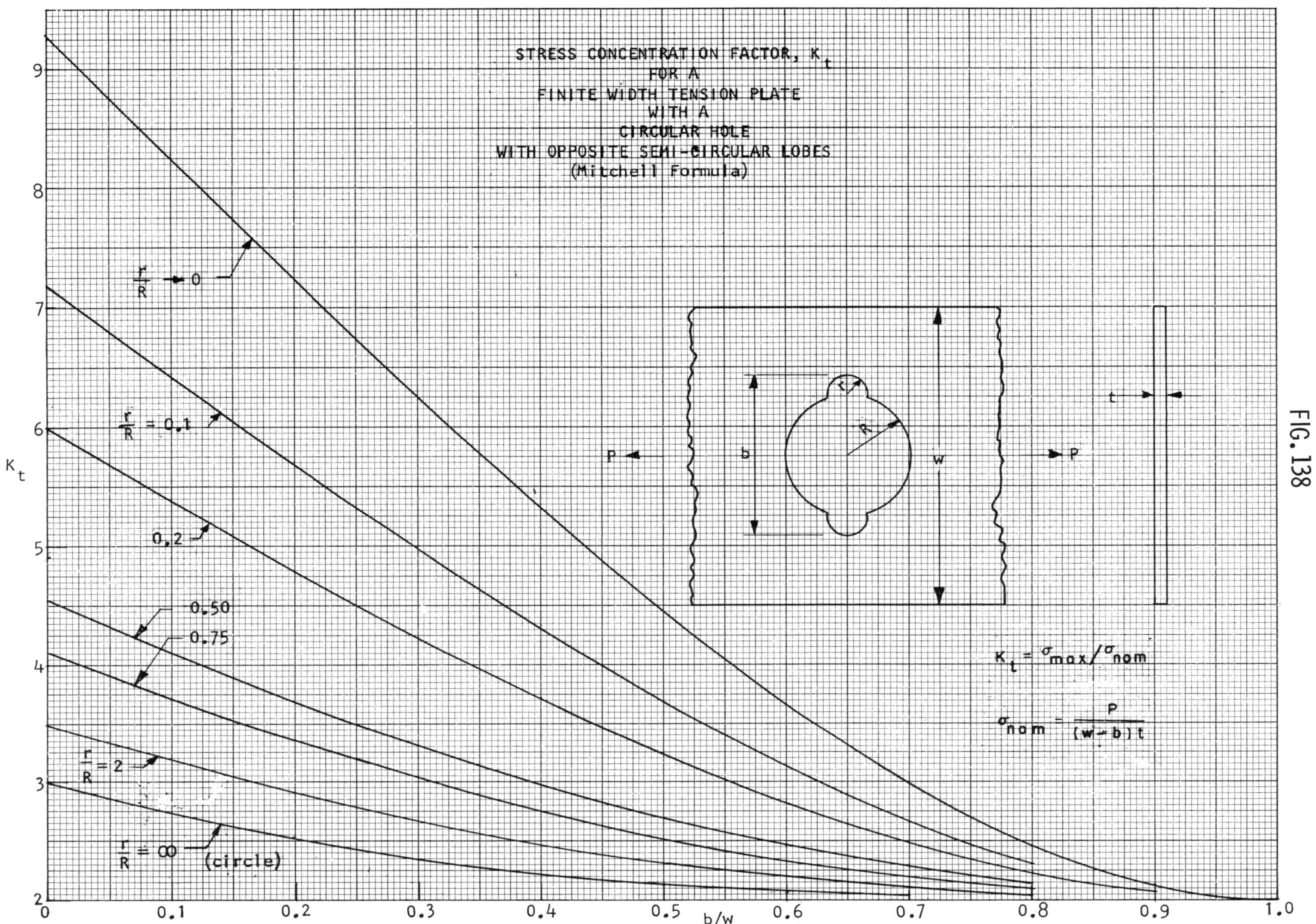

STRESS CONCENTRATION FACTOR, K_t
FOR A
FINITE WIDTH TENSION PLATE
WITH A
CIRCULAR HOLE
WITH OPPOSITE SEMI-CIRCULAR LOBES
(Mitchell Formula)
$\frac{r}{R} \to 0$
$\frac{r}{R} = 0.1$
0.2
0.50
0.75
$\frac{r}{R} = 2$
$\frac{r}{R} = \infty$ (circle)
P
P
b
r
R
w
t
$K_t = \sigma_{max}/\sigma_{nom}$
$\sigma_{nom} = \frac{P}{(w-b)t}$
K_t
b/w
2
3
4
5
6
7
8
9
0
0.1
0.2
0.3
0.4
0.5
0.6
0.7
0.8
0.9
1.0

FIG. 139

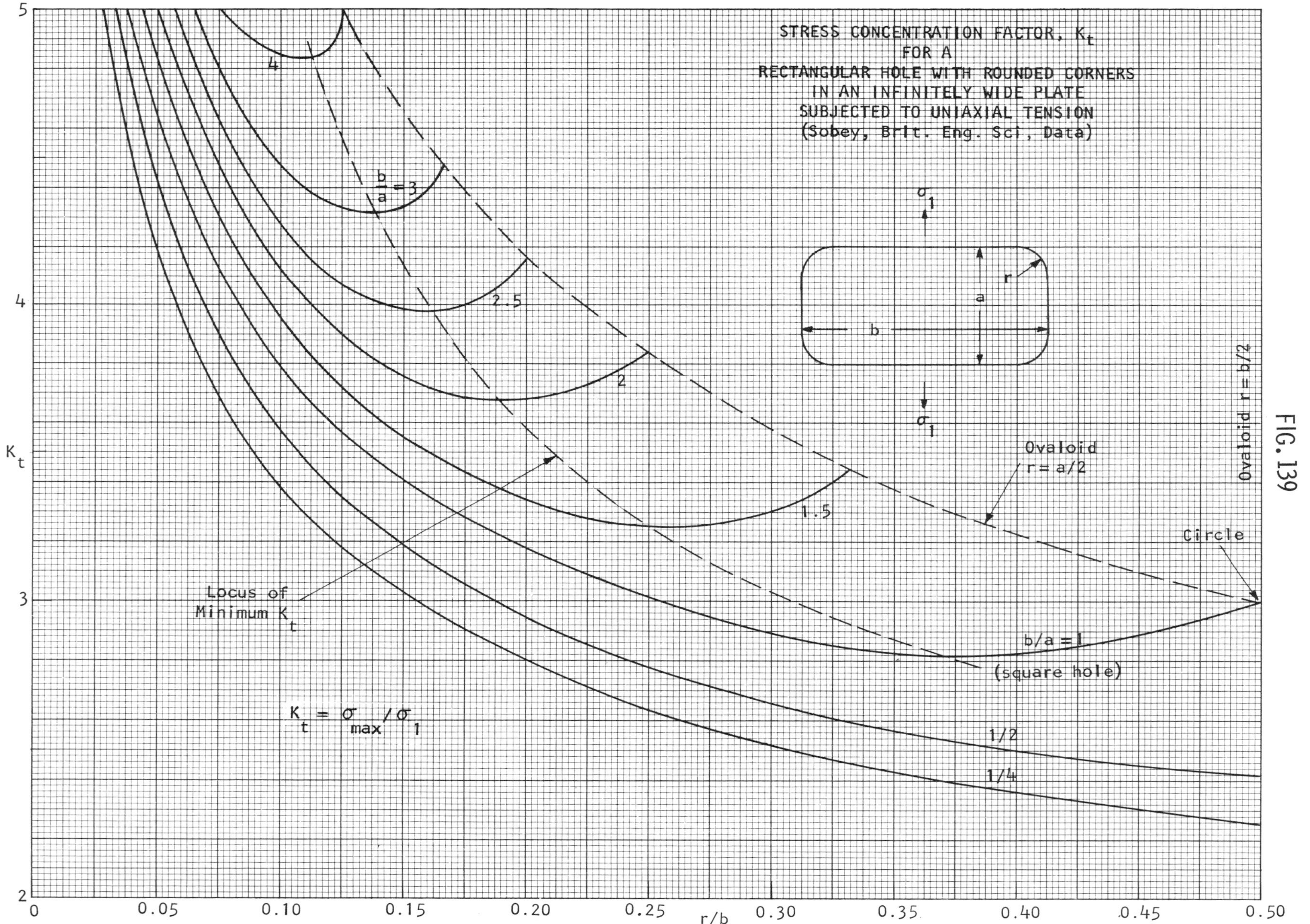

STRESS CONCENTRATION FACTOR, K_t
FOR A
RECTANGULAR HOLE WITH ROUNDED CORNERS
IN AN INFINITELY WIDE PLATE
SUBJECTED TO UNIAXIAL TENSION
(Sobey, Brit. Eng. Sci, Data)
σ_1
σ_1
r
a
b
4
$\frac{b}{a} = 3$
2.5
2
1.5
Ovaloid
r = a/2
Ovaloid r = b/2
Circle
b/a = 1
(square hole)
1/2
1/4
Locus of
Minimum K_t
$K_t = \sigma_{max}/\sigma_1$
K_t
5
4
3
2
r/b
0
0.05
0.10
0.15
0.20
0.25
0.30
0.35
0.40
0.45
0.50

FIG. 140

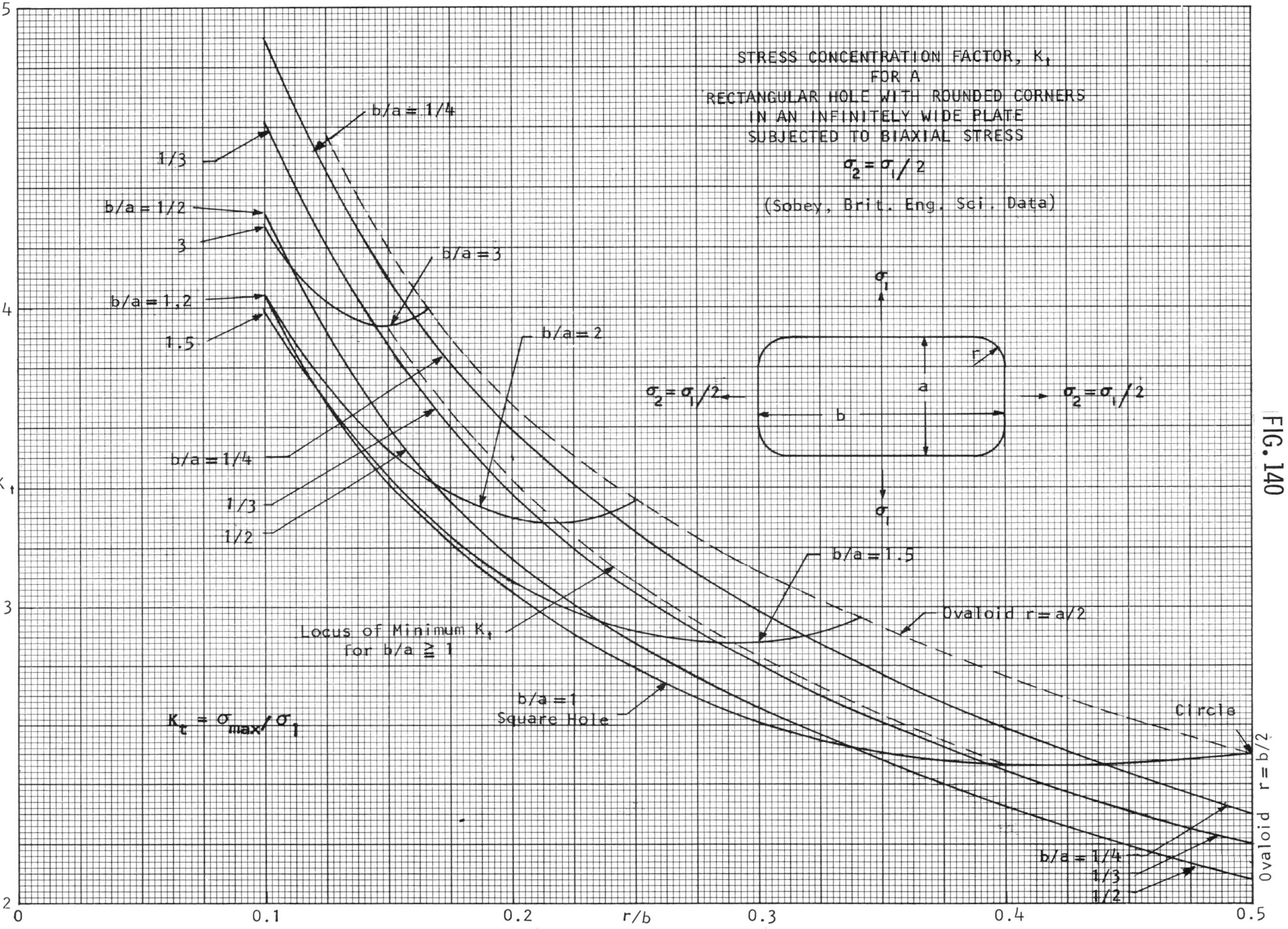

STRESS CONCENTRATION FACTOR, K_t
FOR A
RECTANGULAR HOLE WITH ROUNDED CORNERS
IN AN INFINITELY WIDE PLATE
SUBJECTED TO BIAXIAL STRESS
$\sigma_2 = \sigma_1/2$
(Sobey, Brit. Eng. Sci. Data)
σ_1
$\sigma_2 = \sigma_1/2$
$\sigma_2 = \sigma_1/2$
σ_1
a
b
r
b/a = 1/4
1/3
b/a = 1/2
3
b/a = 1.2
1.5
b/a = 3
b/a = 2
b/a = 1/4
1/3
1/2
b/a = 1.5
Ovaloid r = a/2
Locus of Minimum K_t
for b/a ≥ 1
b/a = 1
Square Hole
$K_t = \sigma_{max}/\sigma_1$
Circle
Ovaloid r = b/2
b/a = 1/4
1/3
1/2
K_t
r/b
0
0.1
0.2
0.3
0.4
0.5
2
3
4
5

FIG. 141

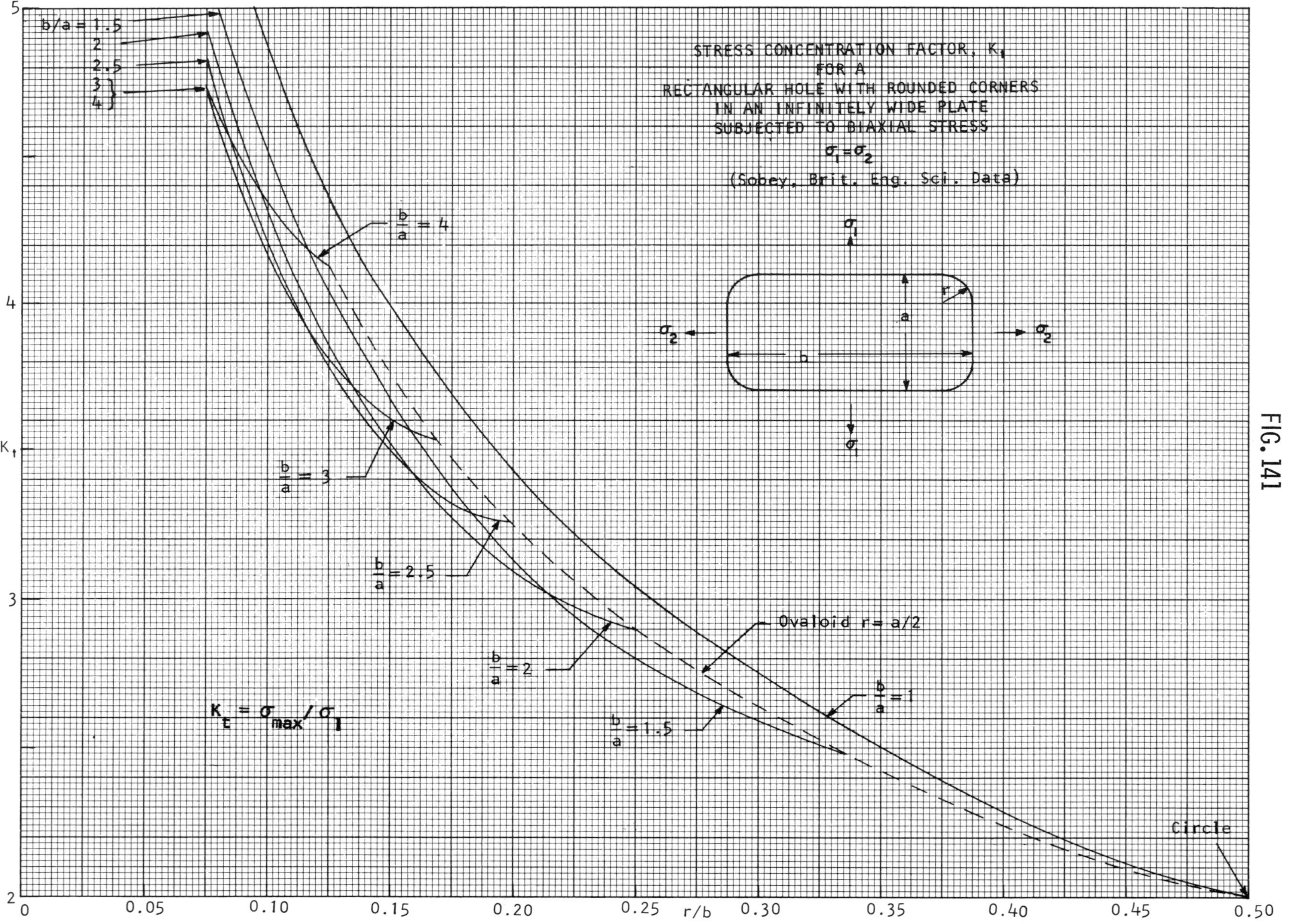
STRESS CONCENTRATION FACTOR, K_t
FOR A
RECTANGULAR HOLE WITH ROUNDED CORNERS
IN AN INFINITELY WIDE PLATE
SUBJECTED TO BIAXIAL STRESS
$\sigma_1 = \sigma_2$
(Sobey, Brit. Eng. Sci. Data)
σ_1
σ_2
r
a
b
$K_t = \sigma_{max}/\sigma_1$
Ovaloid r = a/2
Circle
$\frac{b}{a} = 1$
$\frac{b}{a} = 1.5$
$\frac{b}{a} = 2$
$\frac{b}{a} = 2.5$
$\frac{b}{a} = 3$
$\frac{b}{a} = 4$
b/a = 1.5
2
2.5
3
4
K_t
r/b
0
0.05
0.10
0.15
0.20
0.25
0.30
0.35
0.40
0.45
0.50
2
3
4
5

FIG. 142

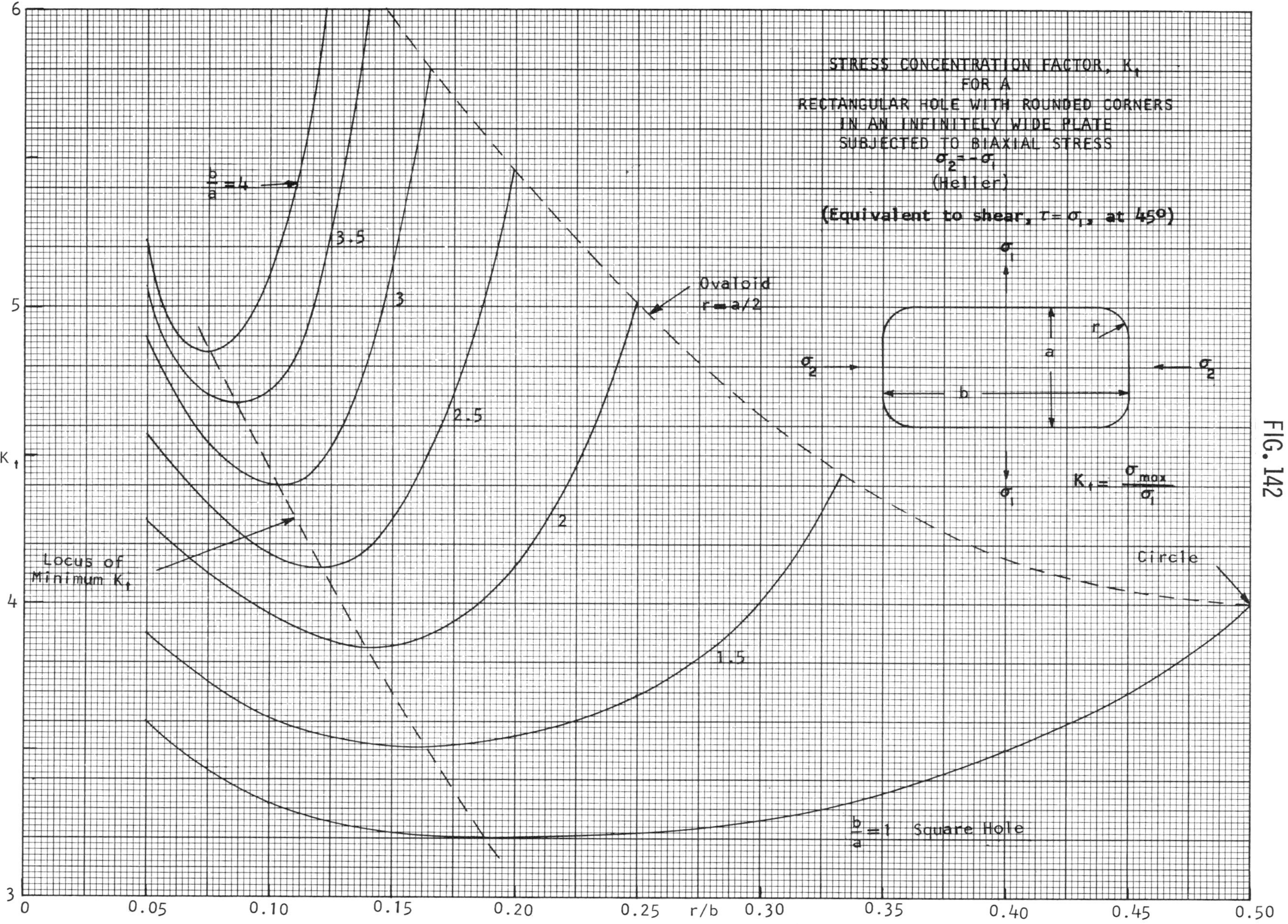

STRESS CONCENTRATION FACTOR, K_t
FOR A
RECTANGULAR HOLE WITH ROUNDED CORNERS
IN AN INFINITELY WIDE PLATE
SUBJECTED TO BIAXIAL STRESS
$\sigma_2 = -\sigma_1$
(Heller)
(Equivalent to shear, $\tau = \sigma_1$, at 45°)
σ_1
σ_2
a
b
r
$K_t = \frac{\sigma_{max}}{\sigma_1}$
Ovaloid
$r = a/2$
Circle
$\frac{b}{a} = 4$
3.5
3
2.5
2
1.5
$\frac{b}{a} = 1$ Square Hole
Locus of
Minimum K_t
K_t
r/b
0
0.05
0.10
0.15
0.20
0.25
0.30
0.35
0.40
0.45
0.50
3
4
5
6

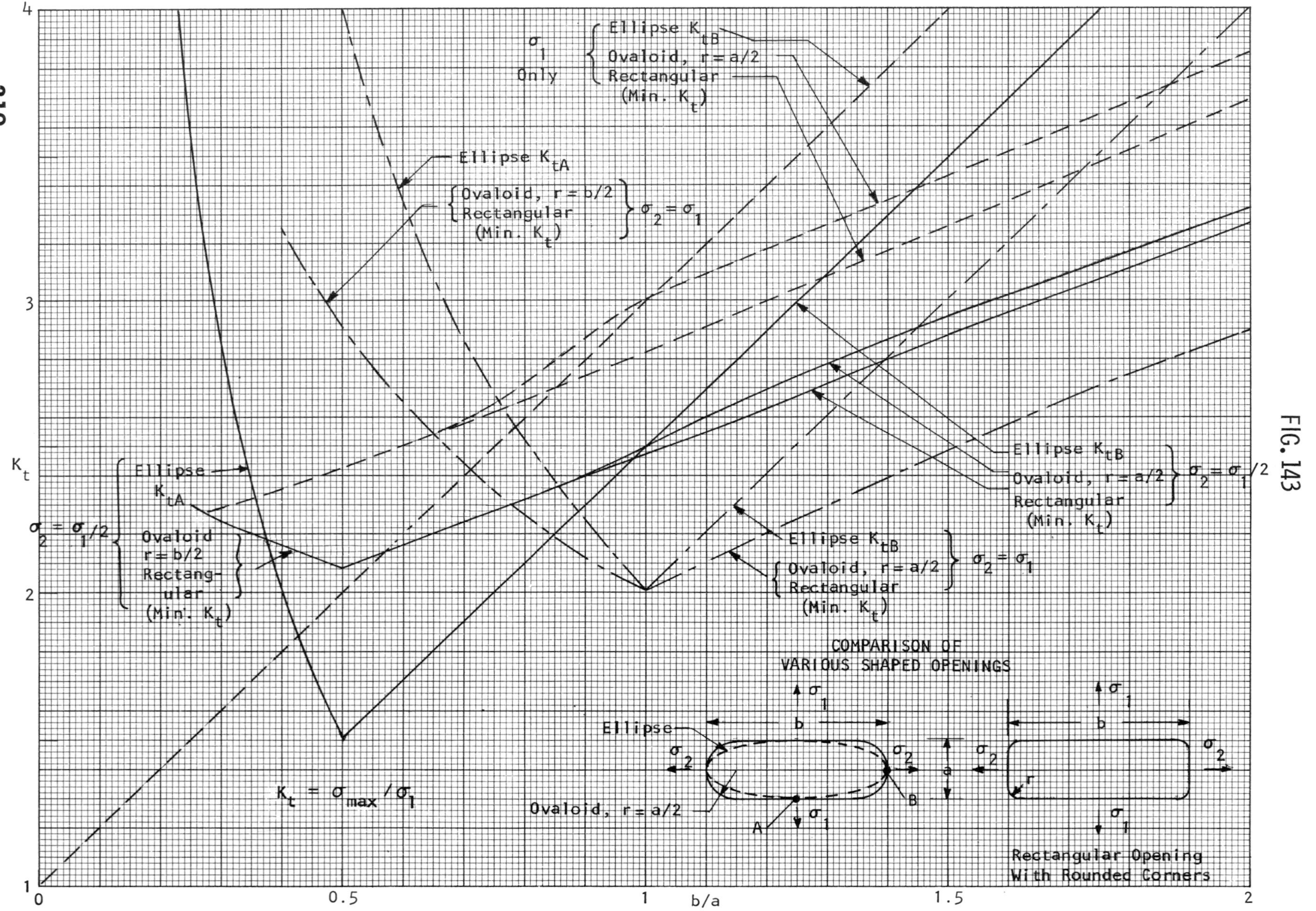

σ_1 Only
Ellipse K_{tB}
Ovaloid, r = a/2
Rectangular (Min. K_t)
Ellipse K_{tA}
Ovaloid, r = b/2
Rectangular (Min. K_t)
$\sigma_2 = \sigma_1$
Ellipse K_{tA}
Ovaloid r = b/2
Rectang- ular (Min. K_t)
$\sigma_2 = \sigma_1/2$
Ellipse K_{tB}
Ovaloid, r = a/2
Rectangular (Min. K_t)
$\sigma_2 = \sigma_1/2$
Ellipse K_{tB}
Ovaloid, r = a/2
Rectangular (Min. K_t)
$\sigma_2 = \sigma_1$
COMPARISON OF
VARIOUS SHAPED OPENINGS
Ellipse
Ovaloid, r = a/2
$K_t = \sigma_{max}/\sigma_1$
Rectangular Opening
With Rounded Corners
σ_1
σ_2
a
b
r
A
B
K_t
b/a
1
2
3
4
0
0.5
1
1.5
2

FIG. 143

FIG. 144

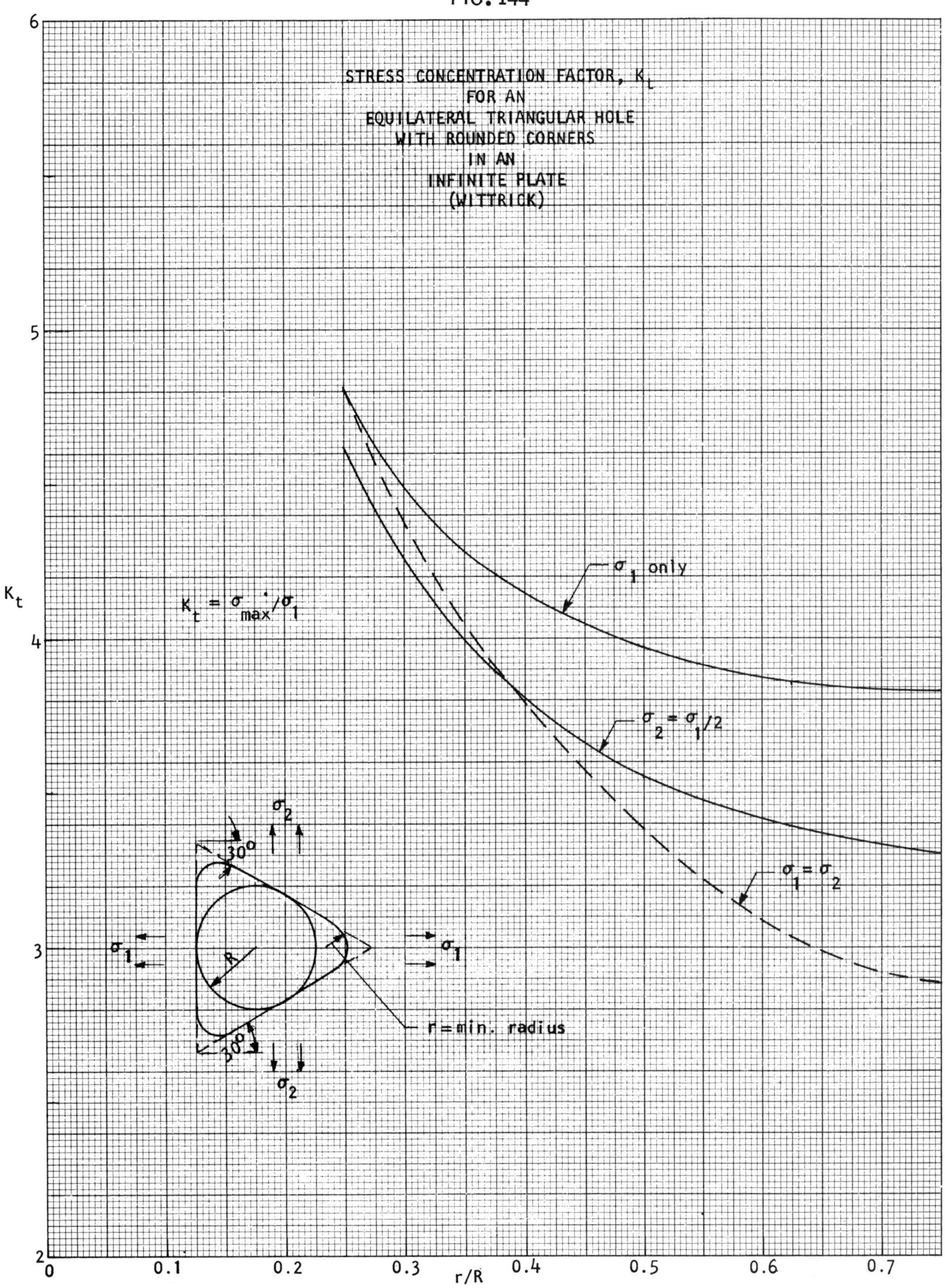

STRESS CONCENTRATION FACTOR, K_t
FOR AN
EQUILATERAL TRIANGULAR HOLE
WITH ROUNDED CORNERS
IN AN
INFINITE PLATE
(WITTRICK)
$K_t = \sigma_{max}/\sigma_1$
K_t
σ_1 only
$\sigma_2 = \sigma_1/2$
$\sigma_1 = \sigma_2$
σ_1
σ_2
30°
R
r = min. radius
r/R
6
5
4
3
2
0
0.1
0.2
0.3
0.4
0.5
0.6
0.7

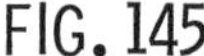

FIG. 145

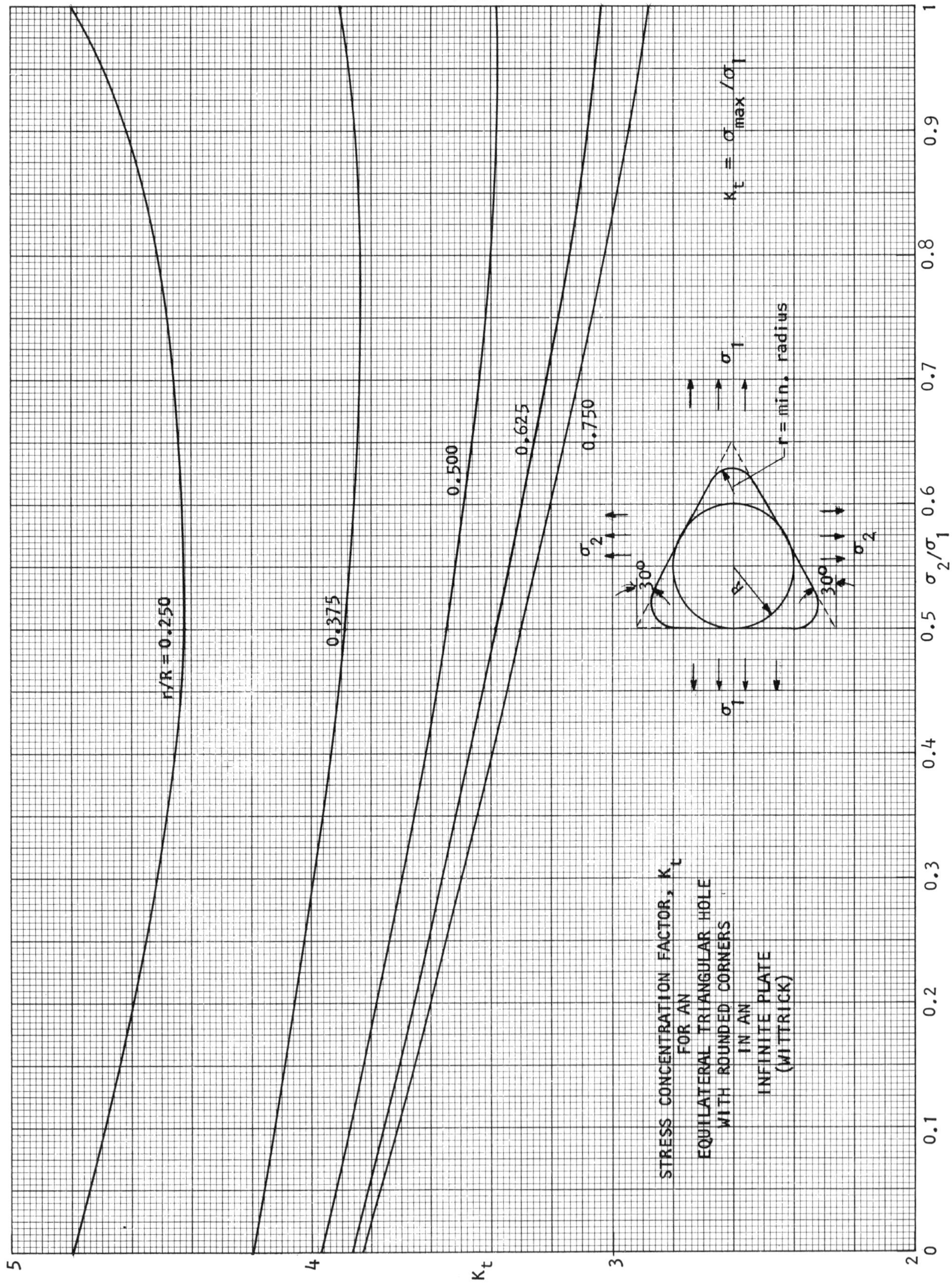
STRESS CONCENTRATION FACTOR, K_t
FOR AN
EQUILATERAL TRIANGULAR HOLE
WITH ROUNDED CORNERS
IN AN
INFINITE PLATE
(WITTRICK)
$K_t = \sigma_{max}/\sigma_1$
r/R = 0.250
0.375
0.500
0.625
0.750
r = min. radius
R
30°
σ_1
σ_2
K_t
σ_2/σ_1
0
0.1
0.2
0.3
0.4
0.5
0.6
0.7
0.8
0.9
1
2
3
4
5

FIG. 145a

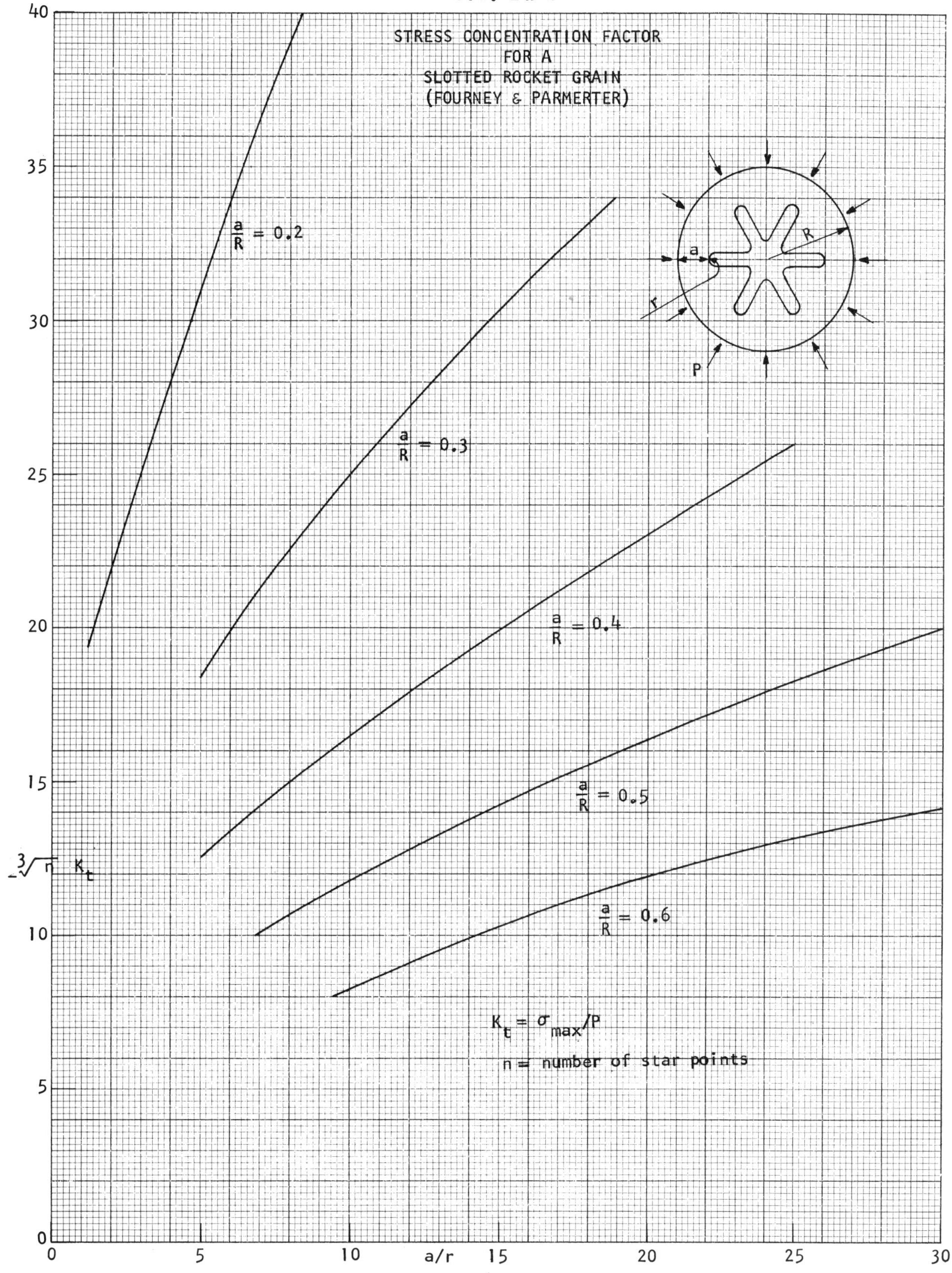

STRESS CONCENTRATION FACTOR
FOR A
SLOTTED ROCKET GRAIN
(FOURNEY & PARMERTER)
a/R = 0.2
a/R = 0.3
a/R = 0.4
a/R = 0.5
a/R = 0.6
a
R
r
P
$\sqrt[3]{n}$ K_t
$K_t = \sigma_{max}/P$
n = number of star points
a/r
0
5
10
15
20
25
30
35
40

FIG. 146

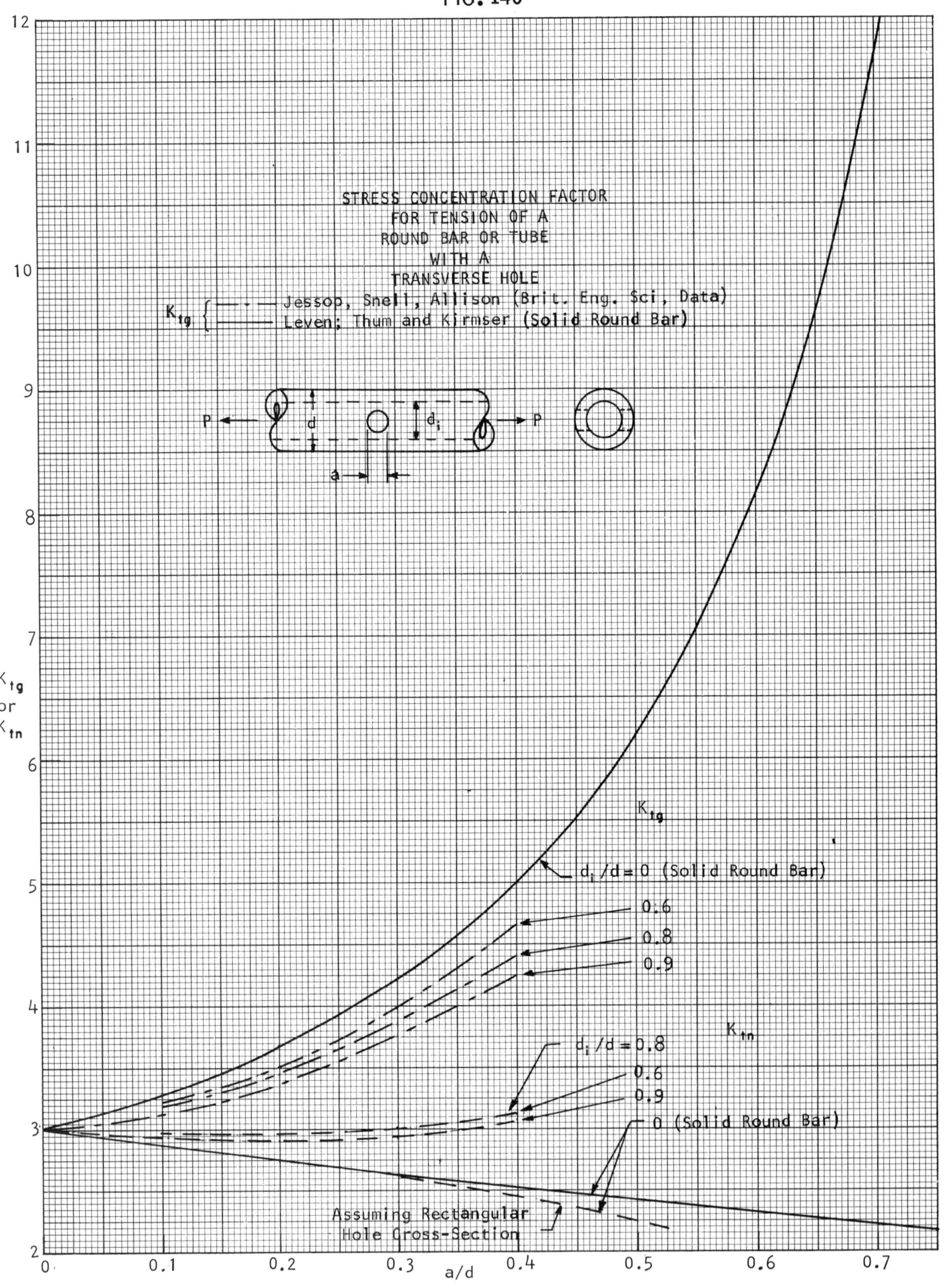

STRESS CONCENTRATION FACTOR
FOR TENSION OF A
ROUND BAR OR TUBE
WITH A
TRANSVERSE HOLE
Ktg
Jessop, Snell, Allison (Brit. Eng. Sci, Data)
Leven; Thum and Kirmser (Solid Round Bar)
P
d
di
a
Ktg
di/d = 0 (Solid Round Bar)
0.6
0.8
0.9
Ktn
di/d = 0.8
0.6
0.9
0 (Solid Round Bar)
Assuming Rectangular
Hole Cross-Section
Ktg
or
Ktn
a/d
12
11
10
9
8
7
6
5
4
3
2
0
0.1
0.2
0.3
0.4
0.5
0.6
0.7

FIG. 147

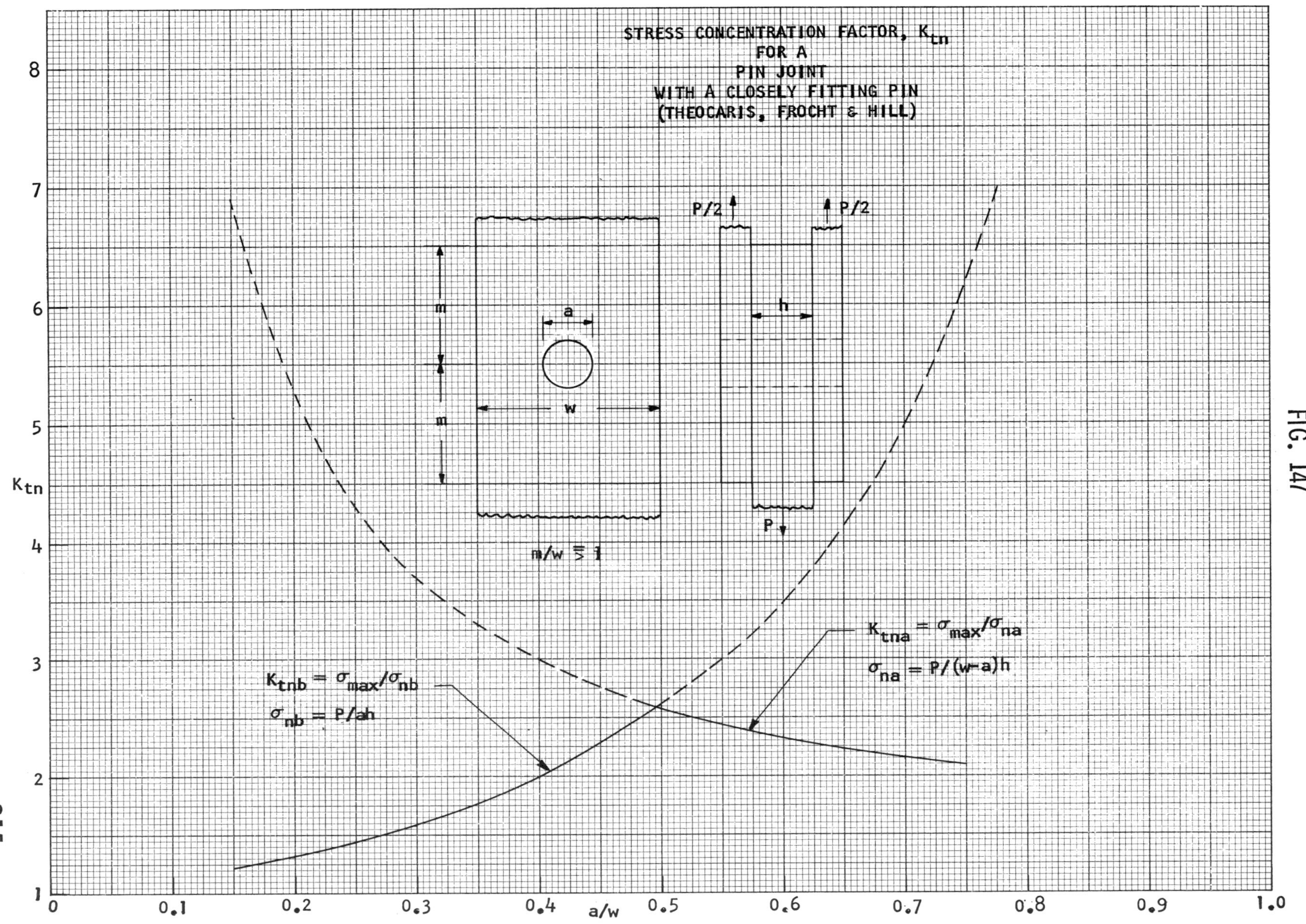
STRESS CONCENTRATION FACTOR, K_{tn}
FOR A
PIN JOINT
WITH A CLOSELY FITTING PIN
(THEOCARIS, FROCHT & HILL)
P/2
P/2
h
P
a
w
m
m
m/w ≅ 1
$K_{tna} = \sigma_{max}/\sigma_{na}$
$\sigma_{na} = P/(w-a)h$
$K_{tnb} = \sigma_{max}/\sigma_{nb}$
$\sigma_{nb} = P/ah$
K_{tn}
a/w
0
0.1
0.2
0.3
0.4
0.5
0.6
0.7
0.8
0.9
1.0
1
2
3
4
5
6
7
8

FIG. 147a

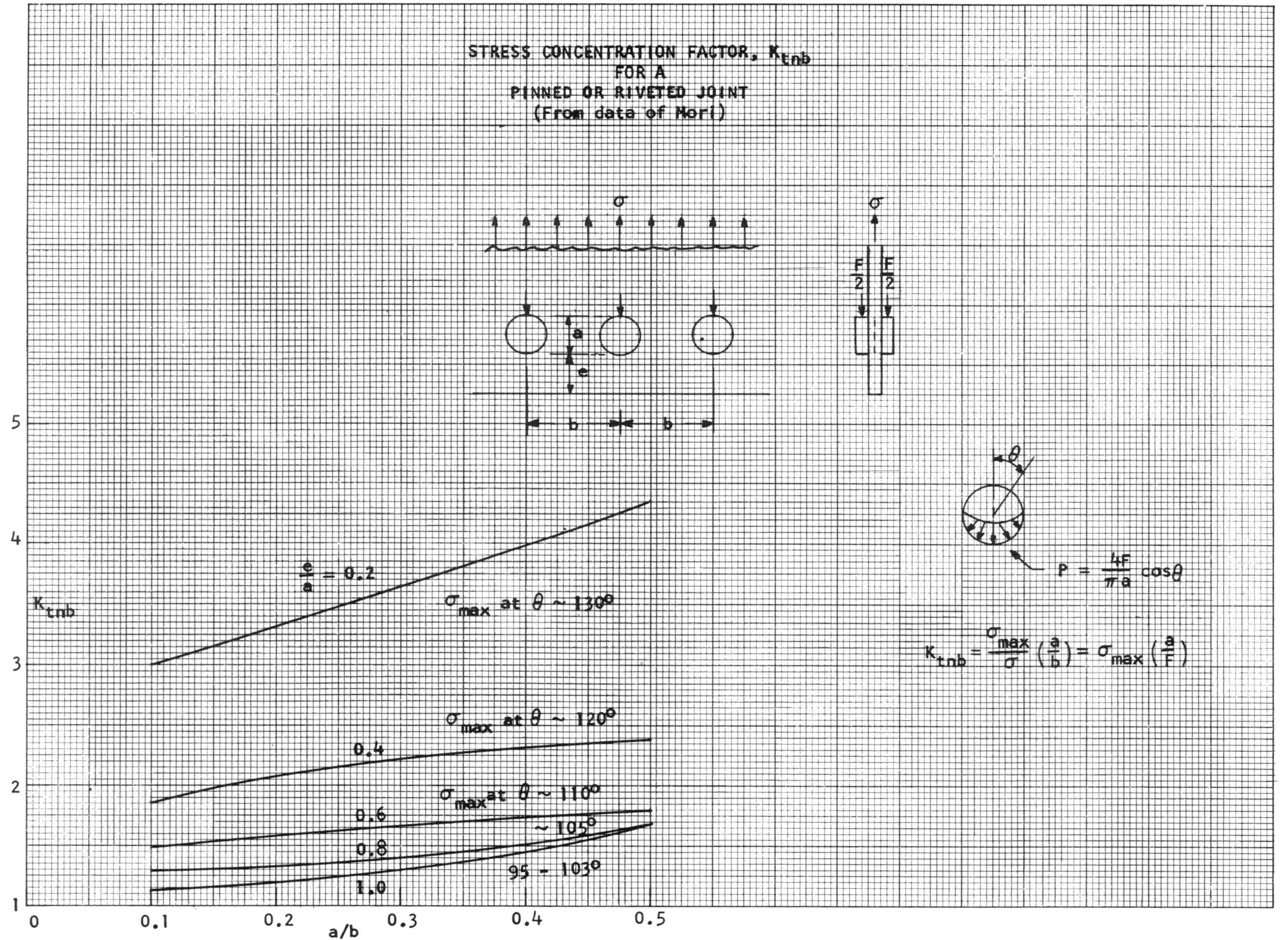

STRESS CONCENTRATION FACTOR, K_{tnb}
FOR A
PINNED OR RIVETED JOINT
(From data of Mori)
σ
F/2
a
e
b
θ
$P = \frac{4F}{\pi a}\cos\theta$
$K_{tnb} = \frac{\sigma_{max}}{\sigma}\left(\frac{a}{b}\right) = \sigma_{max}\left(\frac{a}{F}\right)$
$\frac{e}{a} = 0.2$
σ_{max} at $\theta \sim 130^o$
σ_{max} at $\theta \sim 120^o$
σ_{max} at $\theta \sim 110^o$
$\sim 105^o$
95 - 103°
0.4
0.6
0.8
1.0
K_{tnb}
1
2
3
4
5
0
0.1
0.2
0.3
0.4
0.5
a/b

FIG. 148

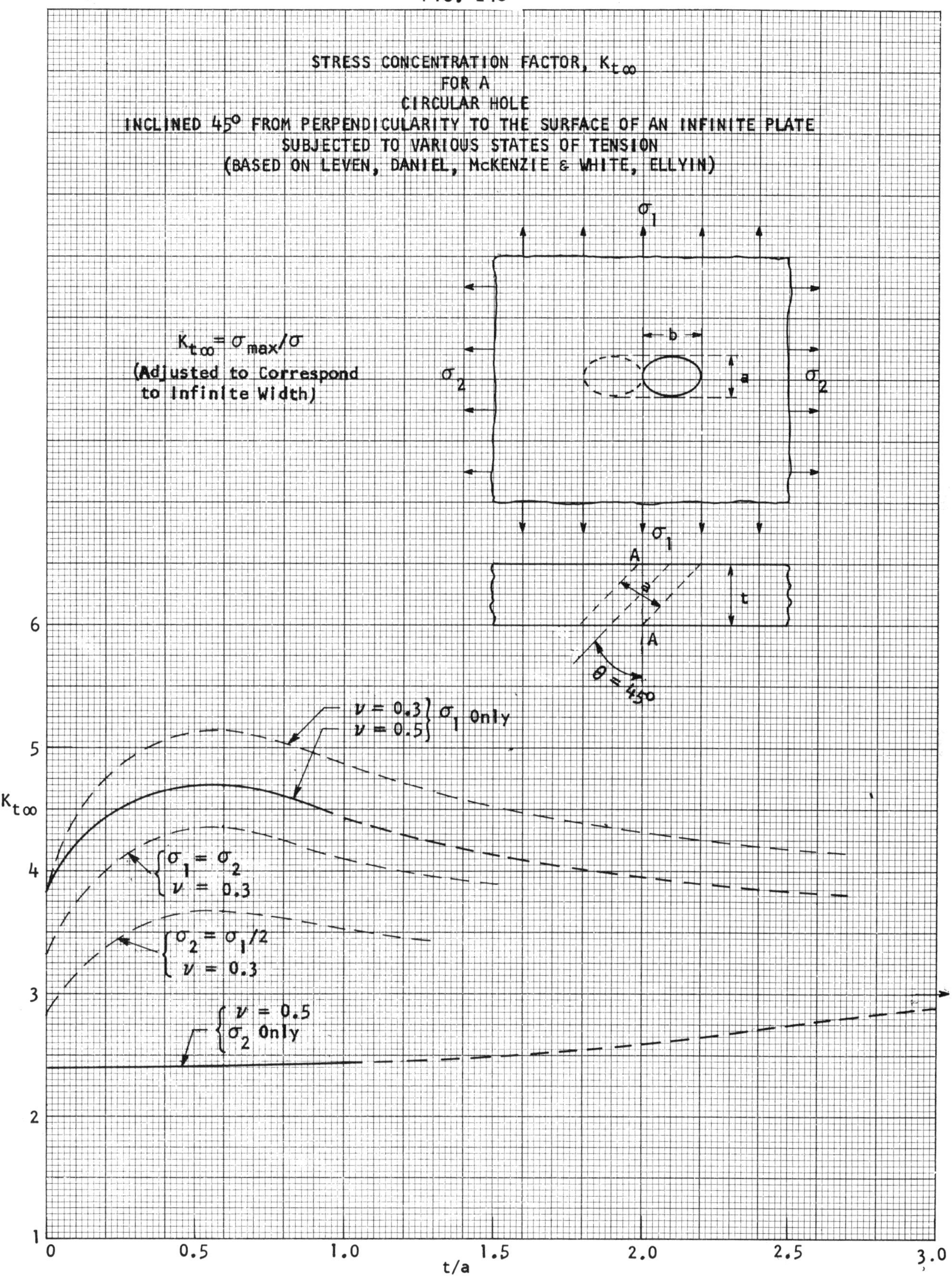

STRESS CONCENTRATION FACTOR, $K_{t\infty}$
FOR A
CIRCULAR HOLE
INCLINED 45° FROM PERPENDICULARITY TO THE SURFACE OF AN INFINITE PLATE
SUBJECTED TO VARIOUS STATES OF TENSION
(BASED ON LEVEN, DANIEL, McKENZIE & WHITE, ELLYIN)
$K_{t\infty} = \sigma_{max}/\sigma$
(Adjusted to Correspond to Infinite Width)
σ_1
σ_2
b
a
A
t
$\theta = 45°$
$\nu = 0.3$ } σ_1 Only
$\nu = 0.5$
$\sigma_1 = \sigma_2$, $\nu = 0.3$
$\sigma_2 = \sigma_1/2$, $\nu = 0.3$
$\nu = 0.5$, σ_2 Only
$K_{t\infty}$
t/a
1
2
3
4
5
6
0
0.5
1.0
1.5
2.0
2.5
3.0

FIG. 149

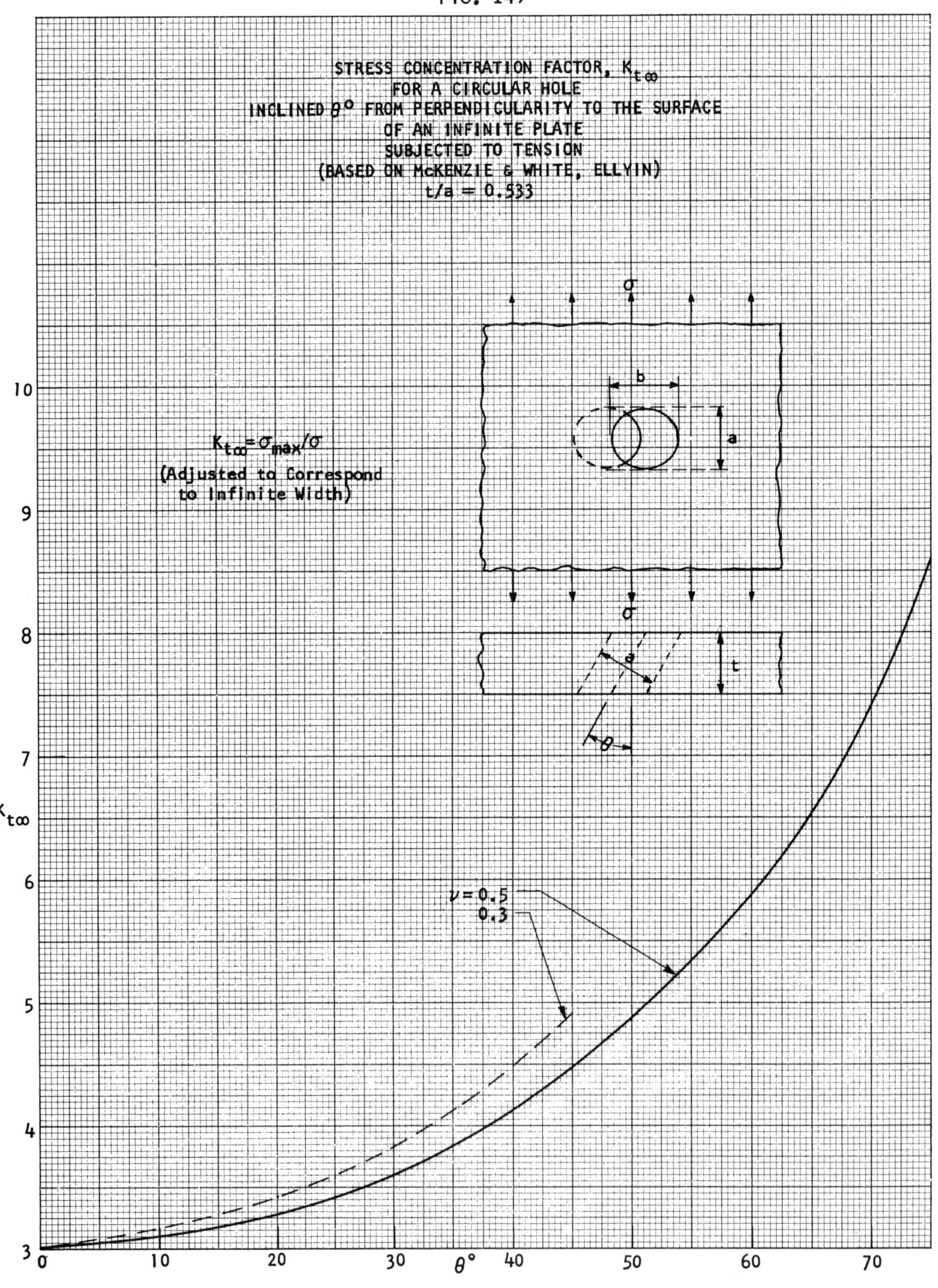

STRESS CONCENTRATION FACTOR, $K_{t\infty}$
FOR A CIRCULAR HOLE
INCLINED $\theta°$ FROM PERPENDICULARITY TO THE SURFACE
OF AN INFINITE PLATE
SUBJECTED TO TENSION
(BASED ON McKENZIE & WHITE, ELLYIN)
t/a = 0.533
$K_{t\infty} = \sigma_{max}/\sigma$
(Adjusted to Correspond to Infinite Width)
σ
b
a
t
θ
$\nu = 0.5$
0.3
$K_{t\infty}$
$\theta°$
3
4
5
6
7
8
9
10
0
10
20
30
40
50
60
70

FIG. 150

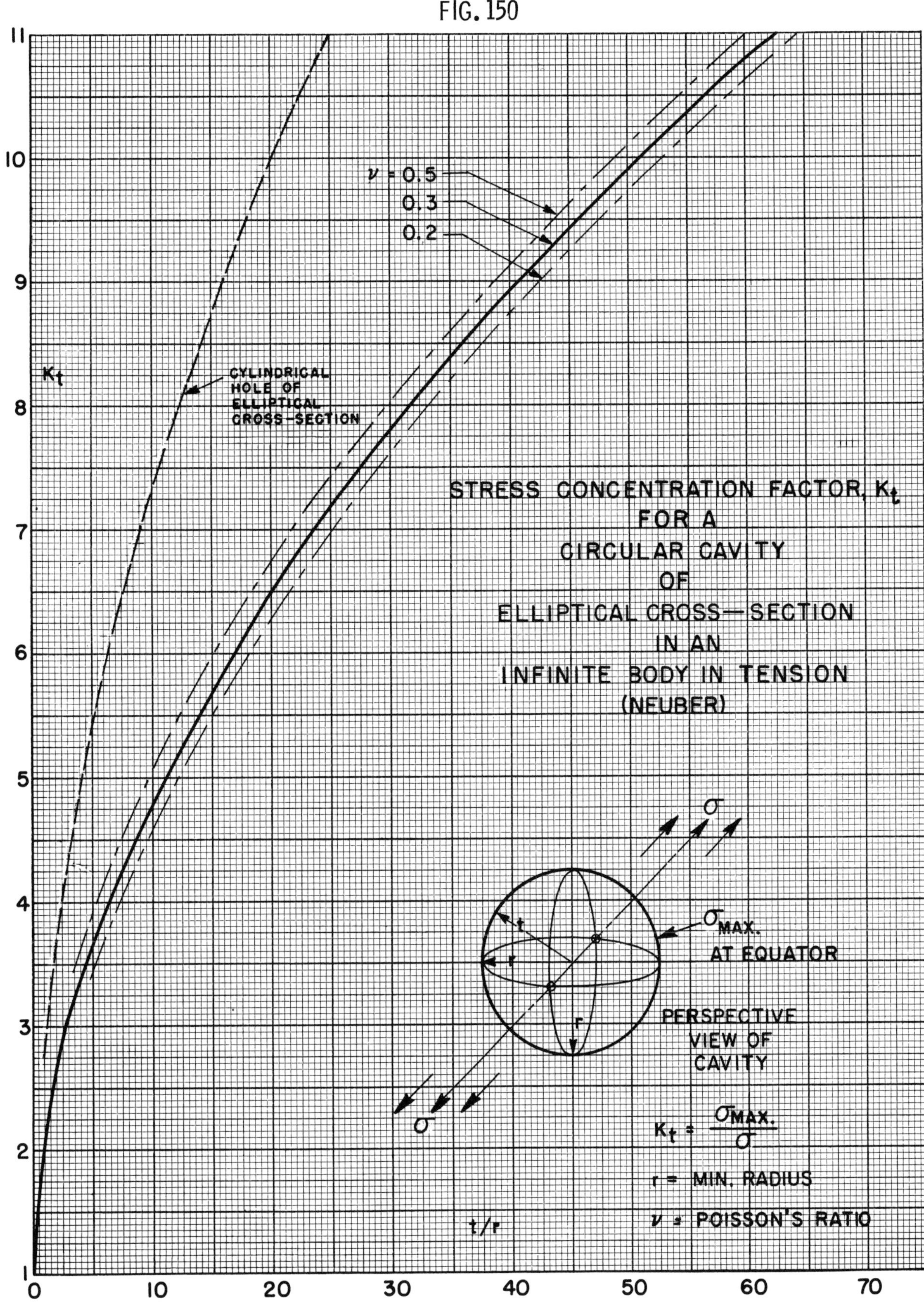

ν = 0.5
0.3
0.2
CYLINDRICAL HOLE OF ELLIPTICAL CROSS-SECTION
Kt
STRESS CONCENTRATION FACTOR, Kt FOR A CIRCULAR CAVITY OF ELLIPTICAL CROSS—SECTION IN AN INFINITE BODY IN TENSION (NEUBER)
σ
σ MAX. AT EQUATOR
t
r
r
PERSPECTIVE VIEW OF CAVITY
σ
Kt = σ MAX. / σ
r = MIN. RADIUS
ν = POISSON'S RATIO
t/r
1
2
3
4
5
6
7
8
9
10
11
0
10
20
30
40
50
60
70

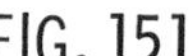

FIG. 151

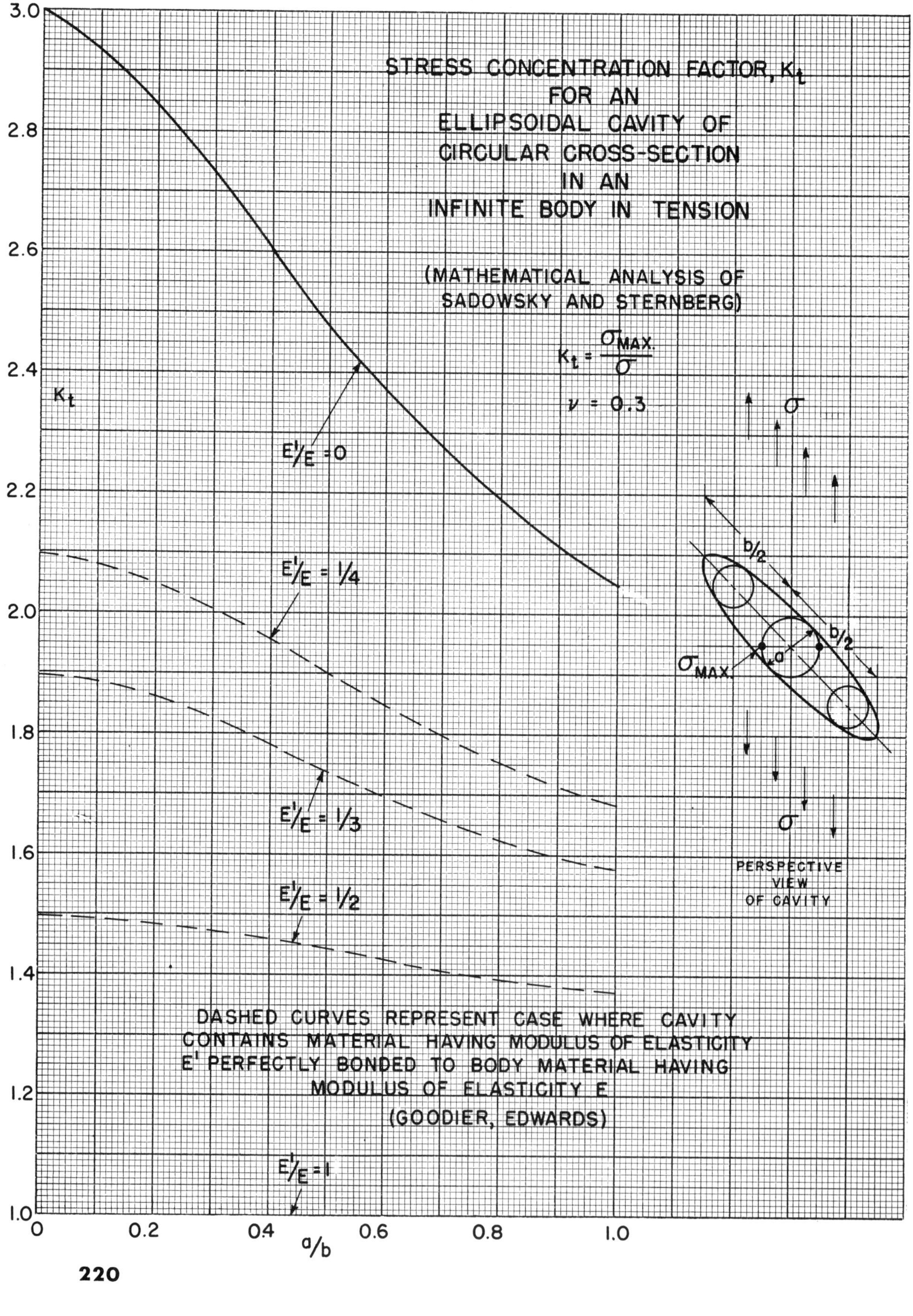

STRESS CONCENTRATION FACTOR, K_t
FOR AN
ELLIPSOIDAL CAVITY OF
CIRCULAR CROSS-SECTION
IN AN
INFINITE BODY IN TENSION

(MATHEMATICAL ANALYSIS OF
SADOWSKY AND STERNBERG)

$K_t = \frac{\sigma_{MAX.}}{\sigma}$

$\nu = 0.3$

$E'/E = 0$

$E'/E = 1/4$

$E'/E = 1/3$

$E'/E = 1/2$

$E'/E = 1$

σ

b/2

b/2

a

$\sigma_{MAX.}$

σ

PERSPECTIVE
VIEW
OF CAVITY

DASHED CURVES REPRESENT CASE WHERE CAVITY
CONTAINS MATERIAL HAVING MODULUS OF ELASTICITY
E' PERFECTLY BONDED TO BODY MATERIAL HAVING
MODULUS OF ELASTICITY E
(GOODIER, EDWARDS)

K_t

3.0
2.8
2.6
2.4
2.2
2.0
1.8
1.6
1.4
1.2
1.0

0 0.2 0.4 0.6 0.8 1.0

a/b

FIG. 152

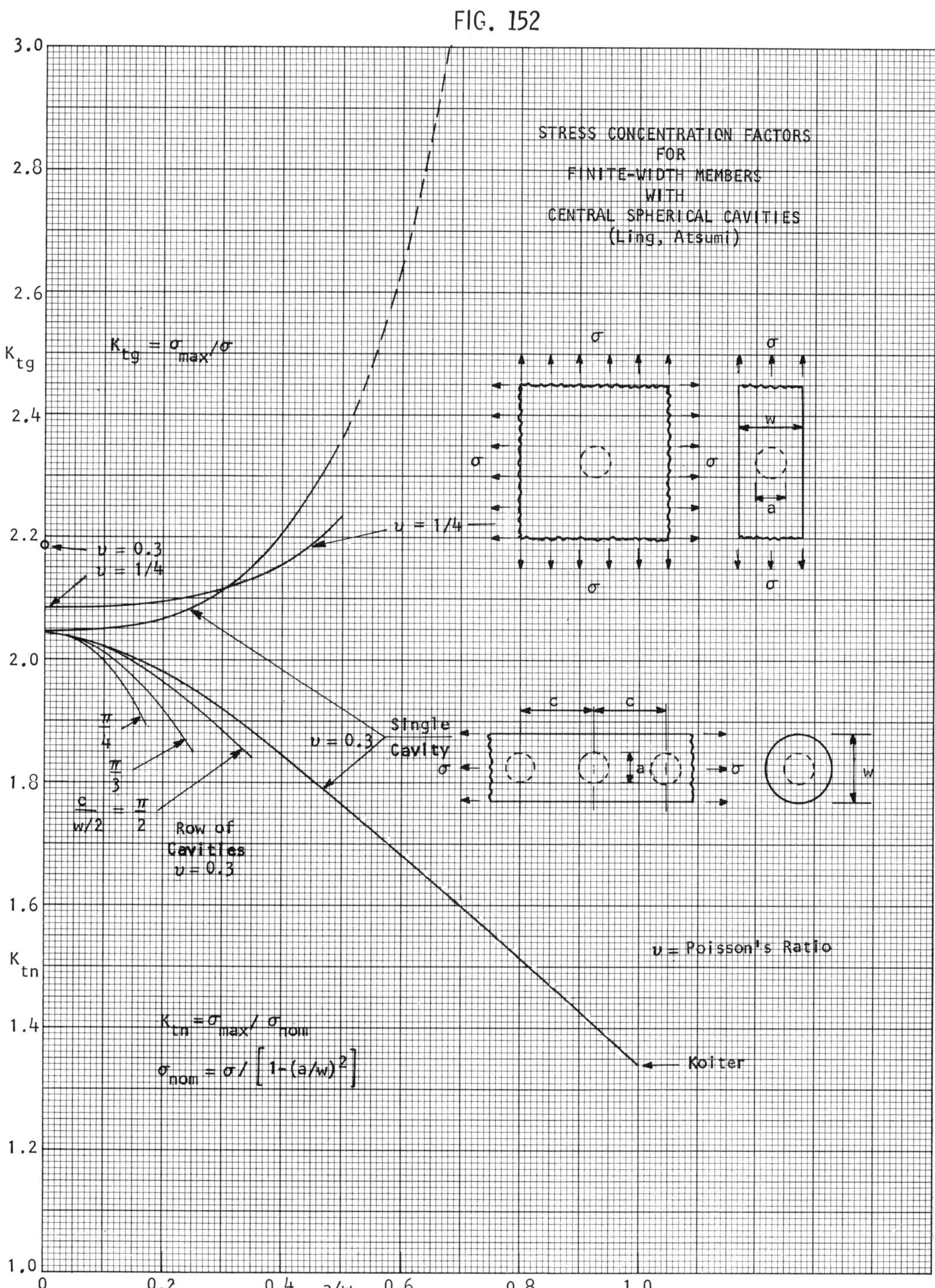

STRESS CONCENTRATION FACTORS
FOR
FINITE-WIDTH MEMBERS
WITH
CENTRAL SPHERICAL CAVITIES
(Ling, Atsumi)
$K_{tg} = \sigma_{max}/\sigma$
K_{tg}
$\nu = 0.3$
$\nu = 1/4$
$\nu = 1/4$
$\frac{\pi}{4}$
$\frac{\pi}{3}$
$\frac{c}{w/2} = \frac{\pi}{2}$
Row of Cavities $\nu = 0.3$
$\nu = 0.3$
Single Cavity
ν = Poisson's Ratio
K_{tn}
$K_{tn} = \sigma_{max}/\sigma_{nom}$
$\sigma_{nom} = \sigma/\left[1-(a/w)^2\right]$
Koiter
σ
w
a
c
a/w
3.0
2.8
2.6
2.4
2.2
2.0
1.8
1.6
1.4
1.2
1.0
0
0.2
0.4
0.6
0.8
1.0

FIG. 153

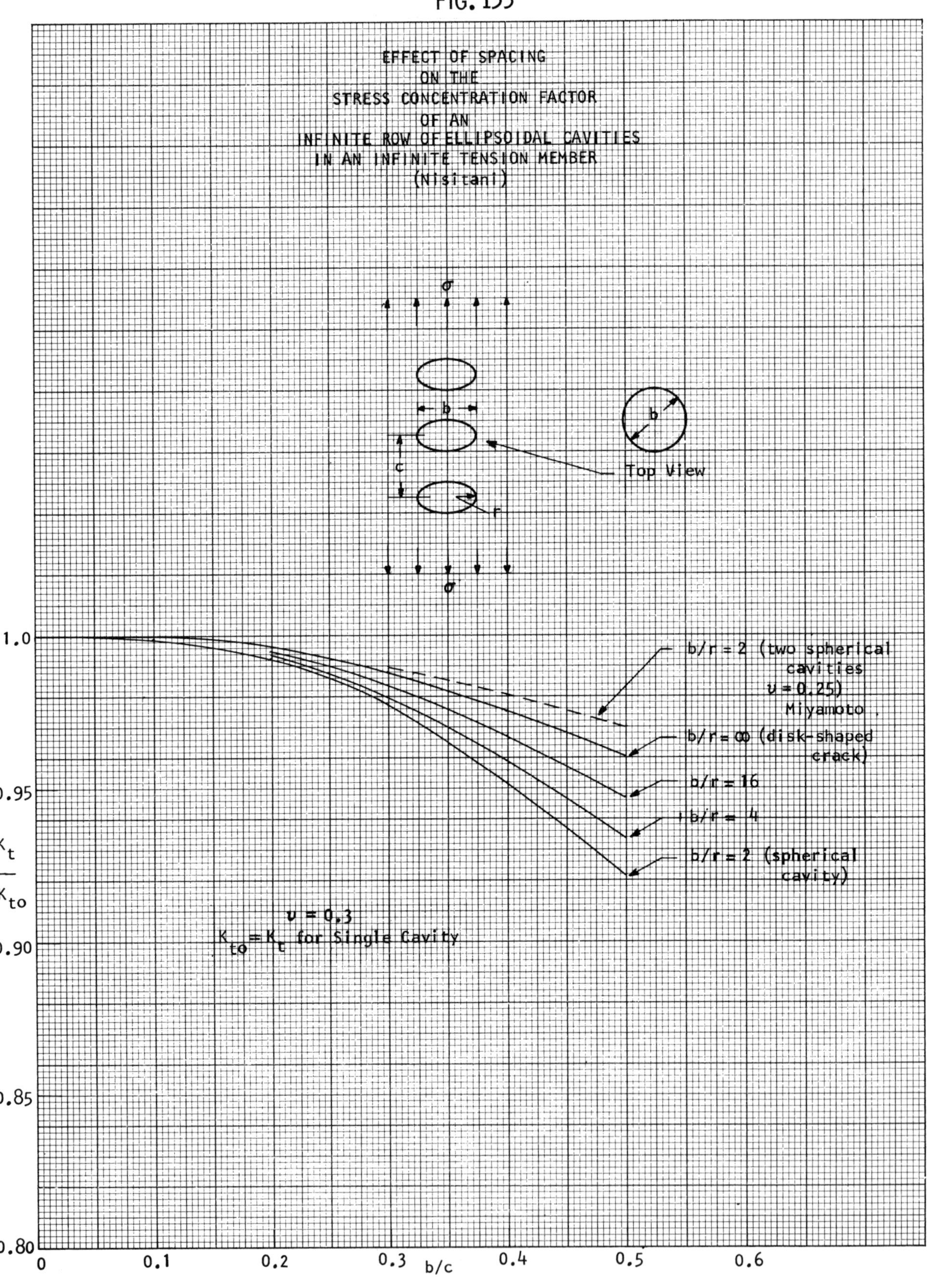
EFFECT OF SPACING
ON THE
STRESS CONCENTRATION FACTOR
OF AN
INFINITE ROW OF ELLIPSOIDAL CAVITIES
IN AN INFINITE TENSION MEMBER
(Nisitani)
σ
b
c
r
σ
b
Top View
b/r = 2 (two spherical cavities ν = 0.25) Miyamoto
b/r = ∞ (disk-shaped crack)
b/r = 16
b/r = 4
b/r = 2 (spherical cavity)
ν = 0.3
Kto = Kt for Single Cavity
1.0
0.95
0.90
0.85
0.80
Kt / Kto
0
0.1
0.2
0.3
0.4
0.5
0.6
b/c

FIG. 154

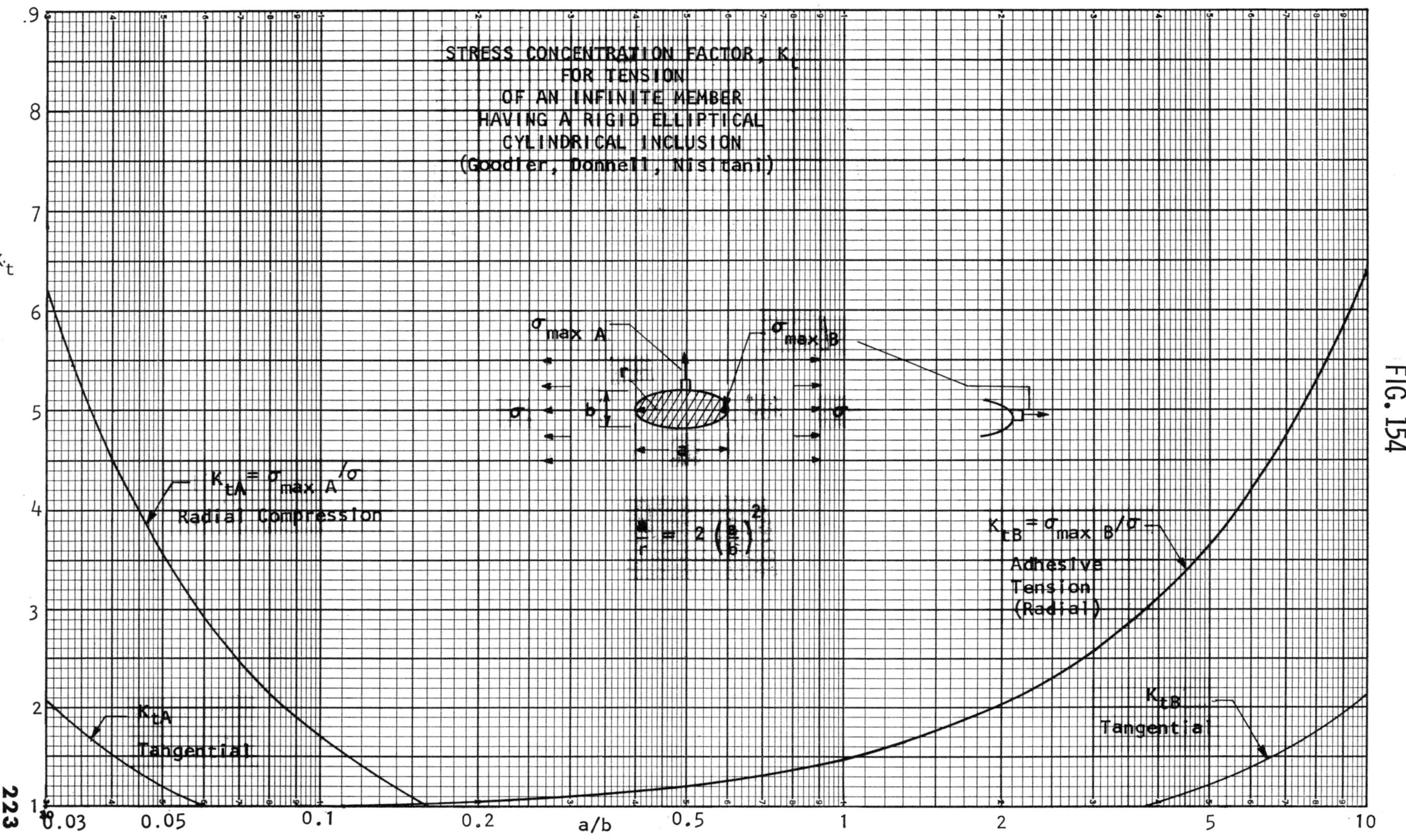
STRESS CONCENTRATION FACTOR, K_t
FOR TENSION
OF AN INFINITE MEMBER
HAVING A RIGID ELLIPTICAL
CYLINDRICAL INCLUSION
(Goodier, Donnell, Nisitani)
$\sigma_{max\,A}$
$\sigma_{max\,B}$
σ
σ
r
b
a
$\frac{a}{r} = 2\left(\frac{a}{b}\right)^2$
$K_{tA} = \sigma_{max\,A}/\sigma$
Radial Compression
K_{tA}
Tangential
$K_{tB} = \sigma_{max\,B}/\sigma$
Adhesive
Tension
(Radial)
K_{tB}
Tangential
K_t
9
8
7
6
5
4
3
2
1
a/b
0.03
0.05
0.1
0.2
0.5
1
2
5
10

FIG. 155

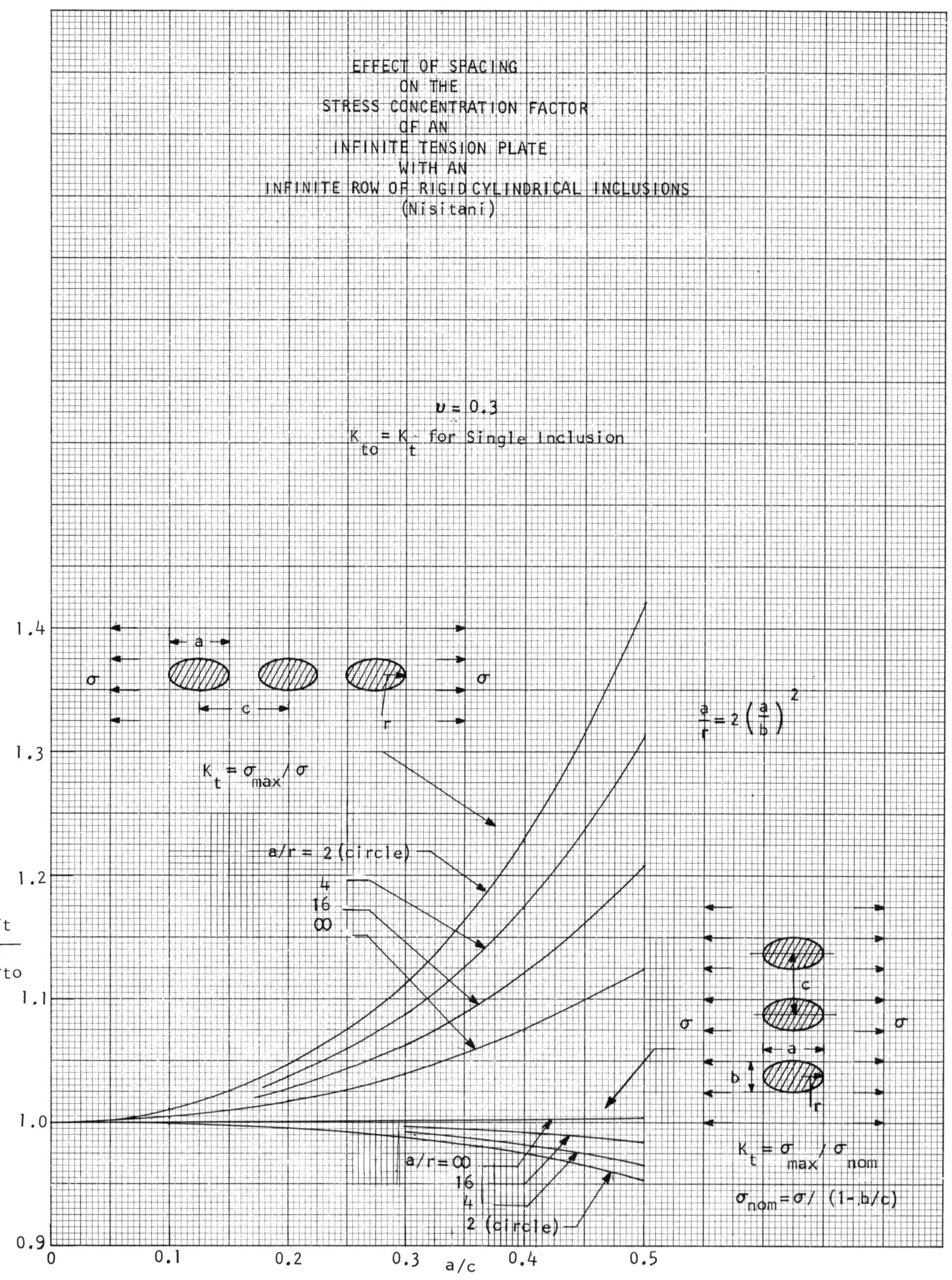

EFFECT OF SPACING
ON THE
STRESS CONCENTRATION FACTOR
OF AN
INFINITE TENSION PLATE
WITH AN
INFINITE ROW OF RIGID CYLINDRICAL INCLUSIONS
(Nisitani)
ν = 0.3
K_to = K_t for Single Inclusion
K_t = σ_max / σ
a/r = 2(a/b)^2
a/r = 2 (circle)
4
16
∞
a/r = ∞
16
4
2 (circle)
K_t = σ_max / σ_nom
σ_nom = σ / (1 - b/c)
K_t / K_to
a/c
a
b
c
r
σ
0.9
1.0
1.1
1.2
1.3
1.4
0
0.1
0.2
0.3
0.4
0.5

FIG. 155a

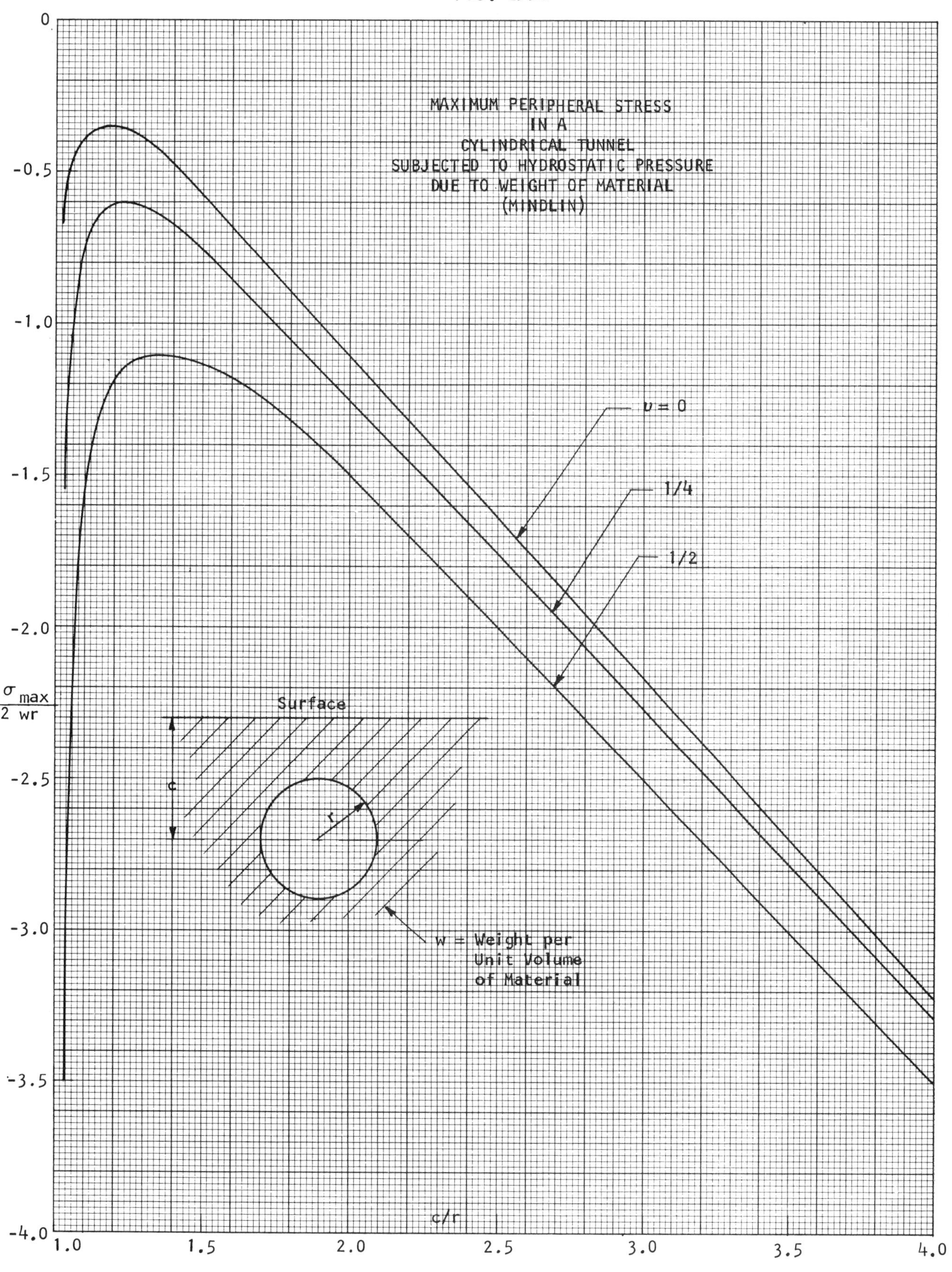

MAXIMUM PERIPHERAL STRESS
IN A
CYLINDRICAL TUNNEL
SUBJECTED TO HYDROSTATIC PRESSURE
DUE TO WEIGHT OF MATERIAL
(MINDLIN)
υ = 0
1/4
1/2
Surface
c
r
w = Weight per
Unit Volume
of Material
σ max
2 wr
c/r
0
-0.5
-1.0
-1.5
-2.0
-2.5
-3.0
-3.5
-4.0
1.0
1.5
2.0
2.5
3.0
3.5
4.0

FIG. 155b

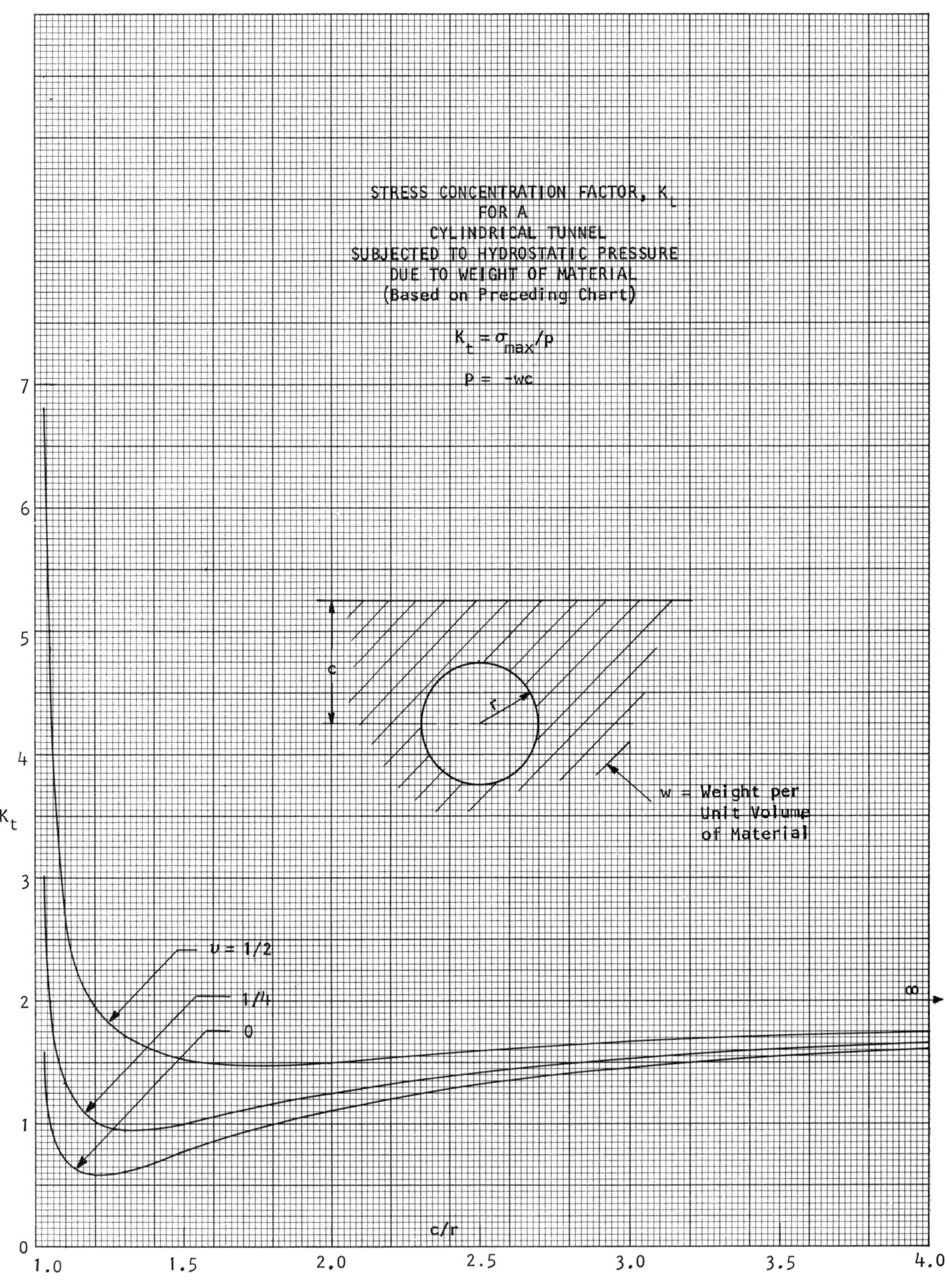

STRESS CONCENTRATION FACTOR, K_t
FOR A
CYLINDRICAL TUNNEL
SUBJECTED TO HYDROSTATIC PRESSURE
DUE TO WEIGHT OF MATERIAL
(Based on Preceding Chart)
$K_t = \sigma_{max}/p$
$p = -wc$
c
r
w = Weight per Unit Volume of Material
K_t
$\nu = 1/2$
1/4
0
∞
c/r
0
1
2
3
4
5
6
7
1.0
1.5
2.0
2.5
3.0
3.5
4.0

FIG. 156

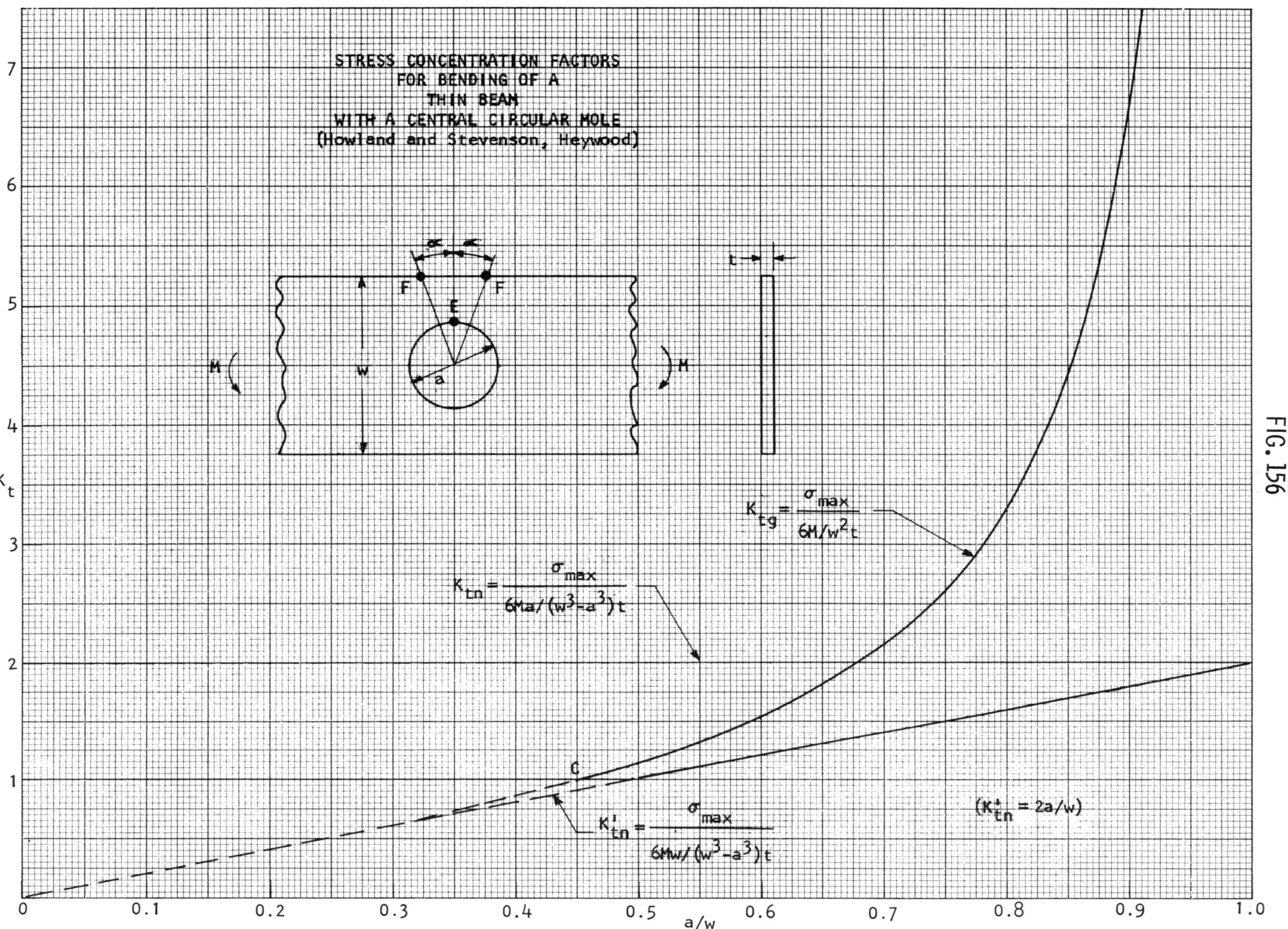
STRESS CONCENTRATION FACTORS
FOR BENDING OF A
THIN BEAM
WITH A CENTRAL CIRCULAR HOLE
(Howland and Stevenson, Heywood)
α
α
F
F
E
M
M
w
a
t
$K_{tg} = \frac{\sigma_{max}}{6M/w^2t}$
$K_{tn} = \frac{\sigma_{max}}{6Ma/(w^3-a^3)t}$
C
$K'_{tn} = \frac{\sigma_{max}}{6Mw/(w^3-a^3)t}$
$(K'_{tn} = 2a/w)$
K_t
7
6
5
4
3
2
1
0
0.1
0.2
0.3
0.4
0.5
0.6
0.7
0.8
0.9
1.0
a/w

FIG. 157

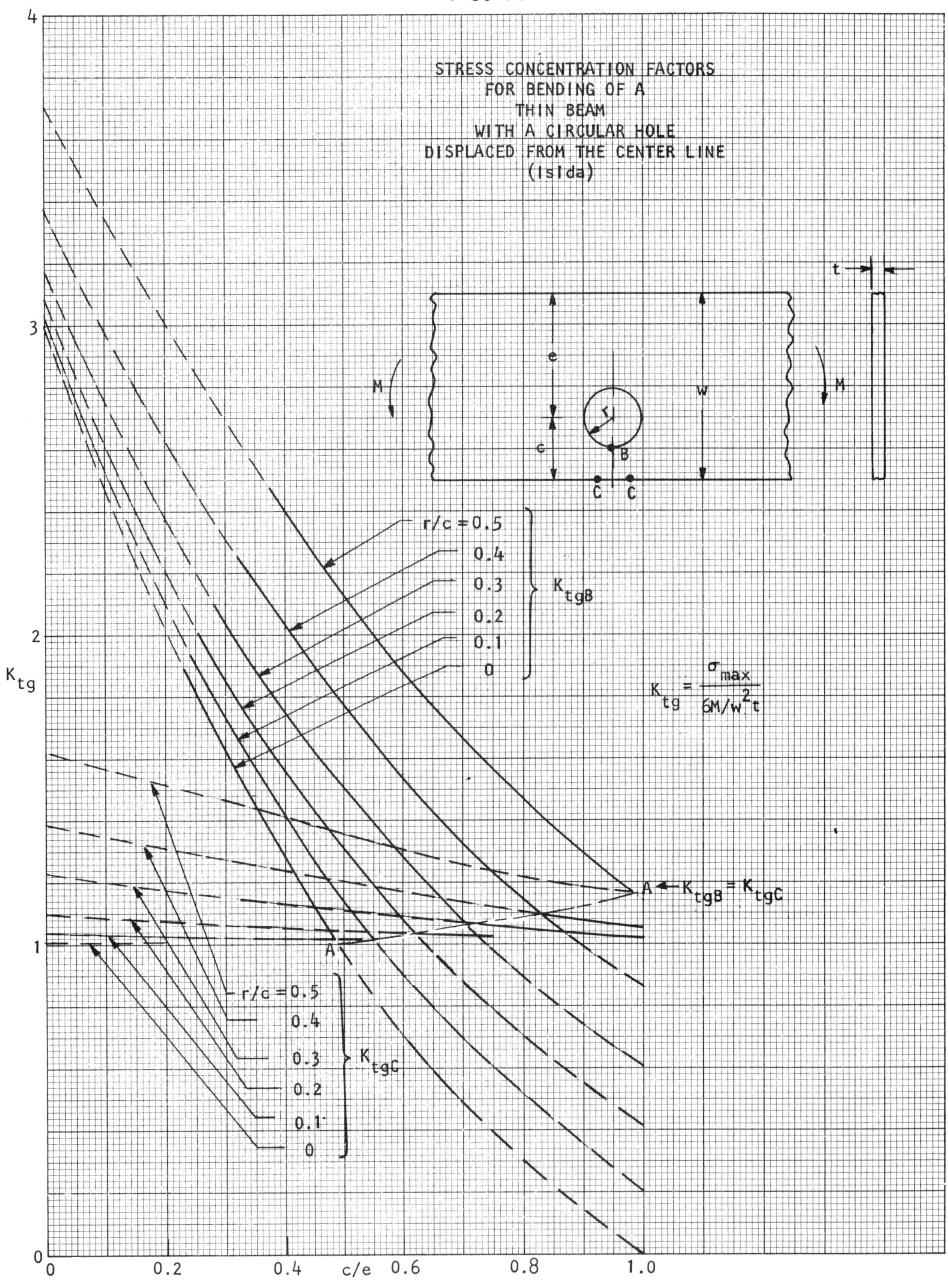
STRESS CONCENTRATION FACTORS
FOR BENDING OF A
THIN BEAM
WITH A CIRCULAR HOLE
DISPLACED FROM THE CENTER LINE
(Isida)
r/c = 0.5
0.4
0.3
0.2
0.1
0
K_{tgB}
$K_{tg} = \frac{\sigma_{max}}{6M/w^2t}$
A ← $K_{tgB} = K_{tgC}$
A
r/c = 0.5
0.4
0.3
0.2
0.1
0
K_{tgC}
K_{tg}
c/e
0
0.2
0.4
0.6
0.8
1.0
1
2
3
4
M
e
w
c
r
B
C
C
t

FIG. 158

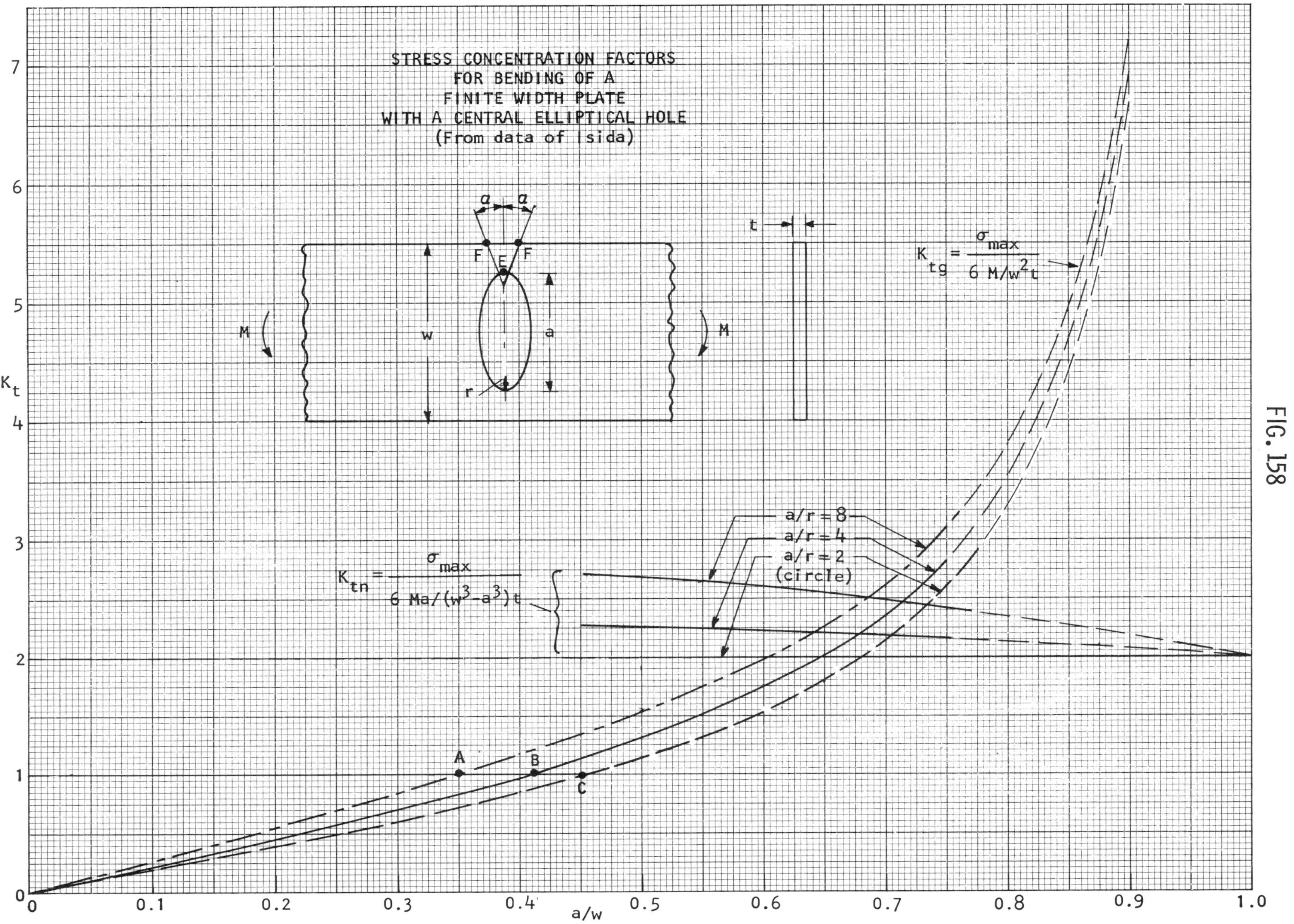

STRESS CONCENTRATION FACTORS
FOR BENDING OF A
FINITE WIDTH PLATE
WITH A CENTRAL ELLIPTICAL HOLE
(From data of Isida)
$K_{tg} = \frac{\sigma_{max}}{6\,M/w^2 t}$
$K_{tn} = \frac{\sigma_{max}}{6\,Ma/(w^3 - a^3)\,t}$
a/r = 8
a/r = 4
a/r = 2
(circle)
M
M
w
a
r
t
F
F
E
α
α
A
B
C
a/w
K_t

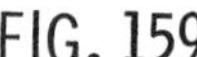

FIG. 159

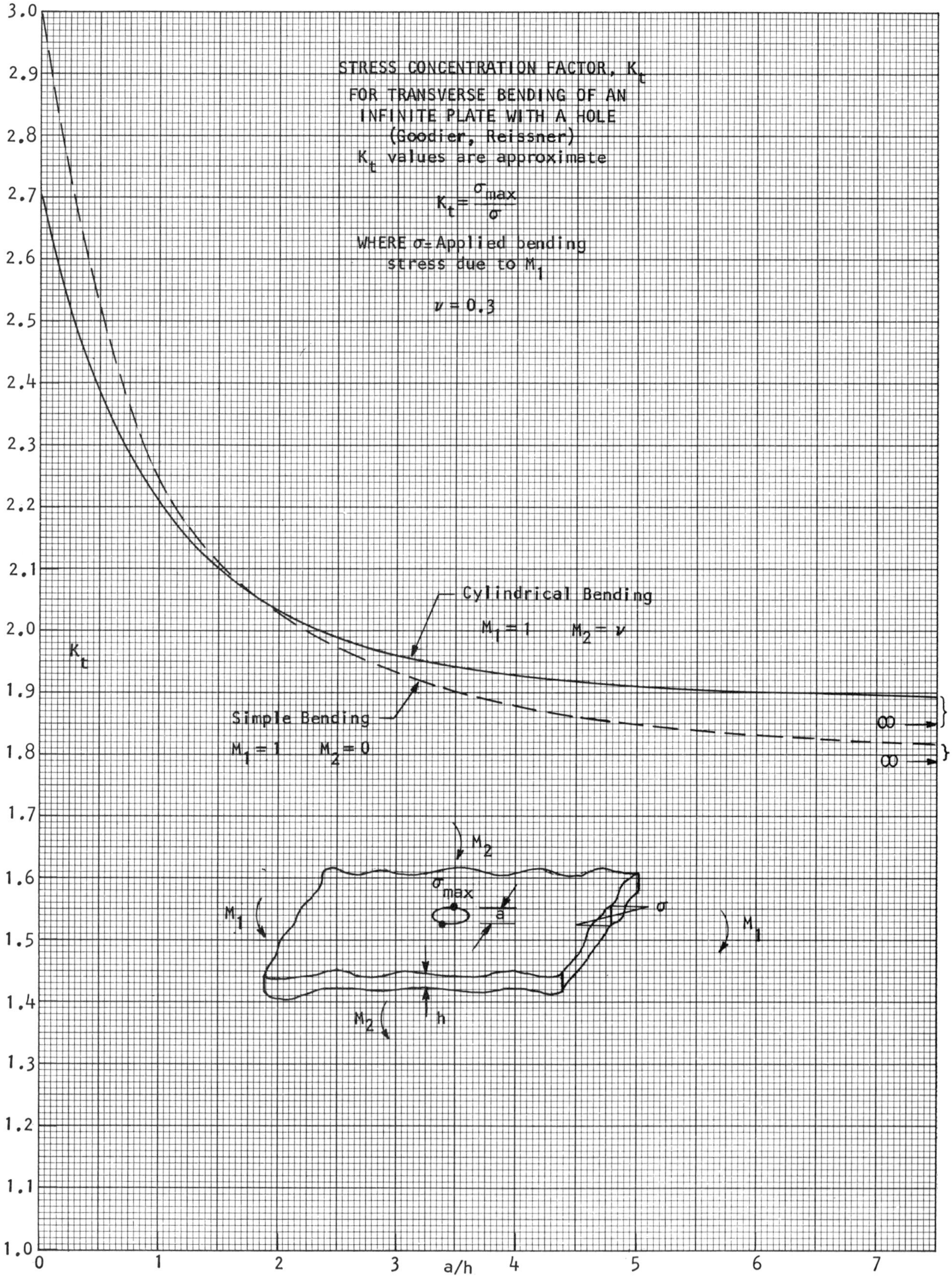

STRESS CONCENTRATION FACTOR, K_t
FOR TRANSVERSE BENDING OF AN
INFINITE PLATE WITH A HOLE
(Goodier, Reissner)
K_t values are approximate
K_t = σ_max / σ
WHERE σ = Applied bending
stress due to M_1
ν = 0.3
Cylindrical Bending
M_1 = 1   M_2 = ν
Simple Bending
M_1 = 1   M_2 = 0
K_t
∞
∞
M_2
σ_max
a
σ
M_1
M_1
h
M_2
a/h
3.0
2.9
2.8
2.7
2.6
2.5
2.4
2.3
2.2
2.1
2.0
1.9
1.8
1.7
1.6
1.5
1.4
1.3
1.2
1.1
1.0
0
1
2
3
4
5
6
7

FIG. 160

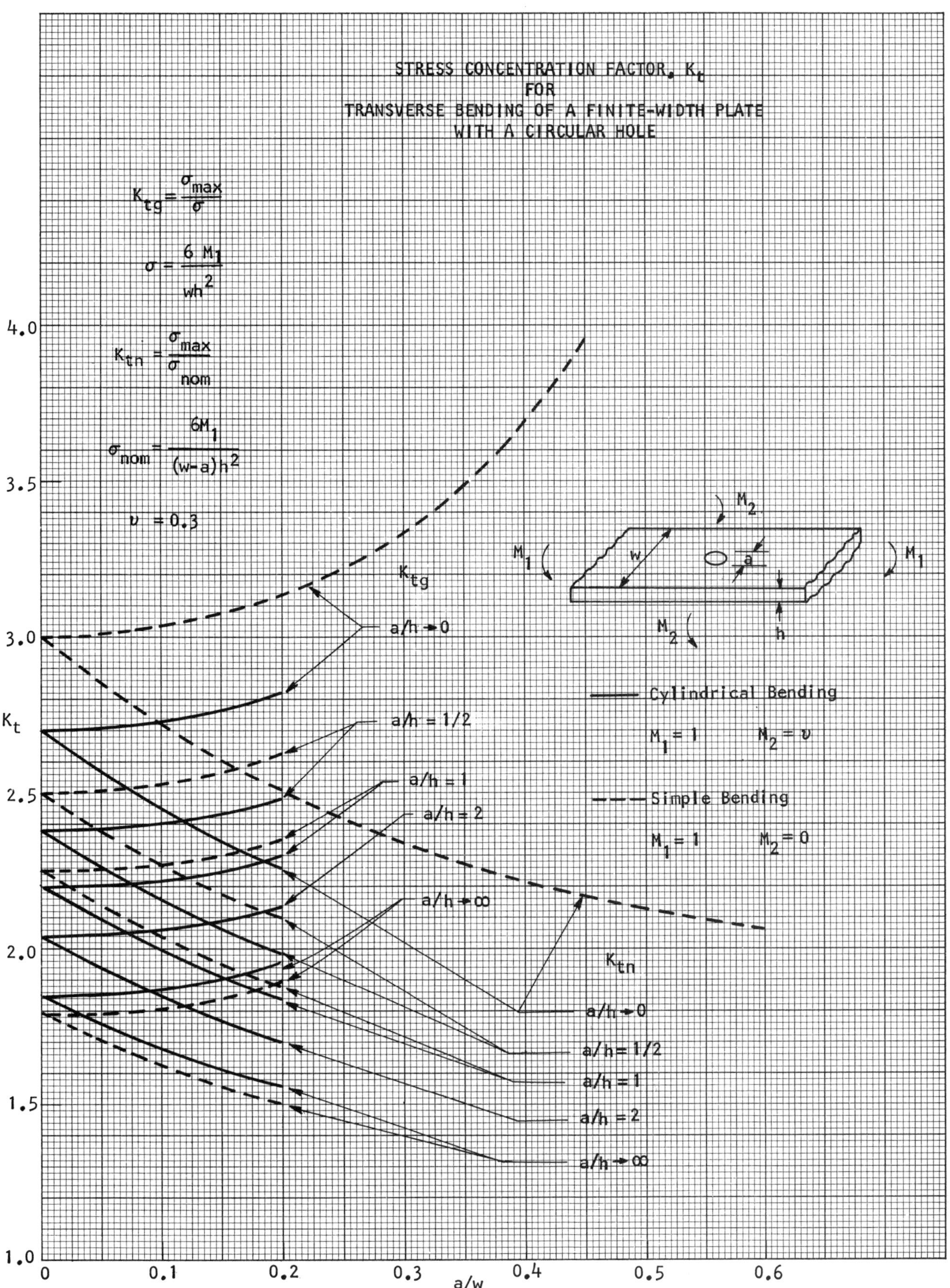

STRESS CONCENTRATION FACTOR, K_t
FOR
TRANSVERSE BENDING OF A FINITE-WIDTH PLATE
WITH A CIRCULAR HOLE
$K_{tg} = \frac{\sigma_{max}}{\sigma}$
$\sigma = \frac{6 M_1}{wh^2}$
$K_{tn} = \frac{\sigma_{max}}{\sigma_{nom}}$
$\sigma_{nom} = \frac{6M_1}{(w-a)h^2}$
$\upsilon = 0.3$
K_{tg}
K_{tn}
$a/h \rightarrow 0$
$a/h = 1/2$
$a/h = 1$
$a/h = 2$
$a/h \rightarrow \infty$
M_1
M_2
w
a
h
Cylindrical Bending
$M_1 = 1$ $M_2 = \upsilon$
Simple Bending
$M_1 = 1$ $M_2 = 0$
K_t
a/w
1.0
1.5
2.0
2.5
3.0
3.5
4.0
0
0.1
0.2
0.3
0.4
0.5
0.6

FIG. 161

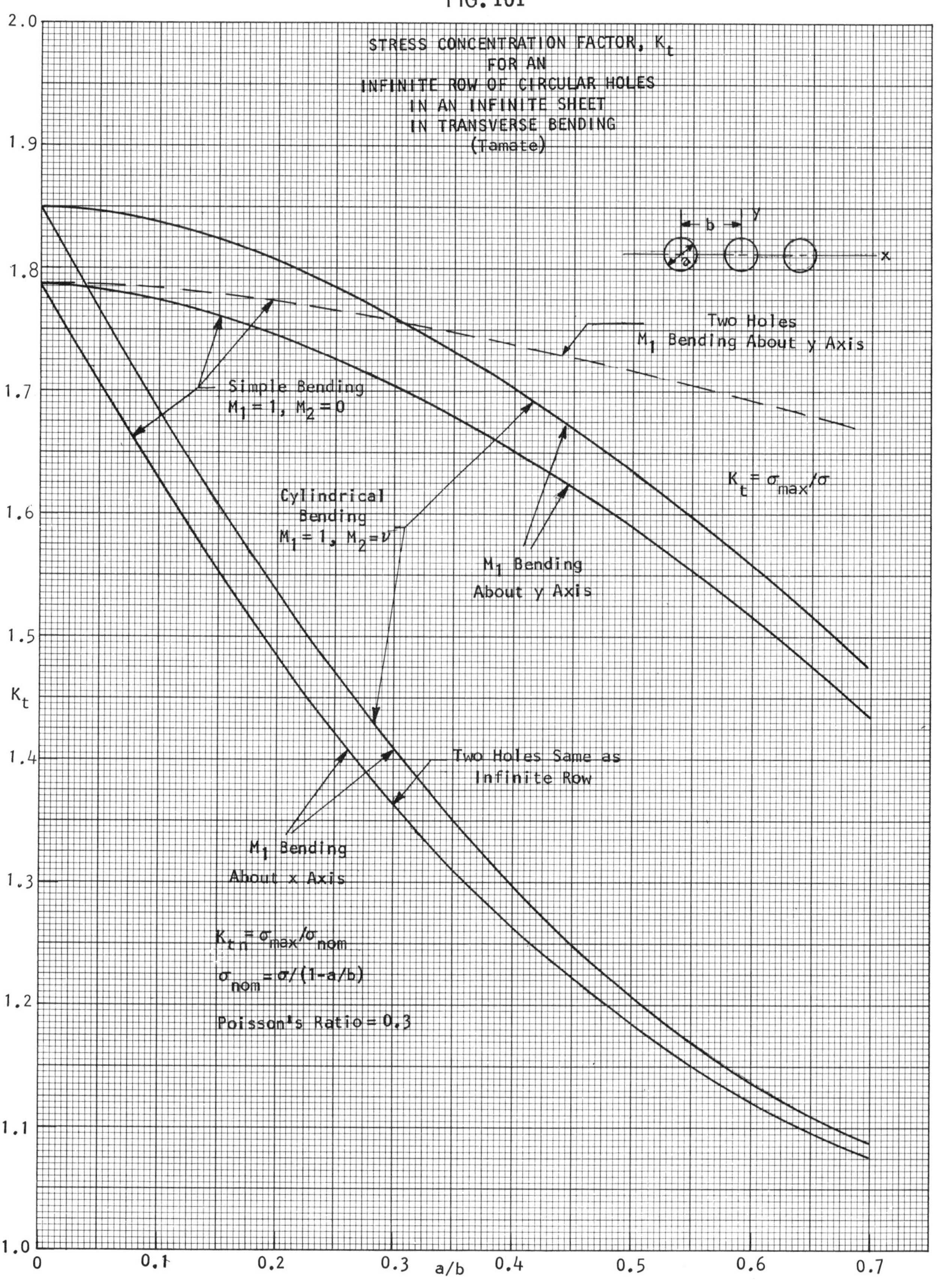

STRESS CONCENTRATION FACTOR, K_t
FOR AN
INFINITE ROW OF CIRCULAR HOLES
IN AN INFINITE SHEET
IN TRANSVERSE BENDING
(Tamate)
b
y
x
a
Two Holes
M_1 Bending About y Axis
Simple Bending
$M_1 = 1$, $M_2 = 0$
$K_t = \sigma_{max}/\sigma$
Cylindrical
Bending
$M_1 = 1$, $M_2 = \nu$
M_1 Bending
About y Axis
Two Holes Same as
Infinite Row
M_1 Bending
About x Axis
$K_{tn} = \sigma_{max}/\sigma_{nom}$
$\sigma_{nom} = \sigma/(1-a/b)$
Poisson's Ratio = 0.3
2.0
1.9
1.8
1.7
1.6
1.5
K_t
1.4
1.3
1.2
1.1
1.0
0
0.1
0.2
0.3
a/b
0.4
0.5
0.6
0.7

FIG. 162

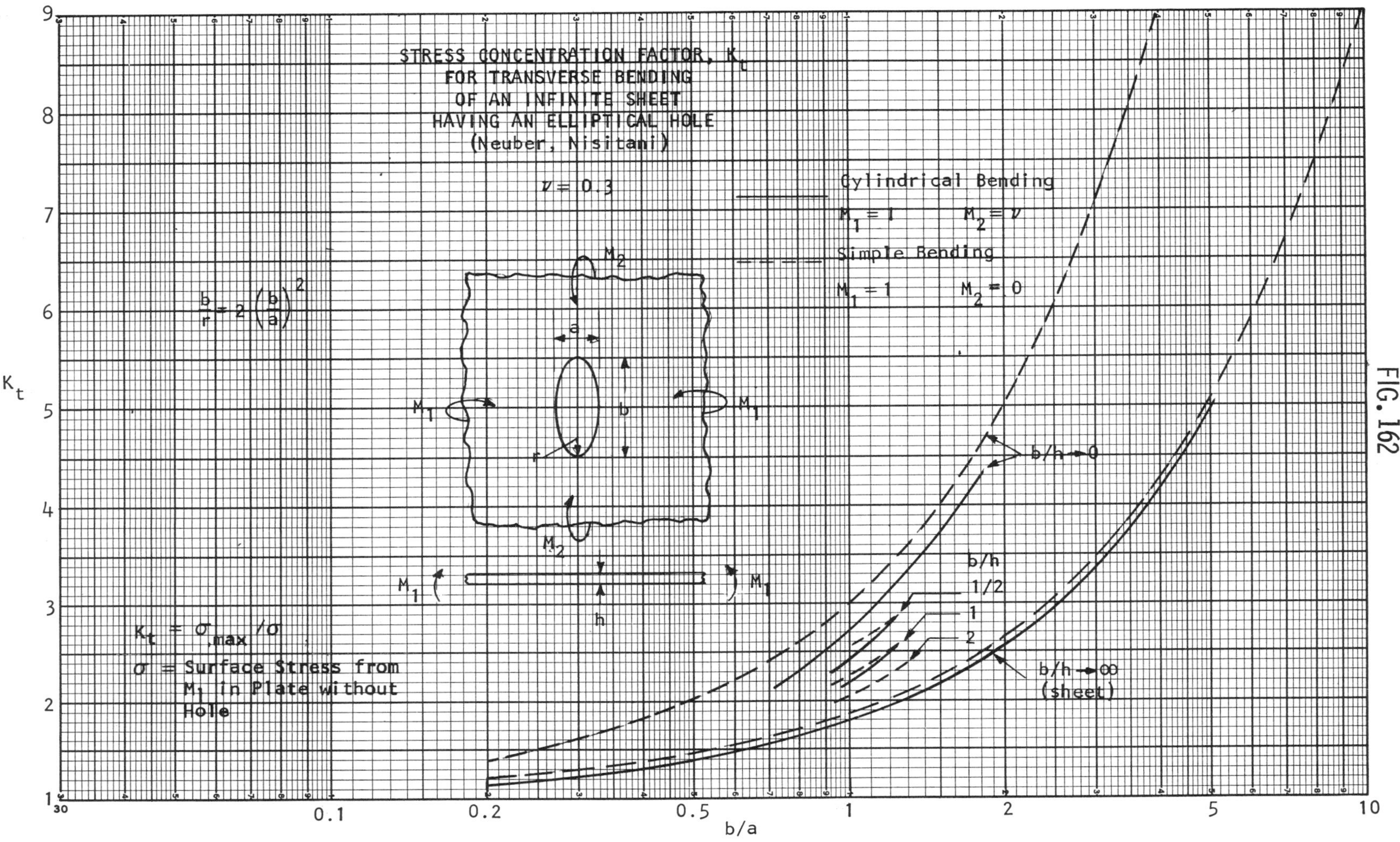
STRESS CONCENTRATION FACTOR, K_t
FOR TRANSVERSE BENDING
OF AN INFINITE SHEET
HAVING AN ELLIPTICAL HOLE
(Neuber, Nisitani)
$\nu = 0.3$
Cylindrical Bending
$M_1 = 1$ $M_2 = \nu$
Simple Bending
$M_1 = 1$ $M_2 = 0$
$\frac{b}{r} = 2\left(\frac{b}{a}\right)^2$
M_1
M_2
a
b
r
h
b/h → 0
b/h
1/2
1
2
b/h → ∞
(sheet)
$K_t = \sigma_{max}/\sigma$
σ = Surface Stress from M_1 in Plate without Hole
K_t
b/a
1
2
3
4
5
6
7
8
9
0.1
0.2
0.5
1
2
5
10

FIG. 163

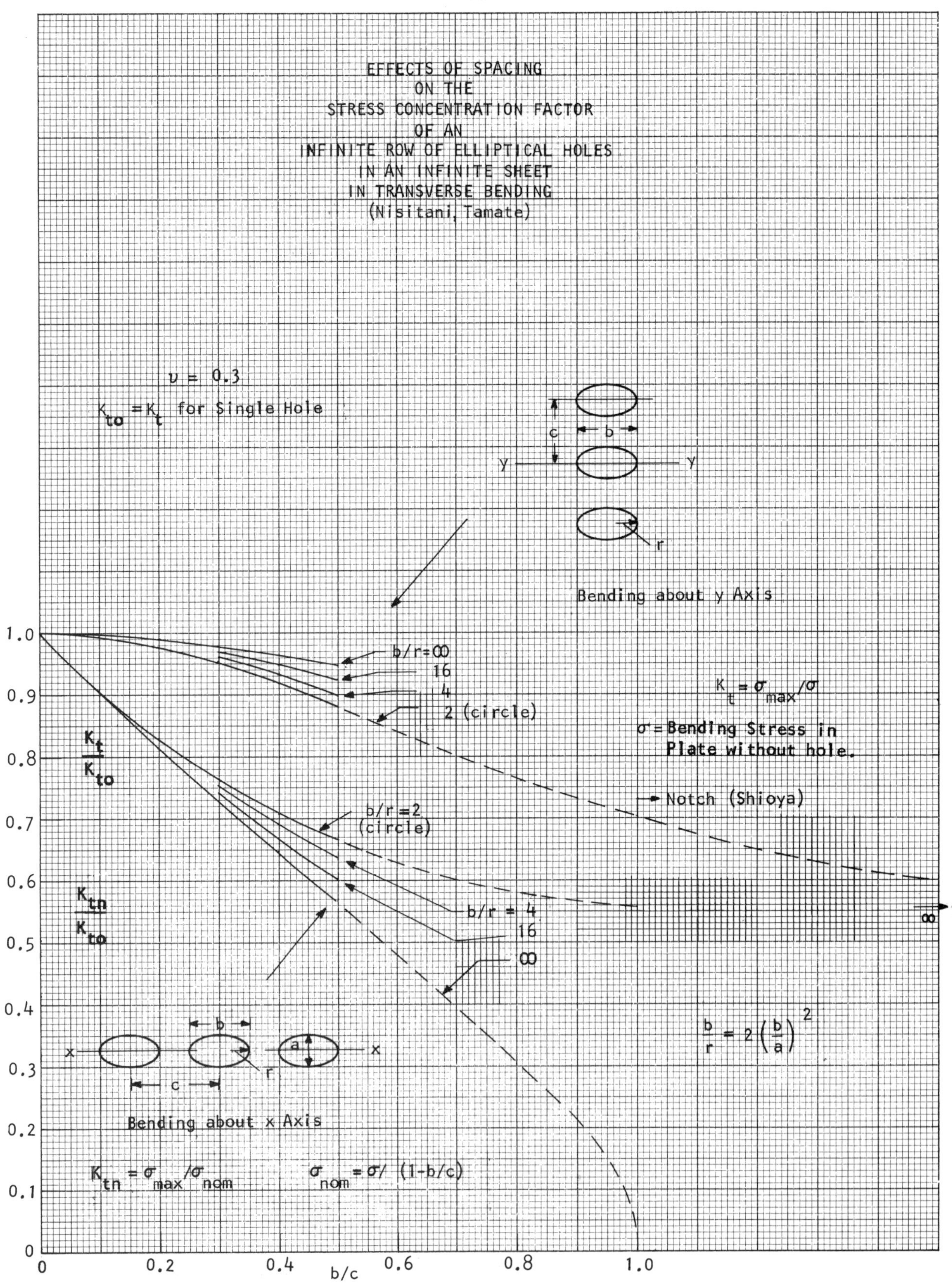

EFFECTS OF SPACING
ON THE
STRESS CONCENTRATION FACTOR
OF AN
INFINITE ROW OF ELLIPTICAL HOLES
IN AN INFINITE SHEET
IN TRANSVERSE BENDING
(Nisitani, Tamate)
ν = 0.3
$K_{to} = K_t$ for Single Hole
c
b
y
y
r
Bending about y Axis
b/r=∞
16
4
2 (circle)
$K_t = \sigma_{max}/\sigma$
σ = Bending Stress in Plate without hole.
Notch (Shioya)
b/r=2 (circle)
b/r = 4
16
∞
$\frac{K_t}{K_{to}}$
$\frac{K_{tn}}{K_{to}}$
$\frac{b}{r} = 2\left(\frac{b}{a}\right)^2$
b
x
a
x
r
c
Bending about x Axis
$K_{tn} = \sigma_{max}/\sigma_{nom}$
$\sigma_{nom} = \sigma/(1-b/c)$
1.0
0.9
0.8
0.7
0.6
0.5
0.4
0.3
0.2
0.1
0
0
0.2
0.4
0.6
0.8
1.0
b/c

FIG. 164

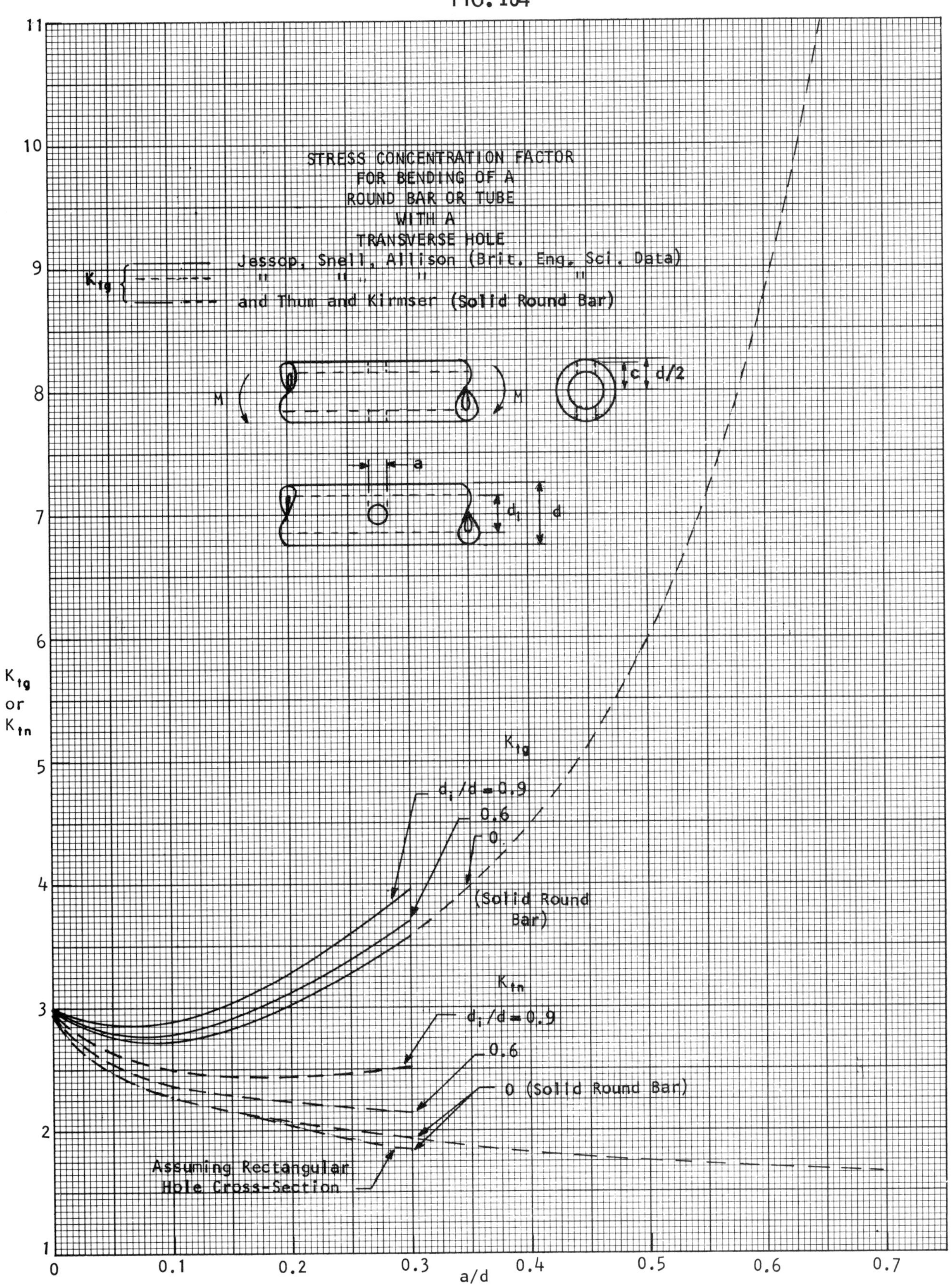
STRESS CONCENTRATION FACTOR
FOR BENDING OF A
ROUND BAR OR TUBE
WITH A
TRANSVERSE HOLE
K_{tg}
Jessop, Snell, Allison (Brit. Eng. Sci. Data)
and Thum and Kirmser (Solid Round Bar)
M
c
d/2
a
d_i
d
K_{tg}
$d_i/d = 0.9$
0.6
0
(Solid Round Bar)
K_{tn}
$d_i/d = 0.9$
0.6
0 (Solid Round Bar)
Assuming Rectangular Hole Cross-Section
K_{tg} or K_{tn}
a/d
1
2
3
4
5
6
7
8
9
10
11
0
0.1
0.2
0.3
0.4
0.5
0.6
0.7

FIG. 165

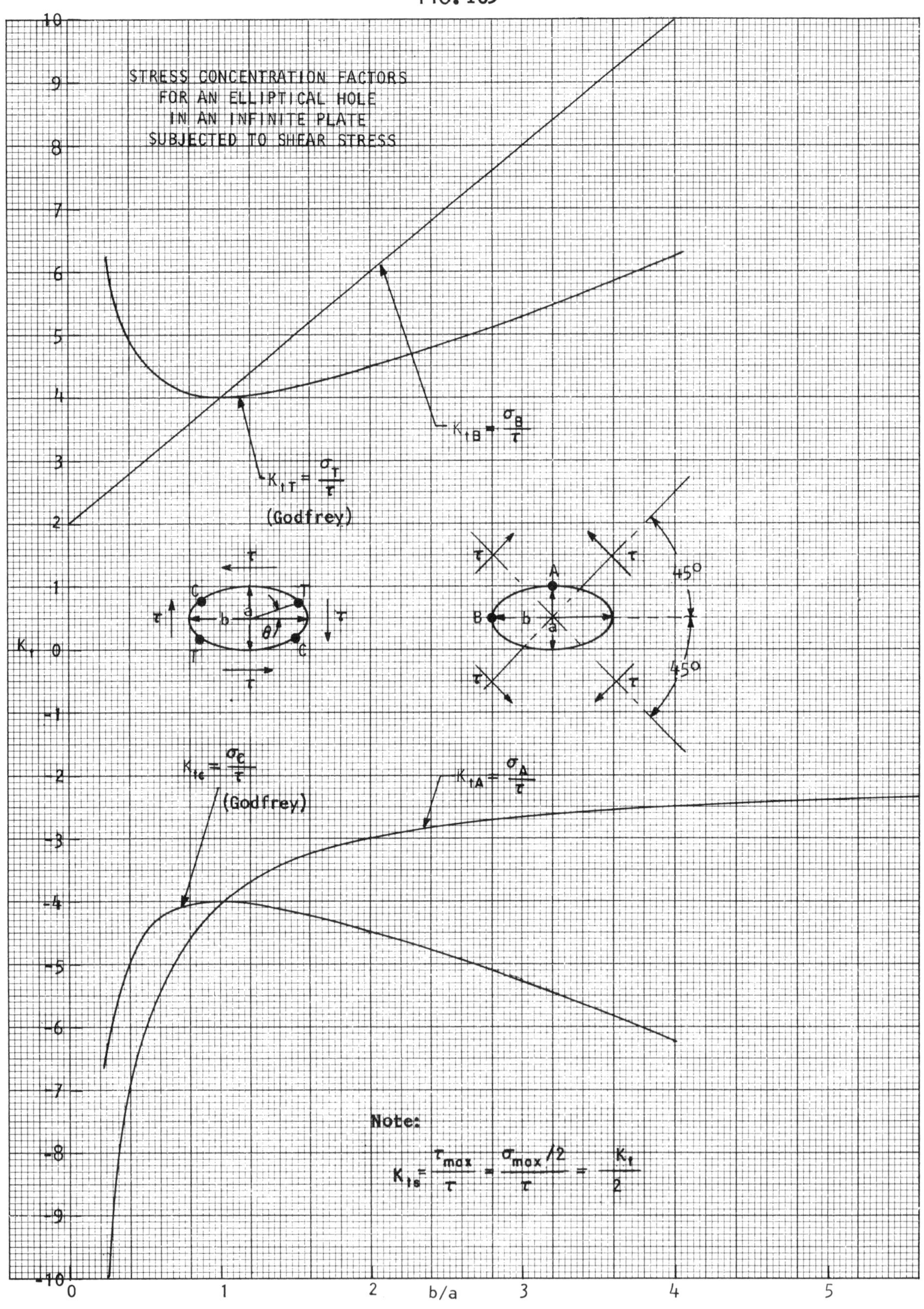
STRESS CONCENTRATION FACTORS
FOR AN ELLIPTICAL HOLE
IN AN INFINITE PLATE
SUBJECTED TO SHEAR STRESS
$K_{tB} = \frac{\sigma_B}{\tau}$
$K_{tT} = \frac{\sigma_T}{\tau}$
(Godfrey)
$K_{tC} = \frac{\sigma_C}{\tau}$
(Godfrey)
$K_{tA} = \frac{\sigma_A}{\tau}$
45°
45°
A
B
b
a
C
T
θ
τ
K_t
b/a
Note:
$K_{ts} = \frac{\tau_{max}}{\tau} = \frac{\sigma_{max}/2}{\tau} = \frac{K_t}{2}$

FIG. 166

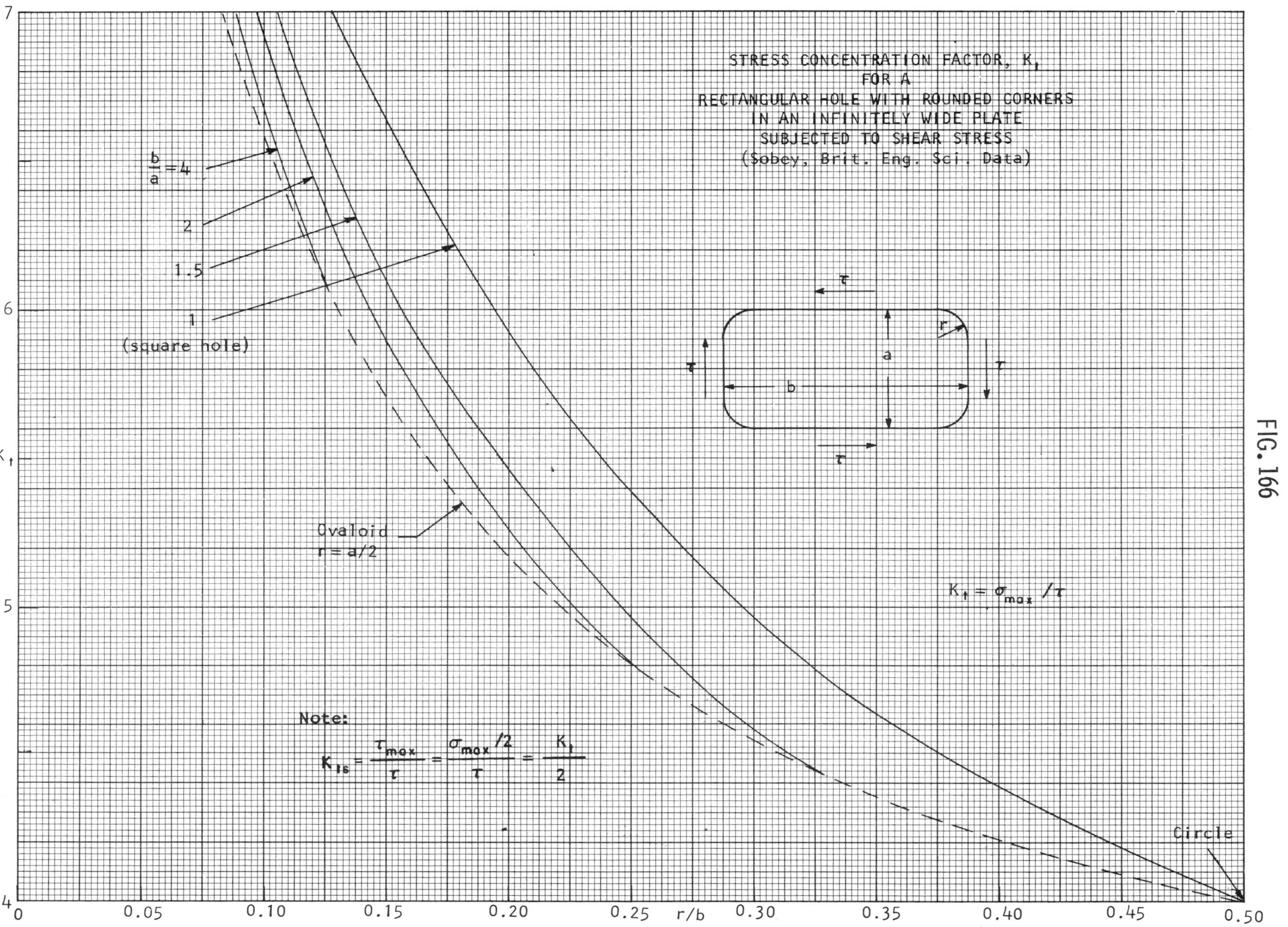

STRESS CONCENTRATION FACTOR, K_t
FOR A
RECTANGULAR HOLE WITH ROUNDED CORNERS
IN AN INFINITELY WIDE PLATE
SUBJECTED TO SHEAR STRESS
(Sobey, Brit. Eng. Sci. Data)
$\frac{b}{a} = 4$
2
1.5
1
(square hole)
Ovaloid
r = a/2
$K_t = \sigma_{max} / \tau$
Note:
$K_{ts} = \frac{\tau_{max}}{\tau} = \frac{\sigma_{max}/2}{\tau} = \frac{K_t}{2}$
Circle
τ
r
a
b
K_t
r/b
7
6
5
4
0
0.05
0.10
0.15
0.20
0.25
0.30
0.35
0.40
0.45
0.50

FIG. 167

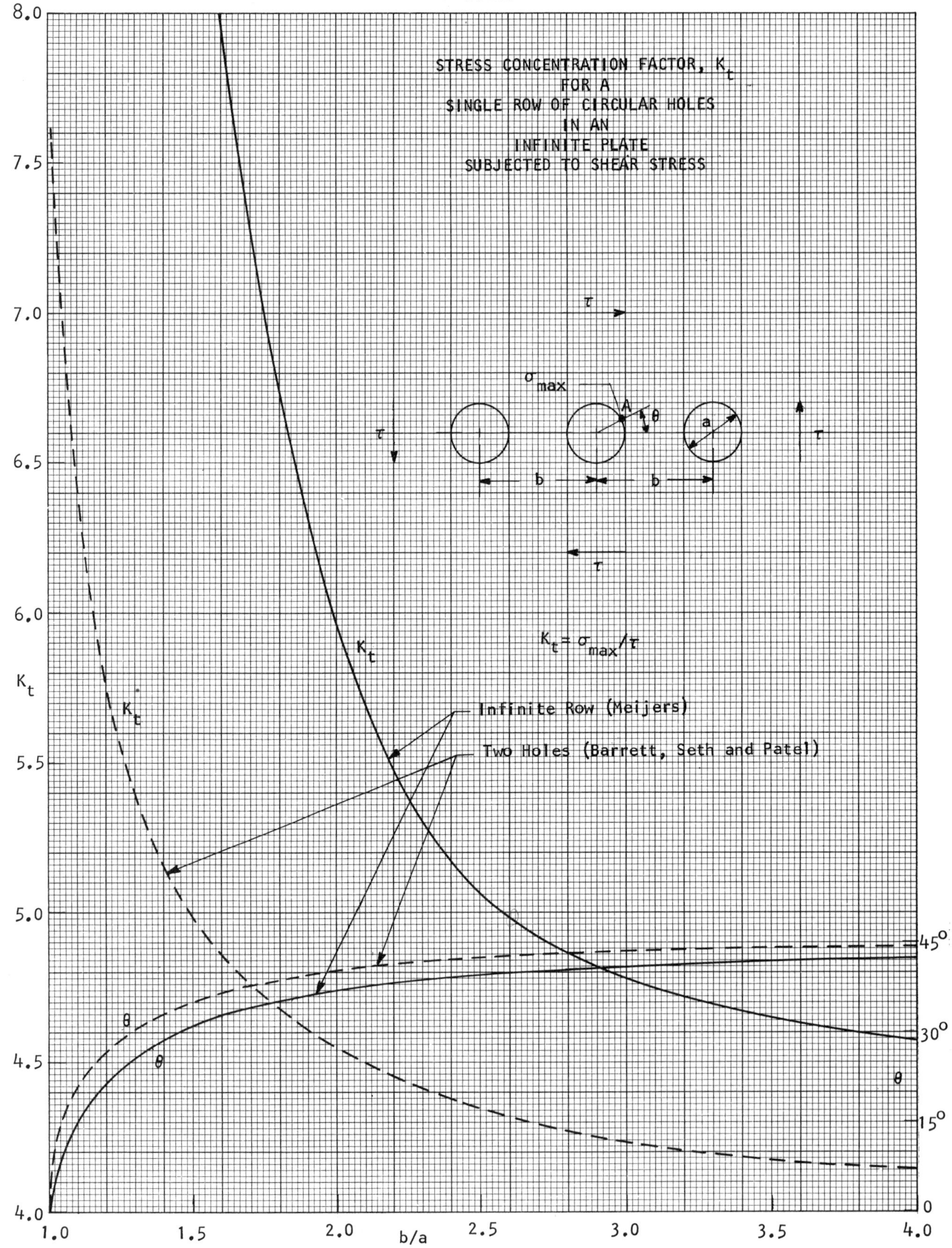

STRESS CONCENTRATION FACTOR, K_t
FOR A
SINGLE ROW OF CIRCULAR HOLES
IN AN
INFINITE PLATE
SUBJECTED TO SHEAR STRESS
τ
σ_{max}
A
θ
a
b
b
$K_t = \sigma_{max}/\tau$
K_t
Infinite Row (Meijers)
Two Holes (Barrett, Seth and Patel)
θ
45°
30°
15°
0
8.0
7.5
7.0
6.5
6.0
5.5
5.0
4.5
4.0
1.0
1.5
2.0
2.5
3.0
3.5
4.0
b/a

FIG. 168

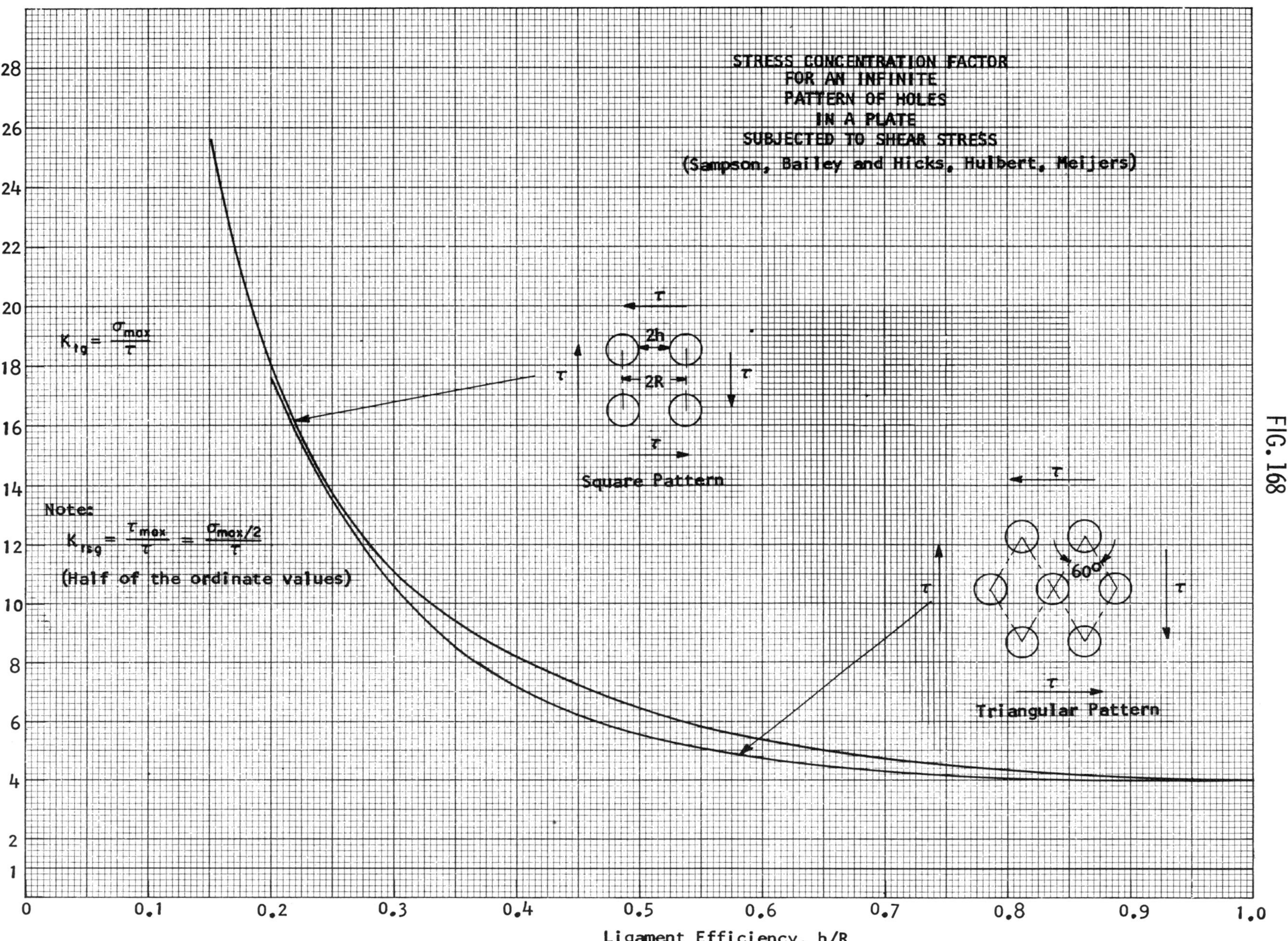

STRESS CONCENTRATION FACTOR
FOR AN INFINITE
PATTERN OF HOLES
IN A PLATE
SUBJECTED TO SHEAR STRESS
(Sampson, Bailey and Hicks, Hulbert, Meijers)
$K_{tg} = \frac{\sigma_{max}}{\tau}$
Note:
$K_{tsg} = \frac{\tau_{max}}{\tau} = \frac{\sigma_{max}/2}{\tau}$
(Half of the ordinate values)
τ
2h
2R
Square Pattern
60°
Triangular Pattern
Ligament Efficiency, h/R
0
0.1
0.2
0.3
0.4
0.5
0.6
0.7
0.8
0.9
1.0
1
2
4
6
8
10
12
14
16
18
20
22
24
26
28

FIG. 169

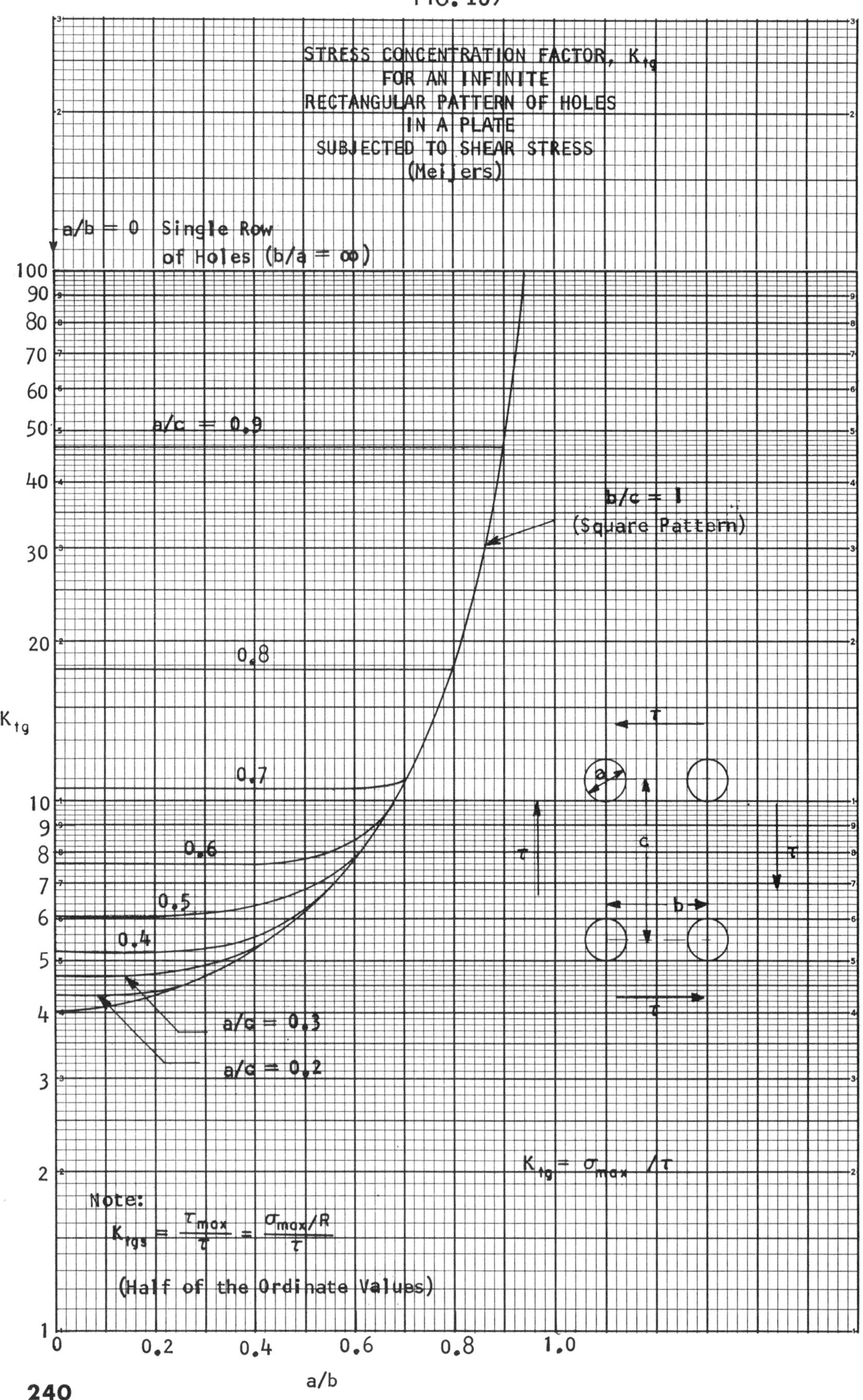

STRESS CONCENTRATION FACTOR, K_{tg}
FOR AN INFINITE
RECTANGULAR PATTERN OF HOLES
IN A PLATE
SUBJECTED TO SHEAR STRESS
(Meijers)
a/b = 0 Single Row of Holes (b/a = ∞)
a/c = 0.9
0.8
0.7
0.6
0.5
0.4
a/c = 0.3
a/c = 0.2
b/c = 1 (Square Pattern)
K_{tg}
a
c
b
τ
$K_{tg} = \sigma_{max} / \tau$
Note:
$K_{tgs} = \frac{\tau_{max}}{\tau} = \frac{\sigma_{max}/R}{\tau}$
(Half of the Ordinate Values)
a/b
0
0.2
0.4
0.6
0.8
1.0
1
2
3
4
5
6
7
8
9
10
20
30
40
50
60
70
80
90
100

FIG. 170

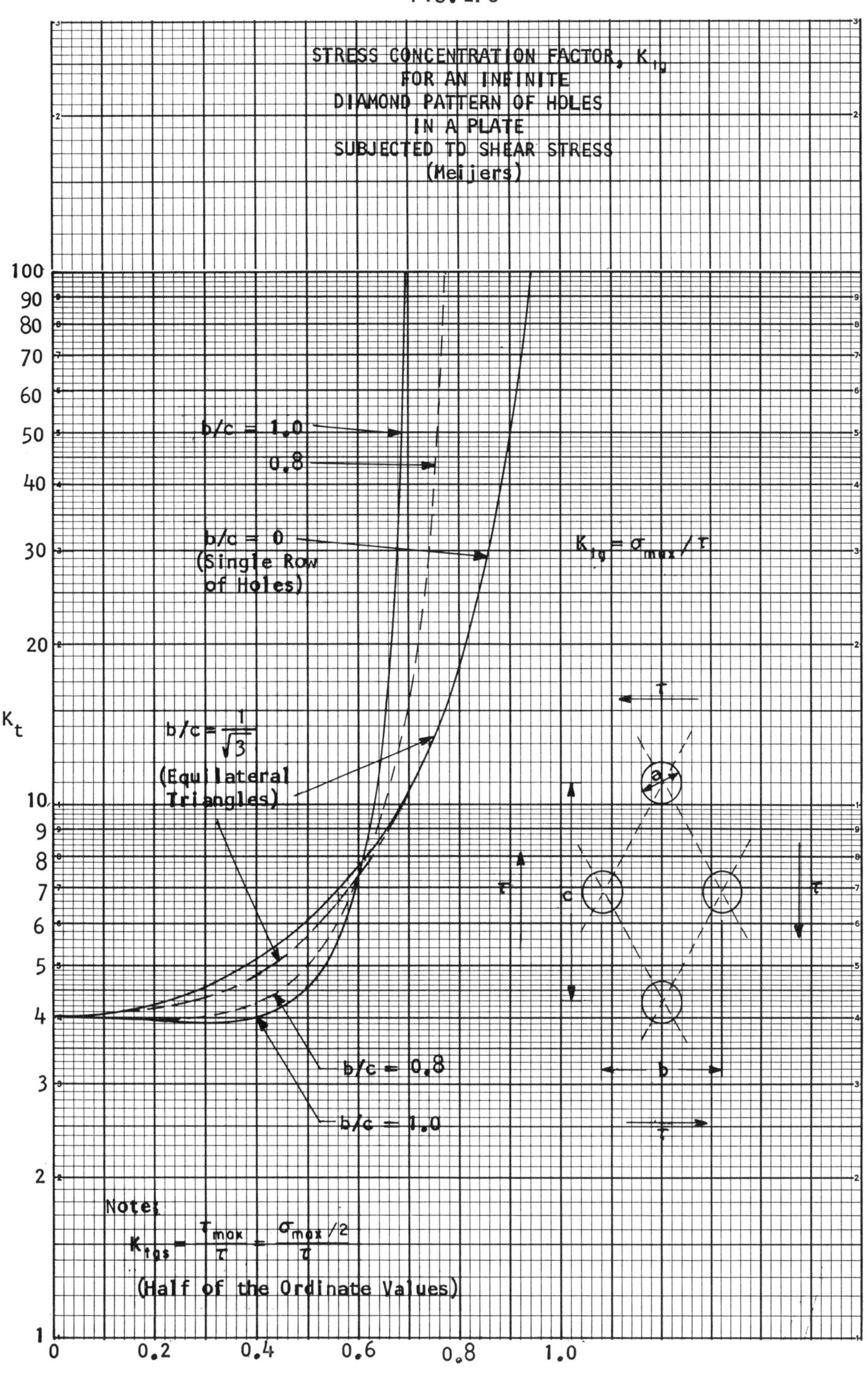

STRESS CONCENTRATION FACTOR, K_{tg}
FOR AN INFINITE
DIAMOND PATTERN OF HOLES
IN A PLATE
SUBJECTED TO SHEAR STRESS
(Meijers)
b/c = 1.0
0.8
b/c = 0
(Single Row
of Holes)
$K_{tg} = \sigma_{max}/\tau$
b/c = $\frac{1}{\sqrt{3}}$
(Equilateral
Triangles)
b/c = 0.8
b/c = 1.0
Note:
$K_{tgs} = \frac{\tau_{max}}{\tau} = \frac{\sigma_{max}/2}{\tau}$
(Half of the Ordinate Values)
K_t
a/b

FIG. 171

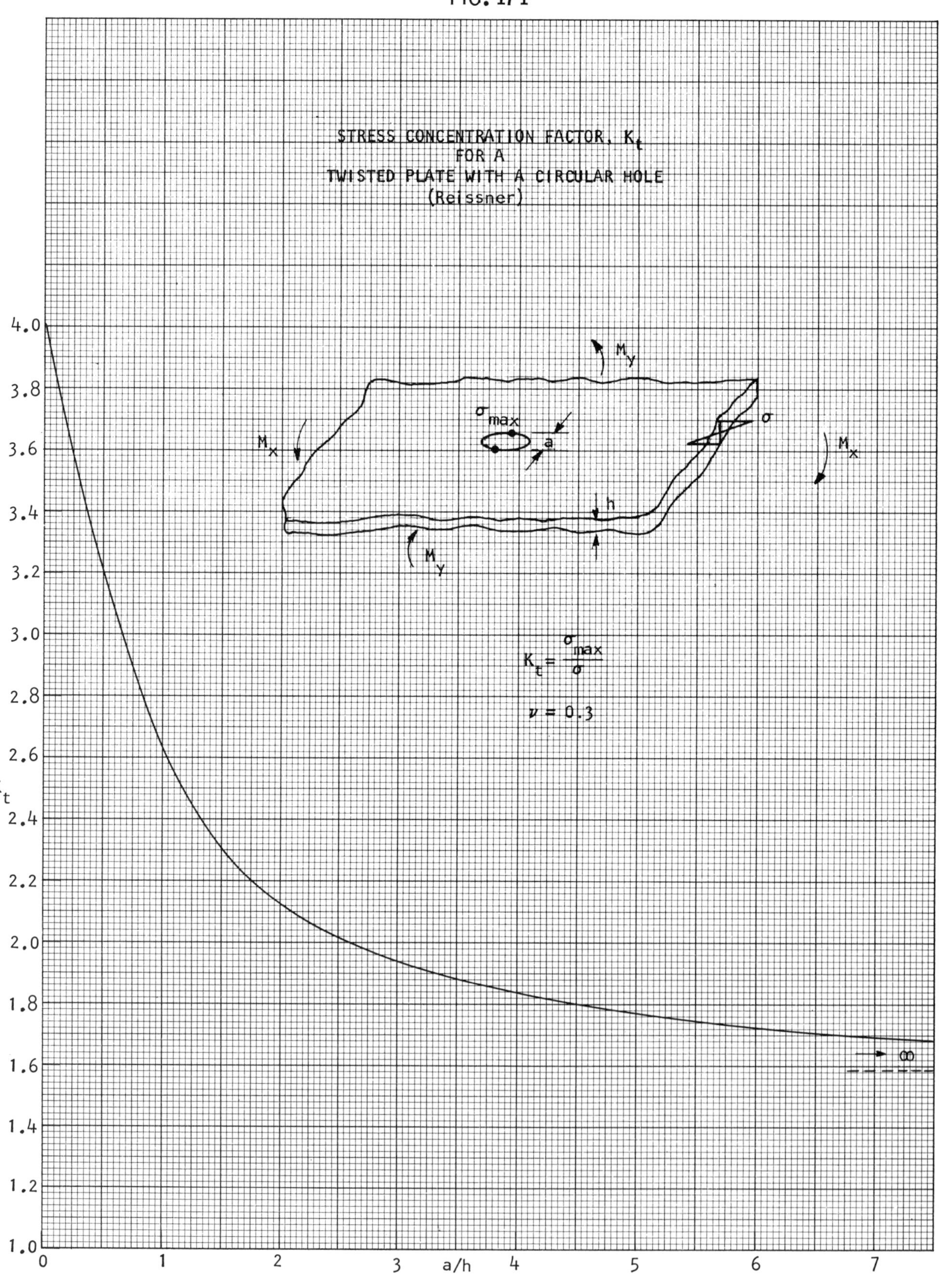

STRESS CONCENTRATION FACTOR, K_t
FOR A
TWISTED PLATE WITH A CIRCULAR HOLE
(Reissner)
M_y
σ_{max}
a
M_x
σ
M_x
h
M_y
$K_t = \frac{\sigma_{max}}{\sigma}$
$\nu = 0.3$
$\rightarrow \infty$
4.0
3.8
3.6
3.4
3.2
3.0
2.8
2.6
K_t
2.4
2.2
2.0
1.8
1.6
1.4
1.2
1.0
0
1
2
3
a/h
4
5
6
7

FIG. 172

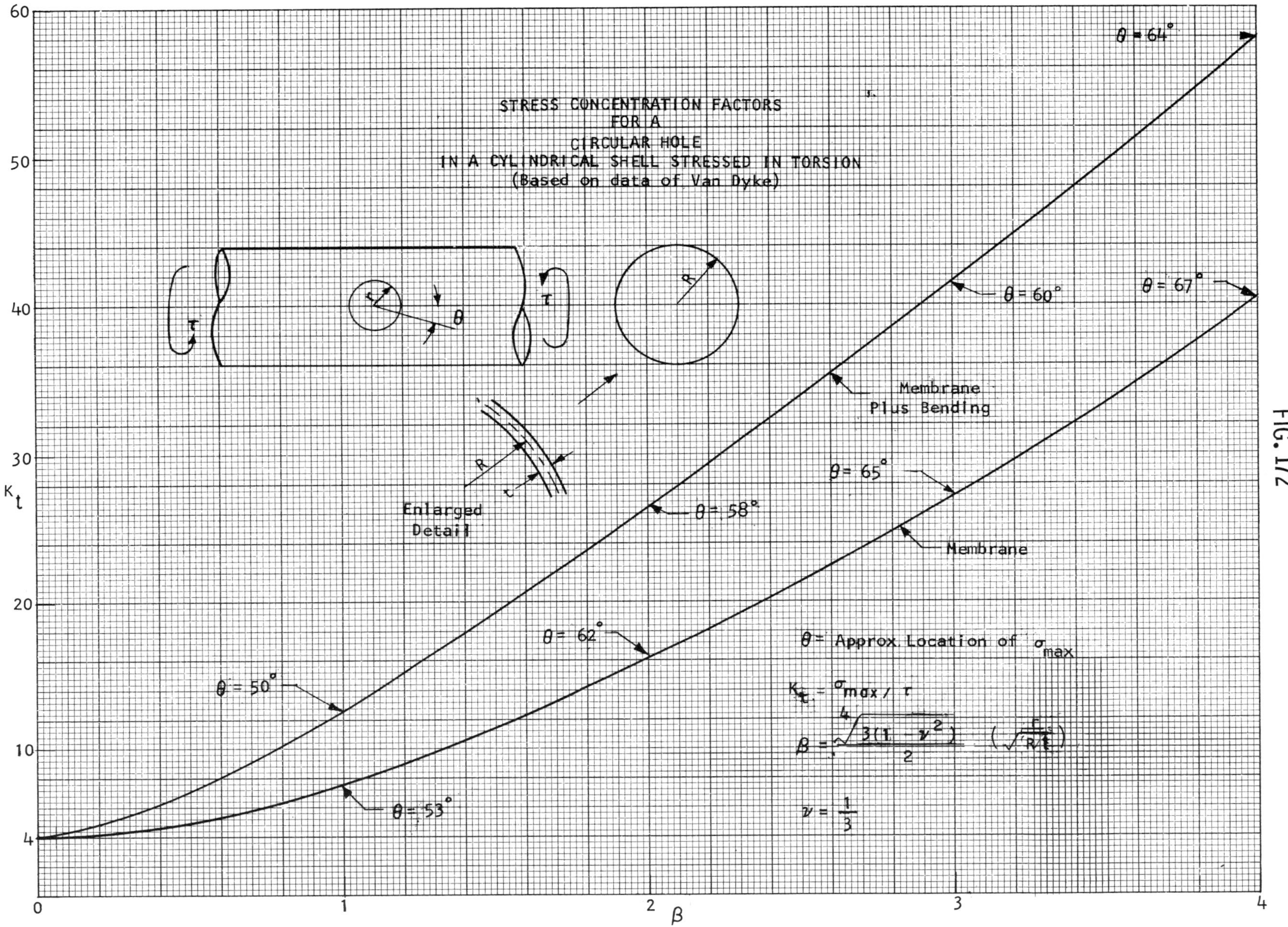
STRESS CONCENTRATION FACTORS
FOR A
CIRCULAR HOLE
IN A CYLINDRICAL SHELL STRESSED IN TORSION
(Based on data of Van Dyke)
Enlarged Detail
Membrane Plus Bending
Membrane
θ = 64°
θ = 67°
θ = 60°
θ = 65°
θ = 58°
θ = 62°
θ = 50°
θ = 53°
θ = Approx. Location of σ max
$K_t = \sigma_{max} / \tau$
$\beta = \frac{\sqrt[4]{3(1-\nu^2)}}{2} \left(\frac{r}{\sqrt{Rt}} \right)$
$\nu = \frac{1}{3}$
K t
β
0
1
2
3
4
10
20
30
40
50
60

FIG. 173

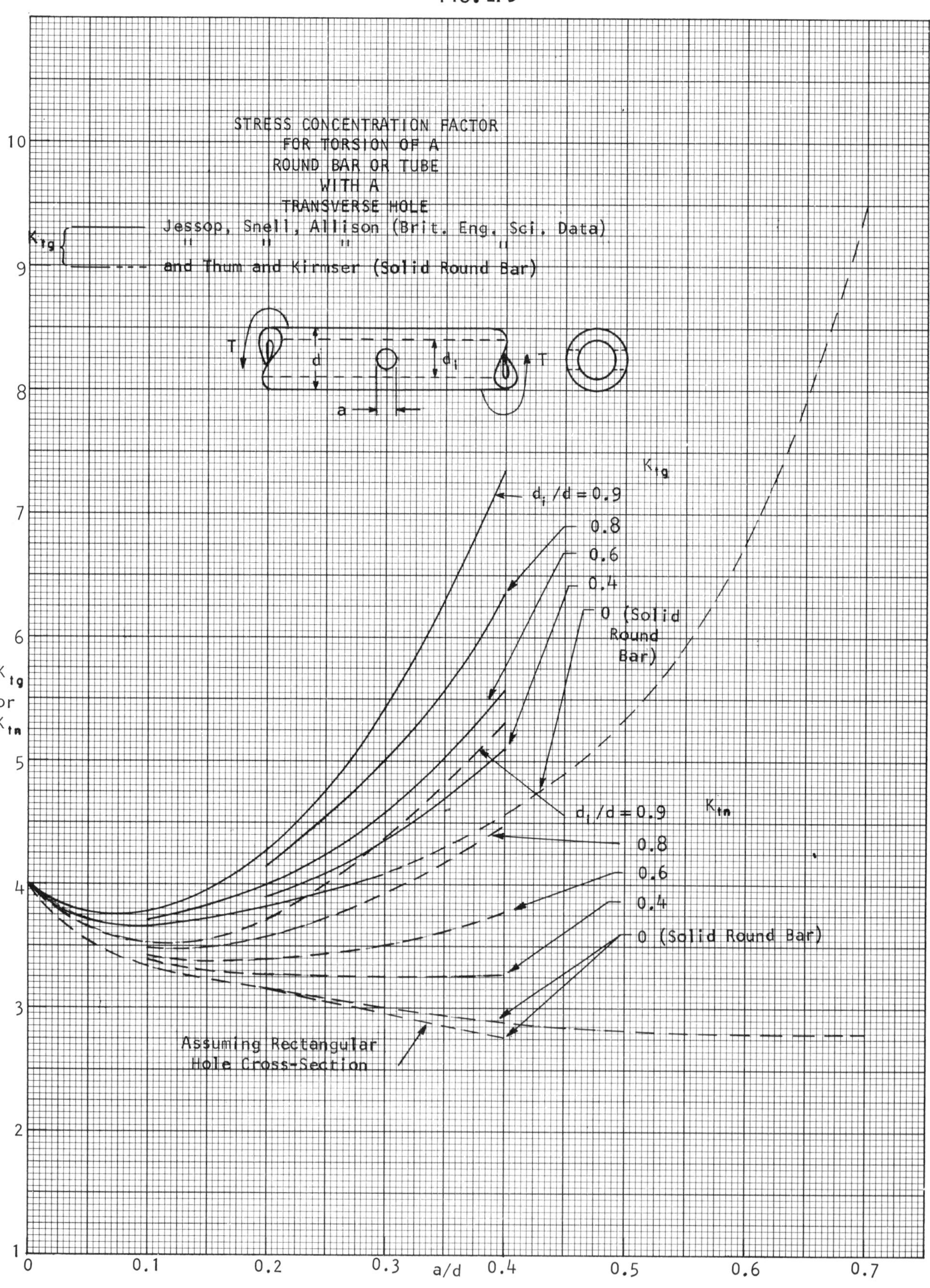

STRESS CONCENTRATION FACTOR
FOR TORSION OF A
ROUND BAR OR TUBE
WITH A
TRANSVERSE HOLE
K tg
Jessop, Snell, Allison (Brit. Eng. Sci. Data)
and Thum and Kirmser (Solid Round Bar)
T
d
d i
T
a
K tg
d i /d = 0.9
0.8
0.6
0.4
0 (Solid Round Bar)
d i /d = 0.9
K tn
0.8
0.6
0.4
0 (Solid Round Bar)
Assuming Rectangular Hole Cross-Section
K tg or K tn
a/d
10
9
8
7
6
5
4
3
2
1
0
0.1
0.2
0.3
0.4
0.5
0.6
0.7

CHAPTER 5

MISCELLANEOUS DESIGN ELEMENTS

(A) SHAFT WITH KEYSEAT

The USA standard keyseat[304] (keyway) has approximate average* values of $b/d = \frac{1}{4}$ and $t/d = \frac{1}{8}$ for shaft diameters up to 6.5 in. (see sketch, Fig. 182). For shaft diameters above 6.5 in., the approximate average values are $b/d = \frac{1}{4}$ and $t/d = 0.09$. The suggested approximate average* fillet radius proportions are $r/d = 1/48 = 0.0208$ for shaft diameters up to 6.5 in. and $r/d = 0.0156$ for shaft diameters above 6.5 in.

In design of a keyed shaft one must also take into consideration the shape of the end of the keyseat. Two types of keyseat end are shown in Fig. 174. The end-milled keyseat is more widely used, perhaps owing to its compactness and definite key positioning longitudinally. However, it is obvious that in bending the stress concentration factor is lower for the sled-runner keyseat.

(a) Bending

A comparison of the surface stresses of the two types of keyseat in bending ($b/d = 0.313$, $t/d = 0.156$) was made photoelastically by Hetényi,[305] who found $K_t = 1.79$ (semicircular end)† and $K_t = 1.38$ (sled-runner). Fatigue tests[306] resulted in K_f factors (fatigue notch factors) having about the same ratio as these two K_t values.

More recently, a comprehensive photoelastic investigation[307] has been made of British standard end-milled keyseats. In Fig. 182, corresponding to USA standards, the British value of $K_{tA} = 1.6$ for the surface has been used, since in both cases $b/d = \frac{1}{4}$ and since it appears that the surface factor is not significantly affected by moderate changes in the keyseat depth ratio.[307] The maximum stress occurs at an angle within about 10° from the point of tangency (see A, Fig. 182). Note that K_{tA} is independent of r/d.

For K_t at the fillet (Fig. 182, location B), the British K_t values for bending have been adjusted for keyseat depth (USA $t/d = \frac{1}{8} = 0.125$; British $t/d = 1/12 = 0.0833$) in accordance with corresponding t/d values from extrapolations in Fig. 65. Also note that the maximum fillet stress is located at the end of the keyway, about 15° up on the fillet.

*The keyseat width, depth, and fillet radius are in multiples of 1/32 in., each size applying to a range of shaft diameters.

†Also denoted "end-milled" or "profiled."

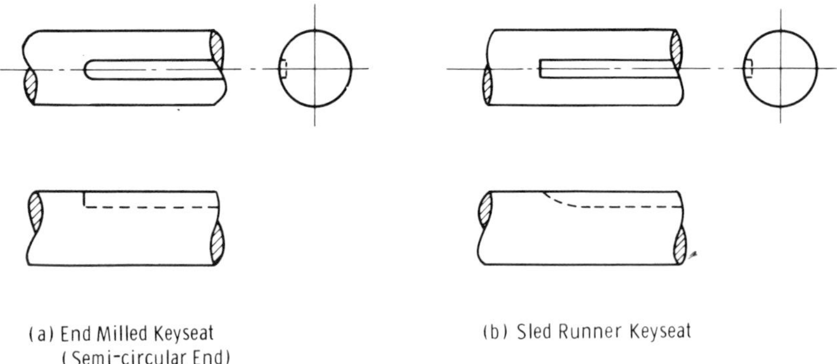

(a) End Milled Keyseat (Semi-circular End)

(b) Sled Runner Keyseat

FIG. 174 – Types of keyseat end

The foregoing discussion refers to shafts below 6.5-in. diameter, for which $t/d = 0.125$. For larger diameter shafts, where $t/d = 0.09$, it would seem that the K_t factor would not differ significantly, taking account of the different values of t/d and r/d. It is suggested that for design, the K_t values for $t/d = 0.125$ and $r/d = 0.0208$ be used for all shaft diameters.

(b) Torsion

For the surface at the semicircular keyseat end Leven[308] and the British investigators[307] found $K_{tA} = \sigma_{\max}/\tau = \backsim 3.4$.The maximum normal stress, tangential to the semicircle, occurs at 50° from the axial direction and is independent of r/d. The maximum shear stress is at 45° to the maximum normal stress and is half its value; $K_{tsa} = \tau_{\max}/\tau = \backsim 1.7$. For the stresses in the fillet of the straight part of a USA standard keyseat, Leven[308] obtained K_{ts} values mathematically and also obtained confirmatory results photoelastically. The maximum shear stresses at B (Fig. 183) are in the longitudinal and perpendicular directions. The maximum normal stresses are of the same magnitude and are at 45° to the shear stress; therefore $K_{tsB} = \tau_{\max}/\tau$ is equal to $K_{tB} = \sigma_{\max}/\sigma = \sigma_{\max}/\tau$, where $\tau = 16T/\pi d^3$.

Nisida[309] made photoelastic tests of models with a keyseat having the same depth ratio ($t/d = 1/8$) but with somewhat greater width ratio ($b/d = 0.3$). Allowing for the difference in keyseat shape, the degree of agreement of K_t factors is good. Griffith and Taylor[310] and Okubo[311] obtained results for cases with other geometrical proportions. For a semicircular[312] groove, $K_t = 2$ for $r/d \to 0$. For $r/d = 0.125$, K_t based on gross section would be somewhat higher, less than 2.1 as an estimate. This fits quite well with an extension of Leven's curve. Recent photoelastic results[307] are in reasonable agreement with Leven's values at $r/d = 0.0052$ and 0.0104, but at $r/d = 1/48 = 0.0208$ the K_t value[307] seems low in comparison with the previously mentioned results and the extension to $r/d = 0.125$.

The K_t values in the fillet of the semicircular keyseat end[307] appear to be lower than, or about the same as, in the straight part if the shape of Leven's curve for the straight part is accepted.

(c) Torque Transmitted Through a Key

The stresses in the keyseat when torque is transmitted through a key have been investigated by two-dimensional photoelasticity;[313,314] the results are not applicable to design since the stresses vary along the key length.

The upper dashed curve of Fig. 183 is an estimate of the fillet K_t when torque is transmitted by a key of length 2.5*d*. The dashed curve has been obtained by use of ratios for K_t values with and without a key, as determined by an "electroplating method," with keyseats of somewhat different cross-sectional proportions.[315] In the tests with a key,[315] shaft friction was held to a minimum. In a design application, the degree of press-fit pressure is an important factor.

(d) Combined Bending and Torsion

In Ref. 307 a chart was developed for obtaining a design K_t for combined bending and torsion for a shaft with the British keyseat proportions ($b/d = 0.25$, $t/d = 1/12 = 0.0833$) for $r/d = 1/48 = 0.0208$. The nominal stress for the chart is defined as

$$\sigma_{nom} = \frac{16M}{\pi d^3}\left[1 + \sqrt{1 + \left(\frac{T^2}{M^2}\right)}\right] \qquad [131a]$$

Figure 183*a* provides a rough estimate for the USA standard keyseat, based on use of straight lines to approximate the results of the British chart. Note that $K_t = \sigma_{max}/\sigma_{nom}$ and $K_{ts} = \tau_{max}/\sigma_{nom}$, with σ_{nom} defined in [131a].

Figure 183*a* is for $r/d = 1/48 = 0.0208$. The effect of a smaller r/d is to move the middle two lines upward in accordance with the values of Figs. 182 and 183; but the top and bottom lines remain fixed.

(e) Effect of Proximity of Keyseat to Shaft Shoulder Fillet

Photoelastic tests[316] were made of shafts with $D/d = 1.5$ (large diameter/small diameter) and with British keyseat proportions ($b/d = 0.25$, $t/d = 0.0833$, $r/d = 0.0208$). With the keyseat end located at the position where the shaft shoulder fillet begins, Fig. 175(*a*), there was no effect on the maximum keyseat fillet stress by varying the shaft shoulder fillet over a r_s/d range of 0.021 to 0.083, where r_s = shaft shoulder fillet radius. In torsion, the maximum surface stress on the semicircular keyseat end terminating at the beginning of the shoulder fillet was increased about 10% over the corresponding stress of a straight shaft with a keyseat; the increase decreases to zero as the keyseat end was moved a distance of $d/10$ away from the beginning of the shaft shoulder fillet radius, Fig. 175(*b*).

For keyseats cut into the shaft shoulder, Fig. 175(*c*), the effect was to reduce K_t for bending (fillet and surface) and for torsion (surface). For torsion (fillet) K_t was reduced, except for an increase when the end of the keyway was located at an axial distance of 0.07*d* to 0.25*d* from the beginning of the fillet.

(f) Fatigue Failures

The designer is interested in applying the foregoing K_t factors for keyseats in attempting to prevent fatigue failure. Although the problem is a complex one, some comments may be helpful.

Fatigue is initiated by shear stress, but the eventual crack propagation is usually by normal stress. Referring to Figs. 182 to 183*a*, two locations are involved: keyseat fillet,

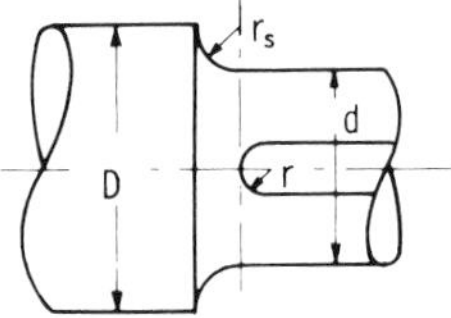

(a) Keyseat End at Beginning of Shoulder Fillet

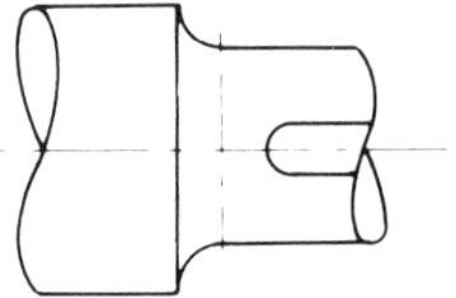

(b) Keyseat End Away From Shoulder Fillet

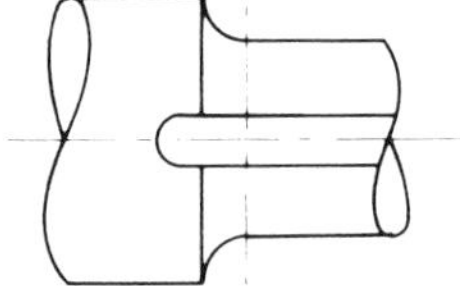

(c) Keyseat End Cut into Shoulder

Fig. 175 – Location of end of keyseat with respect to shaft shoulder

with a small radius; and surface of shaft at semicircular keyway end, with a relatively larger radius, three or more times the fillet radius. Since the development of both the initiation process and the eventual crack are functions of the stress gradient, which is related mainly to the "notch" radius, this consideration must be kept in mind in attempting to predict fatigue failure. It is possible, in certain instances, to have a nonpropagating crack in a fillet having a small radius.

Referring to Fig. 183*a*, for pure torsion ($M/T = 0$), it would be expected that initiation (shear) would start in the fillet, but the stress gradient is so steep that initial failure at the surface is also a possibility. The final crack direction will be determined by the surface stresses, where the normal stresses, associated with K_{tA}, are relatively high. For pure bending ($T/M = 0$), it would seem more likely that failure would occur primarily at the surface. Laboratory fatigue tests[306] and service failures[316a,316b] appear to support the foregoing remarks. In certain instances, torsional fatigue starts at the fillet and develops into a peeling type of failure; this particular type may be influenced by the key and possibly by a smaller than standard keyseat fillet radius. Predictions are difficult, owing to differing geometries, press-fit and key conditions and ratios of steady and alternating bending and torsional stress components.

(B) SPLINED SHAFT IN TORSION

In a three-dimensional photoelastic study by Yoshitake[317] of a particular eight-tooth spline, the tooth fillet radius was varied in three tests; from these data a K_{ts} versus r/d curve was drawn (Fig. 184). A test of an involute spline with a full fillet radius gave a K_{ts} value of 2.8. These values are for an open spline, that is, no mating member.

A test was made with a mated pair wherein the fitted length was slightly greater than the outside diameter of the spline. The maximum longitudinal bending stress of a tooth occurred at the end of the tooth and was about the same numerically as the maximum torsion stress.

A related mathematical analysis has been made by Okubo[318] of the torsion of a shaft with n longitudinal semicircular grooves. The results are on a strain basis. When $r/d \to 0$, $K_{ts} = 2$, as in Ref. 312 for a single groove.

(C) GEAR TEETH

A gear tooth is essentially a short cantilever beam, with the maximum stress occurring at the fillet at the base of the tooth. Owing to the angularity of load application the stresses on the two sides are not the same; the tension side is of design interest since fatigue failures occur there. The photoelastic tests of Dolan and Broghamer[319] provide the accepted stress concentration factors used in design (Figs. 185 and 186). For gear notation see Fig. 175a.

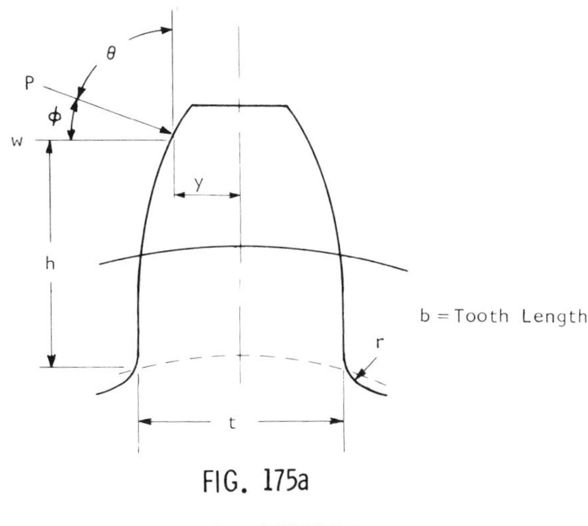

FIG. 175a

GEAR NOTATION

Ordinarily, the fillet is not a constant radius but is a curve produced by the generating hob or cutter.[320] The hobbing tool has straight sided teeth (see sketch, Fig. 187) with a tip radius r_t which has been standardized:[321] $r_t = 0.209/P_d$ for $14\frac{1}{2}°$ pressure angle and $r_t = 0.235/P_d$ for a 20° pressure angle, both for full depth teeth. For stub teeth r_t has not been standardized, but $r_t = 0.3/P_d$ has been used.* The tool radius r_t generates a gear tooth fillet of variable radius; the minimum radius is denoted r_f. Candee[322] has developed a relation between r_f and r_t:

$$r_f = \frac{(b - r_t)^2}{N/2P_d + (b - r_t)} + r_t \qquad [132]$$

where b = dedendum
N = number of teeth
P_d = diametral pitch

*Note that in Figs. 185 and 186, the full curves are for the above r_t values; these curves are approximate in that they were obtained by interpolation of the curves corresponding to the photoelastic tests.

The dedendum $b = 1.157/P_d$ for full-depth teeth and $1/P_d$ for stub teeth. Relation [132] is shown graphically in Fig. 187; the curve for 20° stub teeth is shown dashed, since r_t is not standardized and there is uncertainty regarding application of [132].

The following empirical relations have been developed[319] for the stress concentration factor for the fillet on the tension side:

For 14½° pressure angle:

$$K_t = 0.22 + \frac{1}{(r_f/t)^{0.2}\,(h/t)^{0.4}} \qquad [133]$$

For 20° pressure angle:

$$K_t = 0.18 + \frac{1}{(r_f/t)^{0.15}\,(h/t)^{0.45}} \qquad [134]$$

In certain instances, a specific form grinder has been used to provide a semicircular fillet radius between the teeth. To evaluate the effect of fillet radius, Fig. 188 is constructed from relations [133] and [134]. Note that the lowest K_t factors occur when the load is applied at the tip of the tooth; however, owing to increased moment arm the maximum fillet stress occurs at this position (neglecting load division,[323] which beneficial effect can be reliably taken into account only for extremely accurate gearing[324]). Considering, then, the lowest curves ($h/t = 1$) and keeping in mind that r_f/t values for standard generated teeth lie in the 0.1 to 0.2 region (depends on number of teeth), we see that going to a semicircular fillet $r_t/t \backsim 0.3$ does not result in a very large decrease in K_t. Although the decrease per se represents a definite available gain, this needs to be weighed against other factors, economic and technical (such as decreased effective rim of a pinion with a small number of teeth,[324a] especially where a keyway is present).

In addition to their photoelastic tests of gear tooth models, Dolan and Broghamer[319] also made tests of straight-sided short cantilever beams, varying the distance of load application and fillet radius (Fig. 189). The following empirical formula for K_t for the tension side was developed:[319]

$$K_t = 1.25\left[\frac{1}{(r/t')^{0.2}\,(h/t)^{0.3}}\right] \qquad [135]$$

Results for the compression side (K_t in most cases higher) are also given in Fig. 189; an empirical formula was not developed.

For presenting the effect of fillet radius in gear teeth it was found preferable (see Fig. 188) to use relations [133] and [134], owing mainly to angular application of load. Figure 189 is included for its application to other problems. Results of Weibel[325] and Frocht[326] on longer beams are included in Fig. 189 to represent the case of large h/t.

A subsequent photoelastic investigation of gear teeth by Jacobson[327] resulted in stress concentration factors in good agreement with Ref. 319.

More recently Aida and Terauchi[328] obtained the following analytical solution for the tensile maximum stress at the gear fillet:

$$\sigma_{max} = \left(1 + 0.08\,\frac{t}{r}\right)\left(0.66\,\sigma_{Nb} + 0.40\sqrt{\sigma_{Nt}^2 + 36\,\tau_N^2} + 1.15\,\sigma_{Nc}\right) \qquad [136]$$

where

$$\sigma_{Nb} = \frac{6\,Ph\sin\theta}{bt^2} \quad \text{(see Fig. 175a)}$$

$$\sigma_{Nc} = -\frac{P\cos\theta}{bt} - \frac{6\,Py\cos\theta}{bt^2}$$

$$\tau_N = \frac{P\sin\theta}{bt}$$

Some photoelastic tests were made which resulted in a satisfactory check of the foregoing analytical results, which also were in good agreement with the results of Ref. 319.

(D) PRESS-FITTED OR SHRINK-FITTED MEMBERS

Gears, pulleys, wheels, and similar elements are often assembled on a shaft by means of a press fit or shrink fit. Photoelastic tests of flat models[329] (Fig. 176) have been made with $\sigma_{nom}/p = 1.36$, where σ_{nom} = nominal bending stress in shaft and p = average normal pressure exerted by the member on the shaft. The following values were found; for the plain member, $K_t = 1.95$; for the grooved member, $K_t = 1.34$.

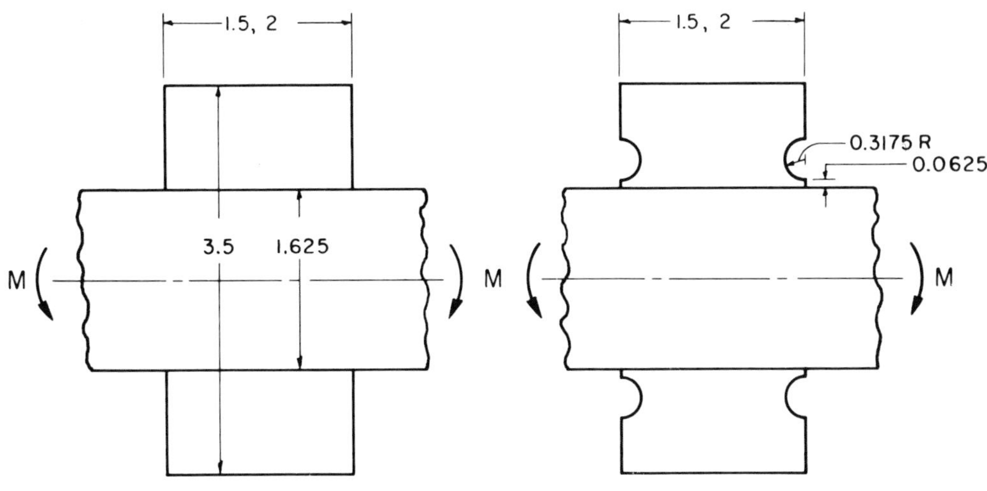

FIG. 176

PRESS FIT MODELS

(Dimensions in inches)

Fatigue tests of the "three-dimensional" case of a collar pressed* on a 1.625-in. diameter medium-carbon (0.42% C) steel shaft, the proportions being approximately the same as for the above-mentioned photoelastic models, gave the following "fatigue-notch factors" in bending: for the plain member, $K_f = 2.0$; for the grooved member, $K_f = 1.7$. It will be noted that the factors for the plain member seem to be in good agreement, but this is not significant since the fatigue result is due to a combination of stress concentration and "fretting

*The calculated radial pressure in this case was 16,000 lb/in.^{2} ($\sigma_{nom}/p = 1$). However, tests[329,330] indicate that, over a wide range of pressures, this variable does not affect K_f, except for very light pressures which result in a lower K_f.

corrosion,"[331–334] the latter producing a weakening effect over and above that produced by stress concentration. Note that the fatigue factor for the grooved member is higher than the stress concentration factor; this is no doubt due to fretting corrosion, which becomes relatively more prominent for lower stress condition cases. The fretting corrosion effect varies considerably with different combinations of materials. A roller-bearing inner race of case-hardened Cr-Ni-Mo steel pressed on a 2-in. diameter shaft gave the following bending fatigue results:[335]

	K_f
1. No external reaction through collar	
a. 0.45% C axle steel shaft	2.3
2. External reaction through collar	
a. 0.45% C axle steel shaft	2.9
b. Cr-Ni-Mo steel, heat-treated to 310 Brinell	3.9
c. 2.6% Ni steel, 57,000 psi fatigue limit	3.3–3.8
d. Same, heat treated to 253 Brinell	3.0

Similar tests were made in Germany[161] on shafts of 0.66-in. diameter. For the case where the reaction was not carried through the inner race the following results were obtained:

	K_f
3. 0.36% C axle steel shaft	
a. Press fit and shoulder fillet (r = 0.04 in., D/d = 1.3)	1.82
b. Same, shoulder fillet only (no inner race present)	1.64
c. Press fit only (no shoulder)	1.49
4. 1.5% Ni-0.5% Cr steel shaft (236 Brinell)	
a. Press fit and shoulder fillet (r = 0.04 in. D/d = 1.3)	2.04
b. Same, shoulder fillet (no inner race)	1.82

Where the reaction was carried through the inner race, somewhat lower values were obtained.

Some tests were made with relief grooves (see Chapter 3, Section D), showing lower K_f values.

Another favorable construction,[336,336a] as shown in Fig. 177, is to enlarge the shaft at the fit and to round out the shoulders in such a way that the critical region A (Fig. 177*a*) is relieved as at B (Fig. 177*b*). The photoelastic tests[336] did not provide quantitative information, but it is clear that, if the shoulder is ample, failure will occur in the fillet, in which case the design can be rationalized in accordance with Chapter 3.

As noted previously, K_f factors are a function of size, increasing toward a limiting value for increasing size of geometrically similar values. For 50 mm ($\backsim$ 2 in.) diameter, K_f = 2.8 was obtained for 0.39% C axle steel.[337] For models 3½ to 5 in. diameter, K_f values of the order of 3 to 4 were obtained for turbine rotor alloy steels.[338] For 7 to 9½ in. wheel fit models,[339–341] K_f values of the order 4 to 5 were obtained for a variety of axle steels, based on the fatigue limit of conventional specimens. Nonpropagating cracks were found, in some instances at about half of the fatigue limit of the press-fitted member.

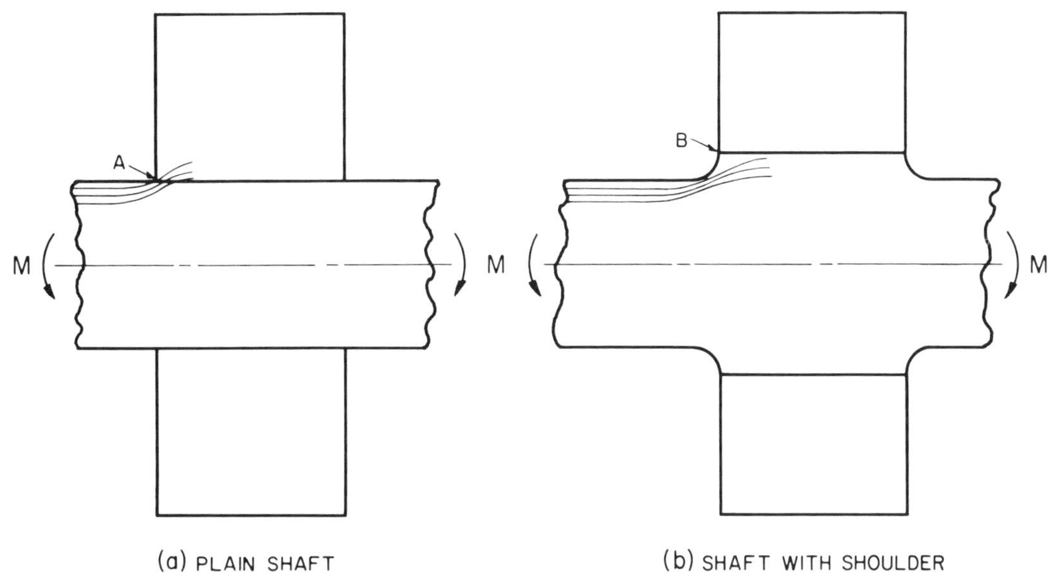

FIG. 177

SHOULDER DESIGN FOR FITTED MEMBER

(STRESS "FLOW LINES" ARE SCHEMATIC)

Photoelastic tests[341a] of a press fitted ring on a shaft with six lands or spokes gave K_t factors on the order of 2 to 4.

The situation with regard to press fits is complicated by the simultaneous presence of stress concentration and fretting corrosion. The relations governing fretting corrosion are not well understood at the present time.

(E) BOLT AND NUT

It has been estimated[342] that bolt failures are distributed about as follows: (*a*) 15% under the head; (*b*) 20% at the end of the thread; and (*c*) 65% in the thread at the nut face.

By using a reduced bolt shank (Fig. 13*c* as compared to 13*b*) the situation with regard to fatigue failures of group *b* type can be improved.[343,344]

With a reduced shank a larger fillet radius can be provided under the head (see Section F), thereby improving design with regard to group *a* type failure.

With regard to failure in the threads at the nut face, group *c* type, Hetényi[115] investigated various bolt-and-nut combinations by means of three-dimensional photoelastic tests. For Whitworth threads, root radius = 0.1373 pitch,[345] he obtained the following results for the designs shown in Fig. 178: for bolt and nut of standard proportions, $K_{tg} = 3.85$; for nut having lip, $K_{tg} = 3.00$, based on the full body (shank) nominal stress. If the factors are calculated for the area at the thread bottom (which is more realistic from a stress concentration standpoint since this corresponds to the location of the maximum stress), the following is obtained: standard nut, $K_{tn} = 2.7$; tapered nut, $K_{tn} = 2.1$.

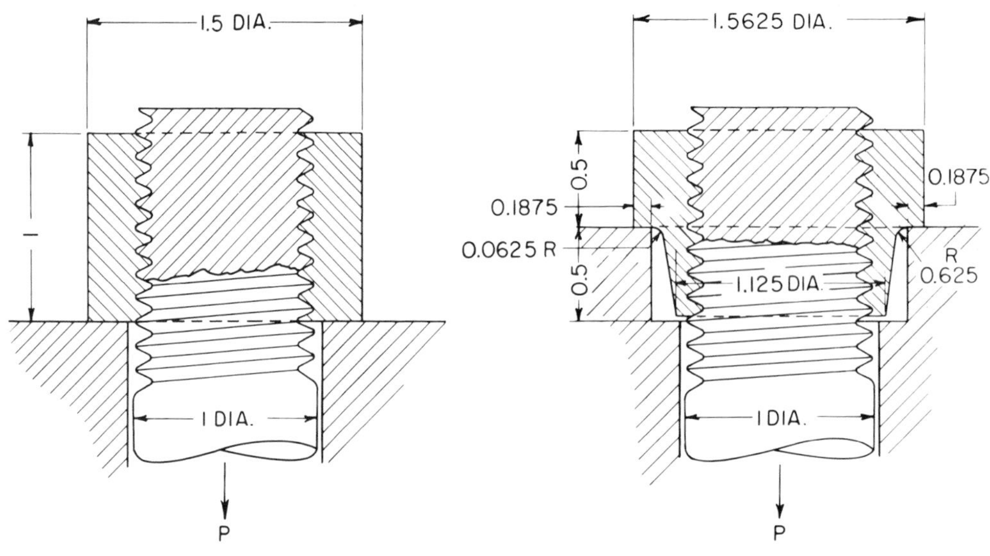

FIG. 178

NUT DESIGNS TESTED PHOTOELASTICALLY
(HETÉNYI)

(Dimensions in inches)

Later tests by Brown and Hickson,[346] using a Fosterite model twice as large and thinner slices, resulted in $K_{tg} = 9$ for the standard nut based on body diameter (see authors' closure).[346] This corresponds to $K_{tn} = 6.7$ for the standard nut, based on root diameter. This compares with the Hetényi value of 2.7. The value of 6.7 should be used in design where fatigue or embrittling is involved, with a correction for notch sensitivity (Fig. 8).

In the discussion of the foregoing paper,[346] Taylor reports a fatigue $K_{fn} = 7$ for a 3-in. diameter bolt with a root contour radius/root diameter half that of the photoelastic model.[346] He estimates that if his fatigue test had been made on a bolt of the same geometry as the photoelastic model, the K_{fn} value might be as low as 4.2.

For a root radius of 0.023 in., a notch sensitivity factor q of about 0.67 is estimated from Fig. 8 for "mild steel." The photoelastic $K_{tn} = 6.7$ would then correspond to $K_{fn} = 4.8$. Although this is in fair agreement with Taylor's estimate, the basis of the estimate has some uncertainties.

A photoelastic investigation[347] of buttress threads showed that by modifying the thread-root contour radius, a reduction of the maximum stress by 22% was achieved.

In a nut designed with a lip (Fig. 179*b*) the peak stress is relieved by virtue of the lip being stressed in the same direction as the bolt. Fatigue tests[344] showed the lip design to be about 30% stronger than the standard nut design (Fig. 179*a*), which is in approximate agreement with photoelastic tests.[115]

In the arrangement shown in Fig. 179*c* the transmitted load is not reversed. Fatigue tests[344] showed a fatigue strength more than double that of the standard bolt-and-nut combination (Fig. 179*a*).

The use of a nut of material having a lower modulus of elasticity is helpful in reducing the peak stress in the bolt threads. Fatigue tests[344,348] have shown gains in strength of 35 to 60% depending on materials.

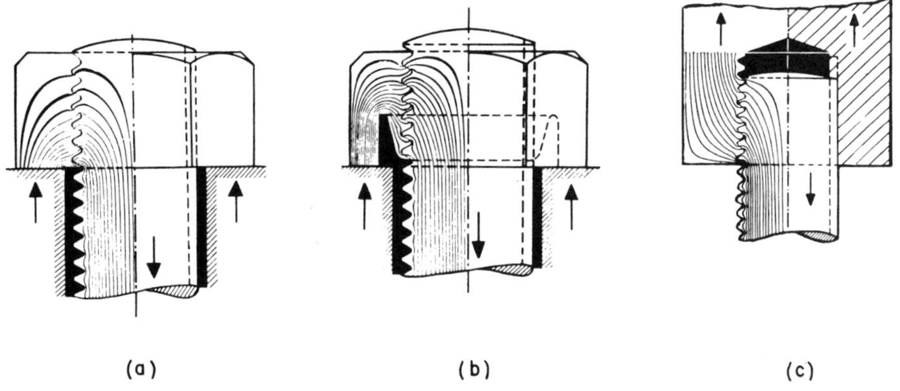

(a) (b) (c)

FIG. 179
NUT DESIGNS FATIGUE TESTED (WIEGAND)
(FLOW LINES-HELE SHAW METHOD)

Other methods can be used for reduction of the K_t factor of a bolt-and-nut combination, such as tapered threads and differential thread spacing, but these methods are not as practical.

(F) BOLT HEAD, TURBINE-BLADE OR COMPRESSOR-BLADE FASTENING (T-HEAD)

A vital difference between the "T-head" case and the case of a bar with shoulder fillets (Chapter 3) is the manner of loading, as illustrated schematically in Fig. 180. Another difference in the above cases is the dimension m, Fig. 180*b*, which is seldom greater than d. As m is decreased, bending of the overhanging portion becomes more prominent.

Figures 190 to 194 present σ_{max}/σ values as determined photoelastically by Hetényi.[349] In this case σ is simply P/A, the load divided by the shank cross-sectional area. Thus σ_{max}/σ values express stress concentration in the simplest form for utilization in design.

However, it is also useful to consider a modified procedure for K_t, so that when comparisons are made between different kinds of fatigue tests, the resulting notch sensitivity values will have a more nearly comparable meaning, as explained in the introduction to Chapter 4. For this purpose, we shall consider two kinds of K_t factor, K_{tA} based on tension and K_{tB} based on bending.

$$\sigma_{\text{nom}A} = \frac{P}{hd} = \sigma$$

where dimensions are given on Figs. 190, 194, and 195:

$$K_{tA} = \frac{\sigma_{\max}}{\sigma} \qquad [137]$$

$$\sigma_{\text{nom}B} = \frac{M}{I/c} = \frac{Pl}{2}\left(\frac{6}{hm^2}\right) = \frac{3}{4}\frac{Pd}{hm^2}\left(\frac{D}{d} - 1\right)$$

where $l = \dfrac{D - d}{4}$.

$$K_{tB} = \frac{\sigma_{max}}{\sigma\left[\dfrac{3(D/d - 1)}{4(m/d)^2}\right]} \qquad [138]$$

Note that for $K_{tA} = K_{tB}$,

$$\frac{(D/d - 1)}{(m/d)^2} = \frac{4}{3}$$

or

$$\frac{ld}{m^2} = \frac{1}{3} \qquad [139]$$

In Fig. 195 values of K_{tA} and K_{tB} are plotted with ld/m^2 as the abscissa variable. For $(ld/m^2) > 1/3$, K_{tB} is used; for $(ld/m^2) < 1/3$, K_{tA} is used. This is a procedure similar to

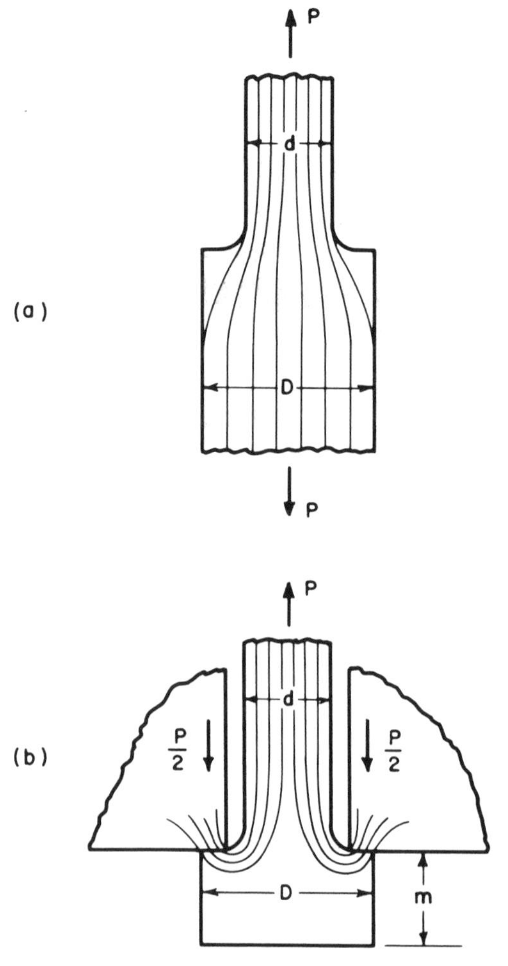

FIG. 180

TRANSMITTAL OF LOAD (SCHEMATIC)
(a) STEPPED TENSION BAR
(b) T-HEAD

that used for the pinned joint (Chapter 4) and, as in that case, not only gets away from extremely high factors but also provides a safer basis for extrapolation (in this case to smaller m/d values).

In Figs. 190 to 193 the dashed line represents equal K_{tA} and K_{tB} values ([137] and [138]). Below this line the σ_{max}/σ values are the same as K_{tA}. Above the dashed line, all the σ_{max}/σ values are higher, usually much higher, than the corresponding K_{tB} values, which in magnitude are all lower than the values represented by the dashed line (i.e., the dashed line represents maximum K_{tA} and K_{tB} values as shown by the peaks in Fig. 195).

The effect of moving concentrated reactions closer to the fillet is shown in Fig. 194. The sharply increasing K_t values are due to a proximity effect,[349] since the nominal bending is decreasing and the nominal tension remains the same.

The T-head factors may be applied directly in the case of a T-shaped blade fastening of rectangular cross section. In the case of the head of a round bolt, somewhat lower factors will result, as can be seen from Chapters 2 and 3. However, the ratios will not be directly comparable to those of Chapter 3, since part of the T-head factor is due to proximity effect. To be on the safe side, it is suggested to use the unmodified T-head factors for bolt heads.

Steam-turbine blade fastenings are often made as a "double T-head." In gas-turbine blades, multiple projections are used in the "fir-tree" type of fastening. Some photoelastic data have been obtained for multiple projections.[350,351]

(G) CURVED BAR

A curved bar subjected to bending will have a higher stress on the inside edge, as shown in Fig. 181. A discussion of the curved bar case is given in advanced textbooks;[352] formulas for the typical cross sections and a graphical method for a general cross section have been published.[353,354] In Fig. 196, values of K_t are given for five cross sections.

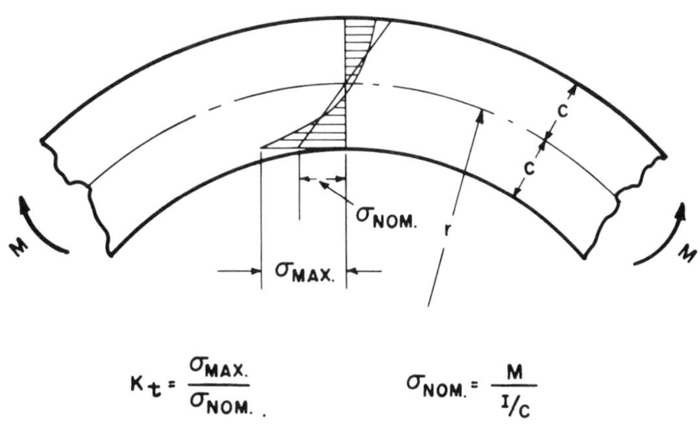

$$K_t = \frac{\sigma_{MAX.}}{\sigma_{NOM.}} \qquad \sigma_{NOM.} = \frac{M}{I/c}$$

FIG. 181

STRESS CONCENTRATION IN CURVED BAR SUBJECTED TO BENDING.

The following formula[353] has been found to be reasonably accurate for ordinary cross sections (not for triangular cross section):

$$K_t = 1.00 + B\left(\frac{I}{bc^2}\right)\left(\frac{1}{r - c} + \frac{1}{r}\right) \qquad [140]$$

where I = moment of inertia of cross section
b = maximum breadth of section
c = distance from centroidal axis to inside edge
r = radius of curvature
B = 1.05 for circular or elliptical cross section; 0.5 for other cross sections.

With regard to notch sensitivity, the q verus r curves of Fig. 8 do not apply to a curved bar; use might be made of the stress gradient concept.[40]

(H) HELICAL SPRING

(a) Round or Square Wire Compression or Tension Spring

A helical spring may be regarded as a curved bar subjected to a twisting moment and a direct shear load.[355] The last sentence in the preceding section applies to helical springs.

For a round wire helical compression or tension spring of small pitch angle, the *Wahl factor*, C_w, a correction factor taking account of curvature and direct shear stress, is generally used in design[355] (Fig. 197).

For round wire:

$$C_w = \frac{\tau_{\max}}{\tau} = \frac{4c - 1}{4c - 4} + \frac{0.615}{c} \qquad [141]$$

$$\tau = \frac{P\,(d/2)}{\pi a^3/16} = \frac{8Pd}{\pi a^3} = \frac{8Pc}{\pi a^2} \qquad [142]$$

where P = axial load
c = spring index d/a
d = mean coil diameter
a = wire diameter

For square wire (Ref. 356):

$$\tau_{\max} = \tau\left(1 + \frac{1.2}{c} + \frac{0.56}{c^2} + \frac{0.5}{c^3}\right) \qquad [143]$$

$$\tau = \frac{Pd}{\alpha a^3} = \frac{2.404Pd}{a^3} = \frac{2.404Pc}{a^2} \qquad [144]$$

From Fig. 199, $\alpha = 0.416$, $1/\alpha = 2.404$, a = width, depth of wire

$$C_w = \frac{\tau_{\max}}{\tau} \qquad [145]$$

The corresponding stress concentration factors, which may be useful for mechanics of materials problems, are obtained as follows:

For round wire:

$$K_{ts} = \frac{\tau_{\max}}{\tau_{\text{nom}}} = \frac{(8Pd/\pi a^3)\,[(4c-1)/(4c-4)] + 4.92P/\pi a^2}{8Pd/\pi a^3 + 4P/\pi a^2}$$

$$K_{ts} = \frac{2c\,[(4c-1)/(4c-4)] + 1.23}{2c+1} \qquad [146]$$

$$\tau_{\max} = K_{ts}\left[\frac{4P}{\pi a^2}(2c+1)\right] \qquad [147]$$

For square wire:

$$K_{ts} = \frac{\tau_{\max}}{\tau_{\text{nom}}} = \frac{(2.404\,Pd/a^3)\,(1 + 1.2/c + 0.56/c^2 + 0.5/c^3)}{2.404\,Pd/a^3 + P/a^2} \qquad [148]$$

$$K_{ts} = \frac{2.404c\,(1 + 1.2/c + 0.56/c^2 + 0.5/c^3)}{2.404\,c + 1} \qquad [149]$$

$$\tau_{\max} = K_{ts}\left[\frac{P}{a^2}(2.404c+1)\right] \qquad [150]$$

Values of C_w and K_{ts} are shown in Fig. 197. K_{ts} is lower than the correction factor C_w. For design calculations it is recommended that the simpler Wahl factor be used; the same value of $\tau_{\max}$ will be obtained whether one uses C_w or K_{ts}.

The effect of pitch angle has been determined by Ancker and Goodier;[357] up to 10° the effect of pitch angle is small, but at 20° the stress increases sufficiently that a correction should be made.

(b) Rectangular Wire Compression or Tension Spring

For wire of rectangular cross section, the results of Liesecke[358] have been converted into stress concentration factors in the following way.

The nominal stress is taken as the maximum stress in a straight torsion bar of the corresponding rectangular cross section plus the direct shear stress:

$$\tau_{\text{nom}} = \frac{Pd}{\alpha x y^2} + \frac{P}{xy} \qquad [151]$$

where x = long side of rectangular cross section
y = short side of rectangular cross section

According to Liesecke,[358]

$$\tau_{\max} = \frac{\beta\,Pd}{ab\sqrt{ab}} \qquad [152]$$

where a = side of rectangle perpendicular to axis
b = side of rectangle parallel to axis
β = Liesecke factor

$$K_{ts} = \frac{\tau_{max}}{\tau_{nom}}$$

$$\tau_{max} = K_{ts}\left(\frac{Pd}{\alpha x y^2} + \frac{P}{xy}\right) \qquad [153]$$

where K_{ts} is given in Fig. 198 and α is given in Fig. 199.

(c) Helical Torsion Spring

Torque is applied in a plane perpendicular to the axis of the spring[355] (see Fig. 200):

$$K_t = \frac{\sigma_{max}}{\sigma_{nom}}$$

For circular wire:

$$\sigma_{nom} = \frac{32Fl}{\pi a^3} \qquad [154]$$

For rectangular wire:

$$\sigma_{nom} = \frac{6Fl}{a^2 b} \qquad [155]$$

where Fl = moment (see Fig. 200)
a = side of rectangle perpendicular to axis
b = side of réctangle parallel to the axis

The effect of pitch angle has been studied by Ancker and Goodier;[357] the correction is small for pitch angles less than 15°.

(l) CRANKSHAFT

The maximum stresses in the fillets of the pin and journal of a series of crankshafts in bending were determined by use of the strain gage by Arai.[359] Design parameters were systematically varied in a comprehensive manner involving 178 tests. The stress concentration factor is defined as $\sigma_{max}/\sigma_{nom}$, where $\sigma_{nom} = M/(\pi d^3/32)$. Strains were measured in the fillet in the axial plane; the smaller circumferential strain in the fillet was not measured.

It was found that the K_t values were in good agreement whether the moment was uniform or applied by means of concentrated loads at the middle of the bearing areas. The K_t values for the pin and journal fillets were sufficiently close that the average value was used.

From the standpoint of stress concentration the most important design variables (see Fig. 201 for notation) are the web thickness ratio, t/d, and the fillet radius ratio, r/d (see Figs. 201 and 202).

It was found that K_t is relatively insensitive to changes in the web width ratio, b/d, and the crank "throw" as expressed* by s/d, over practical ranges of these parameters. It was also found that cutting the corners of the web had no effect on K_t.

*When the inside of the crankpin and the outside of the journal are in line, $s = 0$ (see sketch, Fig. 201). When the crankpin is closer, s is positive (as shown in the sketch); when the crankpin inner surface is farther away than $d/2$, s is negative.

Arai points out that as the web thickness t increases "extremely," K_t should agree with that of a straight stepped shaft. He refers to the author's book[1] (Fig. 65) and an extended t/d value of 1 to 2. This is an enormous extrapolation (see Fig. 201); it seems that all that can be said is that smooth curves can be drawn to the shaft values, but this does not constitute a verification.

Referring to the sketch in Fig. 201, it is sometimes beneficial to recess the fillet fully or partially. It was found that as δ is increased K_t increases. However, the designer should evaluate the increase against the possibility of using a larger fillet radius and increasing the bearing area or decreasing the shaft length.

An empirical formula was developed by Arai to cover the entire range of tests:

$$K_t = 4.84\, C_1\, C_2\, C_3\, C_4\, C_5 \qquad [156]$$

where $C_1 = 0.420 + 0.160\sqrt{[1/(r/d)] - 6.864}$

$C_2 = 1 + 81\,\{0.769 - [0.407 - (s/d)]^2\}\,(\delta/r)\,(r/d)^3$

$C_3 = 0.285\,[2.2 - (b/d)]^2 + 0.785$

$C_4 = 0.444/(t/d)^{1.4}$

$C_5 = 1 - [(s/d) + 0.1]^2/[4(t/d) - 0.7]$

No corresponding investigation of the crankshaft in torsion has been published.

(J) CRANE HOOK

A crane hook is another curved bar case. Since the proportions vary considerably, no generally applicable curve or formula is available. Wahl[360] has developed a simple numerical method and has applied this to a typical example of a crane hook with an approximately trapezoidal cross section, obtaining a K_t value of 1.56.

(K) MEMBER OF U-SHAPE

The case of a U-shaped member subjected to a spreading type of loading has been investigated photoelastically[361] (Figs. 203 and 204).

The location of the maximum stress depends on the proportions of the U member and the position of the load. For variable back depth, d; for $w = r$; and for loads applied at distances one to three times r from the center of curvature (Fig. 203), the maximum stress occurs at position 1 for the smaller values of d and at 2 for the larger values of d. The K_t values were defined[361] as follows:

For position 1:

$$K_{t1} = \frac{\sigma_{\max} - P/a_1}{M_1 c_1/I_1} = \frac{\sigma_{\max} - P/hd}{6P(m + r + d/2)/hd^2} \qquad [157]$$

where d = back depth (Fig. 203)

w = arm width $= r$

h = thickness

r = inside radius

m = distance from line of application of load to center of curvature

For position 2:

$$K_{t2} = \frac{\sigma_{max}}{M_2 c_2/I_2} = \frac{\sigma_{max}}{P(m + b)c_2/I_2} \qquad [158]$$

where I_2 = moment of inertia of section through 2
c_2 = distance from inside edge to centroid of section through 2
b = axial distance from center of radius to centroid of section through 2

In the case of position 2, angle θ (Fig. 203) was found to be approximately 20°.

Where the outside dimensions are constant, $w = d$ and r varies, causing w and d to vary correspondingly (Fig. 204), the maximum stress occurs at position 1, except for very large values of r/d. Values of K_t are given in Fig. 204 for a condition where the line of load application remains the same.

(L) ANGLE AND BOX SECTIONS

Considerable work has been done on beam sections in torsion.[362] In Fig. 205 mathematical results of Huth[363] are given for angle and box sections. For box sections the values given are valid only when a is large compared to c (15 to 20 times as great). An approximation of bending of angle sections can be obtained from results on knee frames[364] and from the references.[364]

(M) ROTATING DISK WITH HOLE

For a rotating disk with a central hole, the maximum stress is tangential, occurring at the edge of the hole:[365]

$$\sigma_{max} = 2\frac{\gamma v^2}{g}\left(\frac{3 + \nu}{8}\right)\left[1 + \left(\frac{1 - \nu}{3 + \nu}\right)\left(\frac{R_1}{R_2}\right)^2\right] \qquad [159]$$

where γ = weight per unit volume
v = peripheral velocity
g = gravitational acceleration
ν = Poisson ratio
R_1 = hole radius
R_2 = disk radius

Note that for $R_1/R_2 = 1$, $\sigma_{max} = \gamma v^2/g$.

The K_t factor can be defined in several ways, depending on the choice of nominal stress.

(*a*) σ_{Na} = stress at center of disk without hole:

$$\sigma_{Na} = \frac{\gamma v^2}{g}\left(\frac{3 + \nu}{8}\right) \qquad [160]$$

Use of this nominal stress results in the top curve of Fig. 206. This curve gives a reasonable result for a small hole (i.e., for $R_1/R_2 \rightarrow 0$, $K_{tA} = 2$), but as R_1/R_2 approaches 1.0 (thin ring), the higher factor is not realistic.

Another basis of σ_N would be to use the tangential stress in a solid disk at the same radial distance as the radius of the hole. Since this would result in a higher K_t for a thin ring than from (*a*), there is no point in showing this on Fig. 206.

(*b*) σ_{Nb} = average tangential stress:

$$\sigma_{Nb} = \frac{\gamma v^2}{3g}\left(1 + \frac{R_1}{R_2} + \frac{R_1^2}{R_2^2}\right) \qquad [161]$$

Use of this nominal stress results in a more reasonable relation, giving $K_t = 1$ for the thin ring. However, for a small hole (*a*) appears preferable.

(*c*) Curve (*b*) adjusted linearly to fit end conditions (*a*) at $R_1/R_2 = 0$ and (*b*) at $R_1/R_2 = 1.0$. For this case σ_{Nb} becomes

$$\sigma_{Nc} = \frac{\gamma v^2}{3g}\left(1 + \frac{R_1}{R_2} + \frac{R_1^2}{R_2^2}\right)\left[3\left(\frac{3+\nu}{8}\right)\left(1 - \frac{R_1}{R_2}\right) + \frac{R_1}{R_2}\right] \qquad [162]$$

For a small central hole, [160] will be satisfactory for most purposes. For larger holes and in cases where notch sensitivity (Chapter 1, Section C) is involved [162] is suggested.

For a rotating disk with a noncentral hole, photoelastic results are available for variable radial location for two sizes of hole[366] (Fig. 207). Here the nominal stress, σ_N, is taken as the tangential stress in a solid disk at a point corresponding to the outermost point (marked *A*, Fig. 207) of the hole. Since the holes in this case are small relative to the disk diameter, this is a reasonable procedure.

$$\sigma_N = \frac{\gamma v^2}{g}\left(\frac{3+\nu}{8}\right)\left[1 - \left(\frac{1+3\nu}{3+\nu}\right)\left(\frac{R_A}{R_2}\right)^2\right] \qquad [163]$$

The same investigation[366] covered the cases of a disk with six to ten noncentral holes located on a common circle, the disk also containing a central hole. Hetényi[305] investigated the special cases of a rotating disk containing a central hole plus two or eight symmetrically disposed noncentral holes.

Similar investigations[367–370] have been made for a disk with a large number of symmetrical noncentral holes, such as is used in gas turbine disks. The optimum number of holes was found[369] for various geometrical ratios; reinforcement bosses did not reduce peak stresses by a significant amount,[370] but use of a tapered disk did lower the peak stresses at the noncentral holes.

(N) RING OR HOLLOW ROLLER

The case of a ring subjected to concentrated loads acting along a diametral line (Fig. 209) has been solved mathematically for $R_1/R_2 = \frac{1}{2}$ by Timoshenko[371] and for $R_1/R_2 = \frac{1}{3}$ by Billevicz.[372] An approximate theoretical solution is given by Case.[373] Photoelastic investigations have been made by Horger[336] and by Leven.[374] The values shown in Figs. 208 and 209 represent the average of the photoelastic data and mathematical results, all of which are in good agreement. For $K_t = \sigma_{max}/\sigma_{nom}$, the maximum tensile stress is used for σ_{max}, and for σ_{nom} the basic bending and tensile components as given by Timoshenko[375] for a thin ring are used.

For the ring loaded internally (Fig. 208):

$$K_t = \frac{\sigma_{\max A}\,[2h(R_2 - R_1)]}{P\left[1 + \dfrac{3(R_2 + R_1)(1 - 2/\pi)}{R_2 - R_1}\right]} \qquad [164]$$

For the ring loaded externally (Fig. 209):

$$K_t = \frac{\sigma_{\max B}\,[\pi h(R_2 - R_1)^2]}{3P(R_2 + R_1)} \qquad [165]$$

The case of a round-cornered square hole in a cylinder subjected to opposite concentrated loads has been analyzed by Seika.[376]

(O) PRESSURIZED CYLINDER

The Lamé solution[228] is

$$\sigma_{\max} = \frac{p(R_1^2 + R_2^2)}{(R_2^2 - R_1^2)} \qquad [166]$$

where p = pressure
R_1 = inside radius
R_2 = outside radius

Two K_t relations are

$$K_{t1} = \frac{\sigma_{\max}}{\sigma_{\text{nom}}} = \frac{\sigma_{\max}}{\sigma_{\text{av}}}$$

$$K_{t1} = \frac{(R_1/R_2)^2 + 1}{(R_1/R_2)^2 + R_1/R_2} \qquad [167]$$

and

$$K_{t2} = \frac{\sigma_{\max}}{p}$$

$$K_{t2} = \frac{(R_1/R_2)^2 + 1}{1 - (R_1/R_2)^2} \qquad [168]$$

The two relations are shown in Fig. 210. At $R_1/R_2 = \frac{1}{2}$, the K_t factors are equal, $K_t = 1.666$. The branches of the curves below $K_t = 1.666$ are regarded as more meaningful when applied to analysis of mechanics of materials problems (see comments in the Introduction to Chapter 4).

(P) CYLINDRICAL PRESSURE VESSEL WITH TORISPHERICAL ENDS

Figure 211 is based on photoelastic data of Fessler and Stanley.[377] Since the maximum stresses used in the K_t factors are in the longitudinal (meridional) direction, the nominal stress used in Fig. 211 is $pd/4t$, which is the stress in the longitudinal direction in a closed cylinder subjected to pressure p. Notations are given in Fig. 211. Although the K_t factors (knuckle) are for $t/d = 0.05$ (or adjusted to that value), it was found[377] that K_t increases only slightly with increasing thickness.

Referring to Fig. 211, above lines ABC and CDE the maximum K_t is at the crown; between lines ABC and FC the maximum K_t is at the knuckle; and below line FE the maximum stress is the hoop stress in the straight cylindrical portion.

Design recommendations[378] are indicated on Fig. 211: r_i/d_o not less than 0.06; R_i/d_o not greater than 1.0; r_i/t not less than 3.0.

A critical evaluation of investigations of stresses in pressure vessels with torispherical ends has been made by Fessler and Stanley.[379]

(Q) PRESSURIZED THICK CYLINDER WITH A CIRCULAR HOLE IN THE CYLINDER WALL

This case is encountered frequently in the high-pressure equipment industry. In Fig. 212, K_t factors[380] are defined as follows:

$$K_t = \frac{\sigma_{\max}}{p\left[\dfrac{(b/a)^2 + 1}{(b/a)^2 - 1}\right]} \qquad [169]$$

The denominator is the hoop stress at the inner surface of a cylinder without a hole, as given by the Lamé equation [166].

Reference 380 also gives K_t factors for a press-fitted cylinder on an unpressurized cylinder with a sidehole or with a crosshole.

Strain gage measurements[381] on pressurized thick-walled cylinders with well rounded crossholes resulted in minimum K_t factors (1.0 to 1.1) when the holes were of equal diameter (K_t defined by [169]). Fatigue failures in compressor heads were largely reduced by making the holes of equal diameter and using larger intersection radii.

For members with crossholes and sideholes, with the member subjected to axial stress, see Chapter 4, Section A.z8.

FIG. 182

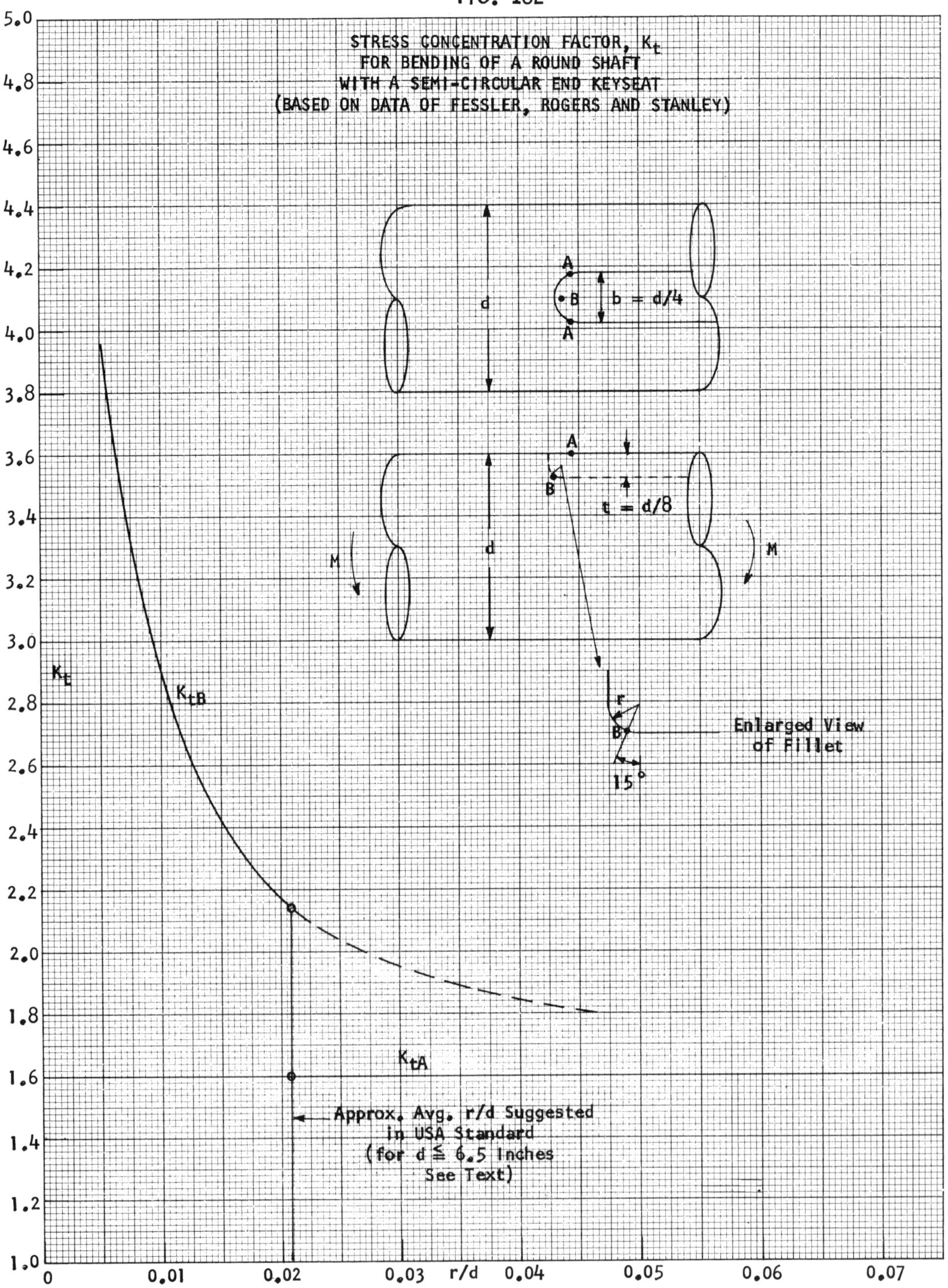

STRESS CONCENTRATION FACTOR, K_t
FOR BENDING OF A ROUND SHAFT
WITH A SEMI-CIRCULAR END KEYSEAT
(BASED ON DATA OF FESSLER, ROGERS AND STANLEY)
K_t
K_{tB}
K_{tA}
A
B
A
d
b = d/4
t = d/8
M
M
r
B
15°
Enlarged View of Fillet
Approx. Avg. r/d Suggested in USA Standard (for d ≦ 6.5 Inches See Text)
5.0
4.8
4.6
4.4
4.2
4.0
3.8
3.6
3.4
3.2
3.0
2.8
2.6
2.4
2.2
2.0
1.8
1.6
1.4
1.2
1.0
0
0.01
0.02
0.03
r/d
0.04
0.05
0.06
0.07

FIG. 183

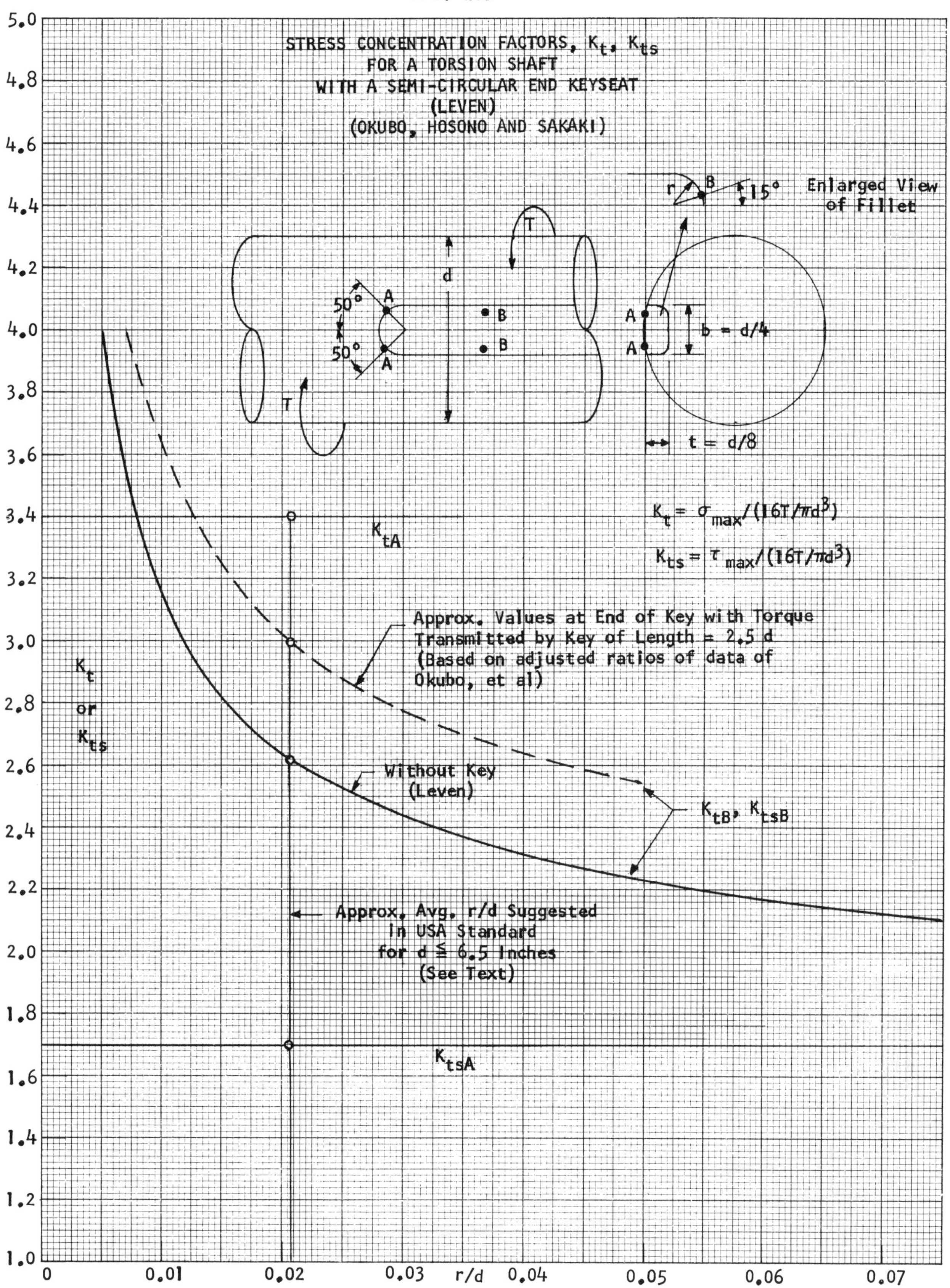

STRESS CONCENTRATION FACTORS, K_t, K_{ts}
FOR A TORSION SHAFT
WITH A SEMI-CIRCULAR END KEYSEAT
(LEVEN)
(OKUBO, HOSONO AND SAKAKI)
Enlarged View of Fillet
r
B
15°
T
d
50°
50°
A
A
B
B
A
A
b = d/4
t = d/8
$K_t = \sigma_{max}/(16T/\pi d^3)$
$K_{ts} = \tau_{max}/(16T/\pi d^3)$
K_{tA}
Approx. Values at End of Key with Torque Transmitted by Key of Length = 2.5 d (Based on adjusted ratios of data of Okubo, et al)
K_t or K_{ts}
Without Key (Leven)
K_{tB}, K_{tsB}
Approx. Avg. r/d Suggested in USA Standard for d ≦ 6.5 Inches (See Text)
K_{tsA}
5.0
4.8
4.6
4.4
4.2
4.0
3.8
3.6
3.4
3.2
3.0
2.8
2.6
2.4
2.2
2.0
1.8
1.6
1.4
1.2
1.0
0
0.01
0.02
0.03
r/d
0.04
0.05
0.06
0.07

FIG. 183a

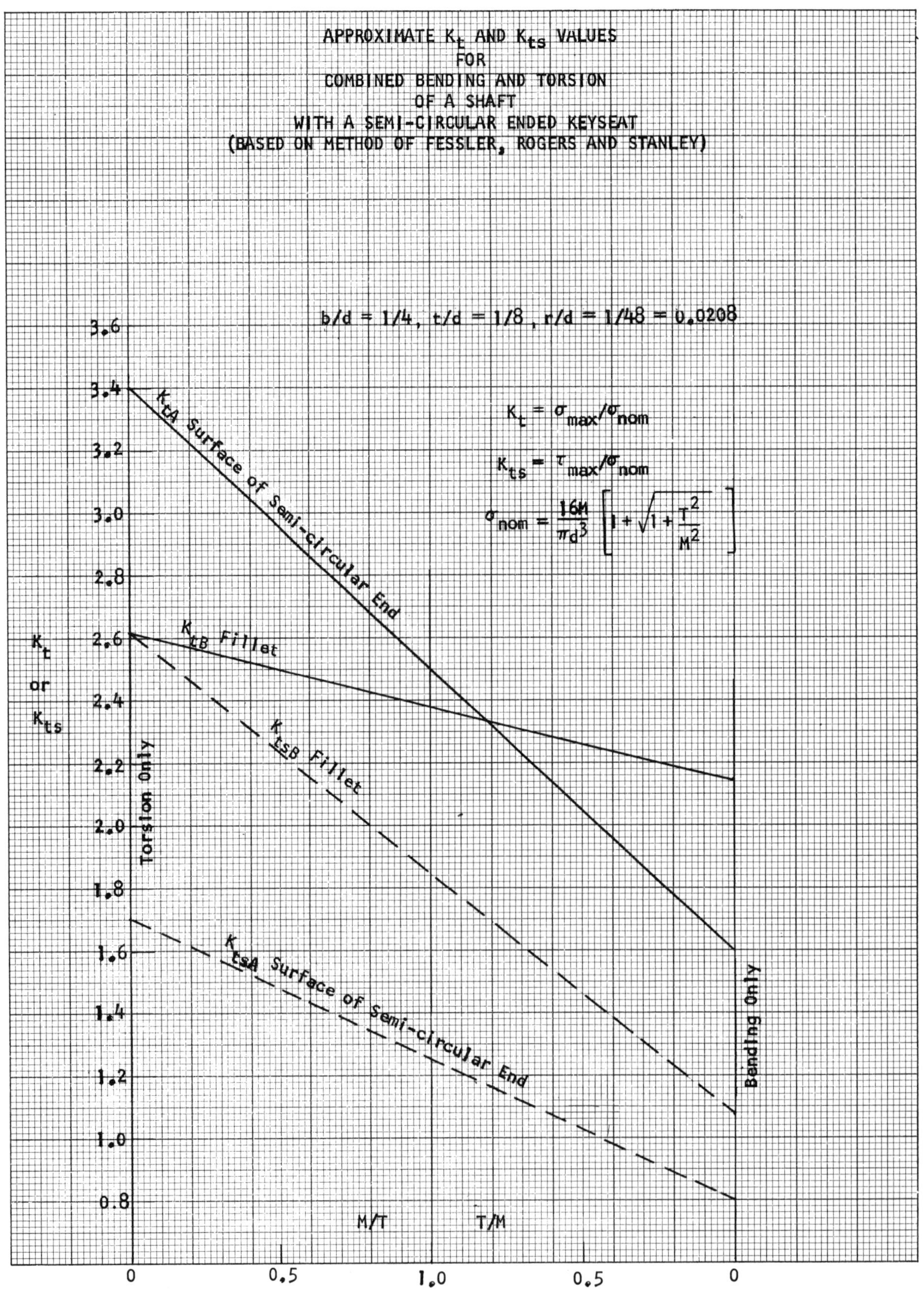
APPROXIMATE K_t AND K_{ts} VALUES
FOR
COMBINED BENDING AND TORSION
OF A SHAFT
WITH A SEMI-CIRCULAR ENDED KEYSEAT
(BASED ON METHOD OF FESSLER, ROGERS AND STANLEY)
b/d = 1/4, t/d = 1/8, r/d = 1/48 = 0.0208
$K_t = \sigma_{max}/\sigma_{nom}$
$K_{ts} = \tau_{max}/\sigma_{nom}$
$\sigma_{nom} = \frac{16M}{\pi d^3}\left[1 + \sqrt{1 + \frac{T^2}{M^2}}\right]$
K_{tA} Surface of Semi-circular End
K_{tB} Fillet
K_{tsB} Fillet
K_{tsA} Surface of Semi-circular End
K_t or K_{ts}
Torsion Only
Bending Only
3.6
3.4
3.2
3.0
2.8
2.6
2.4
2.2
2.0
1.8
1.6
1.4
1.2
1.0
0.8
M/T
T/M
0
0.5
1.0
0.5
0

FIG. 184

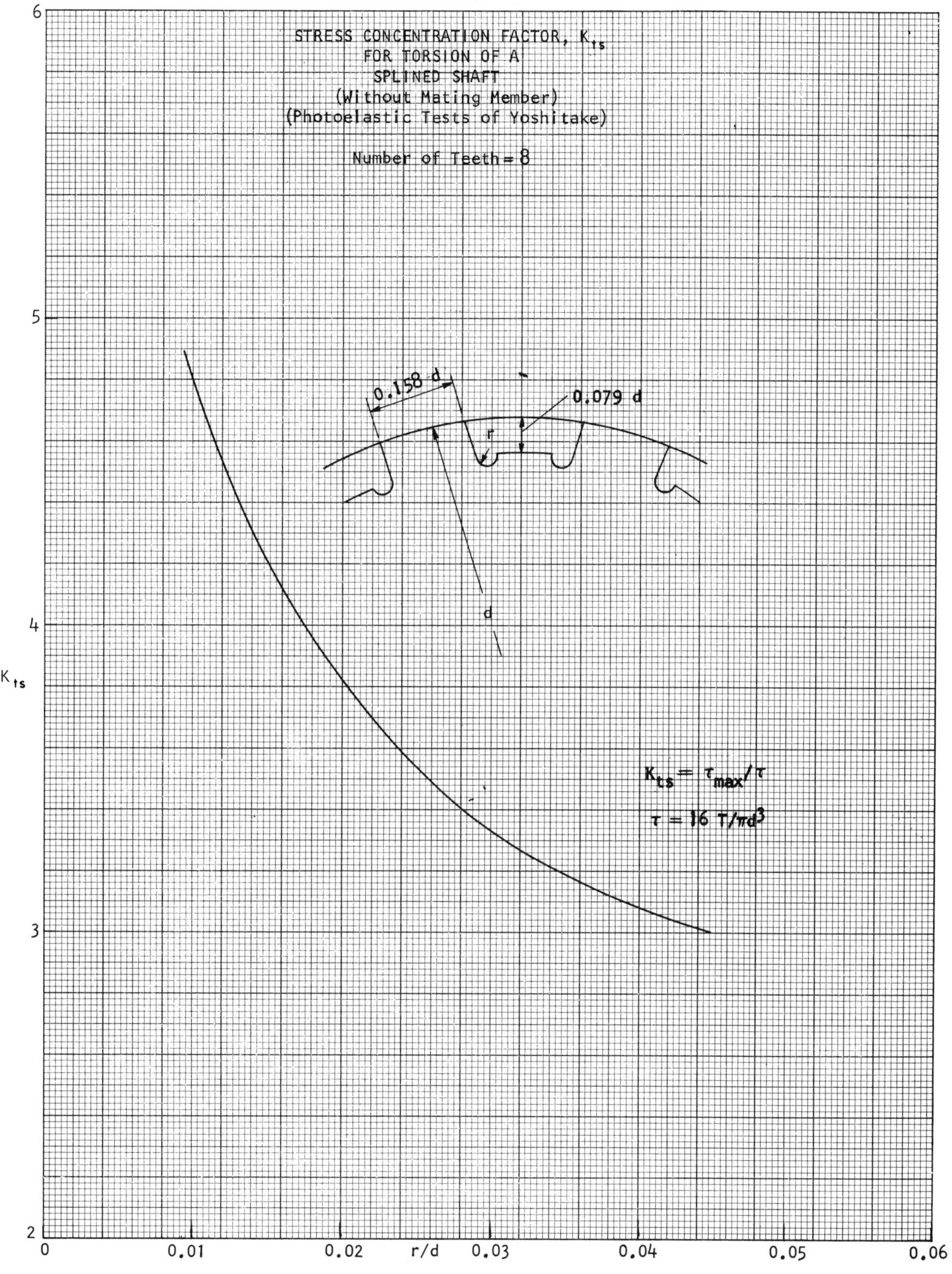

STRESS CONCENTRATION FACTOR, K_{ts}
FOR TORSION OF A
SPLINED SHAFT
(Without Mating Member)
(Photoelastic Tests of Yoshitake)
Number of Teeth = 8
0.158 d
0.079 d
r
d
$K_{ts} = \tau_{max}/\tau$
$\tau = 16\ T/\pi d^3$
K_{ts}
6
5
4
3
2
0
0.01
0.02
r/d
0.03
0.04
0.05
0.06

FIG. 185

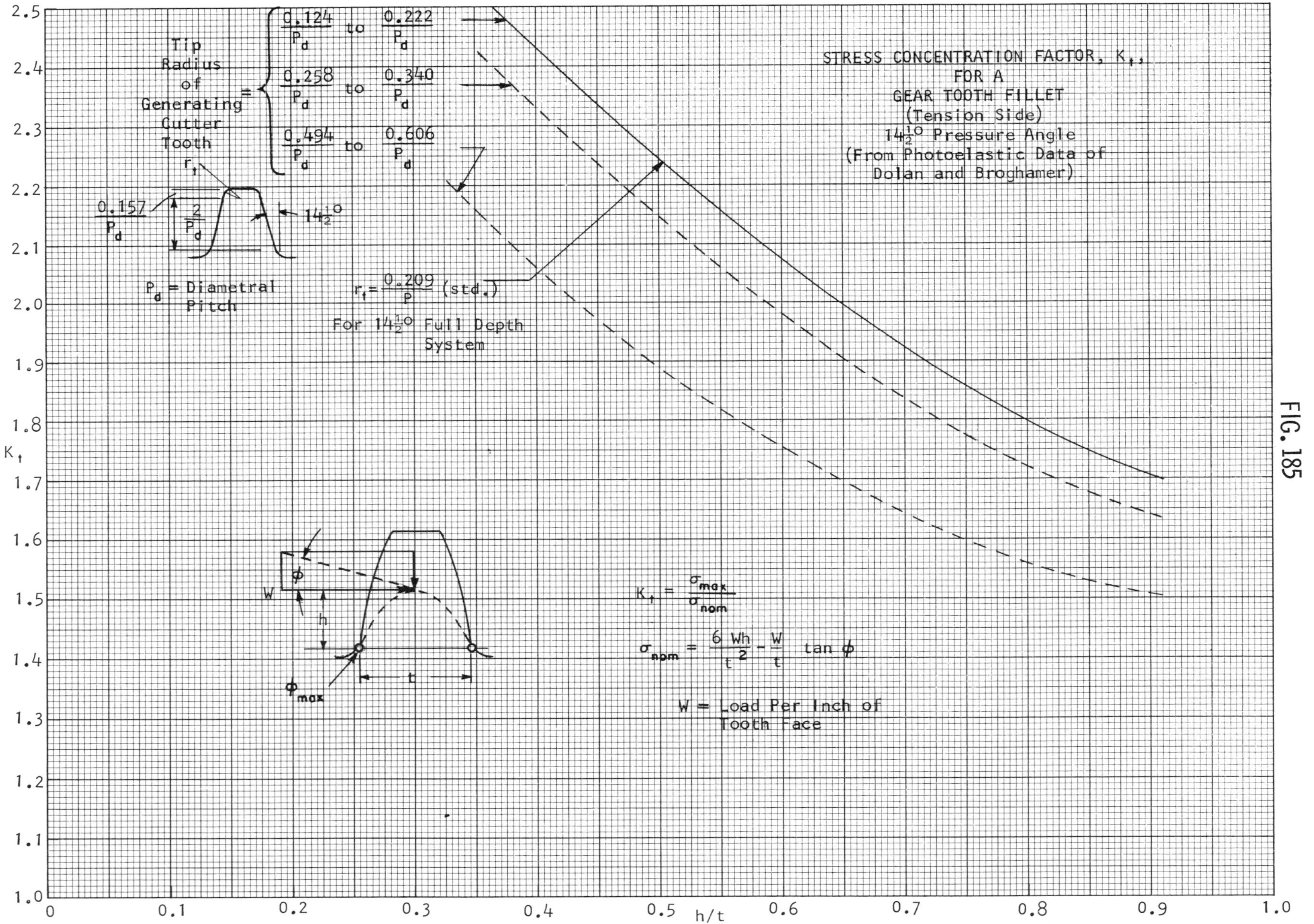

STRESS CONCENTRATION FACTOR, K_t,
FOR A
GEAR TOOTH FILLET
(Tension Side)
$14\frac{1}{2}^\circ$ Pressure Angle
(From Photoelastic Data of
Dolan and Broghamer)
Tip Radius of Generating Cutter Tooth $= \frac{0.124}{P_d}$ to $\frac{0.222}{P_d}$, $\frac{0.258}{P_d}$ to $\frac{0.340}{P_d}$, $\frac{0.494}{P_d}$ to $\frac{0.606}{P_d}$
r_t
$\frac{0.157}{P_d}$
$\frac{2}{P_d}$
$14\frac{1}{2}^\circ$
P_d = Diametral Pitch
$r_t = \frac{0.209}{P}$ (std.)
For $14\frac{1}{2}^\circ$ Full Depth System
ϕ
W
h
t
ϕ_{max}
$K_t = \frac{\sigma_{max}}{\sigma_{nom}}$
$\sigma_{nom} = \frac{6\,Wh}{t^2} - \frac{W}{t}\tan\phi$
W = Load Per Inch of Tooth Face
K_t
2.5
2.4
2.3
2.2
2.1
2.0
1.9
1.8
1.7
1.6
1.5
1.4
1.3
1.2
1.1
1.0
0
0.1
0.2
0.3
0.4
0.5
0.6
0.7
0.8
0.9
1.0
h/t

FIG. 186

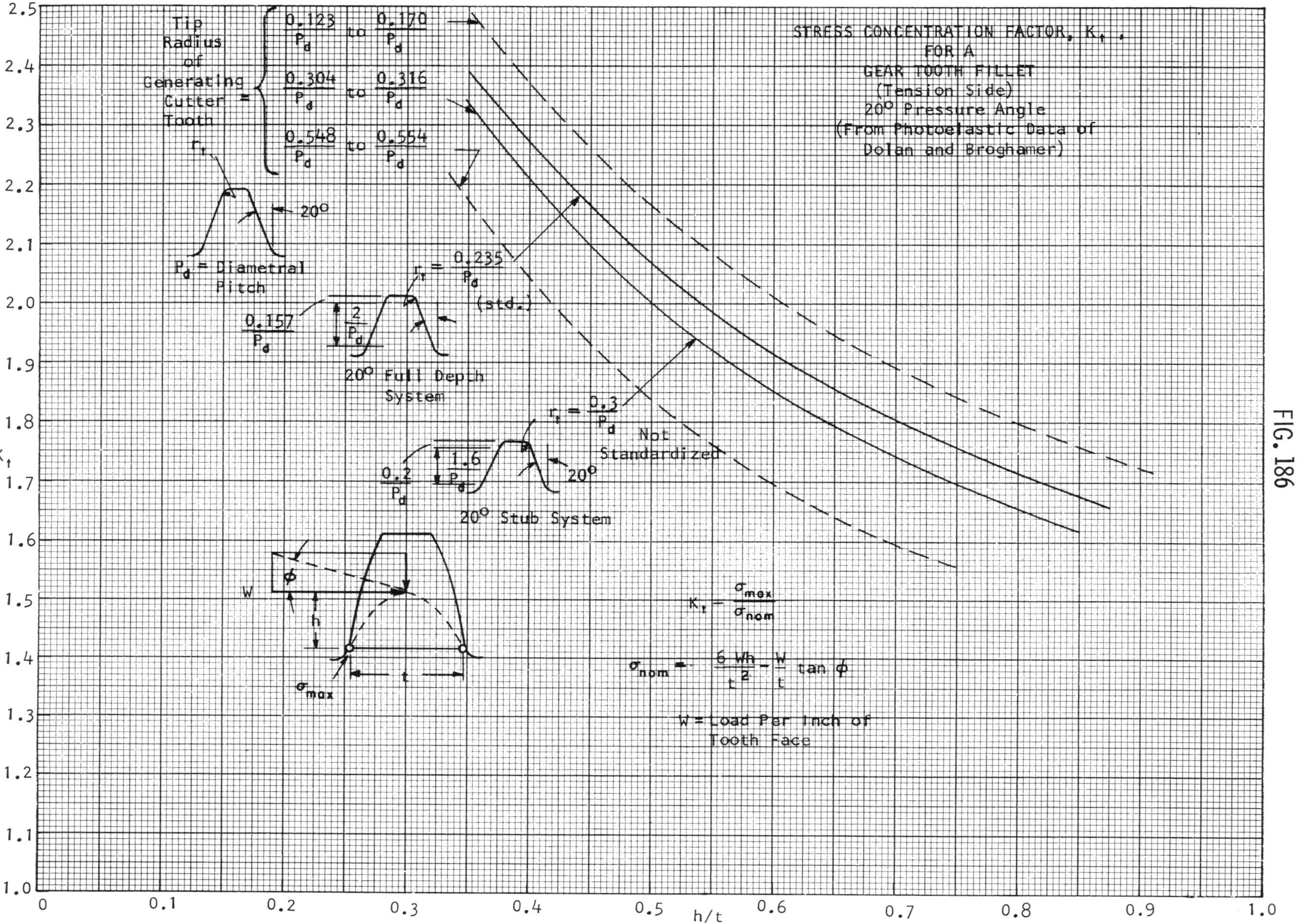

STRESS CONCENTRATION FACTOR, K_t, FOR A GEAR TOOTH FILLET
(Tension Side)
20° Pressure Angle
(From Photoelastic Data of Dolan and Broghamer)
Tip Radius of Generating Cutter Tooth =
0.123/P_d to 0.170/P_d
0.304/P_d to 0.316/P_d
0.548/P_d to 0.554/P_d
r_t
20°
P_d = Diametral Pitch
0.157/P_d
2/P_d
r_t = 0.235/P_d (std.)
20° Full Depth System
r_t = 0.3/P_d Not Standardized
0.2/P_d
1.6/P_d
20°
20° Stub System
W
φ
h
σ_max
t
K_t = σ_max/σ_nom
σ_nom = 6Wh/t² − W/t tan φ
W = Load Per Inch of Tooth Face
K_t
h/t
2.5
2.4
2.3
2.2
2.1
2.0
1.9
1.8
1.7
1.6
1.5
1.4
1.3
1.2
1.1
1.0
0
0.1
0.2
0.3
0.4
0.5
0.6
0.7
0.8
0.9
1.0

FIG. 187

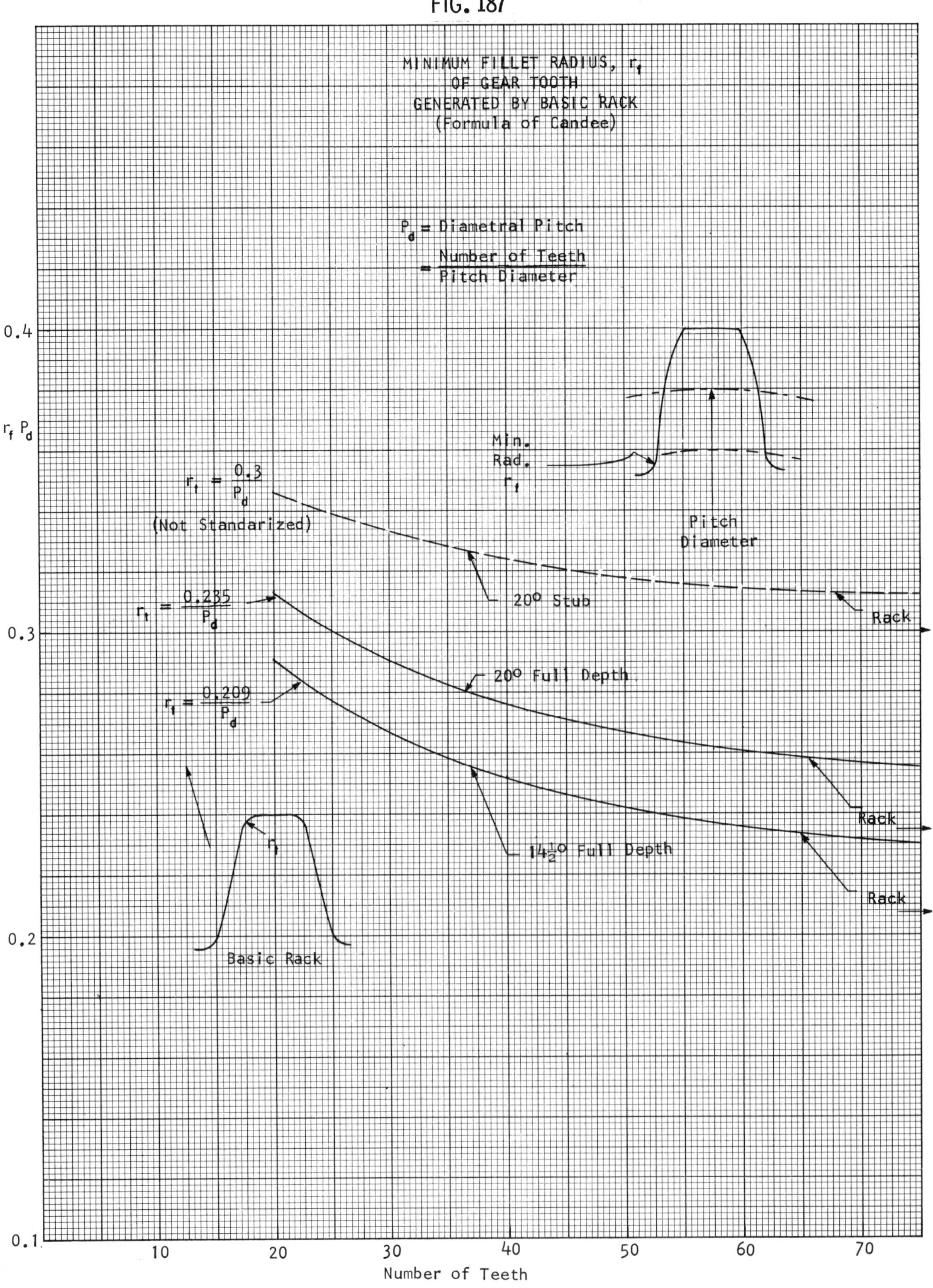

MINIMUM FILLET RADIUS, r_f
OF GEAR TOOTH
GENERATED BY BASIC RACK
(Formula of Candee)
P_d = Diametral Pitch
$= \frac{\text{Number of Teeth}}{\text{Pitch Diameter}}$
Min.
Rad.
r_f
Pitch
Diameter
$r_f = \frac{0.3}{P_d}$
(Not Standarized)
20° Stub
Rack
$r_f = \frac{0.235}{P_d}$
20° Full Depth
$r_f = \frac{0.209}{P_d}$
Rack
$14\frac{1}{2}$° Full Depth
Rack
r_f
Basic Rack
$r_f P_d$
0.4
0.3
0.2
0.1
10
20
30
40
50
60
70
Number of Teeth

FIG. 188

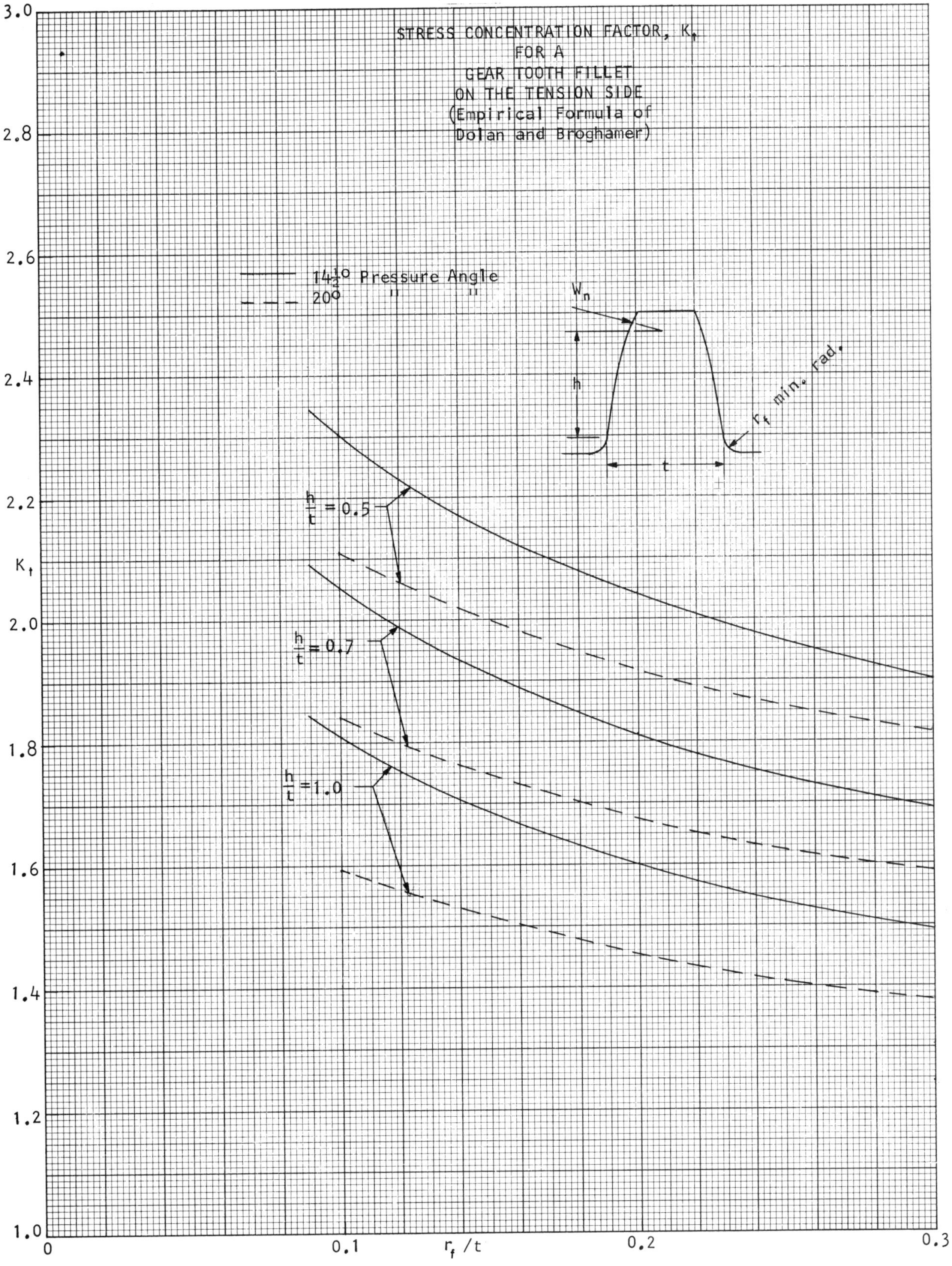

STRESS CONCENTRATION FACTOR, K_t
FOR A
GEAR TOOTH FILLET
ON THE TENSION SIDE
(Empirical Formula of
Dolan and Broghamer)

—— $14\frac{1}{2}^o$ Pressure Angle
– – – 20° " "

W_n

h

r_f min. rad.

t

$\frac{h}{t} = 0.5$

$\frac{h}{t} = 0.7$

$\frac{h}{t} = 1.0$

K_t

3.0
2.8
2.6
2.4
2.2
2.0
1.8
1.6
1.4
1.2
1.0

0
0.1
r_f /t
0.2
0.3

FIG. 189

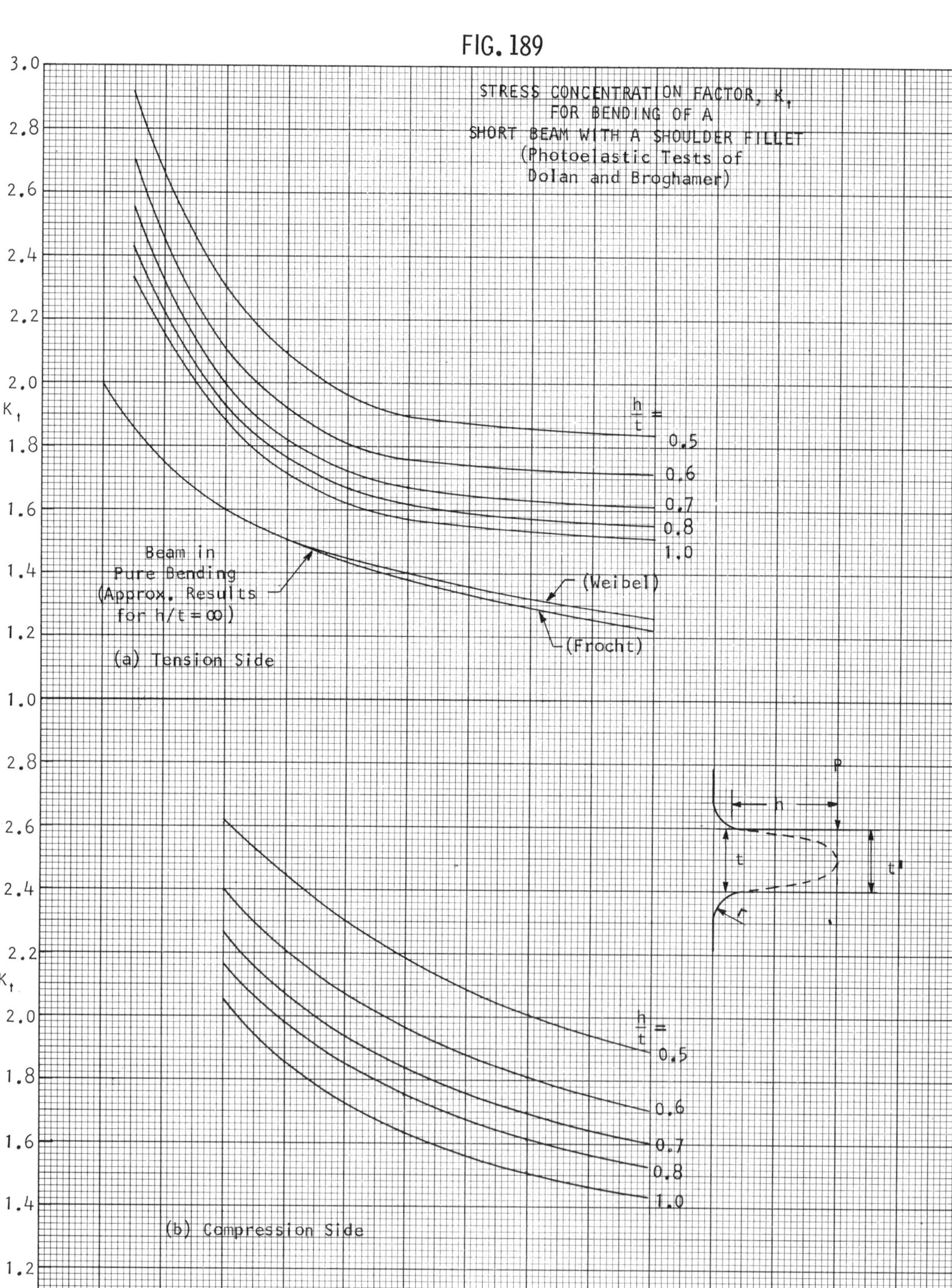

STRESS CONCENTRATION FACTOR, K_t
FOR BENDING OF A
SHORT BEAM WITH A SHOULDER FILLET
(Photoelastic Tests of
Dolan and Broghamer)
3.0
2.8
2.6
2.4
2.2
2.0
K_t
1.8
1.6
1.4
1.2
1.0
$\frac{h}{t}$ =
0.5
0.6
0.7
0.8
1.0
Beam in
Pure Bending
(Approx. Results
for h/t = ∞)
(Weibel)
(Frocht)
(a) Tension Side
P
h
t
t'
r
(b) Compression Side
0
0.1
0.2
r/t'
0.3
0.4
0.5

FIG. 190

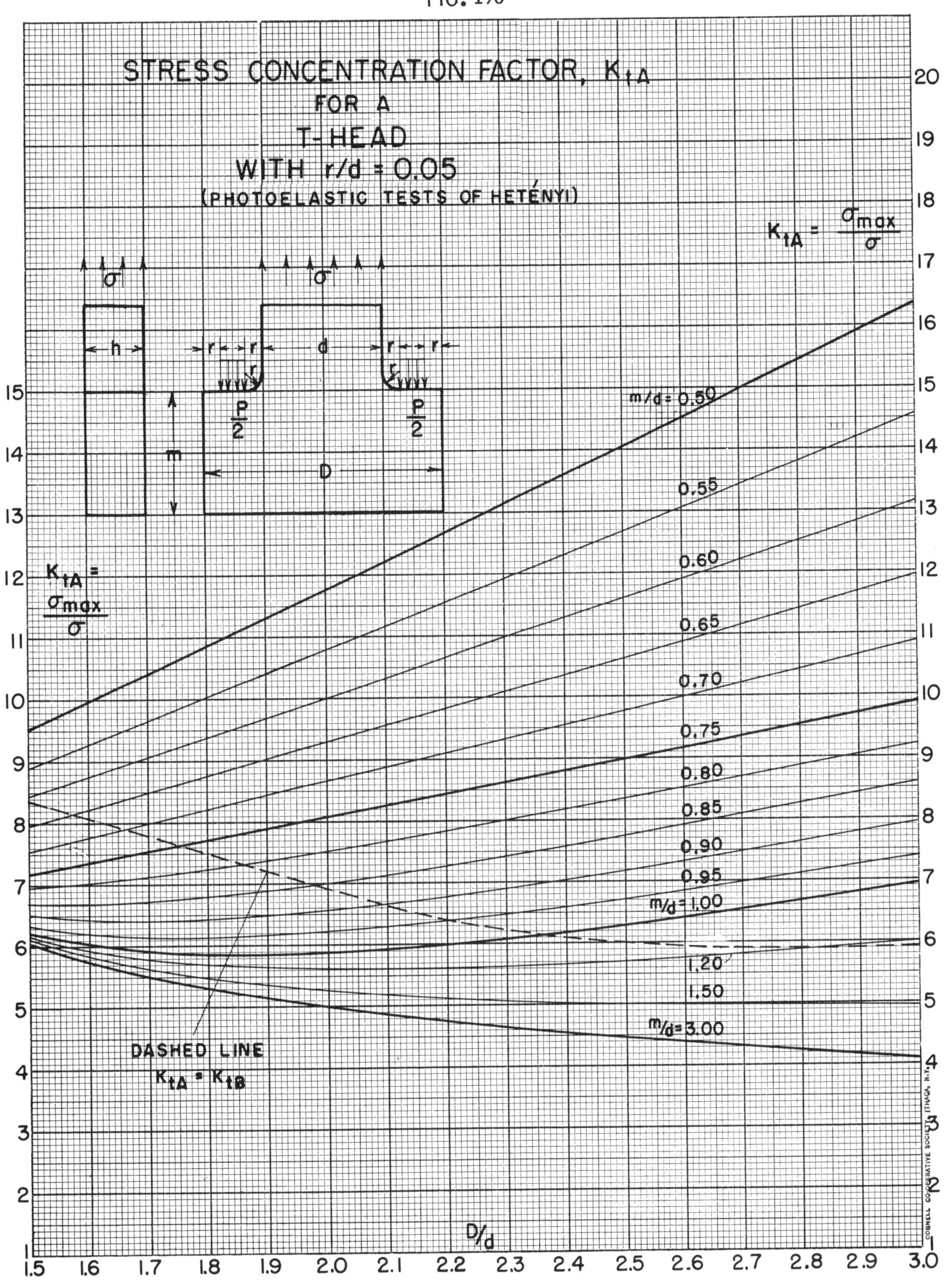

STRESS CONCENTRATION FACTOR, K_{tA}
FOR A
T-HEAD
WITH r/d = 0.05
(PHOTOELASTIC TESTS OF HETÉNYI)
$K_{tA} = \frac{\sigma_{max}}{\sigma}$
σ
h
r
d
m
D
P/2
m/d = 0.50
0.55
0.60
0.65
0.70
0.75
0.80
0.85
0.90
0.95
m/d = 1.00
1.20
1.50
m/d = 3.00
DASHED LINE
$K_{tA} = K_{tB}$
D/d

FIG. 191

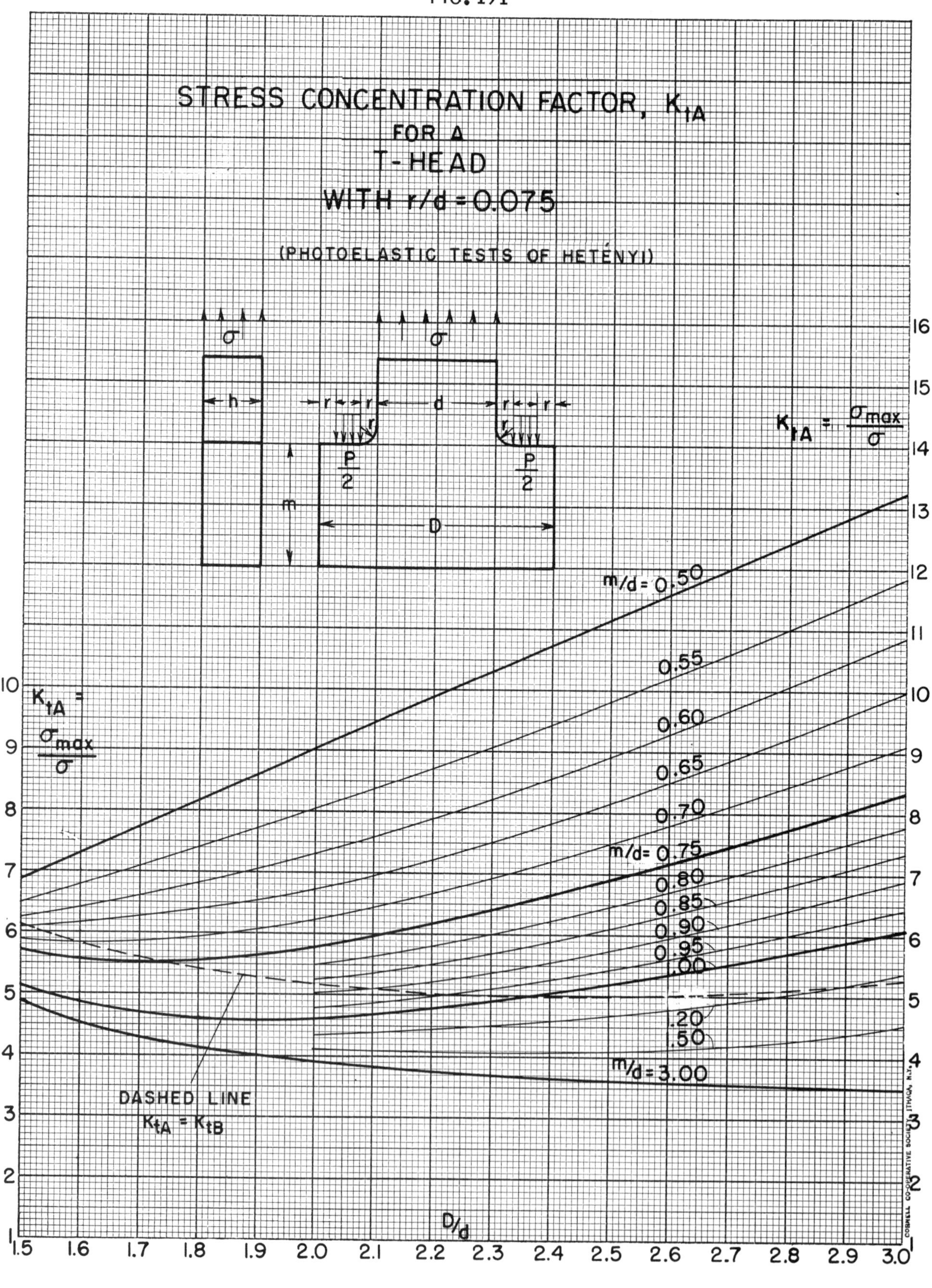

STRESS CONCENTRATION FACTOR, K_{tA}
FOR A
T-HEAD
WITH r/d = 0.075
(PHOTOELASTIC TESTS OF HETÉNYI)
σ
σ
h
r
r
d
r
r
r
r
P/2
P/2
m
D
$K_{tA} = \frac{\sigma_{max}}{\sigma}$
$K_{tA} = \frac{\sigma_{max}}{\sigma}$
m/d = 0.50
0.55
0.60
0.65
0.70
m/d = 0.75
0.80
0.85
0.90
0.95
1.00
1.20
1.50
m/d = 3.00
DASHED LINE
$K_{tA} = K_{tB}$
D/d
1.5
1.6
1.7
1.8
1.9
2.0
2.1
2.2
2.3
2.4
2.5
2.6
2.7
2.8
2.9
3.0
1
2
3
4
5
6
7
8
9
10
11
12
13
14
15
16

FIG. 192

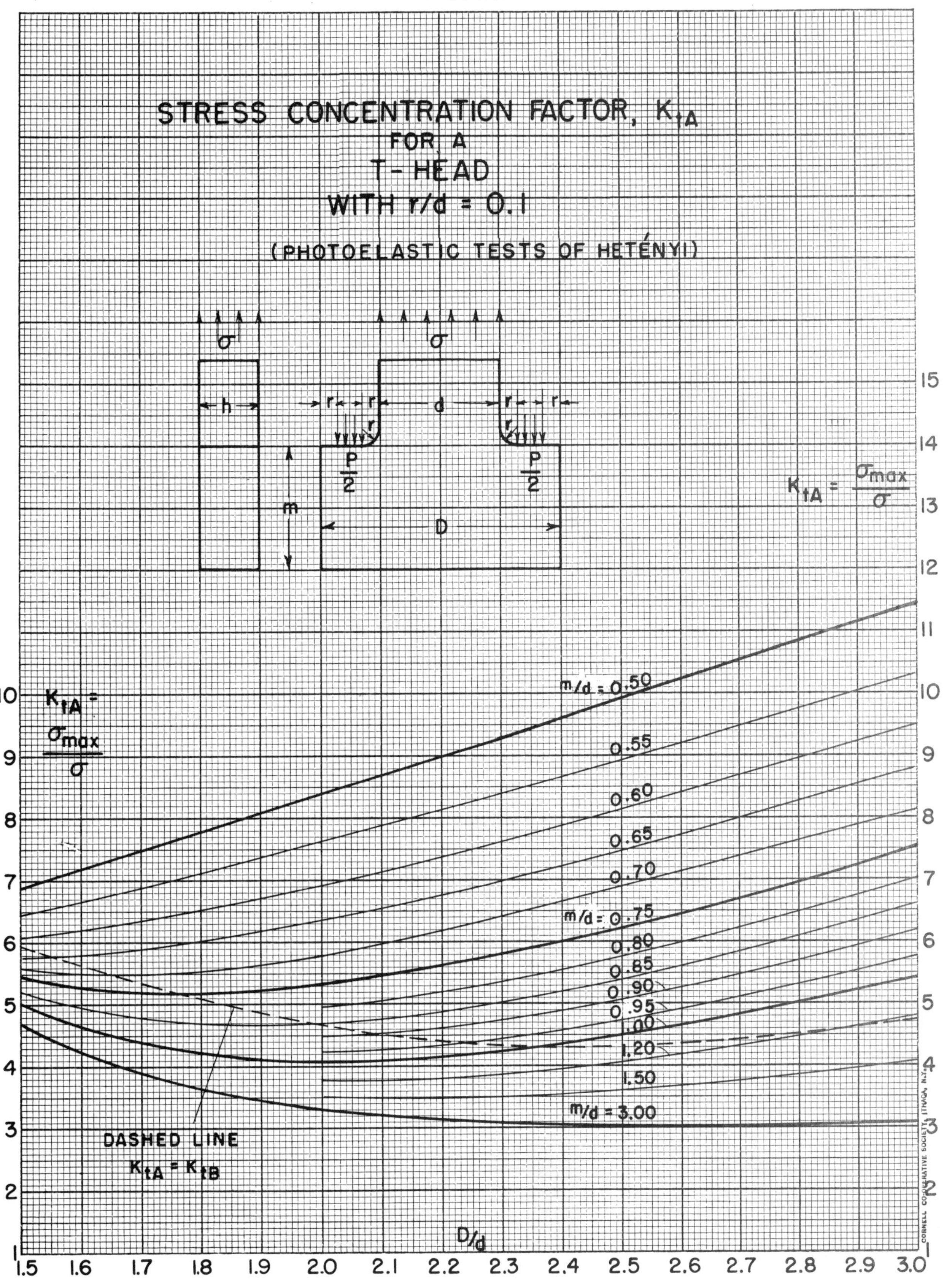
STRESS CONCENTRATION FACTOR, K_{tA}
FOR A
T-HEAD
WITH r/d = 0.1
(PHOTOELASTIC TESTS OF HETÉNYI)
σ
h
σ
r
r
d
r
r
P/2
P/2
m
D
$K_{tA} = \frac{\sigma_{max}}{\sigma}$
m/d = 0.50
0.55
0.60
0.65
0.70
m/d = 0.75
0.80
0.85
0.90
0.95
1.00
1.20
1.50
m/d = 3.00
DASHED LINE
$K_{tA} = K_{tB}$
D/d
1.5
1.6
1.7
1.8
1.9
2.0
2.1
2.2
2.3
2.4
2.5
2.6
2.7
2.8
2.9
3.0

FIG. 193

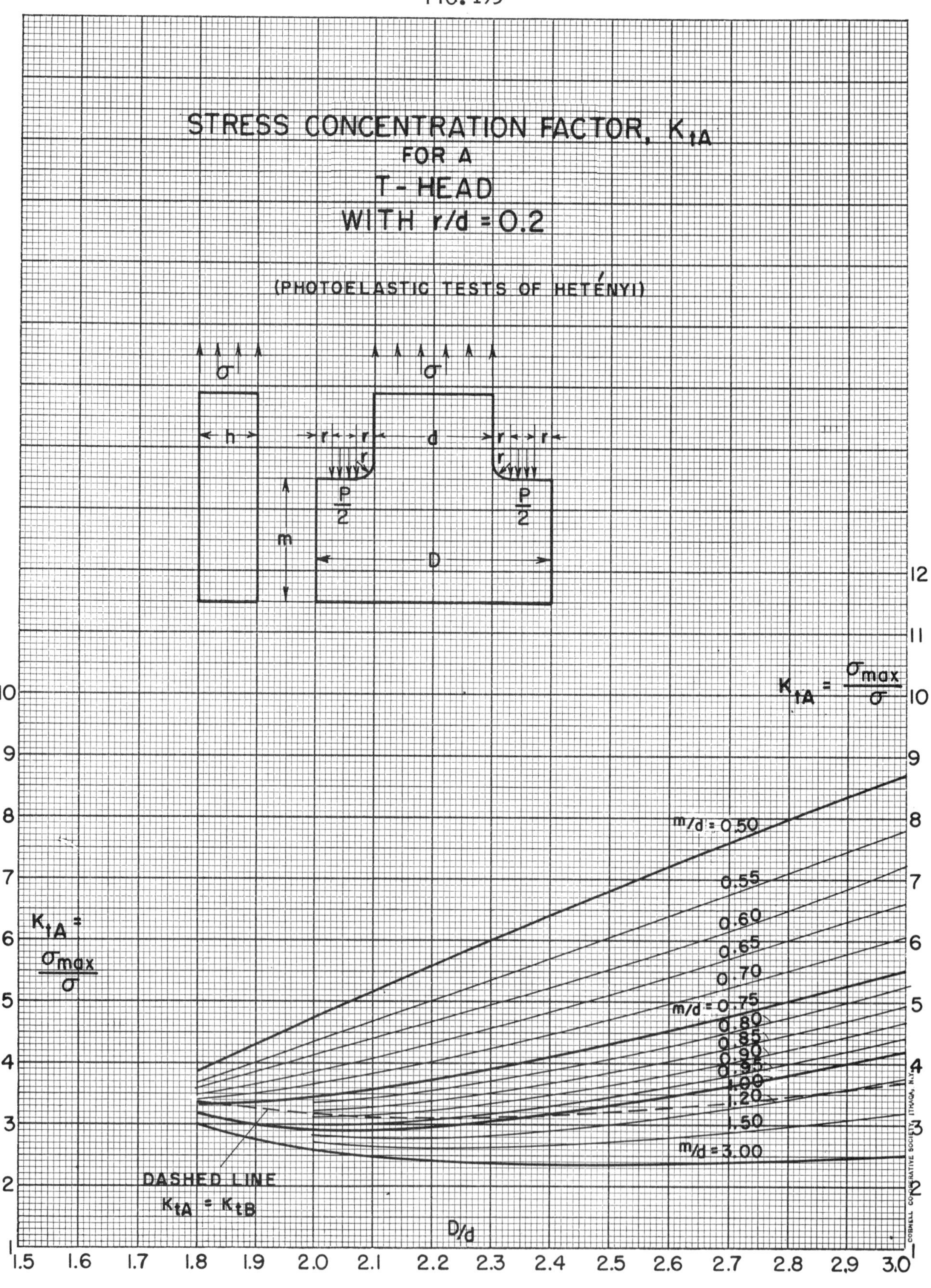

STRESS CONCENTRATION FACTOR, K_{tA}
FOR A
T - HEAD
WITH r/d = 0.2
(PHOTOELASTIC TESTS OF HETÉNYI)
σ
h
r
d
P/2
m
D
$K_{tA} = \frac{\sigma_{max}}{\sigma}$
m/d = 0.50
0.55
0.60
0.65
0.70
m/d = 0.75
0.80
0.85
0.90
0.95
1.00
1.20
1.50
m/d = 3.00
DASHED LINE
$K_{tA} = K_{tB}$
D/d
1.5
1.6
1.7
1.8
1.9
2.0
2.1
2.2
2.3
2.4
2.5
2.6
2.7
2.8
2.9
3.0

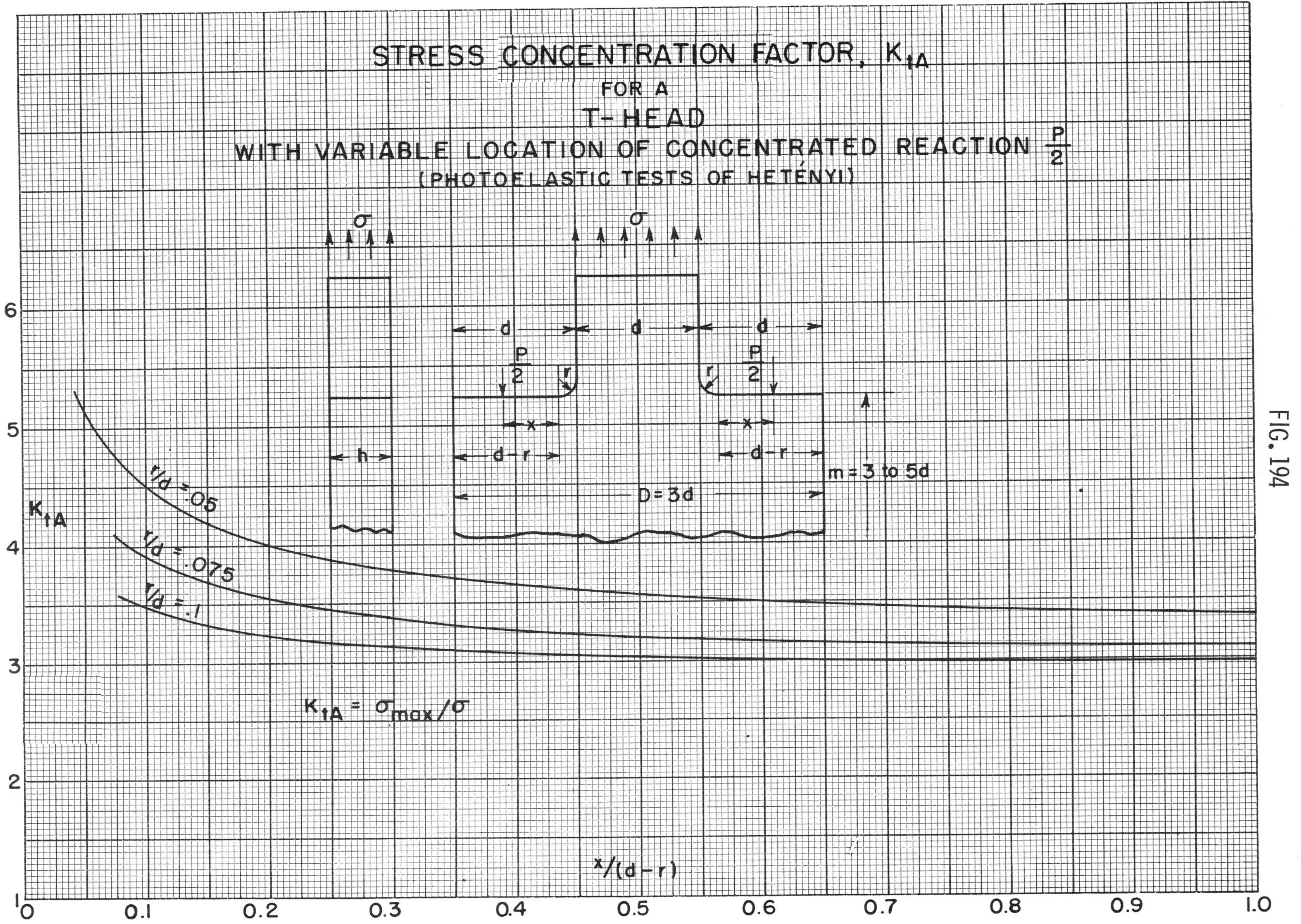

STRESS CONCENTRATION FACTOR, K_{tA}
FOR A
T-HEAD
WITH VARIABLE LOCATION OF CONCENTRATED REACTION $\frac{P}{2}$
(PHOTOELASTIC TESTS OF HETÉNYI)
σ
σ
h
d
d
d
$\frac{P}{2}$
$\frac{P}{2}$
r
r
x
x
d-r
d-r
D=3d
m=3 to 5d
r/d = .05
r/d = .075
r/d = .1
K_{tA}
$K_{tA} = \sigma_{max}/\sigma$
6
5
4
3
2
1
0
0.1
0.2
0.3
0.4
0.5
0.6
0.7
0.8
0.9
1.0
x/(d-r)

FIG. 194

FIG. 195

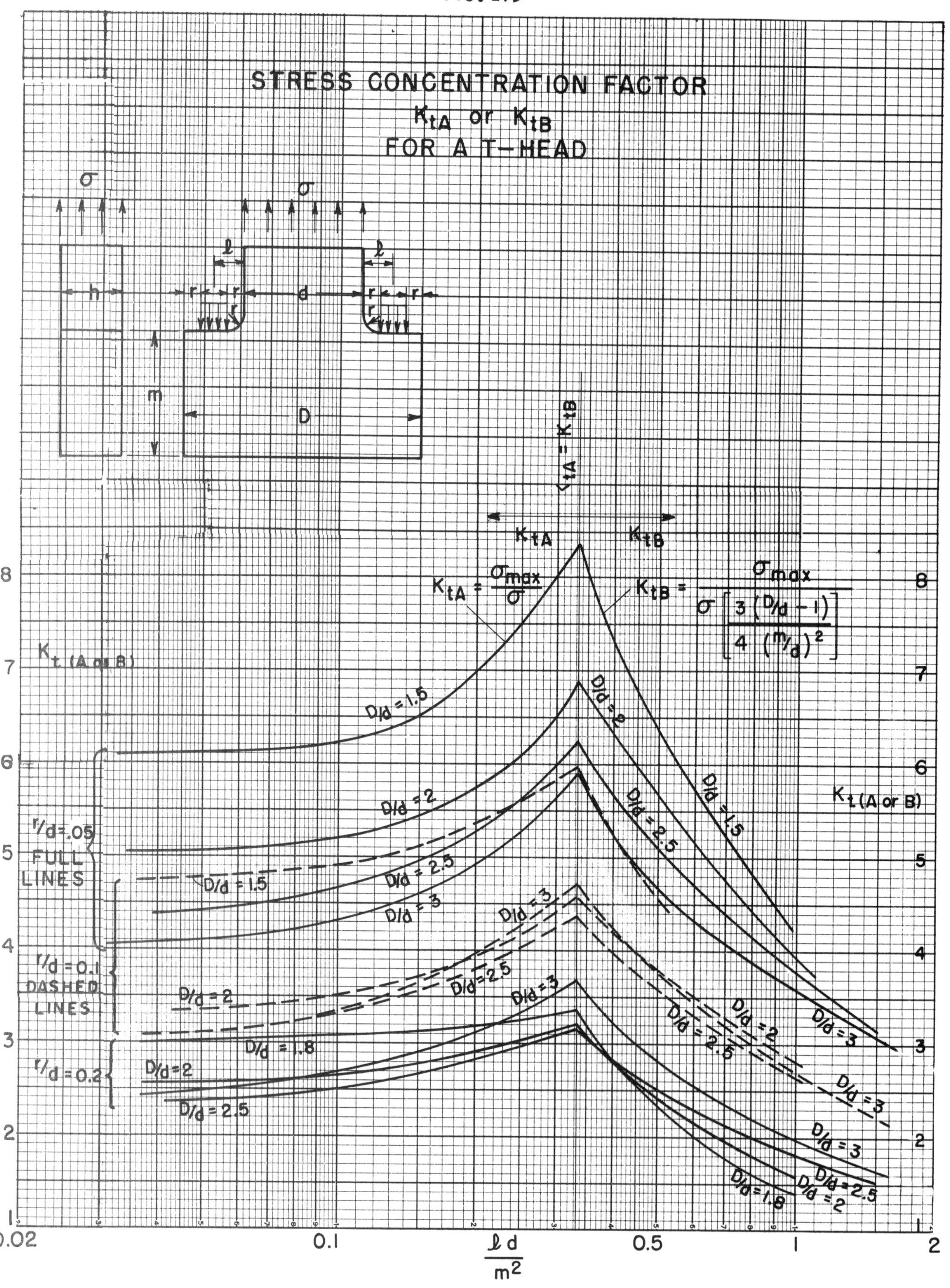
STRESS CONCENTRATION FACTOR
K_{tA} or K_{tB}
FOR A T-HEAD
σ
h
ℓ
r
d
m
D
$K_{tA} = K_{tB}$
K_{tA}
K_{tB}
$K_{tA} = \frac{\sigma_{max}}{\sigma}$
$K_{tB} = \frac{\sigma_{max}}{\sigma\left[\frac{3(D/d - 1)}{4(m/d)^2}\right]}$
$K_{t\,(A\ or\ B)}$
r/d = .05 FULL LINES
r/d = 0.1 DASHED LINES
r/d = 0.2
D/d = 1.5
D/d = 2
D/d = 2.5
D/d = 3
D/d = 1.8
$\frac{\ell d}{m^2}$
0.02
0.1
0.5
1
2

FIG. 196

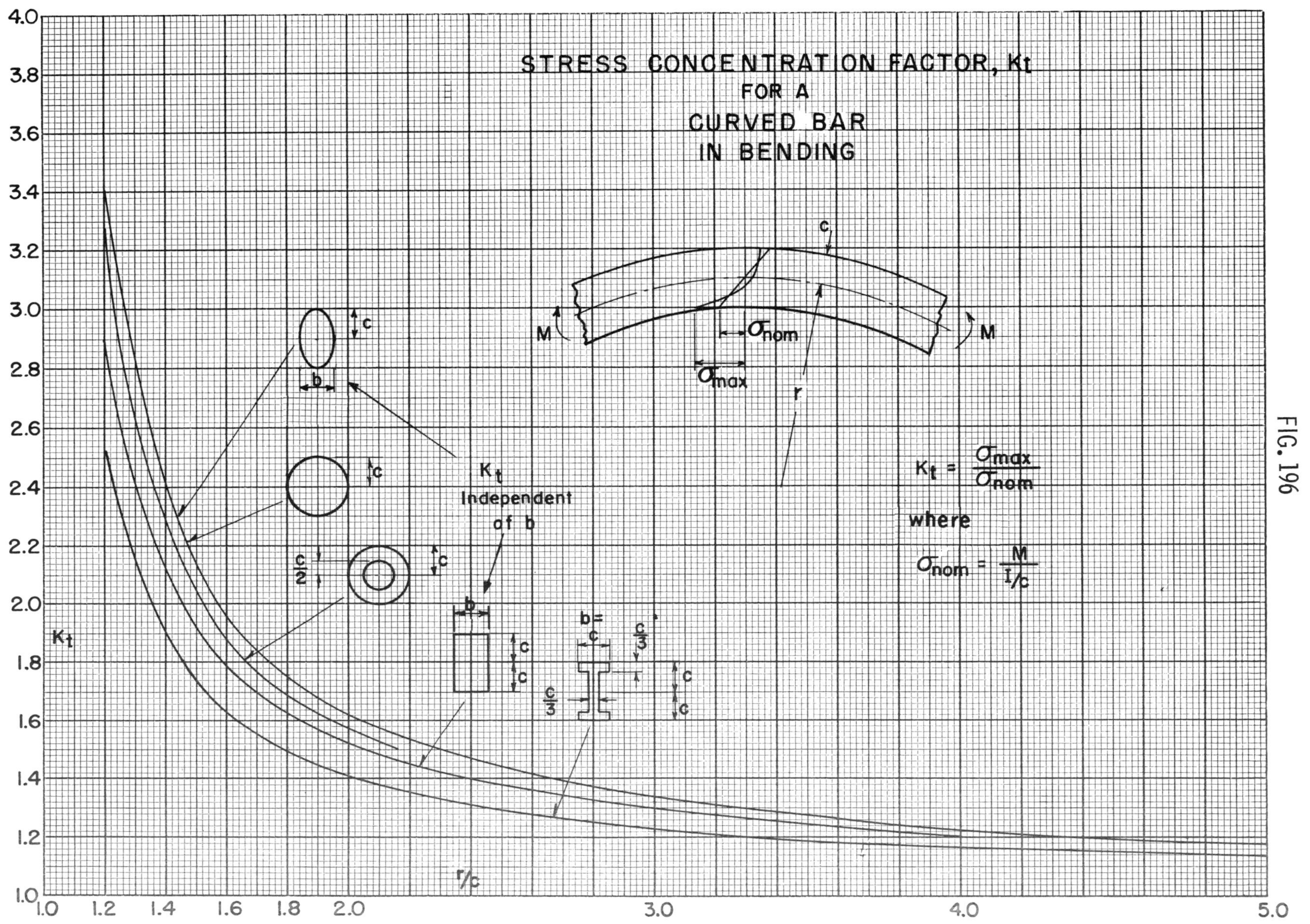

STRESS CONCENTRATION FACTOR, K_t
FOR A
CURVED BAR
IN BENDING
$K_t = \frac{\sigma_{max}}{\sigma_{nom}}$
where
$\sigma_{nom} = \frac{M}{I/c}$
K_t Independent of b
σ_{nom}
σ_{max}
M
c
r
b
b=c
$\frac{c}{3}$
$\frac{c}{2}$
K_t
r/c
1.0
1.2
1.4
1.6
1.8
2.0
2.2
2.4
2.6
2.8
3.0
3.2
3.4
3.6
3.8
4.0
5.0

FIG. 197

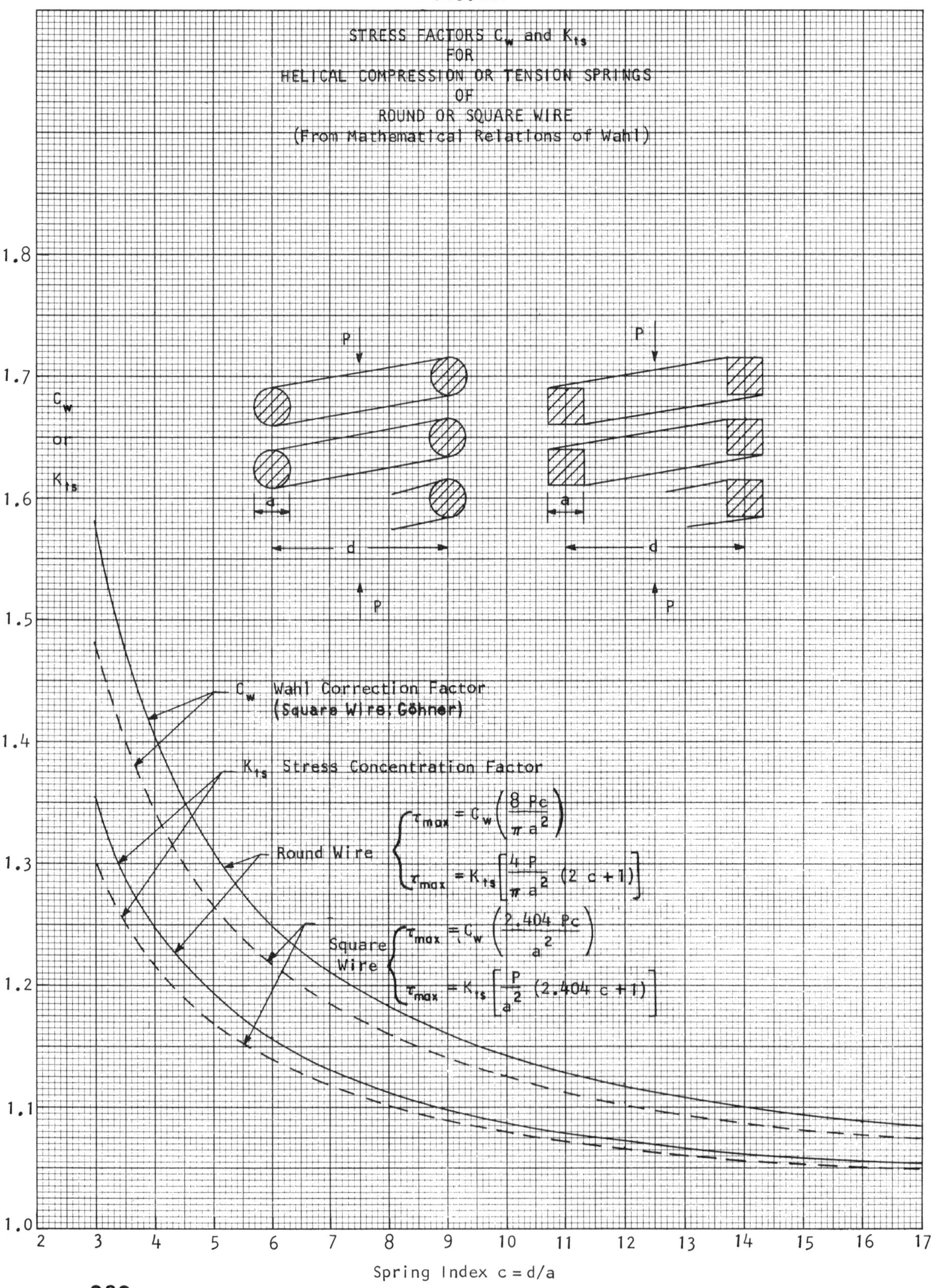
STRESS FACTORS C_w and K_{ts}
FOR
HELICAL COMPRESSION OR TENSION SPRINGS
OF
ROUND OR SQUARE WIRE
(From Mathematical Relations of Wahl)
C_w or K_{ts}
P
a
d
C_w Wahl Correction Factor
(Square Wire; Göhner)
K_{ts} Stress Concentration Factor
Round Wire
$\tau_{max} = C_w\left(\frac{8\ Pc}{\pi\ a^2}\right)$
$\tau_{max} = K_{ts}\left[\frac{4\ P}{\pi\ a^2}\ (2\ c+1)\right]$
Square Wire
$\tau_{max} = C_w\left(\frac{2.404\ Pc}{a^2}\right)$
$\tau_{max} = K_{ts}\left[\frac{P}{a^2}\ (2.404\ c+1)\right]$
1.0
1.1
1.2
1.3
1.4
1.5
1.6
1.7
1.8
2
3
4
5
6
7
8
9
10
11
12
13
14
15
16
17
Spring Index c = d/a

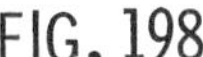

FIG. 198

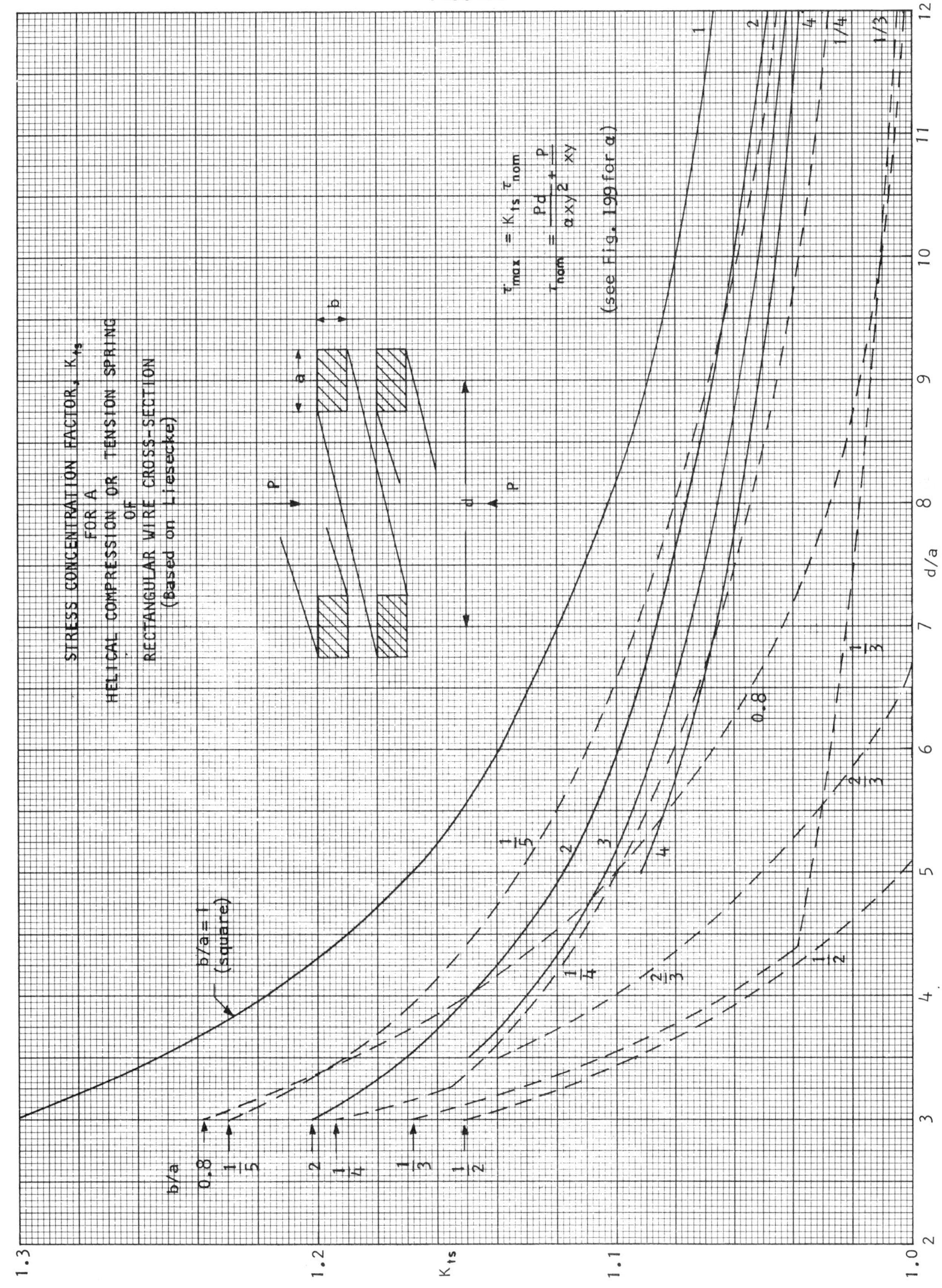
STRESS CONCENTRATION FACTOR, K_{ts}
FOR A
HELICAL COMPRESSION OR TENSION SPRING
OF
RECTANGULAR WIRE CROSS-SECTION
(Based on Liesecke)
$\tau_{max} = K_{ts}\,\tau_{nom}$
$\tau_{nom} = \dfrac{Pd}{\alpha x y^2} + \dfrac{P}{xy}$
(see Fig. 199 for α)
b/a = 1 (square)
b/a
0.8
1/5
2
1/4
1/3
1/2
2/3
3
4
P
d
a
b
d/a
K_{ts}
1.3
1.2
1.1
1.0
2
3
4
5
6
7
8
9
10
11
12

FIG. 199

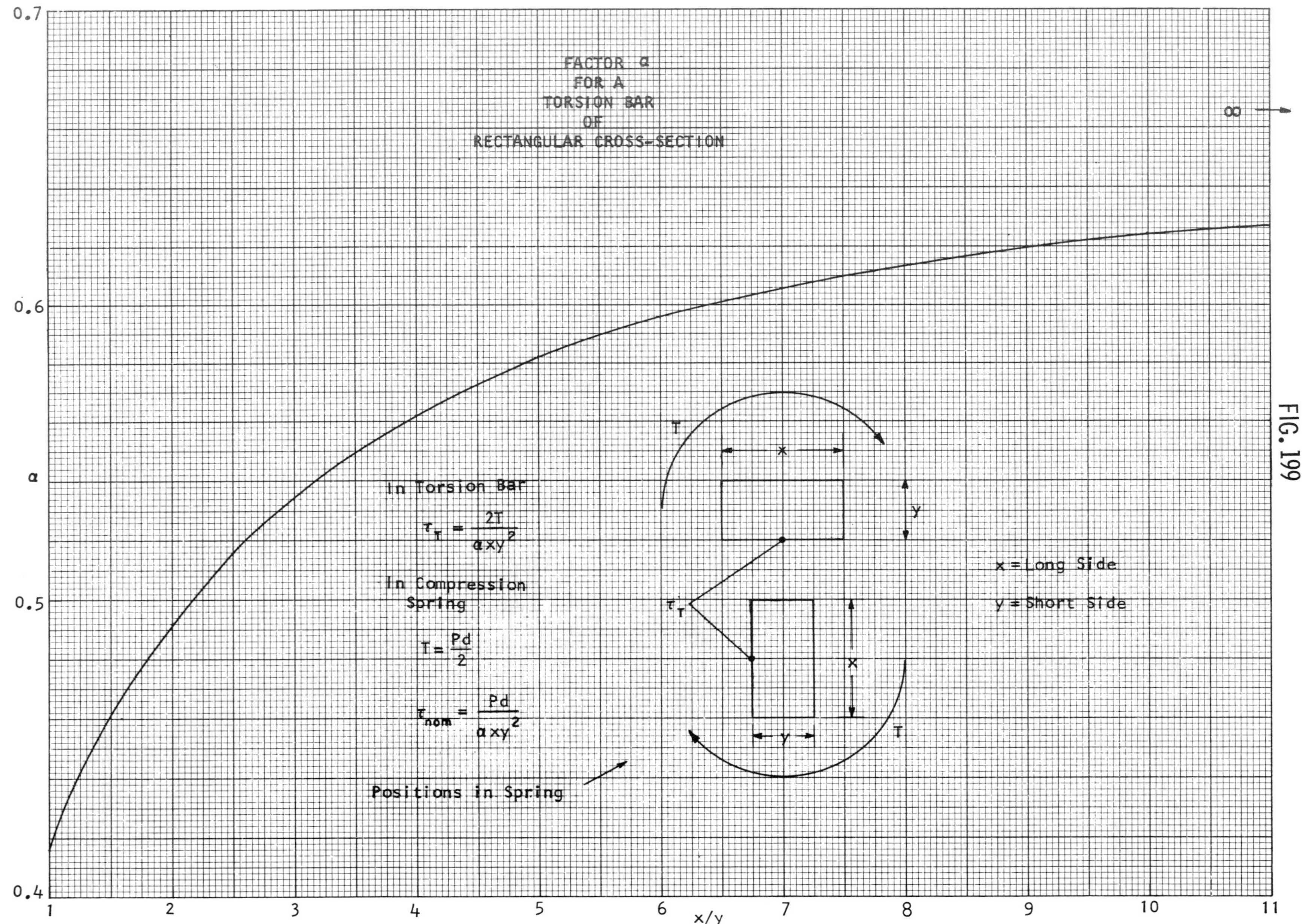

FACTOR α
FOR A
TORSION BAR
OF
RECTANGULAR CROSS-SECTION
In Torsion Bar
$\tau_T = \frac{2T}{\alpha x y^2}$
In Compression Spring
$T = \frac{Pd}{2}$
$\tau_{nom} = \frac{Pd}{\alpha x y^2}$
Positions in Spring
x = Long Side
y = Short Side
x
y
T
τ_T
∞
α
x/y
0.7
0.6
0.5
0.4
1
2
3
4
5
6
7
8
9
10
11

FIG. 200

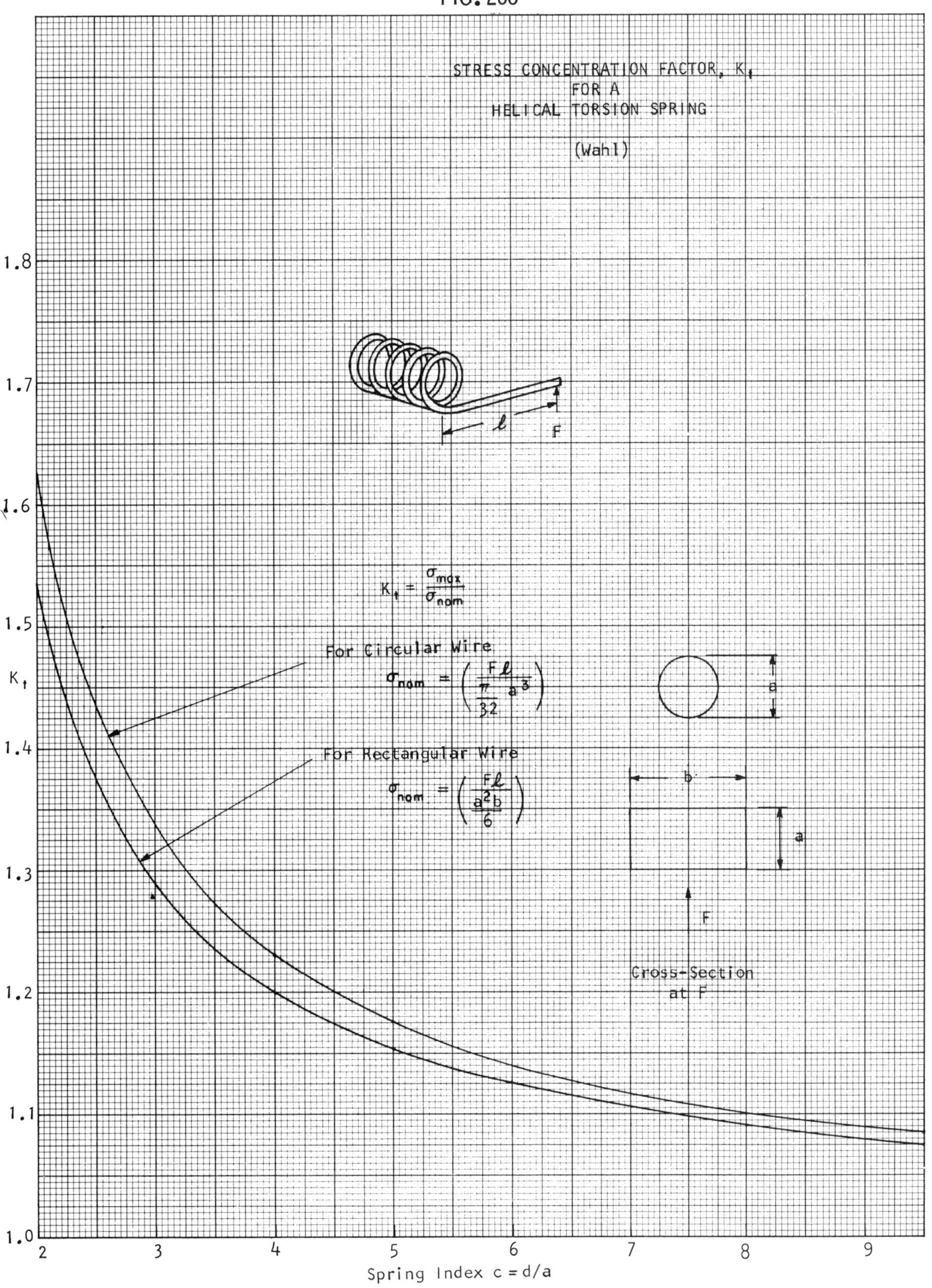
STRESS CONCENTRATION FACTOR, K_t
FOR A
HELICAL TORSION SPRING
(Wahl)
ℓ
F
$K_t = \frac{\sigma_{max}}{\sigma_{nom}}$
For Circular Wire
$\sigma_{nom} = \left(\frac{F\ell}{\frac{\pi}{32} a^3}\right)$
For Rectangular Wire
$\sigma_{nom} = \left(\frac{F\ell}{\frac{a^2 b}{6}}\right)$
d
b
a
F
Cross-Section
at F
1.8
1.7
1.6
1.5
K_t
1.4
1.3
1.2
1.1
1.0
2
3
4
5
6
7
8
9
Spring Index c = d/a

FIG. 201

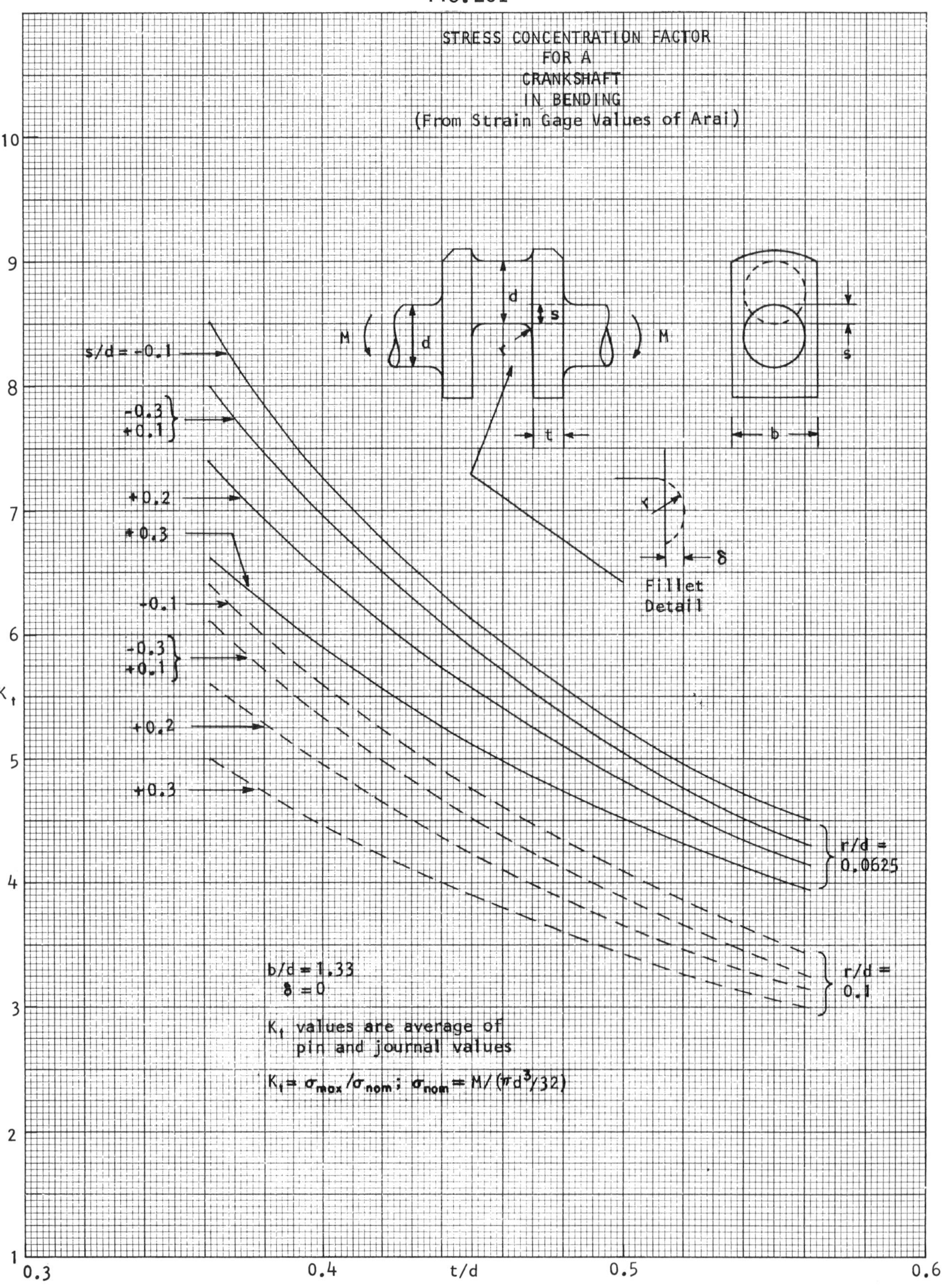

STRESS CONCENTRATION FACTOR
FOR A
CRANKSHAFT
IN BENDING
(From Strain Gage Values of Arai)
s/d = -0.1
-0.3
+0.1
+0.2
+0.3
-0.1
-0.3
+0.1
+0.2
+0.3
M
d
d
s
M
t
b
s
r
δ
Fillet Detail
r/d = 0.0625
r/d = 0.1
b/d = 1.33
δ = 0
K_t values are average of pin and journal values
$K_t = \sigma_{max}/\sigma_{nom}$; $\sigma_{nom} = M/(\pi d^3/32)$
K_t
t/d
0.3
0.4
0.5
0.6
1
2
3
4
5
6
7
8
9
10

FIG. 202

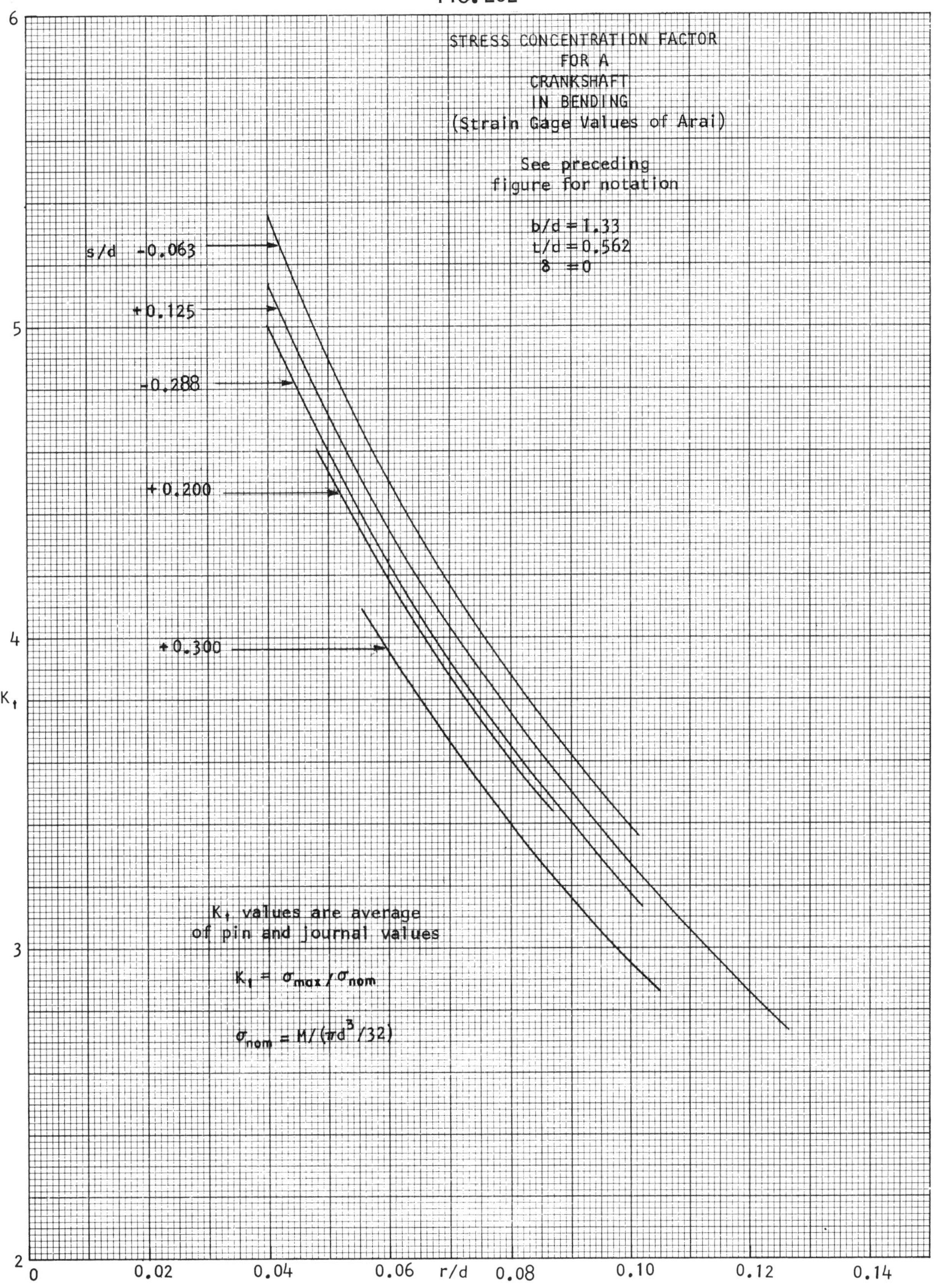
STRESS CONCENTRATION FACTOR
FOR A
CRANKSHAFT
IN BENDING
(Strain Gage Values of Arai)
See preceding
figure for notation
b/d = 1.33
t/d = 0.562
δ = 0
s/d -0.063
+0.125
-0.288
+0.200
+0.300
Kt values are average
of pin and journal values
Kt = σmax / σnom
σnom = M/(πd³/32)
Kt
6
5
4
3
2
0
0.02
0.04
0.06
r/d
0.08
0.10
0.12
0.14

FIG. 203

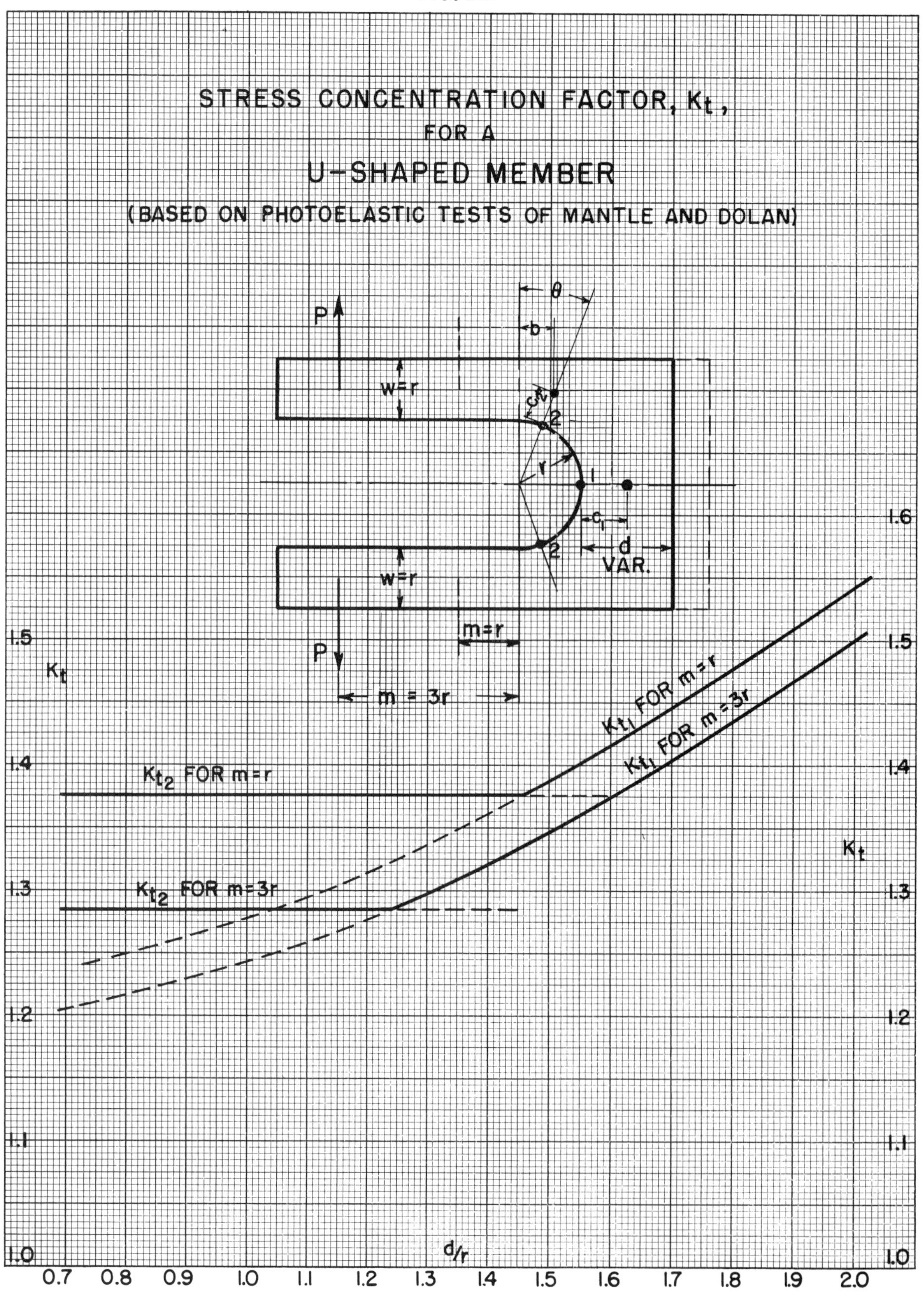

STRESS CONCENTRATION FACTOR, K_t,
FOR A
U-SHAPED MEMBER
(BASED ON PHOTOELASTIC TESTS OF MANTLE AND DOLAN)
P
w=r
θ
b
2
r
1
c_1
d
VAR.
w=r
P
m=r
m = 3r
K_{t_1} FOR m=r
K_{t_1} FOR m=3r
K_{t_2} FOR m=r
K_{t_2} FOR m=3r
K_t
d/r
1.0
1.1
1.2
1.3
1.4
1.5
1.6
0.7
0.8
0.9
1.0
1.1
1.2
1.3
1.4
1.5
1.6
1.7
1.8
1.9
2.0

FIG. 204

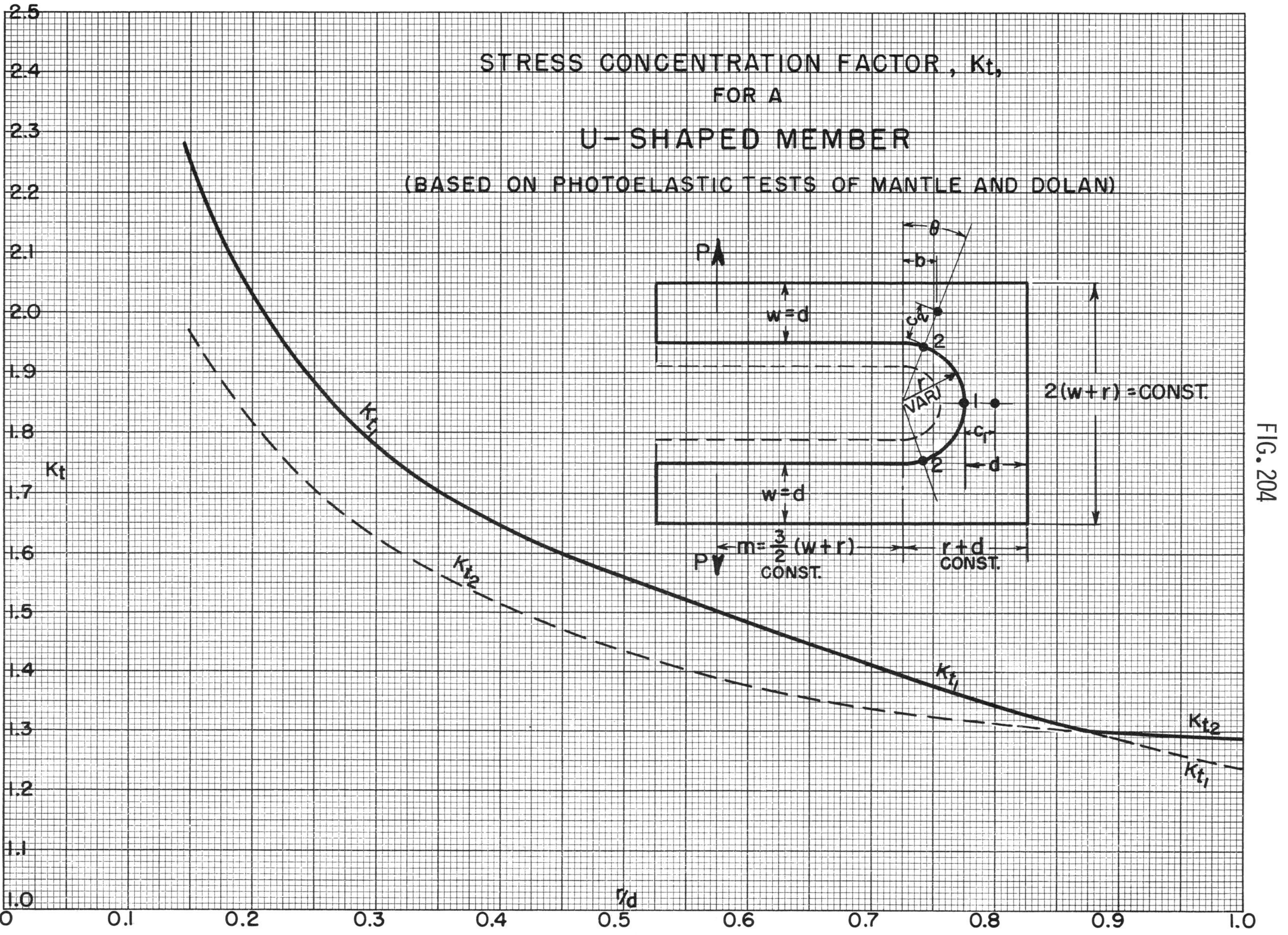

STRESS CONCENTRATION FACTOR, Kt,
FOR A
U-SHAPED MEMBER
(BASED ON PHOTOELASTIC TESTS OF MANTLE AND DOLAN)
P
w=d
w=d
P
θ
b
c
2
2
1
r
VAR.
c
d
2(w+r) = CONST.
m = 3/2 (w+r)
CONST.
r+d
CONST.
Kt1
Kt2
Kt1
Kt2
Kt
r/d
2.5
2.4
2.3
2.2
2.1
2.0
1.9
1.8
1.7
1.6
1.5
1.4
1.3
1.2
1.1
1.0
0
0.1
0.2
0.3
0.4
0.5
0.6
0.7
0.8
0.9
1.0

FIG. 205

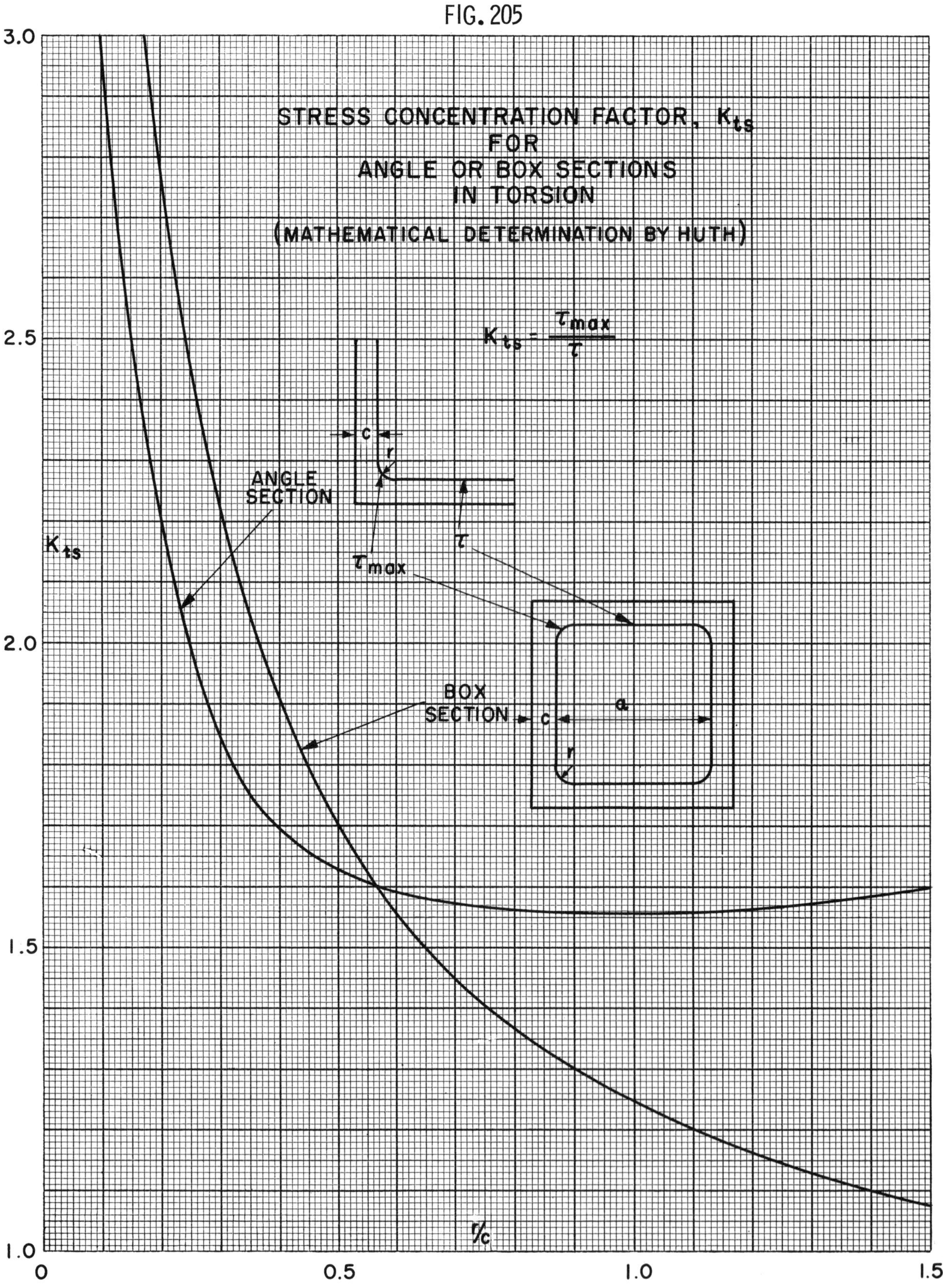

STRESS CONCENTRATION FACTOR, K_{ts}
FOR
ANGLE OR BOX SECTIONS
IN TORSION
(MATHEMATICAL DETERMINATION BY HUTH)
$K_{ts} = \frac{\tau_{max}}{\tau}$
ANGLE SECTION
BOX SECTION
c
r
τ
τ_{max}
a
K_{ts}
r/c
3.0
2.5
2.0
1.5
1.0
0
0.5
1.0
1.5

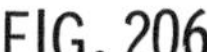

FIG. 206

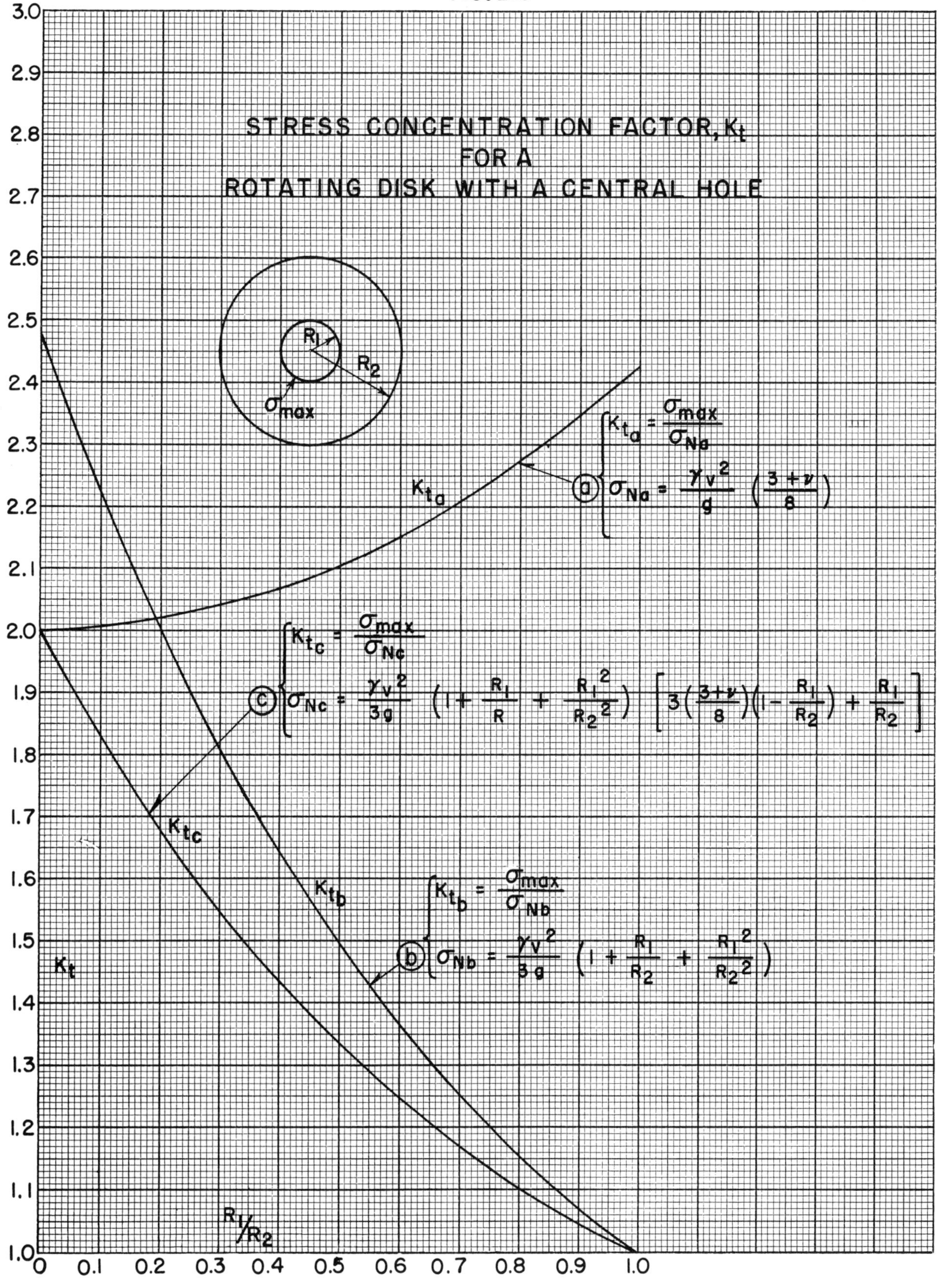

STRESS CONCENTRATION FACTOR, K_t FOR A ROTATING DISK WITH A CENTRAL HOLE

R_1, R_2, σ_{max}

(a) $K_{t_a} = \dfrac{\sigma_{max}}{\sigma_{Na}}$, $\quad \sigma_{Na} = \dfrac{\gamma v^2}{g}\left(\dfrac{3+\nu}{8}\right)$

(b) $K_{t_b} = \dfrac{\sigma_{max}}{\sigma_{Nb}}$, $\quad \sigma_{Nb} = \dfrac{\gamma v^2}{3g}\left(1 + \dfrac{R_1}{R_2} + \dfrac{R_1^2}{R_2^2}\right)$

(c) $K_{t_c} = \dfrac{\sigma_{max}}{\sigma_{Nc}}$, $\quad \sigma_{Nc} = \dfrac{\gamma v^2}{3g}\left(1 + \dfrac{R_1}{R} + \dfrac{R_1^2}{R_2^2}\right)\left[3\left(\dfrac{3+\nu}{8}\right)\left(1 - \dfrac{R_1}{R_2}\right) + \dfrac{R_1}{R_2}\right]$

K_{t_a}, K_{t_b}, K_{t_c}

K_t: 1.0, 1.1, 1.2, 1.3, 1.4, 1.5, 1.6, 1.7, 1.8, 1.9, 2.0, 2.1, 2.2, 2.3, 2.4, 2.5, 2.6, 2.7, 2.8, 2.9, 3.0

R_1/R_2: 0, 0.1, 0.2, 0.3, 0.4, 0.5, 0.6, 0.7, 0.8, 0.9, 1.0

STRESS CONCENTRATION FACTOR, K_t
FOR A
ROTATING DISK WITH A NON-CENTRAL HOLE
(PHOTOELASTIC TESTS OF
BARNHART, HALE AND MERIAM)
$\frac{R_1}{R_2} = 0.1$
$\frac{R_1}{R_2} = 0.04$
K_t
R_2
R_1
A
R_0
$K_t = \frac{\sigma_{max_A}}{\sigma}$
WHERE σ = TANGENTIAL STRESS IN A SOLID DISK
AT RADIUS OF POINT A
R_0/R_2
3.0
2.9
2.8
2.7
2.6
2.5
2.4
2.3
2.2
2.1
2.0
1.9
1.8
1.7
1.6
1.5
1.4
1.3
1.2
1.1
1.0
0
0.10
0.20
0.30
0.40
0.50
0.60
0.70

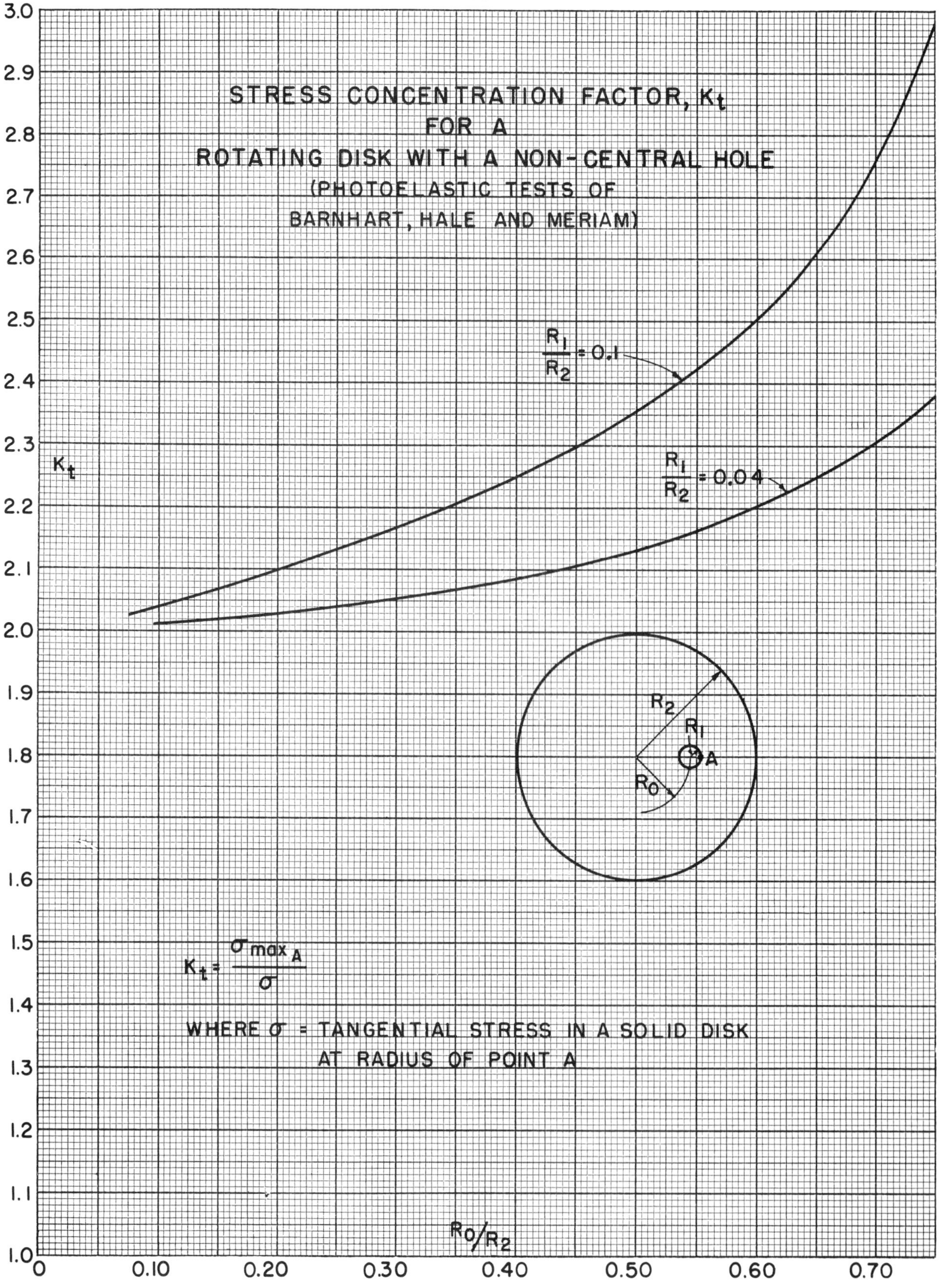

FIG. 207

FIG. 208

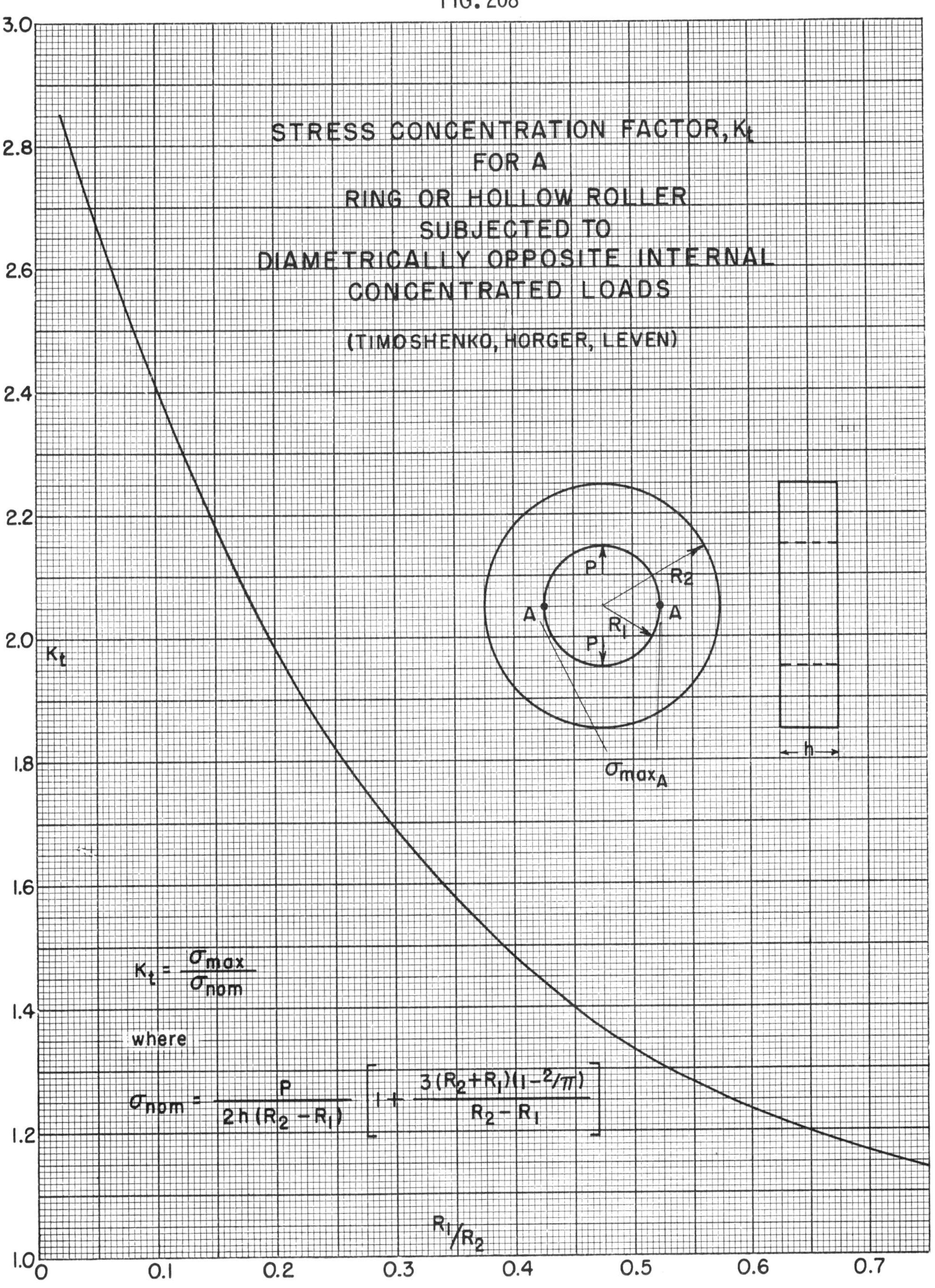

STRESS CONCENTRATION FACTOR, K_t
FOR A
RING OR HOLLOW ROLLER
SUBJECTED TO
DIAMETRICALLY OPPOSITE INTERNAL
CONCENTRATED LOADS
(TIMOSHENKO, HORGER, LEVEN)
K_t
P
P
A
A
R_1
R_2
σ_{max_A}
h
$K_t = \frac{\sigma_{max}}{\sigma_{nom}}$
where
$\sigma_{nom} = \frac{P}{2h(R_2 - R_1)}\left[1 + \frac{3(R_2+R_1)(1-2/\pi)}{R_2 - R_1}\right]$
R_1/R_2
3.0
2.8
2.6
2.4
2.2
2.0
1.8
1.6
1.4
1.2
1.0
0
0.1
0.2
0.3
0.4
0.5
0.6
0.7

FIG. 209

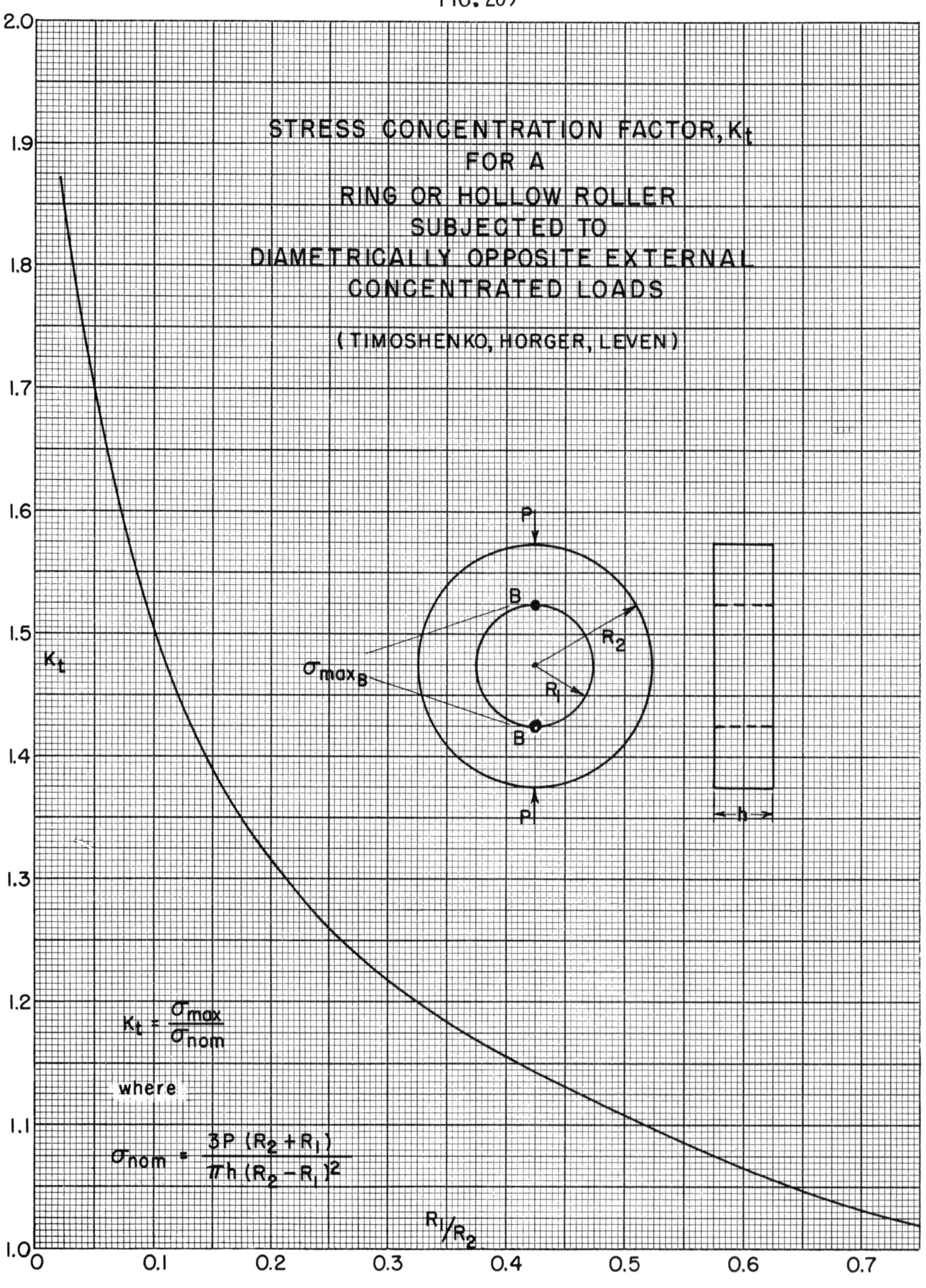

STRESS CONCENTRATION FACTOR, K_t
FOR A
RING OR HOLLOW ROLLER
SUBJECTED TO
DIAMETRICALLY OPPOSITE EXTERNAL
CONCENTRATED LOADS
(TIMOSHENKO, HORGER, LEVEN)
K_t
$K_t = \frac{\sigma_{max}}{\sigma_{nom}}$
where
$\sigma_{nom} = \frac{3P(R_2 + R_1)}{\pi h (R_2 - R_1)^2}$
R_1/R_2
P
B
σ_{max_B}
R_2
R_1
h
2.0
1.9
1.8
1.7
1.6
1.5
1.4
1.3
1.2
1.1
1.0
0
0.1
0.2
0.3
0.4
0.5
0.6
0.7

FIG. 210

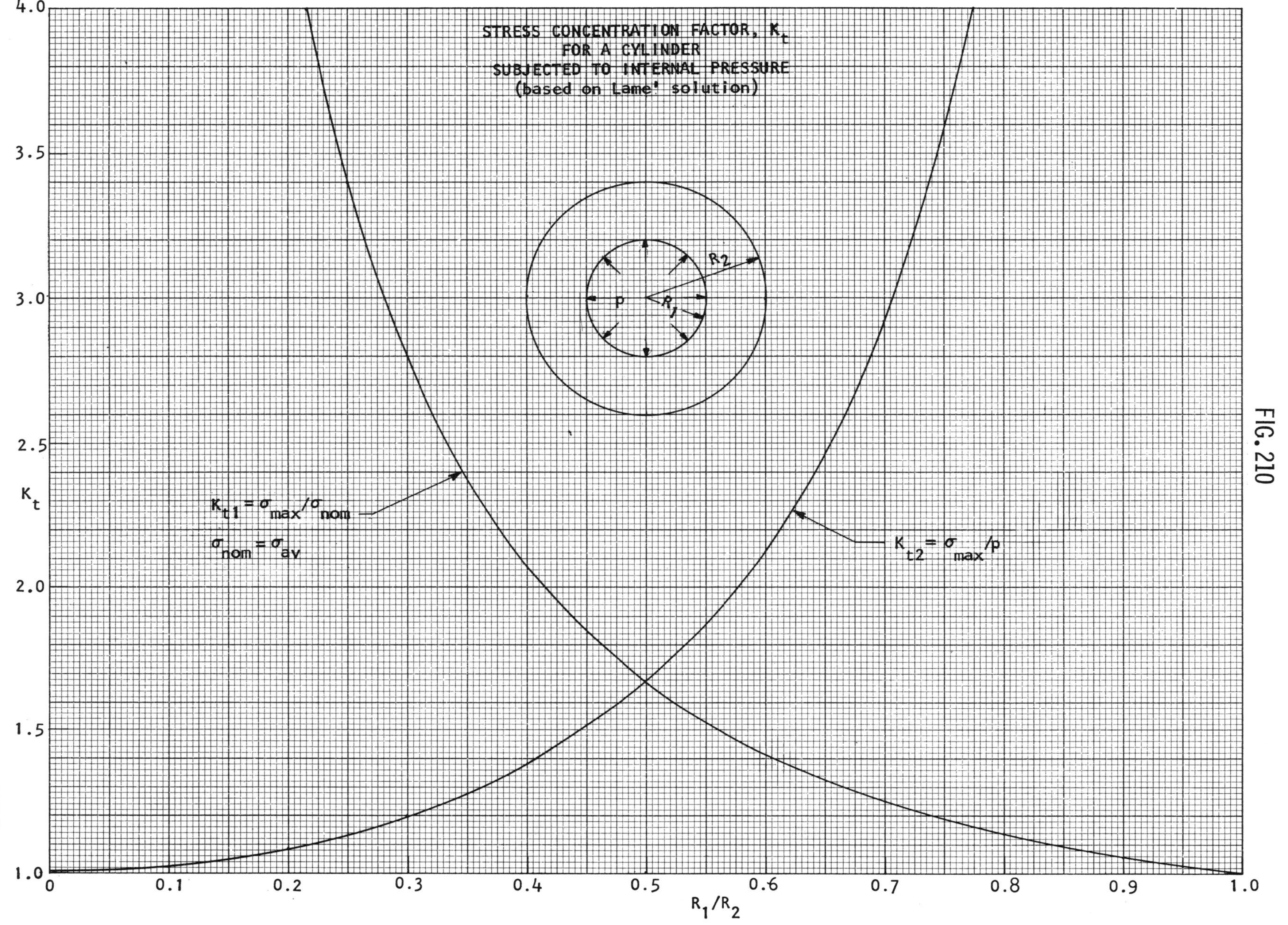

STRESS CONCENTRATION FACTOR, K_t
FOR A CYLINDER
SUBJECTED TO INTERNAL PRESSURE
(based on Lame' solution)
R_2
R_1
p
K_t1 = σ_max/σ_nom
σ_nom = σ_av
K_t2 = σ_max/p
K_t
4.0
3.5
3.0
2.5
2.0
1.5
1.0
0
0.1
0.2
0.3
0.4
0.5
0.6
0.7
0.8
0.9
1.0
R_1/R_2

FIG. 211

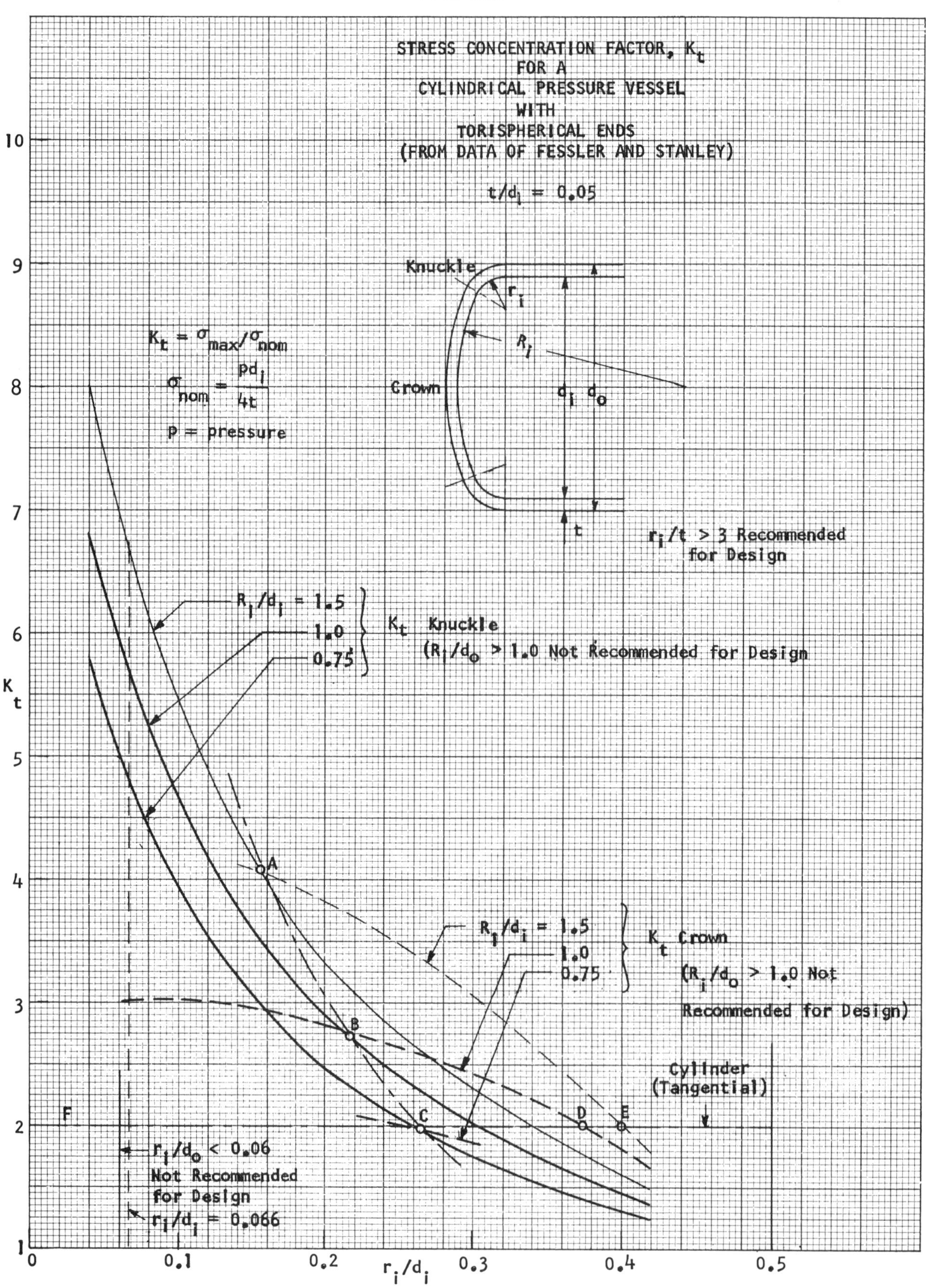
STRESS CONCENTRATION FACTOR, K_t
FOR A
CYLINDRICAL PRESSURE VESSEL
WITH
TORISPHERICAL ENDS
(FROM DATA OF FESSLER AND STANLEY)
$t/d_i = 0.05$
Knuckle
r_i
R_i
Crown
d_i d_o
t
$r_i/t > 3$ Recommended for Design
$K_t = \sigma_{max}/\sigma_{nom}$
$\sigma_{nom} = \frac{pd_i}{4t}$
P = pressure
$R_i/d_i = 1.5$
1.0
0.75
K_t Knuckle
($R_i/d_o > 1.0$ Not Recommended for Design
$R_i/d_i = 1.5$
1.0
0.75
K_t Crown
($R_i/d_o > 1.0$ Not
Recommended for Design)
Cylinder
(Tangential)
A
B
C
D
E
F
$r_i/d_o < 0.06$
Not Recommended
for Design
$r_i/d_i = 0.066$
K_t
r_i/d_i
10
9
8
7
6
5
4
3
2
1
0
0.1
0.2
0.3
0.4
0.5

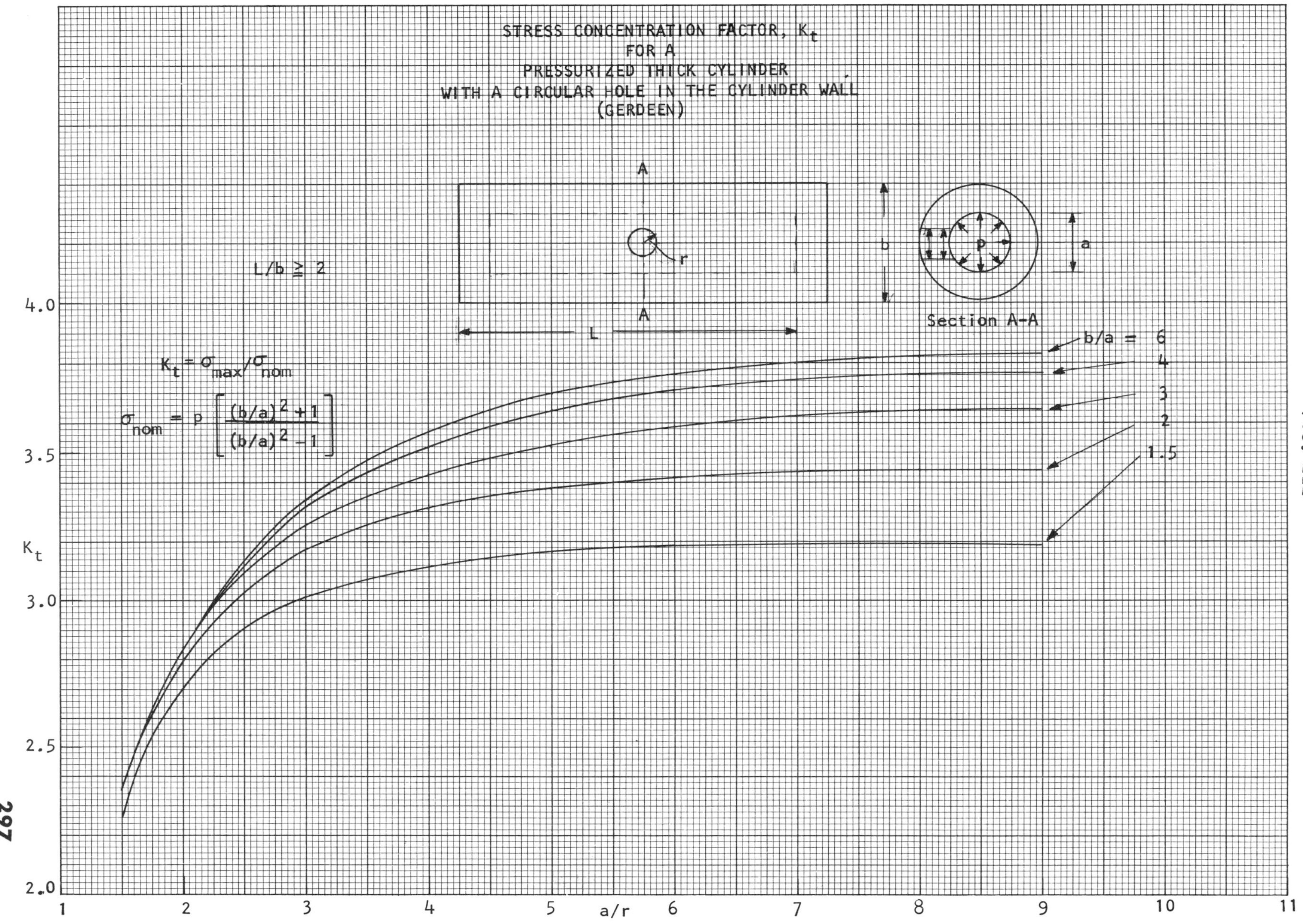

STRESS CONCENTRATION FACTOR, K_t
FOR A
PRESSURIZED THICK CYLINDER
WITH A CIRCULAR HOLE IN THE CYLINDER WALL
(GERDEEN)
A
A
r
L
b
a
p
Section A-A
L/b ≥ 2
$K_t = \sigma_{max}/\sigma_{nom}$
$\sigma_{nom} = p\left[\frac{(b/a)^2 + 1}{(b/a)^2 - 1}\right]$
b/a = 6
4
3
2
1.5
K_t
4.0
3.5
3.0
2.5
2.0
a/r
1
2
3
4
5
6
7
8
9
10
11

FIG. 212

REFERENCES

1. R. E. Peterson, *Stress Concentration Design Factors,* Wiley, New York (1953).
2. *Proc. Intern. Conference on Fatigue of Metals,* Inst. Mech. Eng., London (1956): G. A. Cottell, "Lessons to be Learnt from Failures in Service," p. 563; R. Cazaud, "Fatigue Failure and Service. Experience with Particular Reference to the Shape of the Part," p. 581; T. W. Bunyan, "Service Fatigue Failures in Marine Machinery," p. 713.
3. H. J. Grover, *Fatigue of Aircraft Structures,* NAVAIR 01-1A-13, U.S. Govt. Printing Office, Washington, D.C. (1966).

3a. British Engine Technical Reports, British Engine Boiler and Electrical Insurance Co., Ltd., Longridge House, Manchester 4, England.

4. Oliver Wendell Holmes, "The Deacon's Masterpiece; or the Wonderful One Hoss Shay," *American Poetry,* Scribner, New York (1924), p. 434. Poem originally published in *Atlantic Monthly* (Sept. 1858). The entire poem which is about 1½ pages in length, is most enjoyable.

4a. E. Z. Stowell, "Stress and Strain Concentration at a Circular Hole in an Infinite Plate," Tech. Note 2073, Nat. Advisory Comm. for Aeronautics (now NASA) (1950). See also R. E. Peterson, "Fatigue of Metals in Engineering and Design," Edgar Marburg Lecture (separate publication), ASTM, Philadelphia (1962), Fig. 23; also in *Materials Research and Standards,* ASTM, Vol. 3 (1963), Fig. 23.

5. *Annual Book of Standards, Part 31,* ASTM, Philadelphia, Pa. (1970). Standard E206-66, Definition of Terms Relating to Fatigue Testing and the Statistical Analysis of Fatigue Data, Items 16, 18, 19, 20, p. 572
6. M. Hetényi, Ed. *Handbook of Experimental Stress Analysis,* Wiley, New York (1950).
7. S. Timoshenko, *Strength of Materials,* Part II, 3rd ed., Van Nostrand, Princeton, N.J. (1956), p. 444.
8. C. R. Soderberg, "Working Stresses," Chapter 10 in Ref. 6.
9. C. R. Soderberg, "Working Stresses," *Trans. ASME,* Vol. 52, Part 1 (1930), p. APM 52-2.
10. C. R. Soderberg. Ref. 8, pp. 449–450.
11. J. O. Draffin and W. L. Collins, "Effect of Size and Type of Specimens on the Torsional Properties of Cast Iron," *Proc. ASTM,* Vol. 38 (1938), p. 235.
12. J. Marin, *Engineering Materials,* Prentice-Hall, New York (1952), p. 164.
13. H. J. Gough, "Crystalline Structure in Relation to Failure of Metals," *Proc. ASTM,* Vol. 33, Part 2 (1933), p. 3.
14. R. von Mises, "Mechanik der festen Körper im plastisch deformablen Zustand," *Nachr. Ges. Wiss. Göttingen Jahresber. Geschäftsjahr. Math-phys. Kl.* (1913), p. 582.
15. H. Hencky, "Zur Theorie plastischer Deformationen und der hierdurch im Material hervorgerufenen Nebenspannungen," *Proc. 1st Intern. Congr. Appl. Mech.,* Delft (1924) p. 312. The proposals of both von Mises and Hencky were to a considerable extent anticipated by Huber in 1904, though limited to mean compression and without specifying mode of failure; his paper in the Polish language did not attract international attention until 20 years later.
16. A. Eichinger, "Versuche zur Klärung der Frage der Bruchgefahr," *Proc. 2nd Intern. Congr. Appl. Mech.,* Zurich (1926), p. 325.
17. A. Nadai, "Plastic Behavior of Metals in the Strain Hardening Range," *J. Appl. Phys.,* Vol. 8 (1937), p. 203.

18. G. Sachs, "Zur Ableitung einer Fliessbedingung," *Z. VDI*, Vol. 72 (1928), p. 734.

19. H. L. Cox and D. G. Sopwith, "The Effect of Orientation on Stresses in Single Crystals and of Random Orientation on the Strength of Polycrystalline Aggregates," *Proc. Phys. Soc. (London)*, Vol. 49 (1937), p. 134.

20. R. Beeching, Discussion, *Proc. Inst. Mech. Engrs. (London)*, Vol. 159 (1948), p. 113.

21. R. Hill, *The Mathematical Theory of Plasticity,* Oxford University Press (1950).

22. W. Prager and P. G. Hodge, *Theory of Perfectly Plastic Solids,* Wiley, New York (1951).

23. R. E. Peterson and A. M. Wahl, Closure to discussion of "Two and Three Dimensional Cases of Stress Concentration, and Comparison with Fatigue Tests," *Trans. ASME*, Vol. 58 (1936), Applied Mechanics Section, p. A-149.

24. H. Majors, B. D. Mills, and C. W. MacGregor, "Fatigue under Combined Pulsating Stresses," *Trans. ASME,* Vol. 71 (1949), Applied Mechanics Section, p. 269.

25. W. Sawert, "Verhalten der Baustähle bei wechselnder mechrachsiger Beanspruchung," *Z. VDI*, Vol. 87 (1943), p. 609.

26. J. Marin, Ref. 12, p. 163.

27. P. Ludwik, "Kerb- und Korrosionsdauerfestigkeit," *Metall*, Vol. 10 (1931), p. 705.

28. H. J. Gough and H. V. Pollard, "Strength of Materials under Combined Alternating Stresses," *Proc. Inst. Mech. Engrs. (London)*, Vol. 131 (1935), p. 1, Vol. 132 (1935), p. 549.

29. T. Nisihara and A. Kawamoto, "The Strength of Metals under Combined Alternating Stresses," *Trans. Soc. Mech. Engrs. Japan, Vol. 6, No. 24* (1940), p. S-2.

30. N. M. Newmark, R. J. Mosborg, W. H. Munse, and R. E. Elling, "Fatigue Tests in Axial Compression," *Proc. ASTM*, Vol. 51 (1951), p. 792. See also Discussion.

31. T. Nishihara and K. Kojima, "Diagram of Endurance Limit of Duralumin for Repeated Tension and Compression," *Trans. Soc. Mech. Engrs. Japan*, Vol. 5, No. 20 (1939), p. I-1. See also T. Nishihara and T. Sakurai, "Fatigue Strength for Steel for Repeated Tension and Compression," *Trans. Soc. Mech. Engrs. Japan,* Vol. 5, No. 8 (1939), p. 25.

32. M. Rôs and A. Eichinger, "Die Bruchgefahr fester Körper," *Eidgenöss. Materialprüf. Ber.* 173, Zurich (1950).

33. W. N. Findley, Discussion of "Engineering Steels under Combined Cyclic and Static Stresses," by H. J. Gough, *Trans, ASME*, Vol. 73 (1951), Applied Mechanics Section, p. 211. See also "Fatigue of 76S-T61 Aluminum Alloy under Combined Bending and Torsion," *Proc. ASTM*, Vol. 52 (1952), p. 818.

34. R. E. Peterson, "Stress Concentration Phenomena in Fatigue of Metals," *Trans. ASME*, Vol. 55 (1933), Applied Mechanics Section, p. 157. See Figs. 6, 7, and 8. Also "Model Testing as Applied to Strength of Materials," *Trans. ASME,* Vol. 55 (1933), Applied Mechanics Section, p. 79.

35. R. E. Peterson and A. M. Wahl, "Two and Three Dimensional Cases of Stress Concentration, and Comparison with Fatigue Tests," *Trans. ASME,* Vol. 57, (1936), Applied Mechanics Section, p. A-15. See Figs. 15 and 16.

36. R. E. Peterson, "Application of Stress Concentration Factors in Design," *Proc. SESA*, Vol. 1, No. 1 (1943), p. 118.

37. R. E. Peterson, "Relation between Life Testing and Conventional Tests of Materials," *ASTM Bull.* (March 1945), p. 13. See Figs. 20, 21.

38. Ref. 35, Discussion by Horger and Maulbetsch, p. A-148.

39. H. Neuber, *Kerbspannungslehre,* Springer, Berlin (1937), p. 142. Translation, *Theory of Notch Stresses,* J. W. Edwards Co., Ann Arbor Mich. (1946), p. 155; 2nd ed., *Kerbspannungslehre,* Springer, Berlin (1958), p. 164. Translation, *Theory of Notch Stresses,* Office of Technical Services, Dept. of Commerce, Washington, D.C. (1961), p. 207.

40. R. E. Peterson, "Methods of Correlating Data from Fatigue Tests of Stress Concentration Specimens," *Stephen Timoshenko Anniversary Volume,* Macmillan, New York (1938), p. 179.

40a. M. M. Leven, "Stress Gradients in Grooved Bars and Shafts," *Proc. SESA*, Vol. 13, No. 1 (1955), p. 207.

41. H. A. Von Phillipp, "Einfluss von Querschittsgrösse und Querschittsform auf die Dauerfestigkeit bei ungleichmässig verteilten Spannungen," *Forschung,* Vol. 13 (1942), p. 99.

42. Ref. 37, Fig. 19.

43. R. E. Peterson, "Relation between Stress Analysis and Fatigue of Metals," *Proc. SESA*, Vol. 11, No. 2 (1950), p. 199. See also Ref. 1, p. 9.

44. R. E. Peterson, "Analytical Approach to Stress Concentration Effect in Aircraft Materials," U.S. Air Force-WADC Symposium on Fatigue of Metals, Technical Report 59–507, Dayton, Ohio (1959), p. 273.

45. H. F. Moore and R. L. Jordan, "Stress Concentration in Steel Shafts with Semi-Circular Notches," *Proc. 5th Intern. Congr. Appl. Mech.*, Wiley, New York (1939), p. 188.

46. H. F. Moore and D. Morkovin, "Second Progress Report on the Effect of Size of Specimen on Fatigue Strength of Three Types of Steel," *Proc. ASTM*, Vol. 43 (1943), p. 109.

47. H. J. Gough, "Engineering Steels under Combined Cyclic and Static Stresses," *Proc. Inst. Mech. Engrs.* (*London*), Vol. 160 (1949), p. 417.

48. M. F. Garwood, H. H. Zurburg, and M. A. Erickson, "Correlation of Laboratory Tests and Service Performance," in *Interpretation of Tests and Correlation with Service*, published by Am. Soc. Metals, Cleveland (1951), p. 1.

49. W. C. Brueggeman, M. Mayer, and W. H. Smith, "Axial Fatigue Tests at Zero Mean Stress of 24 S-T Aluminum Alloy Sheet with and without a Circular Hole," *NACA Tech. Note 955* (1944). Now NASA.

50. W. C. Brueggeman and M. Mayer, "Axial Fatigue Tests at Zero Mean Stress of 24 S-T Aluminum Alloy Strips with a Central Circular Hole," *NACA Tech. Note 1611* (1948). Now NASA.

51. H. J. Grover, S. M. Bishop, and L. R. Jackson, "Fatigue Strengths of Aircraft Materials. Axial Load Fatigue Tests on Notched Sheet Specimens of 24 S-T3 and 75 S-T6 Aluminum Alloys and of SAE 4130 Steel with Stress Concentration Factors of 2.0 and 4.0," *NACA TECH. Note 2389* (1951). Now NASA.

52. W. Illg, "Fatigue Tests on Notched and Unnotched Sheet Specimens of 2024–T3 and 7075-T6 Aluminum Alloys and of SAE 4130 Steel with Special Consideration of the Life Range from 2 to 10,000 Cycles," *NACA Tech. Note 3866* (1956). Now NASA.

53. R. R. Moore, "Effect of Grooves, Threads and Corrosion upon the Fatigue of Metals," *Proc. ASTM*, Vol. 26 (1926), p. 255.

54. H. F. Moore, "The Effect of Size and Notch Sensitivity on Fatigue Characteristics of Two Metallic Materials· Part I—Final Report on Aluminum Alloy 75 S-T," *AF Tech. Report 5726*. USAF Air Material Command, Wright-Patterson Air Force Base, Dayton, Ohio (1948).

55. M. Rôs and A. Eichinger, "Die Bruchgefahr fester Körper bei wiederholter Beanspruchung-Ermüdung," *Eidgenöss. Materialprüfungs und Versuchsanstalt für Industrie, Bauwesen und Gewerbe* (*EMPA*) *Ber. 83*, Zurich (1950), p. 71.

56. N. J. F. Gunn, "Fatigue Properties at Low Temperature on Transverse and Longitudinal Notched Specimens of DTD363A Aluminum Alloy," Tech. Note Met. 163, Royal Aircraft Establishment, Farnborough, England (1952).

57. B. J. Lazan and A. A. Blatherwick, "Strength Properties of Rolled Aluminum Alloys under Various Combinations of Alternating and Mean Axial Stresses," *Proc. ASTM*, Vol. 53 (1953), p. 856.

58. R. L. Templin, "Fatigue of Aluminum," *Proc. ASTM,* Vol. 54 (1954), p. 641.

59. R. W. Fralich, "Experimental Investigation of Effects of Random Loading on the Fatigue Life of Notched Cantilever Beam Specimens of 7075-T6 Aluminum Alloy," NASA Memo 4-12-59L (1959).

60. R. E. Peterson, "Design of Parts Subjected to Variable Stresses," Soc. Promotion Eng. Education, June 13, 1941, Ann Arbor, Mich. Linear function formula also in discussion of paper by H. F. Moore, "A Study of Size Effect and Notch Sensitivity in Steel," *Proc. ASTM*. Vol. 45 (1945), p. 522. Formula q versus r based on Neuber K_N, Ref. 39 (1937), p. 149; (1946), p. 163.

61. H. J. Grover, "Fatigue Notch Sensitivity of Some Aircraft Materials," *Proc. ASTM*, Vol. 50 (1950) p. 731.

62. P. Kuhn and H. F. Hardrath, "An Engineering Method for Estimating Notch-Size Effect in Fatigue Tests of Steel,", *NACA Tech. Note 2805* (1952). Now NASA.

63. C. S. Yen and T. J. Dolan, "A Critical Review of the Criteria for Notch-Sensitivity in Fatigue of Metals," *Univ. Illinois Expt. Sta. Bull. 398* (1952).

64. R. E. Peterson, "Notch Sensitivity," Chapter 13 in G. Sines and J. L. Waisman, *Metal Fatigue*, McGraw-Hill, New York (1959), p. 293. Formula 13.16, p. 300. Note that this formula is based on a linear stress distribution, with a q versus r curve correction to give $q = 0$ at $r = 0$. On p. 299, disregard second sentence and six-line paragraph in middle of page.

65. R. B. Heywood, "Stress Concentration Factors," *Engineering* (*London*), Vol. 179 (1955), p. 146.

65a. H. Neuber, "Theoretical Determination of Fatigue Strength at Stress Concentration," Report AFML-TR-68-20 Air Force Materials Lab, Wright-Patterson Air Force Base Dayton, Ohio. (1968).

66. E. A. Davis and M. J. Manjoine, "Effect of Notch Geometry on Rupture Strength at Elevated Temperature," *Proc. ASTM*, Vol. 52 (1952).

67. G. R. Irwin, "Fracture," *Encyclopedia of Physics,* Vol. 6, Springer, Berlin, (1958), p. 551.

68. *Fracture Toughness Testing and Its Applications,* STP 381, ASTM, Philadelphia, Pa. (1964).

69. *Plane Strain Crack Toughness Testing of High Strength Metallic Materials,* STP 410, ASTM, Philadelphia, Pa. (1966).

70. J. A. Van Den Broek, *Theory of Limit Design*, Wiley, New York (1942). See Timoshenko, Ref. 7. Chapter 9, for specific values.

71. M. C. Steele, C. K. Liu, and J. O. Smith, "Critical Review and Interpretation of the Literature on Plastic (Inelastic) Behavior of Engineering Metallic Materials," *Research Report of Dept. of Theoretical and Applied Mechanics,* Univ. of Illinois, Urbana, Ill. (Sept. 1952).

72. R. J. Roark, R. S. Hartenberg, and R. Z. Williams, "The Influence of Form and Scale on Strength," *Univ. Wisconsin Expt. Sta. Bull. 84* (1938).

73. H. J. Gough and W. J. Clenshaw, "Some Experiments on the Resistance of Metals to Fatigue under Combined Stresses," *Aeronaut. Research Counc. Repts. Memoranda 2522* (1951), London: H. M. Stationery Office.

74. R. Houdremont and H. Bennek, "Federstähle," *Stahl u. Eisen*, Vol. 52 (1932), p. 660. Includes Schenck data.

75. R. E. Peterson, "Brittle Fracture and Fatigue in Machinery," *Fatigue and Fracture of Metals,* Wiley, New York (1952), p. 74.

76. R. W. Nichols, Ed., *A Manual of Pressure Vessel Technology,* Chapter 3, Elsevier Publishing Co., London (1969).

77. B. J. Lazan and A. A. Blatherwick, "Fatigue Properties of Aluminum Alloys at Various Direct Stress Ratios," WADC TR 52-306 Part I, Wright-Patterson Air Force Base, Dayton, Ohio (1952).

78. J. O. Smith, "The Effect of Range of Stress on the Fatigue Strength of Metals," *Univ. Illinois Expt. Sta. Bull. 334* (1942).

79. A. Ono, "Fatigue of Steel under Combined Bending and Torsion," *Mem. Coll. Eng. Kyushu Imp. Univ.*, Vol. 2, No. 2 (1921). See also A. Ono, "Some Results of Fatigue Tests of Metals," *J. Soc. Mech. Engrs. Japan* (Tokyo) Vol. 32 (1929), p. 331. Also discussions in *Engineering* (*London*). Vol. 123 (1926), p. 222, and *Proc. Inst. Mech.* Engrs. (London), Vol. 131 (1935), p. 92.

80. F. C. Lea and H. P. Budgen, "Combined Torsional and Repeated Bending Stresses," *Engineering* (*London*), Vol. 122 (1926), p. 242.

81. V. C. Davies, discussion based on theses of S. K. Nimhanmimie and W. J. Huitt (Battersea Polytechnic), *Proc. Inst. Mech. Engrs.* (*London*), Vol. 131 (1935), p. 66.

82. K. Hohenemser and W. Prager, "Zur Frage der Ermüdungsfestigkeit bei mehrachsigen Spannungszuständen," *Metall*, Vol. 12 (1933), p. 342.

83. P. C. Paris, "The Fracture Mechanics Approach to Fatigue," *Fatigue—An Interdisciplinary Approach,* Syracuse University Press, Syracuse, N.Y. (1964), p. 107.

84. *Fatigue Crack Propagation*, STP 415, ASTM, Philadelphia, Pa. (1966).

85. H. Neuber, Ref. 39, 1st ed., p. 6, 2nd ed., p. 11, Translations, 1st ed., p. 7, 2nd ed., p. 21.

86. Ref. 1, p. 136–138 (Summary of formulas).

86a. S. I. Suzuki, "Stress Analysis of a Semi-Infinite Plate Containing a Reinforced Notch Under Uniform Tension," *Intern. J. Solids and Structures,* Vol. 3 (1967), p. 649.

87. M. Seika, "Stresses in a Semi-Infinite Plate Containing a U-Type Notch Under Uniform Tension," *Ingenieur Archiv.*, Vol. 27 (1960), p. 20.

88. O. L. Bowie, "Analysis of Edge Notches in a Semi-Infinite Region," Army Materials & Mechanics Research Center AMRA TR 66-07 (June 1966).

89. F. I. Barrata and D. M. Neal, "Stress Concentration Factors in U-Shaped and Semi-Elliptical Shaped Edge Notches," *J. Strain Anal.*, Vol. 5 (1970), p. 121.

90. Chi-Bing Ling, "On Stress Concentration at Semicircular Notch," *Trans. ASME,* Vol. 89, Series E (1967) Applied Mechanics Section, p. 522.

91. C. E. Inglis, "Stresses in a Plate due to the Presence of Cracks and Sharp Corners," *Engineering* (*London*), Vol. 95 (1913), p. 415.

92. M. Isida, "On the Tension of the Strip with Semi-Circular Notches," *Trans. Japan Soc. Mech. Eng.*, Vol. 19 (1953), p. 5.

93. Chi-Bing Ling, "On Stress Concentration Factor in a Notched Strip," *Trans. ASME*, Vol. 90, Appl. Mech. (1968), p. 833. Discussion by F. J. Appl, *Trans. ASME,* Vol. 91 (1969), Applied Mechanics Section, p. 654.

94. F. J. Appl and D. R. Koerner, "Numerical Analysis of Plane Elasticity Problems," *Proc. Am. Soc. Civil Eng.*, Vol. 94, No. EM3 (1968), p. 743.

95. C. J. Hooke, "Numerical Solution of Plane Elastostatic Problems by Point Matching," *J. Strain Anal.*, Vol. 3 (1968), p. 109.

95a. T. Slot, *Stress Analysis of Thick Perforated Plates,* Technomic Publ. Co., Westport, Conn. (1972).

96. M. Kikukawa, "Factors of Stress Concentration for Notched Bars Under Tension and Bending," *Proc. 10th Intern. Cong. Appl. Mech.*, Elsevier, New York (1962), p. 337.

97. P. D. Flynn and A. A. Roll, "Re-examination of Stresses in a Tension Bar with Symmetrical U-Shaped Grooves," *Proc. Soc. Exp. Stress Analysis,* Vol. 23, Pt. 1 (1966), p. 93. Also "A Comparison of Stress Concentration Factors in Hyperbolic and U-Shaped Grooves," *Proc. Soc. Exp. Stress Analysis,* Vol. 24, Pt. 1 (1967), p. 272.

98. F. J. Appl and D. R. Koerner, "Stress Concentration Factors for U-Shaped, Hyperbolic and Rounded V-Shaped Notches" ASME Paper 69-DE-2; Eng. Soc. Library, United Eng. Center, New York (1969).

98a. I. H. Wilson and D. J. White, "Stress Concentration Factors for Shoulder Fillets and Grooves in Plates," *J. Strain Anal.*, Vol. 8 (1973), p. 43.

98b. F. I. Barrata, "Comparison of Various Formulae and Experimental Stress-Concentration Factors for Symmetrical U-Notched Plates," *J. Strain Anal.*, Vol. 7 (1972), p. 84.

98c. R. B. Heywood, *Designing by Photoelasticity,* Chapman and Hall, London (1952), p. 163.

98d. H. Liebowitz, H. Vandervelt, and R. J. Sanford, "Stress Concentrations Due to Sharp Notches," *Exper. Mech.*, Vol. 7 (1967), p. 513.

99. J. R. Dixon, "Stress Distribution Around Edge Slits in a Plate Loaded in Tension—The Effect of Finite Width of Plate," *J. Royal Aero. Soc.*, Vol. 66 (1962), p. 320.

100. H. M. Westergaard, "Bearing Pressures and Cracks," *Trans. ASME,* Vol. 61 (1939), Applied Mechanics Section, p. A-49.

101. G. R. Irwin, *Fracture*, Vol. 6, *Encyclopedia of Physics,* Springer, Berlin (1958), p. 565.

102. O. L. Bowie, "Rectangular Tensile Sheet with Symmetric Edge Cracks," Army Materials & Mechanics Research Center, AMRA TR 63–22 (Oct. 1963).

103. W. F. Brown and J. E. Srawley, "Plane Strain Crack Toughness Testing of High Strength Metallic Materials," STP 410, Amer. Soc. Testing Mtls., Philadelphia, Pa. (1966), p. 11.

104. W. T. Koiter, "Note on the Stress Intensity Factors for Sheet Strips with Crack under Tensile Loads," Rpt. 314 of Laboratory of Engr. Mechanics, Technological University, Delft, Holland (1965).

105. G. R. Irwin, "Fracture Mechanics," in *Structural Mechanics,* Pergamon, New York (1960).

106. P. C. Paris and G. C. Sih, "Stress Analysis of Cracks," *ASTM Special Tech. Publ.* 381 (1965), p. 34.

107. M. M. Leven and M. M. Frocht, "Stress Concentration Factors for a Single Notch in a Flat Plate in Pure and Central Bending," *Proc. SESA,* Vol. 11, No. 2 (1953), p. 179. Also *Trans. ASME,* Vol. 74 (1952), Applied Mechanics Section, p. 560.

107a. H. Neuber, Ref. 39, Translation 2nd ed., p. 69.

108. A. G. Cole and A. F. Brown, "Photoelastic Determination of Stress Concentration Factors Caused by a Single U-notch on One Side of a Plate in Tension," *J. Royal Aero. Soc.*, Vol. 62 (1958), p. 597.

109. A. Atsumi, "Stress Concentrations in a Strip under Tension and Containing an Infinite Row of Semicircular Notches," *Q. J. Mech. & Appl. Math.*, Vol. 11, Part 4 (1958), p. 478.

110. C. Weber, "Halbebene mit periodisch gewelltem Rand," *Z. angew. Math. u. Mech.*, Vol. 22 (1942), p. 29.

111. K. J. Schulz, "Over den Spannungstoestand in doorborde Platen" (On the State of Stress in Perforated Plates), Doctoral Thesis, Techn. Hochschule, Delft (1941) (in Dutch).

111a. K. J. Schulz, "On the State of Stress in Perforated Strips and Plates," *Proc. Koninklÿke Nederlandsche Akadamie van Wetenschappen* (Netherlands Royal Academy of Science), Amsterdam, Vol. 45 (1942), p. 233, 341, 457 524; Vol. 46–48 (1943–1945), p. 282, 292 (in English).
The foregoing six papers were summarized by C. B. Biezeno, "Survey of Papers on Elasticity Published in Holland 1940–1946," *Advances in Applied Mechanics,*. Vol. 1, Academic Press, New York (1948), p. 105. Biezeno comments, "The calculations are performed only for a strip with one row of holes, though an extensive program of further investigation was projected, which unfortunately could not be executed; the Jewish author, who lived in Holland, was arrested and marched off to Germany or Poland, which meant his destruction."

112. H. Neuber, Ref. 39, 2nd ed., p. 163.

113. A. J. Durelli, R. L. Lake, and E. Phillips, "Stress Concentrations Produced by Multiple Semi-Circular Notches in Infinite Plates under Uniaxial State of Stress," *Proc. SESA*, Vol. 10, No. 1 (1952), p. 53. See also "Stress Distri-

bution in Plates under a Uniaxial State of Stress, with Multiple Semicircular and Flat-Bottom Notches," *Proc. 1st Nat. Congr. Appl. Mech.*, (1952), p. 309.

114. A. Atsumi, "Stress Concentrations in a Strip Under Tension and Containing Two Pairs of Semicircular Notches Placed on the Edges Symmetrically," *Trans. ASME*, Vol. 89, Series E (1967), Applied Mechanics Section, p. 565.

115. M. Hetényi, "The Distribution of Stress in Threaded Connections," *Proc. SESA*, Vol. 1, No. 1 (1943), p. 147. Also *Trans. ASME*, Vol. 65 (1943), Applied Mechanics Section, p. A-93.

116. R. R. Moore, "Effect of Grooves, Threads and Corrosion upon the Fatigue of Metals," *Proc. ASTM*, Vol. 26, Part 2 (1926), p. 255.

117. R. A. Eubanks, "Stress Concentration Due to a Hemispherical Pit at a Free Surface," *Trans. ASME*, Vol. 76 (1954), Applied Mechanics Section, p. 57.

118. B. P. Denardo, "Projectile Shape Effects on Hypervelocity Impact Craters in Aluminum," NASA TN D-4953, Washington, D.C. (1968).

119. R. E. Reed and P. R. Wilcox, "Stress Concentration Due to a Hyperboloid Cavity in a Thin Plate," NASA TN D-5955, Washington, D.C. (1970).

120. G. R. Cowper, "Stress Concentrations Around Shallow Spherical Depressions in a Flat Plate," Aero Report LR-340, National Research Laboratories, Ottawa, Canada (1962).

120a. Y. F. Cheng, "Stress at Notch Root of Shafts under Axially Symmetric Loading," *Exp. Mechanics*, Vol. 10 (1970), p. 534.

121. M. M. Frocht, "Factors of Stress Concentration Photoelastically Determined," *Trans. ASME*, Vol. 57 (1935), Applied Mechanics Section, p. A-67.

121a. H. Kitagawa and K. Nakade, "Stress Concentrations in Notched Strip Subjected to In-Plane Bending," *Technology Reports of Osaka University*, Vol. 20 (1970), p. 751.

121b. H. Neuber, Ref. 39, Translation 2nd ed., p. 71.

121c. W. K. Wilson, "Stress Intensity Factors for Deep Cracks in Bending and Compact Tension Specimens," *Engineering Fracture Mechanics*, Vol. 2, Pergamon Press, London (1970), p. 169.

122. C. H. Tsao, A. Ching, and S. Okubo, "Stress Concentration Factors for Semi-elliptical Notches in Beams Under Pure Bending," *Exp. Mechanics*, Vol. 5 (1965) p. 19A.

123. K. Nishioka and N. Hisamitsu, "On the Stress Concentration in Multiple Notches," *Trans. ASME*, Vol. 84, Series E (1962), Applied Mechanics Section, p. 575.

124. A. Ching, S. Okubo, and C. H. Tsao, "Stress Concentration Factors for Multiple Semi-Elliptical Notches in Beams Under Pure Bending," *Exp. Mechanics*, Vol. 8 (1968), p. 19N.

125. H. Neuber, Ref. 39, 2nd ed., p. 84.

126. G. H. Lee, "The Influence of Hyperbolic Notches on the Transverse Flexure of Elastic Plates," *Trans. ASME*, Vol. 62, (1940), Applied Mechanics Section, p. A-53.

127. S. Shioya, "The Effect of Square and Triangular Notches with Fillets on the Transverse Flexure of Semi-Infinite Plates," *Z. angew. Math. u. Mech.*, Vol. 39 (1959), p. 300.

128. S. Shioya, "On the Transverse Flexure of a Semi-Infinite Plate with an Elliptic Notch," *Ingenieur-Archiv*, Vol. 29 (1960), p. 93.

129. S. Shioya, "The Effect of an Infinite Row of Semi-Circular Notches on the Transverse Flexure of a Semi-Infinite Plate," *Ingenieur-Archiv*, Vol. 32 (1963), p. 143.

130. H. Neuber, Ref. 39, 2nd ed., p. 45.

131. K. R. Rushton, "Stress Concentrations Arising in the Torsion of Grooved Shafts," *J. Mech. Sci.*, Vol. 9 (1967), p. 697.

132. G. J. Matthews and C. J. Hooke, "Solution of Axisymmetric Torsion Problems by Point Matching," *J. Strain Anal.*, Vol. 6 (1971), p. 124.

133. M. Hamada and H. Kitagawa, "Elastic Torsion of Circumfentially Grooved Shafts," *Bull. Japan Soc. Mech. Eng.*, Vol. 11 (1968), p. 605.

134. M. M. Leven, "Quantitative Three-Dimensional Photoelasticity," *Proc. SESA*, Vol. 12, No. 2 (1955), p. 167.

135. H. Okubo, "Approximate Approach for Torsion Problem of a Shaft with a Circumferential Notch," *Trans. ASME*, Vol. 74 (1952), Applied Mechanics Section, p. 436.

136. H. Okubo, "Determination of Surface Stress by Means of Electroplating," *J. Appl. Physics*, Vol. 24 (1953), p. 1130.

137. H. Fessler, C. C. Rogers, and P. Stanley, "Shouldered Plates and Shafts in Tension and Torsion," *J. Strain Anal.*, Vol. 4 (1969), p. 169.

138. R. V. Baud, "Study of Stresses by Means of Polarized Light and Transparencies," *Proc. Engrs. Soc. West. Penn.* Vol. 44 (1928), p. 199.

139. M. M. Leven and J. B. Hartman, "Factors of Stress Concentration for Flat Bars with Centrally Enlarged Section," *Proc. SESA,* Vol. 19, No. 1 (1951), p. 53.

140. K. Kumagai and H. Shimada, "The Stress Concentration Produced by a Projection under Tensile Load," *Bull. Japan Soc. Mech. Eng.*, Vol. 11 (1968), p. 739.

141. B. Scheutzel and D. Gross, *Konstruktion, Vol. 18* (1966), p. 284.

142. D. Spangenberg, *Konstruktion,* Vol. 12 (1960), p. 278.

143. A. T. Derecho and W. H. Munse, "Stress Concentration at External Notches in Members Subjected to Axial Loading," *Univ. Illinois Eng. Exp. Eng. Sta. Bulletin No. 494* (1968).

144. R. V. Baud, "Beiträge zur Kenntnis der Spannungsverteilung in Prismatischen und Keilförmigen Konstruktionselementen mit Querschnittsübergängen," *Eidgenöss. Materialprüf. Ber 83,* Zurich (1934). See also *Product Eng.*, Vol. 5 (1934), 133.

145. A. Thum and W. Bautz, "Der Entlastungsübergang—Günstigste Ausbildung des Überganges an abgesetzten Wellen u. dgl." *Forsch. Ingwes.*, Vol. 6 (1934), p. 269.

146. K. Lurenbaum, *Ges. Vortrage der Hauptvers. der Lilienthal Gesell.* (1937), p. 296.

147. P. Grodzinski, "Investigations on Shaft Fillets," *Engineering* (London), Vol. 152 (1941), p. 321.

148. W. Morgenbrod, "Die Gestaltftestigkeit von Walzen und Achsen mit Hohlkehlen," *Stahl u. Eisen,* Vol. 59 (1939), p. 511.

149. D. J. McAdam, "Endurance Properties of Steel," *Proc. ASTM,* Vol. 23, Part II (1923), p. 68

150. R. E. Peterson, "Fatigue Tests of Small Specimens with Particular Reference to Size Effect," *Proc. Am. Soc. Steel Treatment,* Vol. 18 (1930), p. 1041.

151. L. S. Clock, "Reducing Stress Concentration with an Elliptical Fillet," *Design News,* Rogers Publishing Co., Detroit, Mich. (May 15, 1952).

152. R. B. Heywood, *Photoelasticity for Designers.* Pergamon, New York (1969), Chapter 11.

153. I. M. Allison, "The Elastic Concentration Factors in Shouldered Shafts, Part III: Shafts Subjected to Axial Load," *Aeronautical Q.*, Vol. 13, (1962), p. 129.

153a. D. S. Griffin and A. L. Thurman, "Comparison of DUZ Solution with Experimental Results for Uniaxially and Biaxially Loaded Fillets and Grooves," WAPD TM-654, Clearinghouse for Scientific and Technical Information, Springfield, Va. (1967).

153b. D. S. Griffin and R. B. Kellogg, "A Numerical Solution for Axially Symmetrical and Plane Elasticity Problems," *Intern. J. Solids and Structures,* Vol. 3 (1967), p. 781.

153c. M. M. Leven, "Stress Distribution in a Cylinder with an External Circumferential Fillet Subjected to Internal Pressure," Res. Memo 65-9D7-520-M1, Westinghouse Research Lab. (1965).

153d. J. H. Heifetz and I. Berman, "Measurements of Stress Concentration Factors in the External Fillets of a Cylindrical Pressure Vessel," *Exp. Mechanics.* Vol. 7 (1967), p. 518.

153e. R. C. Gwaltney, J. M. Corum, and W. L. Greenstreet, "Effect of Fillets on Stress Concentration in Cylindrical Shells with Step Changes in Outside Diameter," *Trans. ASME,* Vol. 93, *J. Eng. for Industry* (1971), p. 986.

154. D. C. Berkey, "Reducing Stress Concentration with Elliptical Fillets," *Proc. SESA,* Vol. 1, No. 2 (1944), p. 56.

155. I. M. Allison, "The Elastic Concentration Factors in Shouldered Shafts, Part II: Shafts Subjected to Bending," *Aeronautical Q.*, Vol. 12 (1961), p. 219.

156. L. S. Jacobsen, "Torsional Stress Concentrations in Shafts of Circular Cross-Section and Variable Diameter," *Trans. ASME,* Vol. 47 (1925), p. 619.

157. A. Weigand, "Ermittung der Formziffer der auf Verdrehung beanspruchten abgesetzten Welle mit Hilfe von Feindehnungsmessungen," *Luftfahrt-Forsch.*, Vol. 20 (1943), p. 217 (NACA *Translation* No. 1179). Now NASA.

158. I. M. Allison, "The Elastic Stress Concentration Factors in Shouldered Shafts," *Aeronautical Q.*, Vol. 12 (1961), p. 189.

159. K. R. Rushton, "Elastic Stress Concentrations for the Torsion of Hollow Shouldered Shafts Determined by an Electrical Analogue," *Aeronautical Q.*, Vol. 15 (1964), p. 83.

160. H. Oschatz, "Gesetzmässigkeiten des Dauerbruches und Wege zur Steigerung der Dauerhaltbarkeit," *Mitt. der Materialprüfungsanstalt an den Technischen Hochschule Darmstadt,* Vol. 2 (1933).

161. A. Thum and E. Bruder, "Dauerbruchgefahr an Hohlkehlen von Wellen und Achsen und ihre Minderung," No. 11, *Deutsche Kraftfahrtforschung im Auftrag des Reichs-Verkehrsministeriums,* VDI Verlag, Berlin (1938).

162. E. Sternberg and M. A. Sadowsky, "Three-Dimensional Solution for the Stress Concentration Around a Circular Hole in a Plate of Arbitrary Thickness," *Trans. ASME,* Vol. 71 (1949), Applied Mechanics Section, p. 27.

163. C. K. Youngdahl and E. Sternberg, "Three-Dimensional Stress Concentration Around a Cylindrical Hole in a Semi-Infinite Elastic Body," *Trans. ASME,* Vol. 88, Series E (1966), Applied Mechanics Section, p. 855.

164. B. Kirsch, *Z. VDI* (July 16, 1898), p. 797. See S. Timoshenko, Ref. 7, p. 301.

165. R. C. J. Howland, "On the Stresses in the Neighborhood of a Circular Hole in a Strip under Tension." *Phil. Trans. Roy. Soc. (London) A,* Vol. 229 (1929–30), p. 67.

166. *Design Data,* booklet published by ASME (1939). A collection of data was also made by G. H. Neugebauer "Stress Concentration Factors and Their Effect on Design," *Product Eng.,* Vol. 14 (1943), p. 82.

167. A. M. Wahl and R. Beeuwkes, "Stress Concentration Produced by Holes and Notches," *Trans. ASME,* Vol. 56 (1934), Applied Mechanics Section, p. 617.

167a. S. Christiansen, "Numerical Determination of Stresses in a Finite or Infinite Plate with Several Holes of Arbitrary Form," *Z. angew. Math. u. Mech.,* Vol. 48 (1968), p. T131.

168. T. Udoguti, "Solutions of Some Plane Elasticity Problems by Using Dipole Coordinates—Part II," *Trans. Japan Soc. Mech. Eng.,* Vol. 15 (1949), p. I-80 (in Japanese).

169. M. Isida, "Form Factors of a Strip with an Elliptic Hole in Tension and Bending," *Scientific Papers of Faculty of Engineering, Tokushima University,* Vol. 4 (1953), p. 70. Factors for b/w from 0.5 to 1.0 from private communication from Isida.

169a. E. G. Coker and L. N. G. Filon. *A Treatise on Photoelasticity,* Cambridge University Press, Cambridge, England (1931), p. 486.

169b. W. T. Koiter, "An Elementary Solution of Two Stress Concentration Problems in the Neighborhood of a Hole," *Q. Appl. Math.,* Vol. 15 (1957), p. 303.

170. R. B. Heywood, Ref. 98c, p. 268.

170a. A. Hennig, "Polarizationsoptische Spannungsuntersuchungen am gelochten Zugstab und am Nietloch," *Forsch. Gebiete Ingenieur., VDI,* Vol. 4, No. 2 (1933), p. 53.

170b. R. G. Belie and F. J. Appl., "Stress Concentrations in Tensile Strips with Large Circular Holes," *Exp. Mechanics,* Vol. 12 (1972), p. 190. Discussion, Vol. 13 (1973), p. 255.

171. R. D. Mindlin, "Stress Distribution Around a Hole near the Edge of a Plate in Tension," *Proc. SESA,* Vol. 5, No. 2 (1948), p. 56.

172. T. Udoguti, "Solutions of Some Plane Elasticity Problems by Using Dipole Coordinates—Part I," *Trans. Japan Soc. Mech. Eng.,* Vol. 13 (1947), p. 17 (in Japanese).

173. M. Isida, "On the Tension of a Semi-Infinite Plate with an Elliptic Hole," *Scientific Papers of Faculty of Engineering, Tokushima University,* Vol. 5 (1955), p. 75 (in Japanese).

174. S. Sjöström, "On the Stresses at the Edge of an Eccentrically Located Circular Hole in a Strip under Tension," Report No. 36, *Aeronaut. Research Inst. Sweden* (Stockholm) (1950).

175. J. G. Lekkerkerker, "Stress Concentration Around Circular Holes in Cylindrical Shells," *Proc. 11th Internat. Congr. Appl. Mech.,* Springer, Berlin (1964), p. 283.

176. A. C. Eringen, A. K. Naghdi, and C. C. Thiel, "State of Stress in a Circular Cylindrical Shell with a Circular Hole," *Welding Research Council Bulletin 102* (1965).

177. P. Van Dyke, "Stresses about a Circular Hole in a Cylindrical Shell," *AIAA J.,* Vol. 3 (1965), p. 1733.

178. N. C. Lind, "Stress Concentration of Holes in Pressurized Cylindrical Shells," *AIAA J.,* Vol. 6 (1968), p. 1397.

179. D. S. Houghton and A. Rothwell, "The Effect of Curvature on the Stress Concentration around Holes in Shells," College of Aeronautics, Cranfield, England, Rpt. 156 (1962).

180. J. T. Jessop, C. Snell, and I. M. Allison, "The Stress Concentration Factors in Cylindrical Tubes with Transverse Cylindrical Holes." *Aeronautical Q.,* Vol. 10 (1959), p. 326.

180a. A. J. Durelli, C. J. del Rio, V. J. Parks, and H. Feng, "Stresses in a Pressurized Cylinder with a Hole," *Proc. Am. Soc. Civil Eng.,* Vol. 93 (1967), p. 383.

181. D. N. Pierce and S. I. Chou, "Stress State Around an Elliptic Hole in a Circular Cylindrical Shell Subjected to Axial Loads," Presented at SESA Meeting, Los Angeles, Cal., May 16, 1973.

181a. H. Hanzawa, M. Kishida, M. Murai, and K. Takashina, "Stresses in a Circular Cylindrical Shell Having Two Circular Cutouts," *Bull. Japan Soc. Mech. Eng.,* Vol. 15 (1972), p. 787.

181b. M. Hamada, K. Yokoya, M. Hamamoto, and T. Masuda, "Stress Concentration of a Cylindrical Shell With One or Two Circular Holes," *Bull. Japan Soc. Mech. Eng., Vol. 15* (1972), p. 907.

181c. A. J. Durelli, et. al., "Stresses in a Pressurized Ribbed Cylindrical Shell with a Reinforced Circular Hole Interrupting a Rib," *Trans. ASME,* Vol. 93, (1971) Jl. Eng. for Industry, p. 897.

182. M. V. V. Murthy, "Stresses Around an Elliptic Hole in a Cylindrical Shell," *Trans. ASME,* Vol. 91, Series E (1969) Applied Mechanics Section, p. 39.

182a. M. V. V. Murthy and M. N. Bapu Rao, "Stresses in a Cylindrical Shell Weakened by an Elliptic Hole with Major Perpendicular to Shell Axis," *Trans. ASME,* Vol. 92, Series E (1970), Applied Mechanics Section, p. 539.

182b. O. Tingleff, "Stress Concentration in a Cylindrical Shell with an Elliptical Cutout," *AIAA J.,* Vol. 9 (1971), p. 2289.

182c. F. A. Leckie, D. J. Paine, and R. K. Penny, "Elliptical Discontinuities in Spherical Shells," *J. Strain Anal.,* Vol. 2 (1967), p. 34.

183. M. Seika and M. Ishii, "Photoelastic Investigation of the Maximum Stress in a Plate with a Reinforced Circular Hole under Uniaxial Tension," *Trans. ASME,* Vol. 86, Series E (1964), Applied Mechanics Section, p. 701.

184. M. Seika and A. Amano, "The Maximum Stress in a Wide Plate with a Reinforced Circular Hole under Uniaxial Tension—Effects of a Boss with Fillet," *Trans. ASME,* Vol. 89, Series E (1967), Applied Mechanics Section, p. 232.

185. S. Timoshenko, "On Stresses in a Plate with a Circular Hole," *J. Franklin Inst.,* Vol. 197 (1924), p. 505.

186. K. Lingaiah, W. P. T. North, and J. B. Mantle, "Photoelastic Analysis of an Asymmetrically Reinforced Circular Cutout in a Flat Plate Subjected to Uniform Unidirectional Stress," *Proc. SESA,* Vol. 23, No. 2 (1966), p. 617.

187. C. Gurney, "An Analysis of the Stresses in a Flat Plate with Reinforced Circular Hole under Edge Forces," *Aeronautical Research Comm. R&M 1834,* London (1938).

188. L. Beskin, "Strengthening of Circular Holes in Plates under Edge Forces," *Trans. ASME,* Vol. 66, (1944), Applied Mechanics Section, p. A-140.

189. S. Levy, A. E. McPherson, and F. C. Smith, "Reinforcement of a Small Circular Hole in a Plane Sheet under Tension," *Trans. ASME,* Vol. 70 (1948), Applied Mechanics Section, p. 160.

190. H. Reissner and M. Morduchow, "Reinforced Circular Cutouts in Plane Sheets, "*NACA TN 1852* (1949). Now NASA.

191. A. A. Wells, "On the Plane Stress Distribution in an Infinite Plate with Rim-Stiffened Elliptical Opening," Q. *J Mech. Appl. Math.,* Vol. 2, Part 1 (1950), p. 23.

192. E. H. Mansfield, "Neutral Holes in Plane Sheet—Reinforced Holes which are Elastically Equivalent to the Uncut Sheet," Q. *J. Mech. Appl. Math.,* Vol. 6 (1953), p. 370.

193. R. Hicks, "Reinforced Elliptical Holes in Stressed Plates," *J. Royal Aero. Soc.,* Vol. 61 (1957), p. 688.

194. W. H. Wittrick, "The Stresses Around Reinforced Elliptical Holes in Plane Sheet," *Aero. Res. Lab. (Australia) Report 267* (1959).

195. G. N. Savin, *Stress Concentration Around Holes,* Pergamon Press New York (1961), p. 234.

196. D. S. Houghton and A. Rothwell, "The Analysis of Reinforced Circular and Elliptical Cutouts under Various Loading Conditions," College of Aeronautics, Cranfield, England, Report 151 (1961).

196a. G. A. O. Davies, "Plate-Reinforced Holes," *Aeronautical* Q. Vol. 18 (1967), p. 43.

197. A. Kaufman, P. T. Bizon, and W. C. Morgan, "Investigation of Circular Reinforcements of Rectangular Cross-Section Around Central Holes in Flat Sheets under Biaxial Loads in the Elastic Range," NASA TN D-1195 (1962).

197a. British Science Data 70005, Engineering Science Data Unit, 4 Hamilton Pl., London W1 (1970), p. 15.

198. *ASME Boiler and Pressure Vessel Code—Pressure Vessels, Section VIII,* ASME, New York (1971), p. 27.

199. A. Kaufman, P. T. Bizon, and W. C. Morgan, "Investigation of Tapered Circular Reinforcements Around Central Holes in Flat Sheets Under Biaxial Loads in the Elastic Range," NASA TN D 1101 (1962).

199a. S. K. Dhir and J. S. Brock, "A New Method of Reinforcing a Hole Effecting Large Weight Savings," *Intern. J. Solids and Structures,* Vol. 6 (1970), p. 259.

200. E. H. Mansfield, "Analysis of a Class of Variable Thickness Reinforcement Around a Circular Hole in a Flat Sheet," *Aeronaut. Q.,* Vol. 21 (1970), p. 303.

201. W. F. Riley, A. J. Durelli, and P. S. Theocaris, "Further Stress Studies on a Square Plate with a Pressurized Central Circular Hole," *Proc. 4th Ann. Conf. on Solid Mechanics,* Univ. of Texas, Austin (1959).

202. A. J. Durelli and W. F. Riley, *Introduction to Photomechanics,* Prentice-Hall, Englewood Cliffs, N.J. (1965), p. 233.

203. T. Sekiya, "An Approximate Solution in the Problems of Elastic Plates with an Arbitrary External Form and a Circular Hole," *Proc. 5th Japan Nat. Congr. Appl. Mech.* (1955), p. 95.

204. G. A. O. Davies, "Stresses in a Square Plate Having a Central Circular Hole," *J. Royal Aero. Soc.,* Vol. 69 (1965), p. 410.

204a. A. J. Durelli and A. S. Kobayashi, "Stress Distributions around Hydrostatically Loaded Circular Holes in the Neighborhood of Corners," *Trans. ASME,* Vol. 80 Applied Mechanics Section, (1958), p. 178.

205. Chi-Bing Ling, "On the Stresses in a Plate Containing Two Circular Holes," *J. Appl. Physics,* Vol. 19 (1948), p. 77. Also "The Stresses in a Plate Containing an Overlapped Circular Hole," *J. Appl. Physics,* Vol. 19 (1948), p. 405.

206. R. A. W. Haddon, "Stresses in an Infinite Plate with Two Unequal Circular Holes," *Q. J. Mech. Math.,* Vol. 20 (1967), p. 277

207. W. E. North, "A Photoelastic Study of the Interaction Effect of Two Neighboring Holes in a Plate Under Tension," M. S. Thesis, University of Pittsburgh (1965).

208. V. L. Salerno and J. B. Mahoney, "Stress Solution for an Infinite Plate Containing Two Arbitrary Circular Holes Under Equal Biaxial Stresses," *Trans. ASME,* Vol. 90, Series B, (1968), Industry Section, p. 656.

209. R. C. J. Howland, "Stresses in a Plate containing an Infinite Row of Holes," *Proc. Roy. Soc. (London) A,* Vol. 148 (1935), p. 471

210. P. Meijers, "Doubly-Periodic Stress Distributions in Perforated Plates," Dissertation, Tech. Hochschule Delft, Netherlands (1967).

211. A. Hütter, "Die Spannungsspitzen in gelochten Blechscheiben und Streifen," *Z. angew. Math. Mech.,* Vol. 22 (1942), p. 322.

212. W. J. O'Donnell and B. F. Langer, "Design of Perforated Plates," *Trans. ASME,* Vol. 84, Series B (1962) Industry Section, p. 307.

213. G. Horvay, "The Plane-Stress Problem of Perforated Plates," *Trans. ASME,* Vol. 74, Series E (1952), Applied Mechanics Section, p. 355.

214. R. C. Sampson, "Photoelastic Analysis of Stresses in Perforated Material Subject to Tension or Bending," *Bettis Technical Review,* WAP-BT-18 (April 1960).

215. M. M. Leven, "Effective Elastic Constants in Plane Stressed Perforated Plates of Low Ligament Efficiency," Westinghouse Research Report 63-917-520-R1 (August 28, 1963).

216. M. M. Leven, "Stress Distribution in a Perforated Plate of 10% Ligament Efficiency Subjected to Equal Biaxial Stress," Westinghouse Research Report 64-917-520-R1 (March 17, 1964).

216a. E. I. Grigolyuk and L. A. Fil'shtinskii, *Perforirorannye Plastiny i Obolocki* (Perforated Plates and Shells), Nauka Publishing House, Moscow (1970).

216b. J. E. Goldberg and K. N. Jabbour, "Stresses and Displacements in Perforated Plates," *Nuclear Structural Engineering,* North Holland Publ. Co., Amsterdam (1965), p. 360.

217. R. Bailey and R. Hicks, "Behavior of Perforated Plates under Plane Stress," *J. Mech. Eng. Sci.,* Vol. 2, No. 2 (1960), p. 143.

218. L. E. Hulbert and F. W. Niedenfuhr, "Accurate Calculation of Stress Distributions in Multiholed Plates," *Trans. ASME,* Vol. 87, Series B, (1965) Industry Section, p. 331.

219. H. Nuno, T. Fujie, and K. Ohkuma, "Experimental Study on the Elastic Properties of Perforated Plates with Circular Holes in Square Patterns," MAPI Laboratory Research Report 74, Mitsubishi Atomic Power Industries Laboratory (March 1964).

220. W. J. O'Donnell, "A Study of Perforated Plates with Square Penetration Patterns," WAPD-T-1957, Bettis Atomic Power Laboratory (August 1966).

221. M. M. Leven, "Photoelastic Analysis of a Plate Perforated with a Square Pattern of Penetrations," Westinghouse Research Memo 67-1D7-TAADS-MI (January 1967).

222. W. J. O'Donnell, "A Study of Perforated Plates with Square Penetration Patterns," *Welding Research Council Bulletin 124* (September 1967).

223. L. E. Hulbert, *The Numerical Solution of Two-Dimensional Problems of the Theory of Elasticity*, Ohio State Univ., *Eng. Exp. Sta. Bull. 198*, Columbus, Ohio, (1965).

224. V. N. Buivol, "Experimental Investigations of the Stressed State of Multiply Connected Plates" (in Ukranian), *Prikladna Mekhanika*, Vol. 6 (3) (1960), p. 328

225. V. N. Buivol, "Action of Concentrated Forces on a Plate with Holes," (in Ukranian), *Prikladna Mehkanika*, Vol. 8(1) (1962), p. 42.

226. H. Kraus, "Stress Concentration Factors for Perforated Annular Bodies Loaded in Their Plane," Unpublished Report, Pratt and Whitney Co., E. Hartford, Conn. (1963).

227. H. Kraus, P. Rotondo, W. D. Haddon, "Analysis of Radially Deformed Perforated Flanges," *Trans. ASME*, Vol. 88, Series E (1966), Applied Mechanics Section, p. 172.

228. S. Timoshenko, Ref. 7, p. 208.

229. G. N. Savin, *Stress Concentrations Around Holes* (English Translation), Pergaman Press, New York (1961).

230. H. Kraus, "Pressure in Multibore Bodies," *Intern. J. Mech. Sci.*, Vol. 4 (1962), p. 187.

231. G. Kolosoff, Disertation, St. Petersburg, 1910. See S. Timoshenko, Ref. 7, p. 306.

232. M. Isida, "On the Tension of a Strip with a Central Elliptic Hole," *Trans. Japan Soc. Mech. Eng.*, Vol. 21 (1955), p. 514.

233. N. Jones and D. Hozos, "A Study of the Stress Around Elliptical Holes in Flat Plates," *Trans. ASME*, Vol. 93, Series B, *J. Eng. for Industry*, (1971), p. 688.

233a. A. J. Durelli, V. J. Parks, and H. C. Feng, "Stresses Around an Elliptical Hole in a Finite Plate Subjected to Axial Loading," *Trans. ASME*, Vol. 88 Series E (1966), Applied Mechanics Section, p. 192.

234. M. Isida, "On the Stress Distribution around an Elliptic-sectioned Tunnel Excavated under the Horizontal Surface," *Trans. Japan Soc. Mech. Eng.*, Vol. 23 (1957), p. 474 (in Japanese).

235. J. R. Dixon, "Stress Distribution around a Central Crack in a Plate Loaded in Tension; Effect on Finite Width of Plate," *J. Royal Aero. Soc.* Vol. 64 (1960), p. 141.

236. C. Feddersen, Discussion of "Plane Strain Crack Toughness Testing," *ASTM Special Tech. Publ. 410* (1967), p. 77.

237. M. Isida, "Crack Tip Intensity Factors for the Tension of an Eccentrically Cracked Strip," Lehigh University, Dept. Mechanics Rpt. (1965).

238. D. P. Rooke, *Compendium, Engineering Fracture Mechanics,* Vol. 1 (1970), p. 727.

239. R. Papirno, "Stress Concentrations in Tensile Strips with Central Notches of Varying End Radii," *J. Royal Aero. Soc.*, Vol. 66 (1962), p. 323.

240. M. Isida, "Stress Intensity Factors for the Tension of an Eccentrically Cracked Strip," *Trans. ASME*, Vol. 88, Series E (1966), Applied Mechanics Section, p. 674.

241. H. Nisitani, "Method of Approximate Calculation for Interference of Notch Effect and its Application," *Bull. Japan Soc. Mech. Eng.*, Vol. 11 (1968), p. 725.

242. H. L. Cox, "Four Studies in the Theory of Stress Concentration," *Aero Research Council (London), Rpt. 2704* (1953).

243. A. J. Sobey, "Stress Concentration Factors for Rounded Rectangular Holes in Infinite Sheets," *Aero. Res. Council R&M 3407*. Her Majesty's Stationery Office, London, (1963).

244. F. Hirano, "Study of Shape Coefficients of Two-Dimensional Elastic Bodies," *Trans. Japan Soc. Mech. Eng.*, Vol. 16, No. 55 (1950), p. 52 (in Japanese).

245. M. Isida, "On the Tension of an Infinite Strip Containing a Square Hole with Rounded Corners," *Bull. Japan Soc. Mech. Eng.*, Vol. 3 (1960), p. 254.

246. M. M. Frocht and M. M. Leven, "Factors of Stress Concentration for Slotted Bars in Tension and Bending," *Trans. ASME*, Vol. 73 (1951), Applied Mechanics Section, p. 107.

247. A. J. Durelli, V. J. Parks, and S. Uribe, "Optimization of a Slot End Configuration in a Finite Plate Subjected to Uniformly Distributed Load," *Trans. ASME*, Vol. 90, Series E (1968), Applied Mechanics Section, p. 403.

248. E. Gassner and K. F. Horstmann, "The Effect of Ground to Air to Ground Cycle on the Life of Transport Aircraft Wings which are Subject to Gust Loads," *RAE Translation* 933 (1961).

249. W. Schutz, "Fatigue Test Specimens," Tech. Note TM 10/64, Laboratorium für Betriebsfestigkeit, Darmstadt (1964).

250. L. H. Mitchell, "Stress Concentration Factors at a Doubly-symmetric Hole," *Aeronautical Q.*, Vol. 17 (1966), p. 177.

251. Y. F. Cheng, "A Photoelastic Study of Stress Concentration Factors at a Doubly Symmetric Hole in Finite Strips under Tension," *Trans. ASME*, Vol. 90, Series E (1968), Applied Mechanics Section, p. 188.

251a. K. Miyao, "Stresses in a Plate Containing a Circular Hole with a Notch," *Bull. Japan Soc. Mech. Eng.,* Vol. 13 (1970), p. 483.

252. S. R. Heller, J. S. Brock, and R. Bart, "The Stresses Around a Rectangular Opening with Rounded Corners in a Uniformly Loaded Plate," *Proc. 3rd U. S. Nat. Congr. Appl. Mech.,* publ. by ASME, 1958, p. 357.

253. British Engineering Science Data, 70005, Engineering Science Data Unit, 4 Hamilton Pl., London W1 (1970), p. 23.

254. S. R. Heller, "Stress Concentration Factors for a Rectangular Opening with Rounded Corners in a Biaxially Loaded Plate," *J. Ship Research,* Vol. 13 (1969), p. 178.

255. E. Steneroth, L. Lindau, and B. Önnermark, "Photoelastic Investigations of Stress Concentrations at Hatch Corners," *J. Ship Research,* Vol. 7 (1963), p. 24.

255a. W. O. Richmond, Discussion of Reference 287c, *Proc. ASCE,* Vol. 65 (1939), p. 1465.

255b. R. Mindlin, Ref. 255a, p. 1476.

256. W. H. Wittrick, "Stress Concentrations for Uniformly Reinforced Equilateral Triangular Holes with Rounded Corners," *Aeronautical Q.,* Vol. 14 (1963), p. 254.

256a. M. E. Fourney and R. R. Parmerter, "Photoelastic Design Data for Pressure Stresses in Slotted Rocket Grains," *J. AIAA,* Vol. 1 (1963), p. 697.

256b. H. B. Wilson, "Stresses Owing to Internal Pressure in Solid Propellant Rocket Grains," *J. ARS,* Vol. 31 (1961), p. 309.

256c. M. E. Fourney and R. R. Parmerter, "Stress Concentrations for Internally Perforated Star Grains," Bur. Naval Weapons, NAVWEPS Rep. 7758 (1961); "Parametric Study of Rocket Grain Configurations by Photoelastic Analysis," AFSC Rep. AFRPL-TR-66-52 (1966).

256d. W. C. Jenkins, "Comparison of Pressure and Temperature Stress Concentration Factors for Solid Propellant Grains," *Experimental Mechanics,* Vol. 8 (1968), p. 94.

256e. A. Mondina and M. Falco, "Richerche Sperimentali su Modelli Dell'interno di un Reattore Nuclear PWR," (Experimental Stress Analysis on Models of PWR Internals), *Disegno di Macchine, Palermo,* Vol. 3 (1972), p. 77.

257. M. M. Leven, "Quantitative Three-Dimensional Photoelasticity," *Proc. SESA,* Vol. 12, No. 2 (1955), p. 157.

258. A. Thum and W. Kirmser, "Überlagerte Wechselbeanspruchungen, ihre Erzeugung und ihr Einfluss auf die Dauerbarkeit und Spannungsausbildung quergebohrten Wellen," *VDI-Forschungsheft 419,* Vol 14(b) (1943), p. 1.

259. H. T. Jessop, C. Snell, and I. M. Allison, "The Stress Concentration Factors in Cylindrical Tubes with Transverse Circular Holes," *Aeronautical Q.,* Vol. 10 (1959), p. 326.

260. British Engineering Science Data, 65004, Engineering Science Data Unit, 4 Hamilton Pl., London W1 (1965), p. 29.

261. R. E. Peterson, "Stress Concentration Factors for a Round Bar with a Transverse Hole," Report 68-1D7-TAEUG-R8, Westinghouse Research Labs., Pittsburgh, Pa. (1968).

262. W. G. Bickley, "Distribution of Stress Round a Circular Hole in a Plate," *Phil. Trans. Roy. Soc. (London) A,* Vol. 227 (1928), p. 383.

263. R. C. Knight, "Action of a Rivet in a Plate of Finite Breadth," *Phil. Mag.,* Series 7, Vol. 19 (1935), p. 517.

263a. P. S. Theocaris, "The Stress Distribution in a Strip Loaded in Tension by Means of a Central Pin," *Trans. ASME,* Vol. 78 (1956), Applied Mechanics Section, p. 482.

264. Ref. 169a, p. 525.

265. K. Schaechterle, "On the Fatigue Strength of Riveted and Welded Joints and the Design of Dynamically Stressed Structural Members based on Conclusions Drawn from Fatigue Tests," *Intern. Assn. Bridge Structural Eng.,* Vol. 2 (1934), p. 312. Refers to Stoltenberg test.

266. M. M. Frocht and H. N. Hill, "Stress Concentration Factors around a Central Circular Hole in a Plate Loaded Through a Pin in the Hole," *Trans. ASME,* Vol. 62, (1940), Applied Mechanics Section, p. A-5.

266a. H. T. Jessop, C. Snell and G. S. Holister, "Photoelastic Investigation on Plates with Single Interference-Fit Pins with Load Applied (a) to Pin Only and (b) to Pin and Plate Simultaneously," *Aeronaut.,* Q. Vol. 9 (1958), p. 147. See previous papers by authors listed as References.

266b. H. L. Cox and A. F. C. Brown, "Stresses Round Pins in Holes," *Aeronaut. Q.,* Vol. 15 (1964), p. 357.

266c. British Engineering Science Data, 65004, Engineering Science Data Unit, 4 Hamilton Pl., London W1 (1956), p. 51.

266d. A. Mori, "Stress Distributions in a Semi-Infinite Plate with a Row of Circular Holes," *Bull. Japan Soc. Mech. Eng.,* Vol. 15 (1972), p. 899.

267. W. D. Mitchell and D. Rosenthal, "Influence of Elastic Constants on the Partition of Load between Rivets," *Proc. SESA*, Vol. 7, No. 2 (1949), p. 17.

268. M. M. Leven, "Photoelastic Determination of the Stresses at Oblique Openings in Plates and Shells," *Welding Research Council Bulletin 153* (August 1970), p. 52.

269. I. M. Daniel, "Photoelastic Analysis of Stresses Around Oblique Holes," *Experimental Mechanics*, Vol. 10 (1970), p. 467.

270. H. W. McKenzie and D. J. White, "Stress Concentration Caused by an Oblique Round Hole in a Flat Plate Under Uniaxial Tension," *J. Strain Anal.*, Vol. 3 (1968), p. 98.

271. F. Ellyin, "Experimental Study of Oblique Circular Cylindrical Apertures in Plates," *Experimental Mechanics*, Vol. 10 (1970), p. 195.

272. F. Ellyin, "Elastic Stresses Near a Skewed Hole in a Flat Plate and Applications to Oblique Nozzle Attachments in Shells," *Welding Research Council Bulletin 153* (August 1970), p. 32.

272a. F. Ellyin and U. M. Izmiroglu, "Effect of Corner Shape on Elastic Stress and Strain Concentration in Plates with an Oblique Hole," *Trans. ASME*, Vol. 95 (1973), Jl. of Eng. for Industry, p. 151.

272b. D. E. Hardenberg and S. Y. Zamrik, "Effects of External Loadings on Large Outlets in a Cylindrical Pressure Vessel," *Welding Research Council Bulletin 96* (May 1964).

273. *Welding Research Council Bulletin 113* (April 1966).

273b. M. Seika, K. Isogimi, and K. Inoue, "Photoelastic Study of Stresses in a Cylindrical Pressure Vessel with a Nozzle," *Bull. Japan Soc. Mech. Eng.*, Vol. 14, (1971), p. 1036.

274. M. M. Leven, "Photoelastic Determination of the Stresses in Reinforced Openings in Pressure Vessels," *Welding Research Council Bulletin 113* (April 1966), p. 25 (Fig. 32).

275. *Welding Research Council Bulletin 153* (August 1970).

276. R. E. Peterson, "The Interaction Effect of Neighboring Holes or Cavities, with Particular Reference to Pressure Vessels and Rocket Cases," *Trans. ASME*, Vol. 87, Series D (1965), Basic Engineering Section, p. 879.

277. H. Neuber, Ref. 39, Translation 2nd ed., p. 153.

278. M. A. Sadowsky and E. Sternberg, "Stress Concentration Around an Ellipsoidal Cavity in an Infinite Body under Arbitrary Plane Stress Perpendicular to the Axis of Revolution of Cavity," *Trans. ASME*, Vol. 69 (1947), Applied Mechanics Section, p. A-191.

279. E. Sternberg and M. A. Sadowsky, "On the Axisymmetric Problem of the Theory of Elasticity for an Infinite Region Containing Two Spherical Cavities," *Trans ASME*, Vol. 76 (1952), Applied Mechanics Section, p. 19.

280. Chih-Bing Ling, "Stresses in a Circular Cylinder Having a Spherical Cavity under Tension," *Q. Appl. Math*, Vol. 13 (1956), p. 381.

281. S. Timoshenko and J. N. Goodier, *Theory of Elasticity*, 3rd ed., McGraw-Hill, New York (1970), p. 398.

282. Chih-Bing Ling, "Stressses in a Stretched Slab Having a Spherical Cavity," *Trans. ASME*, Vol. 81, Series E (1959), Applied Mechanics Section, p. 235.

282a. E. Tsuchida and I. Nakahara, "Three-Dimensional Stress Concentration around a Spherical Cavity in a Semi-Infinite Elastic Body," *Bull. Japan Soc. Mech. Eng.*, Vol. 13 (1970), p. 499.

282b. G. E. McGinnis, "Stress Concentration at a Spherical Void Near One Surface on an Infinite Slab in a Uniaxial Field," M. S. Thesis, Univ. of Pittsburgh (1960).

283. Chih-Bing Ling, "Torsion of a Circular Cylinder Having a Spherical Cavity," *Q. Appl. Math.*, Vol. 10 (1952), p. 149.

284. H. Miyamoto, "On the Problem of the Theory of Elasticity for a Region Containing more than Two Spherical Cavities," *Trans. Japan Soc. Mech. Eng.*, Vol. 23 (1957), p. 437.

284a. A. Atsumi, "Stresses in a Circular Cylinder Having an Infinite Row of Spherical Cavities under Tension," *Trans. ASME*, Vol. 82, Series E (1960), Applied Mechanics Section, p. 87.

284b. W. L. Chu and D. H. Conway, "A Numerical Method for Computing the Stresses around an Axisymmetrical Inclusion," *Intern. J. Mech. Sci.*, Vol. 12 (1970), p. 575.

285. L. H. Donnell, "Stress Concentrations due to Elliptical Discontinuities in Plates under Edge Forces," *Von Karman Anniversary Volume*, California Inst. of Tech., Pasadena (1941), p. 293.

286. R. H. Edwards, "Stress Concentrations Around Spheroidal Inclusions and Cavities," *Trans. ASME*, Vol. 75 (1951), Applied Mechanics Section, p. 19.

287. J. N. Goodier, "Concentration of Stress Around Spherical and Cylindrical Inclusions and Flaws," *Trans. ASME*, Vol. 55 (1933), Applied Mechanics Section, p. A-39.

287a. A. J. Durelli and W. F. Riley, Ref. 202, p. 245.

287b. S. Shioya, "On the Tension of an Infinite Thin Plate Containing a Pair of Circular Inclusions," *Bull. Japan Soc. Mech. Eng.*, Vol. 14 (1971), p. 117.

287c. R. Mindlin, "Stress Distribution Around a Tunnel," *Proc. ASCE,* Vol. 65 (1939), p. 619. See also Ref. 234.

287d. Y. Y. Yu, "Gravitational Stresses on Deep Tunnels," *Trans. ASME,* Vol. 74 (1952), Applied Mechanics Section, p. 537.

287e. W. F. Riley, "Stresses at Tunnel Intersections," *Proc. ASCE,* Vol. 90, *J. Eng. Mech. Div* (1964), p. 167.

288. R. C. J. Howland and A. C. Stevenson, "Biharmonic Analysis in a Perforated Strip," *Phil. Trans. Royal Soc. A,* Vol. 232 (1933), p. 155.

289. J. J. Ryan and L. J. Fischer, "Photoelastic Analysis of Stress Concentration for Beams in Pure Bending with Central Hole," *J. Franklin Inst.,* Vol. 225 (1938), p. 513.

290. T. Udoguti, Ref. 168, p. I-82.

290a. R. B. Heywood, Ref. 98c, p. 277.

291. M. Isida, "On the Bending of an Infinite Strip with an Eccentric Circular Hole," *Proc. 2nd Japan Congr. Appl. Mech.* (1952), p. 57.

292. M. Nisida, *Rep. Sci. Res. Inst. Japan,* Vol. 28, No. 1 (1952), p. 30.

293. J. A. Joseph and J. S. Brock, "The Stresses Around a Small Opening in a Beam Subjected to Pure Bending," *Trans. ASME,* Vol. 72 (1950), Applied Mechanics Section, p. 353.

294. E. Reissner, "The Effect of Transverse Shear Deformation on the Bending of Elastic Plates," *Trans. ASME,* Vol. 67 (1945), Applied Mechanics Section, p. A-69.

295. J. N. Goodier, "Influence of Circular and Elliptical Holes on Transverse Flexure of Elastic Plates," *Phil. Mag.,* Vol. 22 (1936), p. 69.

296. J. N. Goodier and G. H. Lee, "An Extension of the Photoelastic Method of Stress Measurement to Plates in Transverse Bending," *Trans. ASME Vol. 63* (1941), Applied Mechanics Section, p. A-27.

297. D. C. Drucker, "The Photoelastic Analysis of Transverse Bending of Plates in the Standard Transmission Polariscope," *Trans. ASME,* Vol. 64 (1942), Applied Mechanics Section, p. A-161.

298. C. Dumont, "Stress Concentration Around an Open Circular Hole in a Plate Subjected to Bending Normal to the Plane of the Plate," *NACA Tech. Note 740* (1939). Now NASA.

299. O. Tamate, "Einfluss einer unendliche Reihe gleicher Kreislöcher auf die Durchbiegung einer dünnen Platte," *Z. angew. Math. u Mech.,* Vol. 37 (1957), p. 431.

300. O. Tamate, "Transverse Flexure of a Thin Plate Containing Two Holes," Vol. 80 (1958), Applied Mechanics Section, p. 1.

301. N. Neuber, Ref. 39, Translation 2nd ed., p. 113.

301a. H. Fessler and E. A. Roberts, "Bending Stresses in a Shaft with a Transverse Hole," *Selected Papers on Stress Analysis, Stress Analysis* Conference, *Delft, 1959,* Reinhold Publ. Co., New York (1961), p. 45.

301b. D. E. R. Godfrey, *Theoretical Elasticity and Plasticity for Engineers,* Thames and Hudson, London (1959), p. 109.

301c. H. Neuber, Ref. 39, Translation, 2nd ed., p. 81.

302. R. F. Barrett, P. R. Seth, and G. C. Patel, "Effect of Two Circular Holes in a Plate Subjected to Pure Shear Stress," *Trans. ASME,* Vol. 93, Series E (1971), Applied Mechanics Section, p. 528.

303. S. I. Chou, "Stress State Around a Circular Cylindrical Shell with an Elliptic Hole under Torsion," *Trans. ASME,* Vol. 93, Series E (1971), Applied Mechanics Section, p. 535.

304. *Keys and Keyseats,* USA Standard ANSI B17.1, ASME, New York (1967).

305. M. Hetényi, "The Application of Hardening Resins in Three-Dimensional Photoelastic Studies," *J. Appl. Phys.,* Vol. 10 (1939), p. 295.

306. R. E. Peterson, "Fatigue of Shafts having Keyways," *Proc. ASTM,* Vol. 32, Part 2 (1932), p. 413.

307. H. Fessler, C. C. Rogers and P. Stanley, "Stresses at End-milled Keyways in Plain Shafts Subjected to Tension, Bending and Torsion," *J. Strain Anal.,* Vol. 4 (1969), p. 180.

308. M. M. Leven, "Stresses in Keyways by Photoelastic Methods and Comparison with Numerical Solution," *Proc. SESA,* Vol. 7, No. 2 (1949), p. 141.

309. M. Nisida, "New Photoelastic Methods for Torsion Problems," *Symposium on Photoelasticity,* M. M. Frocht, Ed., Pergamon, New York (1963), p. 109.

310. G. I. Griffith and A. A. Taylor, "Use of Soap Films in Solving Torsion Problems," *Tech. Rep. Brit. Adv. Comm. Aeronaut.*, Vol. 3 (1917–18), p. 910.

311. H. Okubo, "On the Torsion of a Shaft with Keyways," *Q. J. Mech. Appl. Math.*, Vol. 3 (1950), p. 162.

312. S. Timoshenko and J. N. Goodier, Ref. 281, 3rd ed., p. 327.

313. A. G. Solakian and G. B. Karelitz, "Photoelastic Studies of Shearing Stresses in Keys and Keyways," *Trans. ASME,* Vol. 54 (1932), Applied Mechanics Section, p. 97.

314. W. H. Gibson and P. M. Gilet, "Transmission of Torque by Keys and Keyways," *J. Inst. Engrs. Australia,* Vol. 10 (1938), p. 393.

315. H. Okubo, K. Hosono, and K. Sakaki, "The Stress Concentration in Keyways when Torque is Transmitted through Keys," *Experimental Mechanics,* Vol. 8 (1968), p. 375.

316. H. Fessler, C. C. Rogers, and P. Stanley, "Stresses at Keyway Ends Near Shoulders," *J. Strain Anal.*, Vol. 4 (1969), p. 267.

316a. R. E. Peterson, Chapter 13, "Interpretation of Service Fractures," *Handbook of Experimental Stress Analysis,* M. Hetényi, Ed., Wiley, New York (1950), p. 603, 608, 613.

316b. Ref. 3a. Numerous examples of fatigue failures starting at keyseats are found in the technical reports covering many years.

317. H. Yoshitake, "Photelastic Stress Analysis of the Spline Shaft," *Bull. Japan Soc. Mech. Eng.*, Vol. 5 (1962), p. 195.

318. H. Okubo, "Torsion of a Circular Shaft with a Number of Longitudinal Notches," *Trans. ASME,* Vol. 72 (1950), Applied Mechanics Section, p. 359.

319. T. J. Dolan and E. L. Broghamer, "A Photoelastic Study of Stresses in Gear Tooth Fillets," *Univ. Illinois Expt. Sta. Bull.* 335 (1942).

320. G. W. Michalec, *Precision Gearing—Theory and Practice,* Wiley, New York (1966), p. 466.

321. *Marks' Standard Handbook for Mechanical Engineers,* T. Baumeister Ed., McGraw-Hill, New York (1967), p. (8)133.

322. A. H. Candee, *Geometrical Determination of Tooth Factor,* American Gear Manufacturers Assn., Pittsburgh, Pa. (1941).

323. R. V. Baud and R. E. Peterson, "Load and Stress Cycles in Gear Teeth," *Mechanical Engineering,* Vol. 51 (1929), p. 653.

324. R. E. Peterson, "Load and Deflection Cycles in Gear Teeth," *Proc. 3rd Intern. Appl. Mech. Congr.* (1930), p. 382.

324a. G. DeGregorio, "Ricerca sul Forzamento delle Corone Dentate" (Research on Shrunk-on Gear Rings), *Technica Italiana, Trieste,* Vol. 33 (1968), p. 1 (in Italian).

325. E. E. Weibel, "Studies in Photoelastic Stress Determination," *Trans. ASME,* Vol. 56 (1934), Applied Mechanics Section, p. 637.

326. N. C. Riggs and M. M. Frocht, *Strength of Materials,* Ronald Press, New York (1938), p. 389.

327. M. A. Jacobson, "Bending Stresses in Spur Gear Teeth; Proposed New Design Factors based on a Photoelastic Investigation," *Proc. Inst. Mech. Eng.*, Vol. 169 (1955), p. 587.

328. T. Aida and Y. Terauchi, "On the Bending Stress in a Spur Gear," 3 reports, *Bull. Japanese Soc. Mech. Eng.*, Vol. 5 (1962), p. 161.

329. R. E. Peterson and A. M. Wahl, "Fatigue of Shafts at Fitted Members, with a Related Photoelastic Analysis," *Trans. ASME,* Vol. 57 (1935), Applied Mechanics Section, p. A-1.

330. A. Thum and F. Wunderlich, "Der Einfluss von Einspann-und Kantangriffsstellen auf die Dauerhaltbarkeit der Konstruktionen," *Z. VDI,* Vol. 77 (1933), p. 851.

331. G. A. Tomlinson, "The Rusting of Steel Surface on Contact," *Proc. Roy. Soc. (London) A,* Vol. 115 (1927), p. 472.

332. G. A. Tomlinson, P. L. Thorpe, and H. J. Gough, "An Investigation of the Fretting Corrosion of Closely Fitting Surfaces," *Proc. Inst. Mech. Engrs. (London),* Vol. 141 (1939), p. 223.

333. *Symposium on Fretting Corrosion,* STP 144, ASTM, Philadelphia, Pa. (1952).

334. K. Nishioka, S. Nishimura, and K. Hirakawa, "Fundamental Investigations of Fretting Fatigue," *Bull. Japan Soc. Mech. Eng.*, Vol. 11 (1968), p. 437; Vol. 12 (1969), p. 180, 397, 408.

335. O. J. Horger and J. L. Maulbetsch, "Increasing the Fatigue Strength of Press-Fitted Axle Assemblies by Rolling," *Trans. ASME,* Vol. 58 (1936), Applied Mechanics Section, p. A-91.

336. O. J. Horger and T. V. Buckwalter, "Photoelasticity as Applied to Design Problems," *Iron Age,* Vol. 145, Part II, No. 21 (1940), p. 42. See also *J. Appl. Phys.*, Vol. 9 (1938), p. 457.

336a. D. J. White and J. Humpherson, "Finite-Element Analysis of Stresses in Shafts Due to Interference-Fit Hubs," *J. Strain Anal.*, Vol. 4 (1969), p. 105.

337. K. Nishioka and H. Komatsu, "Researches on Increasing the Fatigue Strength of Press-Fitted Shaft Assembly," *Bull. Japan Soc. Mech. Eng.*, Vol. 10 (1967), p. 880.

338. M. B. Coyle and S. J. Watson, "Fatigue Strength of Turbine Shafts with Shrunk-on Discs," *Proc. Inst. Mech. Eng. (London)*, Vol. 178 (1963–64), p. 147.

339. O. J. Horger, "Press Fitted Assembly," *ASME Metals Engineering Handbook—Design*, McGraw-Hill, New York (1953), p. 178.

340. *Passenger Car Axle Tests*, Fourth Progress Report, Assoc. American Railroads (1950), p. 26.

341. O. J. Horger, "Fatigue of Large Shafts by Fretting Corrosion," *Proc. Int. Conf. Fatigue, Inst. Mech. Eng., London* (1956), p. 352.

341a. B. Adelfio and F. DiBenedetto, "Forzamento su Appoggio Discontinuo" (Shrink-fitted Ring over Discontinuous Support), *Disegnio de Macchine, Palermo*, Vol. 3 (1970), p. 21, (in Italian).

342. L. Martinaglia, "Schraubenverbindungen," *Schweiz. Bauztg.*, Vol. 119 (1942), p. 107.

343. W. Staedel, Dauerfestigkeit von Schrauben," *Mitt. der Materialprüfungsanstalt an der Technischen Hochschule Darmstadt*, No. 4, VD1 Verlag, Berlin (1933).

344. H. Wiegand, "Über die Dauerfestigkeit von Schraubenwerkstoffen und Schraubenverbindungen," Thesis, *Technische Hochschule Darmstadt* (1933). Also published as No. 14—*Wissenschaftliche Veröffentlichungen der Firma*, Bauer & Schaurte A. G., Neuss (1934).

345. T. Baumeister Ed., *Marks' Standard Handbook for Mechanical Engineers, 7th Ed.*, McGraw-Hill, New York (1967), p. (8) 20, 21.

346. A. F. C. Brown and V. M. Hickson, "A Photo-elastic Study of Stresses in Screw Threads," *Proc. IME*, Vol. 1B (1952/3), p. 605. Discussion, p. 608.

347. R. L. Marino and W. F. Riley, "Optimizing Thread-root Contours Using Photoelastic Methods," *Experimental Mechanics*, Vol. 4 (1964), p. 1.

348. F. Kaufmann and W. Jäniche, "Beitrag zur Dauerhaltbarkeit von Schraubenverbindungen," *Tech. Mitt. Krupp, Forschungsber.*, Vol. 3 (1940), p. 147.

349. M. Hetényi, "Some Applications of Photoelasticity in Turbine-Generator Design," *Trans. ASME*, Vol. 61 (1939), Applied Mechanics Section, p. A-151. Also M. Hetényi, "A Comprehensive Report of Photoelastic Tests with T-Head Fastenings Subjected to Axial Pull," Westinghouse Research Report R-94027-AQ (1940).

350. R. B. Heywood, Ref. 152, p. 326; Ref. 98c, p. 205.

351. A. J. Durelli and W. F. Riley, Ref. 202, p. 220.

352. S. Timoshenko, Ref. 7, Part I, p. 362.

353. B. J. Wilson and J. F. Quereau, "A Simple Method of Determining Stress in Curved Flexural Members," *Univ. Illinois Expt. Sta. Circ. 16* (1928).

354. R. J. Roark, *Formulas for Stress and Strain*, 4th ed., McGraw-Hill, New York (1965), p. 164.

355. A. M. Wahl, *Mechanical Springs*, 2nd ed., McGraw-Hill, New York (1963).

356. O. Göhner, "Die Berechnung Zylindrische Schraubenfedern," *Z. Ver. Deutsch. Ing.*, Vol. 76 (1932), p. 269.

357. C. J. Ancker and J. N. Goodier, "Pitch and Curvature Corrections for Helical Springs," *Trans. ASME*, Vol. 80 (1958), Applied Mechanics Section, p. 466, 471, 484.

358. G. Liesecke, "Berechnung Zylindrischer Schraubenfedern mit Rechteckigen Drahtquerschnitt," *Z. VDI*, Vol. 77 (1933), p. 425, 892.

359. J. Arai, "The Bending Stress Concentration Factor of a Solid Crankshaft," *Bull. Japan Soc. Mechan. Eng.*, Vol. 8 (1965), p. 322.

360. A. M. Wahl, "Calculation of Stresses in Crane Hooks," *Trans. ASME*, Vol. 68 (1946), Applied Mechanics Section, p. A-239.

361. J. B. Mantle and T. J. Dolan, "A Photoelastic Study of Stresses in U-Shaped Members," *Proc. SESA*, Vol. 6, No. 1 (1948), p. 66.

362. I. Lyse and B. G. Johnston, "Structural Beams in Torsion," *Lehigh University Publication*, Vol. 9 (1935), p. 477.

363. J. H. Huth, "Torsional Stress Concentration in Angle and Square Tube Fillets," *Trans. ASME*, Vol. 72 (1950), Applied Mechanics Section, p. 388.

364. F. E. Richart, T. A. Olson, and T. J. Dolan, "Tests of Reinforced Concrete Knee Frames and Bakelite Models," *Univ. Illinois Expt. Sta. Bull. 307* (1938).

365. S. Timoshenko, Ref. 7, p. 214. See also E. L. Robinson, "Bursting Tests of Steam-Turbine Disk Wheels." *Trans. ASME*, Vol. 66, Applied Mechanics Section, (1944), p. 380.

366. K. E. Barnhart, A. L. Hale, and J. L. Meriam, "Stresses in Rotating Disks due to Non-Central Holes," *Proc. SESA*, Vol. 9, No. 1 (1951), p. 35.

367. K. Leist and J. Weber, "Optical Stress Distributions in Rotating Discs with Eccentric Holes," Report No. 57, Institute for Jet Propulsion; German Research Institute for Aeronautics, Aachen (1956).

368. W. A. Green, G. T. J. Hooper, and R. Hetherington, "Stress Distribution in Rotating Discs with Non-Central Holes," *Aeronautical Q.* ,Vol. 15 (1964), p. 107.

369. H. Fessler and T. E. Thorpe, "Optimization of Stress Concentrations at Holes in Rotating Discs," *J. Strain Anal.*, Vol. 2 (1967), p. 152.

370. H. Fessler and T. E. Thorpe, "Reinforcement of Non-Central Holes in Rotating Discs," *J. Strain Anal.*, Vol. 2 (1967), p. 317.

371. S. Timoshenko, "On the Distribution of Stresses in a Circular Ring Compressed by Two Forces along a Diameter," *Phil. Mag.*, Vol. 44 (1922), p. 1014. Also S. Timoshenko and J. N. Goodier, Ref. 281, p. 136.

372. V. Billevicz, "Analysis of Stress in Circular Rings," doctor's thesis, Univ. of Michigan (1931).

373. J. Case, *Strength of Materials*, Longmans, Green, London (1925), p. 291.

374. M. M. Leven, Unpublished data obtained at Carnegie Inst. of Technology (1938) and Westinghouse Research Labs (1952).

375. S. Timoshenko, Ref. 7, Part I, p. 380.

376. M. Seika, "The Stresses in a Thick Cylinder Having a Square Hole Under Concentrated Loading," *Trans. ASME*, Vol. 80, Series E (1958), Applied Mechanics Series, p. 571.

377. H. Fessler and P. Stanley, "Stresses in Torispherical Drumheads: A Photoelastic Investigation," *J. Strain Anal.*, Vol. 1 (1965), p. 69.

378. Ref. 198, p. 18.

379. H. Fessler and P. Stanley, "Stresses in Torispherical Drumheads: A Critical Evaluation," *J. Strain Anal.*, Vol. 1 (1966), p. 89.

380. J. C. Gerdeen, "Analysis of Stress Concentrations in Thick Cyclinders with Sideholes and Crossholes," *Trans. ASME*, Vol. 94, Series B, *J. Eng. for Industry*, (1972), p. 815.

381. J. C. Gerdeen and R. E. Smith, "Experimental Determination of Stress Concentration Factors in Thick-walled Cylinders with Crossholes and Sideholes," *Experimental Mechanics*, Vol. 12 (1972), p. 530.

INDEX